New Frontiers in Nanotechnology

New Frontiers in Nanotechnology

Edited by
Andrew Green

WILLFORD PRESS

www.willfordpress.com

Published by Willford Press,
118-35 Queens Blvd., Suite 400,
Forest Hills, NY 11375, USA

ISBN: 978-1-68285-372-6

Cataloging-in-Publication Data

New frontiers in nanotechnology / edited by Andrew Green.
p. cm.
Includes bibliographical references and index.
ISBN 978-1-68285-372-6
1. Nanotechnology. 2. Nanostructured materials. I. Green, Andrew.
T174.7 .N49 2017
620.5--dc23

For information on all Willford Press publications
visit our website at www.willfordpress.com

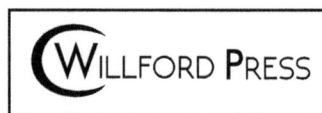

WILLFORD PRESS

Printed in the United States of America.

Contents

Preface

The world is advancing at a fast pace like never before. Therefore, the need is to keep up with the latest developments. This book was an idea that came to fruition when the specialists in the area realized the need to coordinate together and document essential themes in the subject. That's when I was requested to be the editor. Editing this book has been an honour as it brings together diverse authors researching on different streams of the field. The book collates essential materials contributed by veterans in the area which can be utilized by students and researchers alike.

Nanotechnology is the handling of matter on a very minute scale, which cannot be seen by the eye or picked by our hands. It deals with matter at a molecular or atomic scale. It has led to advancements in drug delivery in health sciences, crop protection and livestock productivity in agriculture, water treatment, early detection of diseases, solar energy, etc. This book is essential for readers who wish to broaden their scope of knowledge by learning about the contemporary developments in the field of nanotechnology. Scientists and students actively engaged in this field will find this book full of crucial and unexplored concepts. For all those who are interested in nanotechnology, the case studies included in this text will serve as an excellent guide to develop a comprehensive understanding of the subject.

Each chapter is a sole-standing publication that reflects each author's interpretation. Thus, the book displays a multi-facetted picture of our current understanding of applications and diverse aspects of the field. I would like to thank the contributors of this book and my family for their endless support.

Editor

Ultra-high gain diffusion-driven organic transistor

Fabrizio Torricelli[1,2], Luigi Colalongo[1], Daniele Raiteri[2], Zsolt Miklós Kovács-Vajna[1] & Eugenio Cantatore[2]

Emerging large-area technologies based on organic transistors are enabling the fabrication of low-cost flexible circuits, smart sensors and biomedical devices. High-gain transistors are essential for the development of large-scale circuit integration, high-sensitivity sensors and signal amplification in sensing systems. Unfortunately, organic field-effect transistors show limited gain, usually of the order of tens, because of the large contact resistance and channel-length modulation. Here we show a new organic field-effect transistor architecture with a gain larger than 700. This is the highest gain ever reported for organic field-effect transistors. In the proposed organic field-effect transistor, the charge injection and extraction at the metal–semiconductor contacts are driven by the charge diffusion. The ideal conditions of ohmic contacts with negligible contact resistance and flat current saturation are demonstrated. The approach is general and can be extended to any thin-film technology opening unprecedented opportunities for the development of high-performance flexible electronics.

[1] Department of Information Engineering, University of Brescia, via Branze 38, Brescia 25123, Italy. [2] Department of Electrical Engineering, Eindhoven University of Technology, Groene Loper 19, PO Box 513, Eindhoven 5600MB, The Netherlands. Correspondence and requests for materials should be addressed to F.T. (email: fabrizio.torricelli@unibs.it).

Transistors fabricated with organic, polymeric, amorphous-oxide and carbon-based materials are the basis of emerging technologies for the development of lightweight, large-area and flexible electronics[1–6]. Large-area electronics manufactured at near-to-room temperature on plastic foils aims at enabling new applications where mechanical flexibility, integration in wrapping materials and ultra-low cost are paramount. To fabricate a transistor in flexible technologies, nanometre-thick layers of metals, insulators and semiconductor are stacked together and the semiconductor is directly contacted with the metal electrodes. The overall transistor performance intimately depends on three physical processes: the charge injection from the source electrode to the semiconductor, the charge transport through the semiconductor and the charge extraction at the drain electrode. The impressive development of high-mobility semiconductors[7–9] and short channel-length transistors[10,11] urgently demand high-quality contacts and proper transistor design[12,13]. Unfortunately, the energetic matching between abruptly contacted metal–semiconductor materials is challenging, especially at near-to-room temperature[13]. Electrons and holes must overcome large energy barriers to flow from a material to the other, resulting in a large contact resistance, large device-to-device variations and low transistor amplification[13–18].

The figure-of-merit that determines the intrinsic amplification of a transistor is the gain $= g_m/g_o$, where $g_m = \partial I_D/\partial V_G$ is the transconductance and $g_o = \partial I_D/\partial V_D$ is the output conductance. High-gain transistors are essential for the development of large-scale and robust circuits, high-sensitivity sensors, and adequate signal amplification in sensing systems. Unfortunately, organic field-effect transistors (OFETs) typically show a gain of the order of tens[17,18]. The low gain measured in OFETs is due to the large contact resistance that results in a small g_m and to the channel length modulation that results in a large g_o. Therefore, high-gain OFETs need, at the same time, both high-quality contacts and flat current saturation.

Ohmic contacts with small contact resistance require efficient charge injection and extraction. In organic electronics, the contact optimization is performed on a case-by-case basis, depending on the semiconductor, electrodes and device architecture. Despite ad-hoc approaches[18–25], such as doping, surface treatments and materials blending enable to reduce the contact resistance, a general and simple method is desirable. In addition, the channel length modulation dependents on the specific OFET architecture and geometries, which determine how the charge carriers are extracted at the drain[17,18].

Here we show a new organic transistor with high-quality contacts and flat current saturation. Thanks to the charge diffusion triggered by the transistor architecture, the charge carriers are efficiently injected and extracted from the contacts to the channel, independently of the energy barrier at the contacts. As a prototype and remarkable example, we fabricate Diffusion-driven Organic Field-Effect Transistors (named DOFETs) on flexible plastic substrates with an industrial thin-film technology. The theoretical and experimental analysis unambiguously show that the diffusion-driven contact, proposed in this work for the first time, is fundamental to dramatically improve the charge injection and extraction in organic thin-film field-effect transistors. The ideal conditions of negligible contact resistance and fully flat current saturation are demonstrated. These conditions maximize together the transconductance and the output resistance of the transistors, resulting in OFETs with exceptionally high gain (>700).

Results

Structure and electrical characteristics of the transistor.
The top-view image and the three-dimensional structure of the diffusion-driven organic transistor are shown in Fig. 1a,b. The transistors are bottom-gate co-planar where the gate is patterned first by using photolithography. Thereafter, we deposited by spin coating a photoimageable polymer (polyvinylphenol) used as a gate insulator (named insulator 1) followed by gold source (S) and drain (D) electrodes patterned by a lift-off process. A 100-nm-thick film of pentacene is deposited by spin coating and patterned. A thick layer of polyvinylphenol (named insulator 2) is deposited by spin coating and used as insulator and capping layer. Finally, two electrodes named 'control source' (CS) and 'control drain' (CD) are patterned on the top of the insulator 2 in front of the source and drain electrodes. The transistors are fabricated on a plastic polyethylene naphthalate foil (Fig. 1c) and the overall process temperature is lower than 150 °C. Further details on the transistors fabrication and geometries are shown in Supplementary Fig. 1. The measured transfer and output curves are shown in Fig. 1d–f and Supplementary Fig. 2.

Operation of the transistor. The DOFET operates as follow. An appropriate voltage applied to CS and CD, creates a vertical electric field orthogonal to the S/D contact surface. It triggers a charge injection from the upper surface of S/D into the semiconductor (Fig. 2a). In equilibrium ($V_S = V_D = 0$ V, no current flows), the electric-field below CS/CD is counterbalanced by the injected-charges that are accumulated in the semiconductor region below CS/CD. When a source-drain voltage is applied ($|V_{DS}| > 0$), the charge carriers flow from source to drain despite the contact energy barriers and the potential drop at the contacts is negligible (Fig. 2b).

More in detail (Fig. 2c), the charge carriers accumulated in the CS region move to the right-edge of the CS region itself, attracted by the drain potential (Fig. 2c, arrow 2). As a consequence, the vertical electric field at the left-hand-side of the CS region is not shielded anymore and, despite the energy barrier, other charges can be injected by the source electrode (arrow 1). The excess of the charge carriers at the right-hand-side of the CS region are pushed to the bottom channel by the diffusion against the vertical electric field (arrow 3). As shown in Fig. 2d, few nanometres far from the CS region the vertical electric-field changes direction under the influence of the gate potential and the charge carriers are eventually pulled into the transistor channel (arrow 4). As a result, the CS region acts as an ideal source. The key physical mechanism triggered by the transistor architecture is the charge diffusion, which takes place in less than $L_{diff} = 50$ nm (Fig. 2d) when the semiconductor thickness is $t_S = 100$ nm. We also verified that the diffusion length scales accordingly with the semiconductor thickness (that is, $L_{diff} \cong 25$ nm when $t_S = 50$ nm).

The charge carriers injected into the channel drift to the drain (Fig. 2c, arrow 5) under the force of the longitudinal electric field. When the charge carriers reach the right edge of the channel, they are blocked by the energy barrier at the drain contact, and the local concentration increases. The charges are no more counterbalanced by the gate electric field, and they can diffuse to the CD region (arrow 6) in correspondence of the CD region edge. As shown in Fig. 2e, few nanometres far from the channel the vertical electric field changes direction, the charge carriers are pulled into the CD region (arrow 7) and eventually diffuse (arrow 9) to the drain. The CD region acts as an ideal drain.

The idea is that in the DOFETs the charge injection and extraction do not take place directly from the source and drain metal electrodes as in conventional transistors but, instead, the charge carries are injected by the CS region and are extracted by the CD region. The injection and the extraction are driven by the diffusion triggered by the transistor architecture. As a result, when enough charge carriers are accumulated in the CS and CD regions, the charge injection and extraction are independent of

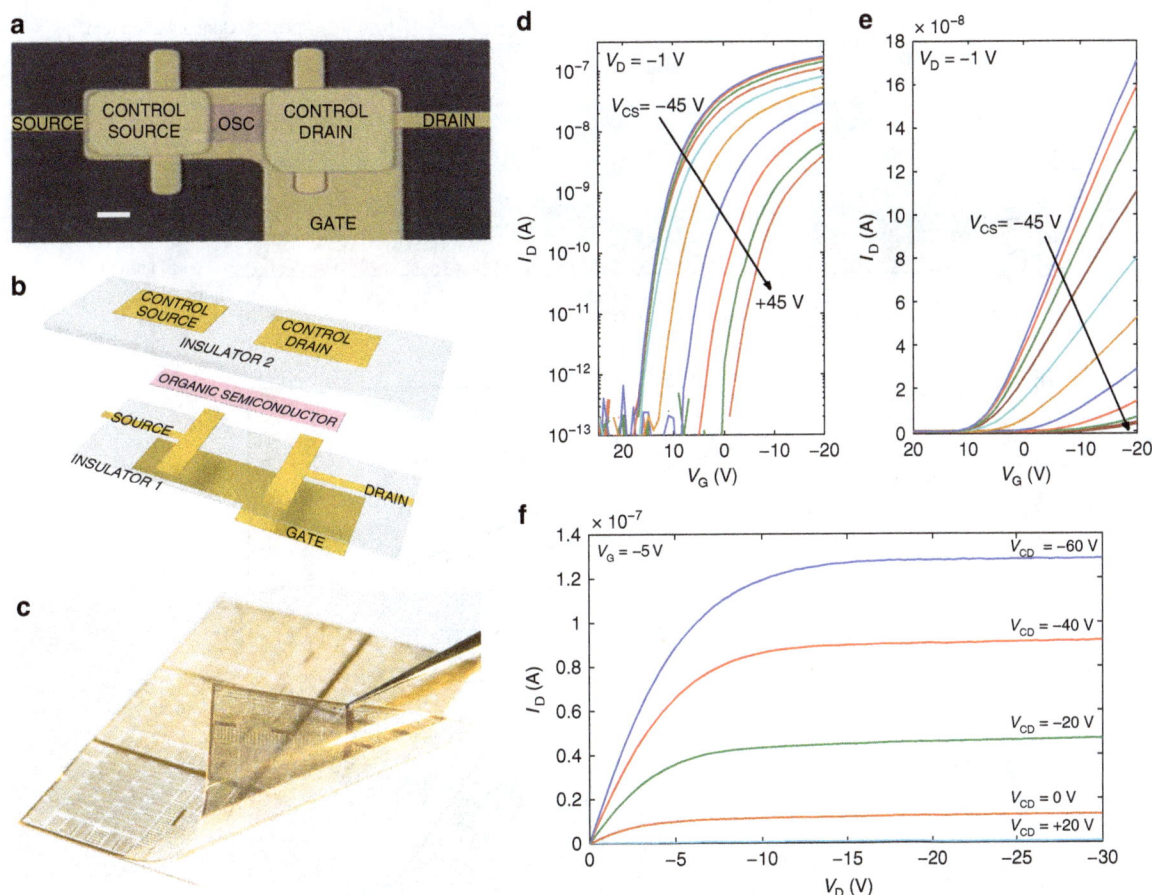

Figure 1 | Transistor architecture and characteristics. (a) Top-view optical image of a diffusion-driven organic field-effect transistor (DOFET) fabricated on plastic foil OSC is the organic semiconductor. Scale bar, 5 μm. **(b)** DOFET components. Photolithographically patterned gold is used for metal electrodes (named gate, source, drain, control source, control drain), the insulators (insulators 1 and 2) are photoimageable polymers (polyvinylphenol), and the organic semiconductor is a solution-processed pentacene. The material thicknesses are detailed in the Supplementary Fig. 1. **(c)** Photograph of the plastic (PEN) foil with the measured transistors detached from the glass substrate. The transistors are fabricated with an industrial thin-film technology with three metal layers. **(d,e)** Measured transfer characteristics at several control source voltages. The V_{CS} step is 10 V, $V_S = 0$ V and $V_{CD} = 0$ V. The DOFET channel width and length are $W = 100$ μm and $L = 12.5$ μm, respectively. **(f)** Measured output characteristics at several control drain voltages.

the applied voltages (viz. V_{CS} and V_{CD}) and the CS and CD regions behave like ideal source and drain for the transistor channel. Therefore, as confirmed by the two-dimensional (2D) numerical simulations shown in Supplementary Fig. 3, the gate electrode is not required to overlap the source and drain electrodes.

The potential at the insulator1–organic interface calculated by means of 2D numerical simulations is shown in Fig. 2b. In the DOFET, the potential drop at the contacts is negligible even if the energy barrier at the metal–semiconductor contacts is 0.5 eV, that is a typical barrier at the metal–organic contacts. In contrast, in a conventional organic transistor (viz. without CS and CD), the charge carriers must overcome the energy barrier flowing from the channel to the S/D electrodes and vice versa. Owing to the energy barrier, the channel is disconnected from the S/D electrodes and more than the half of the drain voltage drops at the contacts (Fig. 2b). The large contact resistance severely limits the transistor performances and this is even worse in case of high-mobility semiconductors and/or short-channel lengths.

Experimental analysis. The effectiveness of the proposed approach is further assessed by means of the experimental results shown in Fig. 3. Figure 3a shows the measured contact resistance

R_P as a function of the gate voltage V_G. The contact resistance of the DOFET biased at $V_{CS} = -40$ V (that corresponds to an electric field $|E_{Y-VCS}| = 0.28$ MV cm^{-1}) is equal to $R_{P[DOFET]} = 20$ kΩ cm, which is lower than the contact resistance in conventional OFETs with Au-pentacene-doped contacts[20] and, more importantly, $R_{P[DOFET]}$ is independent of V_G. In contrast, the contact resistance of an organic transistor without CS/CD (conventional coplanar transistor) fabricated with the same materials and process is V_G dependent. It is up to 24 times larger than that of the DOFET and, even at large gate voltages ($V_G = -25$ V, that is, $|E_{Y-VG}| = 0.7$ MV cm^{-1}), $R_{P[OFET]} > 5 \times R_{P[DOFET]}$. Analogous results are obtained comparing the DOFET with a conventional staggered OFET (Supplementary Fig. 4).

To give more insight, Fig. 3b shows the R_P-V_{CS} characteristic of two nominally identical DOFETs for several V_G. R_P is controlled by V_{CS} despite the gate voltage. Indeed, at low gate voltage ($V_G = -5$ V, that is, $|E_{Y-VG}| = 0.14$ MV cm^{-1}), V_{CS} modulates R_P by more than four orders of magnitude, and at large $V_G = -20$ V ($|E_{Y-VG}| = 0.57$ MV cm^{-1}), R_P still depends on V_{CS}. Interestingly, when $V_{CS} < -20$ V (that is, $|E_{Y-VCS}| = 0.14$ MV cm^{-1}) the contact resistance is negligible compared with the channel resistance (Supplementary Fig. 5) and it is independent of both V_G and V_{CS}. As confirmed by the measurements shown in

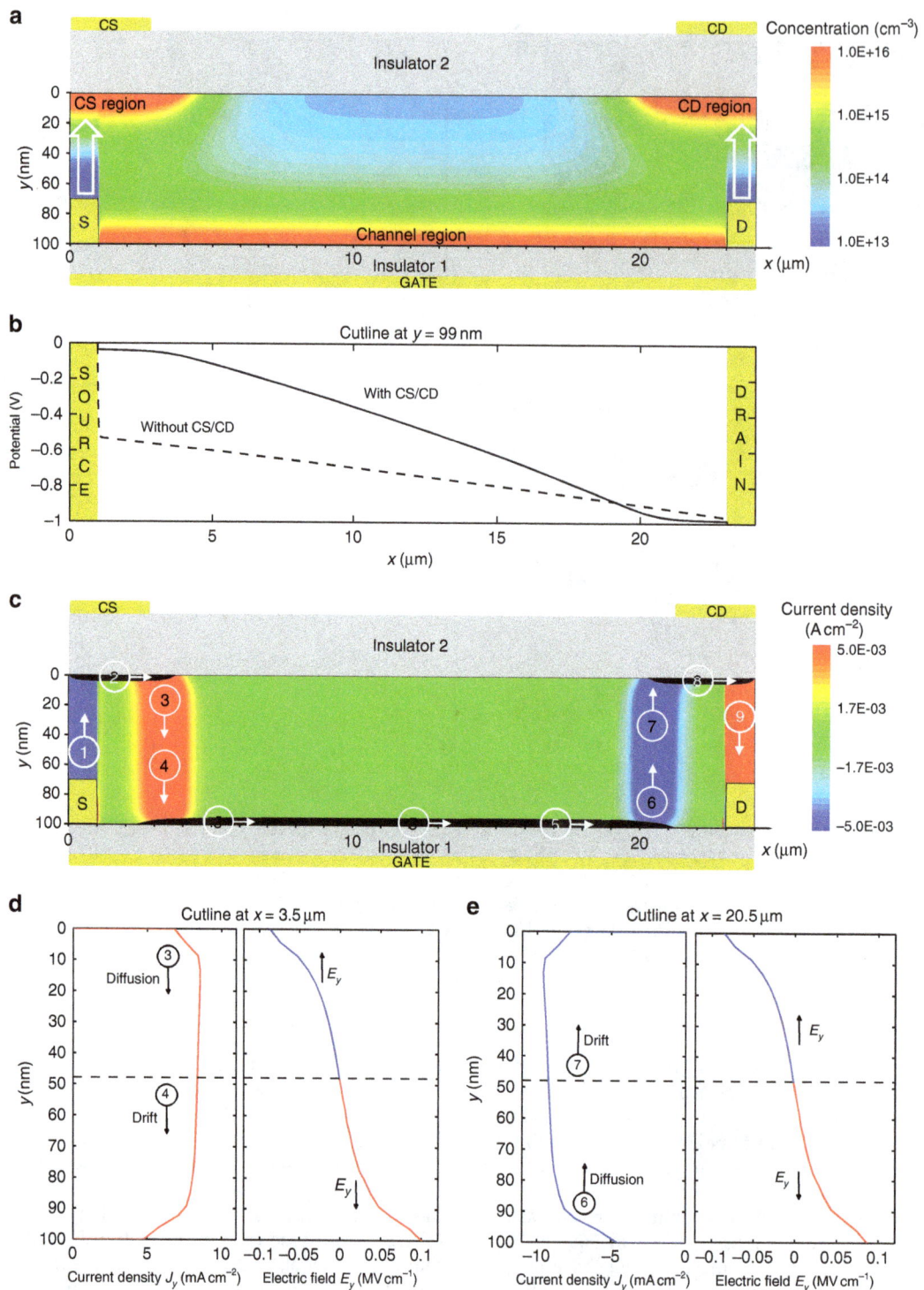

Figure 2 | DOFET operation. Two-dimensional numerical simulations. The applied voltages are $V_G = -5.1$ V, $V_S = 0$ V, $V_D = -1$ V, $V_{CS} = -60$ V, $V_{CD} = -60$ V. Geometrical and physical parameters are listed in the Supplementary Fig. 1 and in the Methods section, respectively. (**a**) Charge concentration in the organic semiconductor. The white arrows depict the charge injection from the source and drain electrodes into the semiconductor when the control source and control drain electrodes are biased. The x-to-y scale ratio is 1:200. (**b**) Quasi-Fermi potential at $y = 99$ nm with (full line) and without (dashed line) CS/CD. Without CS/CD about half of V_{DS} drops at the source and it is required for the charge injection. (**c**) Current density: x-component J_X (black area) is equal to 1 A cm^{-2}, and the y-component J_Y is shown with colour scale levels. (**d**) Current density J_Y and electric field E_Y along the y-direction at $x = 3.5$ µm. In the range $y = [0–47]$ nm, the current is driven by the diffusion, and in the range $y = [47–100]$ nm, the current is driven by the drift. (**e**) J_Y and E_Y along the y-direction at $x = 20.5$ µm. In the range $y = [0–47]$ nm, the current is driven by the drift, and in the range $y = [47–100]$ nm, the current is driven by the diffusion.

the inset of Fig. 3b, this is the experimental evidence that the current enhancement originates from the improved charge injection at the source. According to the physical insight obtained by means of the 2D numerical simulations, at $V_{CS} < -20$ V, the accumulated CS-region is an 'infinite' charge reservoir, the charge diffusion efficiently sustain the charge injection required by the

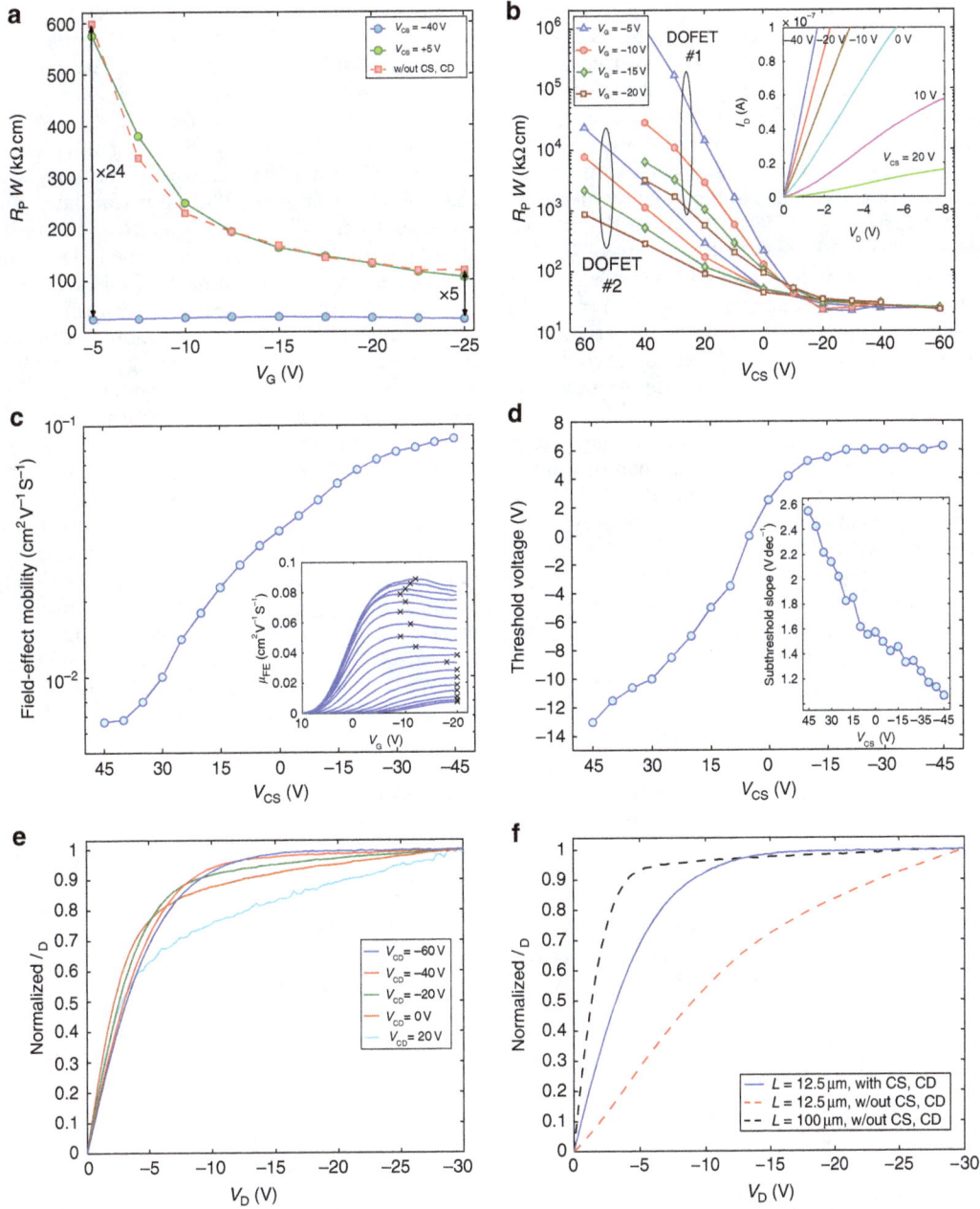

Figure 3 | DOFET measurements and parameters. When it is not specified, the applied voltages are: $V_S = 0\,V$, $V_D = -1\,V$, $V_{CS} = 0\,V$, $V_{CD} = 0\,V$, and the transistors geometries are: $W = 100\,\mu m$, $L = 12.5\,\mu m$. (**a**) Width-normalized contact resistance R_P as a function of the gate voltage V_G. R_P is calculated with the method[26]. In the conventional OFET (viz. without CS and CD), R_P decreases with V_G, whereas in the DOFET, R_P is independent of V_G. When the control source is biased at $V_{CS} = +5\,V$, the DOFET works as a conventional coplanar OFET. (**b**) R_P vs V_{CS} at various V_G measured on two nominally identical DOFETs. When $V_{CS} < -10\,V$, R_P is the same for both the DOFETs and it is independent of both V_G and V_{CS}. Inset: measured output characteristics of a DOFET at several V_{CS}. (**c**) Maximum overall field-effect mobility vs V_{CS}. The inset shows the field-effect mobility as a function of the gate voltage: $\mu_{FE} = (L/W)\,(\partial I_D/\partial V_G)/(C_i\,V_D)$. The × symbol is the maximum value of each curve. (**d**) Threshold voltage (V_{TH}) as a function of V_{CS}. V_{TH} is the intercept to the V_G-axis of the I_D linear fit. Inset: Subthreshold slope as a function of V_{CS}. (**e**) Normalized output characteristics of the DOFET measured at various V_{CD}. I_D is normalized by its maximum value at $V_D = -30\,V$. In saturation, the DOFET is an ideal current generator because the current is diffusion driven. The most important short-channel effect due to the channel-length modulation vanishes. The V_{CD} controls the charge extraction at the drain electrode, which has a strong impact on the output conductance ($g_O = \partial I_D/\partial V_D$). (**f**) Normalized output characteristics of a DOFET and two conventional OFETs (viz. without CS and CD).

channel, and the CS-region behaves like an ohmic contact. On the other hand, at $V_{CS} > +5\,V$, the diffusion-driven charge injection is turned-off, the contact resistance increases, the drain current lowers and it increases super-linearly with V_D as usually obtained in contact limited transistors[13,15,16,19]. We can conclude that it is possible to control (enhance or reduce) the charge injection at the source contact through nanometre-scale charge diffusion.

Comparing the R_P obtained for two nominally identical DOFETs (Fig. 3b), it results that when the virtual-ohmic source contact is not formed, the transistors show different R_P, whereas as soon as the virtual-ohmic source contact is formed ($V_{CS} < -20\,V$), R_P becomes the same for both the DOFETs. According to refs 10,12,13, these measurements suggest that the metal–organic contact is a source of variability. As the DOFET suppresses the contact resistance, the variability due to R_P is reduced as well. This feature is essential for the large-scale integration of flexible circuits. Moreover, the improved charge injection results in a larger overall field-effect mobility (Fig. 3c) as well as in a reduced threshold voltage (Fig. 3d) and steeper subthreshold slope (inset Fig. 3d). Figure 3c shows that the maximum mobility of a DOFET with $L_{[DOFET]} = 12.5 \, \mu m$ is close to $0.1 \, cm^2 \, V^{-1} \, s^{-1}$ and it corresponds to the mobility measured in long-channel OFETs ($L_{[OFET]} = 100 \, \mu m$), where the contact resistance is negligible. Figure 3d shows that by means of V_{CS} the DOFET can be turned into a multi-threshold transistor and the improved DOFET ($V_{CS} < -20 \, V$) operates in depletion-mode. In unipolar technologies, depletion-mode transistors are essential to design high-performance circuits[27,28] and the electrical control of the threshold voltage is extremely important to improve the circuit robustness[27–29].

When the transistor operates in linear region, the energy barrier at the drain side of the channel is smaller than that at the source side. On the other hand, in saturation ($|V_G| < |V_D|$), a wider energy barrier is present at the drain, independently of the metal/semiconductor properties (Fig. 4a). Therefore, we investigated the impact of the control drain in saturation. The output characteristics (I_D–V_D) of the DOFET measured at various V_{CD} are shown in Fig. 1f. As expected, I_D increases with V_{CD} and, more importantly, at large (negative) V_{CD} the DOFET shows fully flat current saturation. The impact of V_{CD} on the current saturation is readily visible in Fig. 3e where the I_D–V_D characteristics are normalized with respect to the maximum I_D measured at $V_D = -30 \, V$. At $V_{CD} < -40 \, V$, the detrimental effect of the channel length modulation on the drain current is completely suppressed and the DOFET behaves like an ideal current generator.

This can be explained in the light of the previous analysis. In saturation, the charge carriers drift to the right-edge of the channel (pinch-off region), and diffuse to the CD region (Fig. 4b, arrow 6). Few nanometres far from the channel edge, the vertical electric-field changes direction because of the control drain

voltage and, in turn, the charge carriers are pulled into the CD region (arrow 7). Now, the excess charges are no more in equilibrium with the vertical electric-field and can diffuse to the drain (arrow 9). As the charge-extraction from the accumulated layer (viz. CD region) is diffusion driven, the drain current is independent of the drain voltage as far as V_{CD} is greater than V_D.

Figure 3f shows the comparison between a DOFET with a channel length $L_{[DOFET]} = 12.5 \, \mu m$ (full line), and two conventional coplanar OFETs with $L_{[OFET1]} = 12.5 \, \mu m$ (red dashed line) and $L_{[OFET2]} = 100 \, \mu m$ (black dashed line). Interestingly, the channel length modulation of the DOFET biased at $V_{CD} = -60$ V is completely suppressed: it is even smaller than that of the long-channel OFET2. This is also more evident when the DOFET is compared with a conventional staggered OFET (Supplementary Fig. 6) where the channel length modulation is very large because the source and drain electrodes are placed at the opposite side of the gate. These results confirm that V_{CD} controls channel length modulation and in turn the output resistance of the DOFET. The channel length modulation is one of the most important short-channel effects and it limits the transistor amplification.

Figure 5 shows the maximum gain measured in a DOFET as a function of V_{CD} (full line with symbols). According to Figs 1f and 3e, the gain depends on V_{CD} because it controls both the contact resistance at the drain and the channel length modulation. When $V_{CD} = -60 \, V$, the gain is larger than 700. This is the largest gain ever reported for OFETs. It is one order of magnitude larger than the gain usually obtained in OFETs[11,16–18,30–35].

Discussion

The ultra-high gain measured in the DOFET is achieved thanks to the diffusion-driven charge injection and extraction. In particular, when the CS and CD regions are accumulated, they act as ideal contacts for the channel and the diffusion enables the efficient and voltage-independent charge injection and extraction. In the DOFET, the CS and CD regions are at the opposite side of the channel and resemble a staggered OFET with ideal ohmic

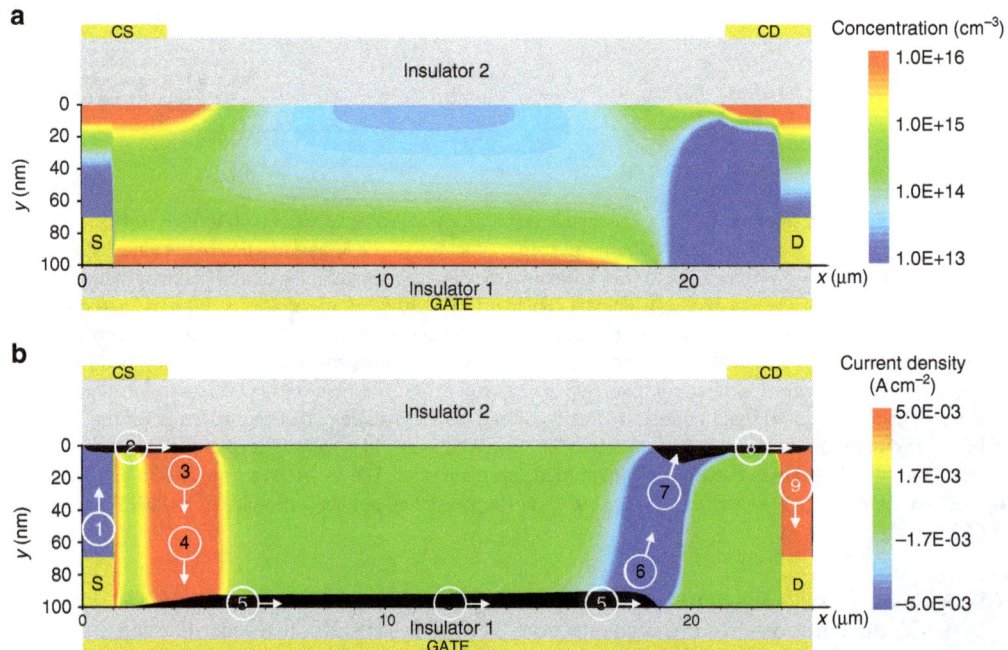

Figure 4 | 2D numerical simulations of a DOFET operating in saturation. The applied voltages are $V_G = -5.1 \, V$, $V_S = 0 \, V$, $V_D = -10 \, V$, $V_{CS} = -60 \, V$, $V_{CD} = -60 \, V$. (**a**) Charge concentration in the organic semiconductor. (**b**) Current density: x-component J_X (black area) is equal to $1 \, A \, cm^{-2}$, and the y-component J_Y is shown with colour scale levels. For the sake of clarity, the positions of control source (CS), control drain (CD) and gate electrodes are shown. Geometrical and physical parameters are listed in the Supplementary Fig. 1.

contacts. It is important to note that in the DOFET this condition is always achieved, thanks to the accumulated CS and CD regions. The charge flow from/to the CS/CD regions and the channel is driven by the charge diffusion, and thus the contact resistance is independent of the gate (Fig. 3a) and drain (inset Fig. 3b) voltages, the saturation current is independent of V_D, and an ultra-high gain is obtained.

As a comparison, the gain measured in the conventional OFET1 ($L_{[OFET1]} = 12.5\,\mu m$, red dashed line) and OFET2 ($L_{[OFET2]} = 100\,\mu m$, black dashed line) are shown in Fig. 5. As expected in both cases, the gain is much lower than that measured in the DOFET at any V_{CD} because in the OFET1 the current is contact limited and the channel modulation is large, whereas in the OFET2 the contact resistance is negligible but the channel length is large and hence g_m is small. In OFETs, the contact resistance can be reduced by means of the contact engineering and optimization[18-25], and the proper choice of the transistor architecture[36,37]. Indeed, staggered OFETs are more tolerant to

the contact resistance with respect to the coplanar OFETs because in the staggered transistors the contact area (of the order of microns) is larger than that of coplanar transistors (of the order of nanometres). On the other hand, in staggered transistors the source and drain electrodes are at the opposite side of the channel and, when operated in saturation, the channel length modulation is larger than that in coplanar OFETs. As an alternative approach, the split-gate OFETs[33,34] are based on a coplanar architecture and lower the contact resistance thanks to the gate bias-assisted charge injection[38]. However, the channel length modulation is not suppressed because the secondary gates are coplanar with the source and drain electrodes, the charge extraction is not diffusion driven and, as a result, the gain is comparable with that typically measured in OFETs (of the order of tens).

In addition to the high-gain, another advantage offered by the DOFET is the possibility to maximize the charge injection/extraction area at the source and drain electrodes, whereas minimizing the overlap between the gate and the electrodes. The 2D numerical simulations in Supplementary Fig. 3 and Fig. 6 show that the gate is not required to overlap the source and drain electrodes because the charge injection/extraction takes place from/to the CS/CD accumulated regions. At the same time, the CS and CD electrodes can be overlapped (without the drawback of extra capacitance) with the source and drain electrodes in order to exploit the full area of the electrodes that is typically in the range 5–10 μm (in our DOFET it is 5 μm). Thanks to the charge diffusion, taking place at the edge of the accumulated CS and CD regions, also the overlap between the gate and the CS and CD electrodes is not required. Moreover, the numerical simulations in Fig. 6 show that the equivalent contact length where the charges are injected/extracted is only $L_C = 0.25\,\mu m$, which is suitable for the megahertz operation[11].

Finally, it is worth noting that the voltages required to form the charge-accumulated CS and CD regions are independent of the DOFET operation. For example, by setting $V_{CS} = V_{CD} = -40\,V$, the DOFET operates as a conventional OFET with ideal ohmic contacts and ultra-high gain. Therefore, the two control electrodes can be connected together and the external circuit design and lines required for the proposed transistor structure is the same of that required for dual-gate transistors. The latter have been successfully used to fabricate an organic microprocessor with 3,381 dual-gate OFETs[39]. Moreover, an alternative approach is to replace the CS and CD electrodes with fixed charges trapped into the insulator 2 (ref. 40). Another very interesting approach

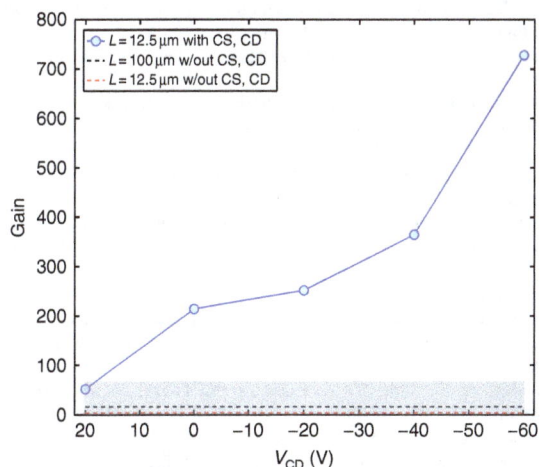

Figure 5 | DOFET gain. Measured gain as a function of V_{CD}. The applied voltages are $V_G = -5\,V$, $V_S = 0\,V$, $V_{CS} = -20\,V$. The transistors width is $W = 100\,\mu m$. The DOFET (full line with symbols) length is $L = 12.5\,\mu m$. The OFET lengths are $L = 12.5\,\mu m$ (red dashed line) and $L = 100\,\mu m$ (black dashed line). The other geometries are the same. The DOFET and OFET are fabricated with the same materials (Supplementary Fig. 1). The grey area shows the gain obtained in OFETs[11,16-18,30-35].

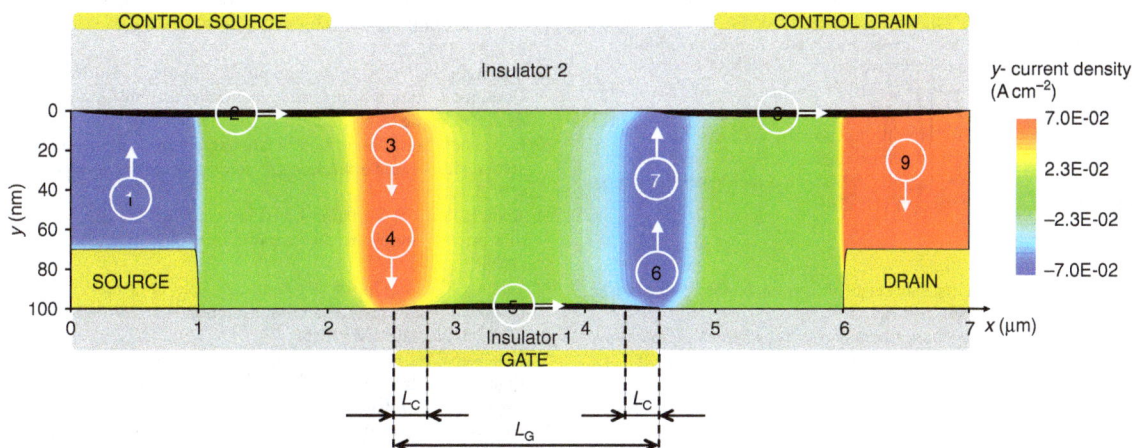

Figure 6 | 2D numerical simulations of a DOFET with minimized capacitances. Current density: x-component J_X (black area) is 10 A cm^{-2}, and the y-component J_Y is shown with colour scale levels. For the sake of clarity, the positions of control source (CS), control drain (CD) and gate electrodes are shown. Geometrical and physical parameters are listed in the Supplementary Fig. 1. The applied voltages are $V_G = -5.1\,V$, $V_S = 0\,V$, $V_D = -1\,V$, $V_{CS} = -60\,V$, $V_{CD} = -60\,V$.

would be the replacement of both the CS and CD electrodes and the insulator 2 with electric dipoles (Supplementary Fig. 7) by local molecular self-assembly functionalization[41,42] of the top surface of the organic semiconductor in front of the source and drain electrodes.

In summary, the DOFET shows that it is possible to dramatically enhance the charge injection and extraction at the metal/semiconductor contacts by means of the nanometre-scale charge diffusion. The enhanced charge injection allowed us to reduce the threshold voltage by more than 15 V, and to increase the field-effect mobility about ten times, approaching the organic semiconductor transport limit also in short-channel transistors. The enhanced charge extraction enables the complete suppression of the channel-length modulation. We show that a short-channel DOFET behaves like an ideal current generator: its channel-length modulation is even smaller than that of an eight times longer organic transistor fabricated in the same technology. These features lead to the fabrication of high performance organic transistors with a unique benefits combination: negligible contact resistance, small device-to-device variability, and exceptionally high gain (>700).

Thanks to the transistor here proposed we theoretically explain and experimentally demonstrate for the first time that the charge diffusion can play a crucial role in organic transistors. Moreover, the ability to independently enhance or reduce the charge injection, transport and extraction in organic semiconductors makes the DOFET the ideal test-bed to study the fundamental physical processes taking place in organic semiconductors and at the metal–organic interfaces.

The proposed approach is a universal method to obtain high-quality contacts without the need of materials or process optimizations. Moreover, according to the approach proposed in ref. 43, the DOFET combined with ambipolar semiconductors could be used to electrically enhance the charge injection of one charge type and to suppress the other. This feature is very relevant for the low-cost fabrication of high-gain and low-power ambipolar complementary electronics.

The diffusion-driven organic transistor opens up new opportunities for the large-scale integration of flexible electronics, high-sensitivity sensors and ultra-large signal amplification in sensing systems.

Methods

Two-dimensional numerical simulations. The coupled drift–diffusion, Poisson and current continuity equations are solved together[43–45]. The simulation parameters are the following: relative permittivity of semiconductor $\varepsilon_{rs} = 3$, relative permittivity of insulators (1 and 2) $\varepsilon_{ri} = 3.757$, highest occupied molecular orbital (HOMO) energy level $E_{HOMO} = 2.8$ eV, lowest unoccupied molecular orbital (LUMO) energy level $E_{LUMO} = 5.2$ eV, effective density of HOMO states $N_{HOMO} = 10^{21}$ cm^{-3}, effective density of LUMO states $N_{LUMO} = 10^{21}$ cm^{-3}, holes effective mobility $\mu_h = 0.1$ cm^2 V^{-1} s^{-1}, electrons effective mobility $\mu_e = 0.1$ cm^2 V^{-1} s^{-1}, metal electrodes work function $\Phi_m = 4.7$ eV (the hole energy barrier at the source/drain metal-semiconductor is $\Phi_B = 0.5$ eV), Schottky barrier lowering $\Delta\Phi_B = e \left[e\, E/(4\,\pi\,\varepsilon_0\,\varepsilon_{rs}) \right]^{\wedge}(1/2)$, where e is the elementary charge, E is the electric field and ε_0 is the vacuum permittivity.

References

1. Kaltenbrunner, M. et al. An ultra-lightweight design for imperceptible plastic electronics. Nature 499, 458–463 (2013).
2. Nomura, K. et al. Room-temperature fabrication of transparent flexible thin-film transistors using amorphous oxide semiconductors. Nature 432, 488–492 (2004).
3. Cao, Q. et al. Medium-scale carbon nanotube thin-film integrated circuits on flexible plastic substrates. Nature 454, 495–500 (2008).
4. Kuribara, K. et al. Organic transistors with high thermal stability for medical applications. Nat. Commun 3, 1–7 (2012).
5. Takei, K. et al. Nanowire active-matrix circuitry for low-voltage macroscale artificial skin. Nat. Mater. 9, 821–825 (2010).
6. Gelinck, G. H. et al. Flexible active-matrix displays and shift registers based on solution-processed organic transistors. Nat. Mater. 3, 106–110 (2005).
7. Minemawari, H. et al. Inkjet printing of single-crystal films. Nature 475, 364–367 (2011).
8. Yan, H. et al. A high-mobility electron-transporting polymer for printed transistors. Nature 457, 679–686 (2009).
9. Yuan, Y. et al. Ultra-high mobility transparent organic thin film transistors grown by an off-centre spin-coating method. Nat. Commun 5, 1–9 (2014).
10. Palfinger, U. et al. Fabrication of n- and p-type organic thin film transistors with minimized gate overlaps by self-aligned nanoimprinting. Adv. Mater. 22, 5115–5119 (2010).
11. Ante, F. et al. Contact resistance and megahertz operation of aggressively scaled organic transistors. Small 8, 73–79 (2012).
12. Arias, A. C., MacKenzie, J. D., McCulloch, I., Rivnay, J. & Salleo, A. Materials and applications for large area electronics: solution-based approaches. Chem. Rev. 110, 2–24 (2010).
13. Natali, D. & Caironi, M. Charge injection in solution-processed organic field-effect transistors: physics, models and characterization methods. Adv. Mater. 24, 1357–1387 (2012).
14. Li, L. et al. The electrode's effect on the stability of organic transistors and circuits. Adv. Mater. 24, 3053–3058 (2012).
15. Léonard, F. & Talin, A. A. Electrical contacts to one- and two-dimensional nanomaterials. Nat. Nanotech 6, 773–783 (2011).
16. Valletta, A. et al. Contact effects in high performance fully printed p-channel organic thin film transistors. Appl. Phys. Lett. 99, 233309 (2011).
17. Raiteri, D., Cantatore, E. & Van Roermund, A. H. M. Circuit Design on Plastic Foils (Springer International Publishing, 2015).
18. Klauk, H. Organic Electronics II: More Materials and Applications (Wiley-VCH, 2012).
19. Ante, F. et al. Contact doping and ultrathin gate dielectrics for nanoscale organic thin-film transistors. Small 7, 1186–1191 (2011).
20. Schaur, S., Stadler, P., Meana-Esteban, B., Neugebauer, H. & Sariciftci, N. S. Electrochemical doping for lowering contact barriers in organic field effect transistors. Org. Electron. 13, 1296–1301 (2012).
21. Gwinner, M. C., Jakubka, F., Gannott, F., Sirringhaus, H. & Zaumseil, J. Enhanced ambipolar charge injection with semiconducting polymer/carbon nanotube thin films for light-emitting transistors. ACS Nano 6, 539–548 (2012).
22. Asadi, K., Gholamrezaie, F., Smits, E. C. P., Blom, P. W. M. & De Boer, B. Manipulation of charge carrier injection into organic field-effect transistors by self-assembled monolayers of alkanethiols. J. Mater. Chem. 17, 1947–1953 (2007).
23. Hamadani, B. H., Corley, D. A., Ciszek, J. W., Tour, J. M. & Natelson, D. Controlling charge injection in organic field-effect transistors using self-assembled monolayers. Nano Lett. 6, 1303–1306 (2006).
24. Liu, Z., Kobayashi, M., Paul, B. C., Bao, Z. & Nishi, Y. Contact engineering for organic semiconductor devices via Fermi level depinning at the metal-organic interface. Phys. Rev. B 82, 035311 (2010).
25. Chai, Y. et al. Low-resistance electrical contact to carbon nanotubes with graphitic interfacial layer. IEEE Trans. Electron Devices 59, 12–19 (2012).
26. Torricelli, F., Ghittorelli, M., Colalongo, L. & Kovacs-Vajna, Z. M. Single-transistor method for the extraction of the contact and channel resistances in organic field-effect transistors. Appl. Phys. Lett. 104, 093303 (2014).
27. Nausieda, I. et al. Dual threshold voltage organic thin-film transistor technology. IEEE Trans. Electron Devices 57, 3027–3032 (2010).
28. Raiteri, D. et al. A 6 b 10 MS/s current-steering DAC manufactured with amorphous Gallium-Indium-Zinc-Oxide TFTs achieving SFDR > 30 dB up to 300 kHz. IEEE Int. Solid-State Circuits Conf. Digest of Technical Papers 55, 314–315 (2012).
29. Cosseddu, P., Vogel, J.-O., Fraboni, B., Rabe, J. P., Koch, N. & Bonfiglio, A. Continuous tuning of organic transistor operation from enhancement to depletion mode. Adv. Mater. 21, 344–348 (2009).
30. Li, Y. et al. Quick fabrication of large-area organic semiconductor single crystal arrays with a rapid annealing self-solution-shearing method. Sci. Rep 5, 13195 (2015).
31. Abdinia, S. et al. Variation-based design of an AM demodulator in a printed complementary organic technology. Org. Electron. 15, 904–912 (2014).
32. Lassnig, R. et al. Optimizing pentacene thin-film transistor performance: temperature and surface condition induced layer growth modification. Org. Electron. 26, 420–428 (2015).
33. Nakayama, K., Hara, K., Tominari, Y., Yamagishi, M. & Takeya, J. Organic single-crystal transistors with secondary gates on source and drain electrodes. Appl. Phys. Lett. 93, 153302 (2008).
34. Hsu, B. B. Y. et al. Split-gate organic field effect transistors: control over charge injection and transport. Adv. Mater. 22, 4649–4653 (2010).
35. Rapisarda, M. et al. Analysis of contact effects in fully printed p-channel organic thin film transistors. Org. Electron. 13, 2017–2027 (2012).
36. Shim, C.-H., Maruoka, F. & Hattori, R. Structural analysis on organic thin-film transistor with device simulation. IEEE Trans. Electron Devices 57, 195–200 (2010).

37. Street, R. A. & Salleo, A. Contact effects in polymer transistors. *Appl. Phys. Lett.* **81,** 2887–2889 (2002).

38. Brondijk, J. J., Torricelli, F., Smits, E. C. P., Blom, P. W. M. & De Leeuw, D. M. Gate-bias assisted charge injection in organic field-effect transistors. *Org. Electron.* **13,** 1526–1531 (2012).

39. Myny, K. *et al.* An 8-Bit, 40-instructions-per-second organic microprocessor on plastic foil. *IEEE J. Solid-State Circuits* **47,** 284–291 (2012).

40. Huang, C., West, J. E. & Katz, H. E. Organic field-effect transistors and unipolar logic gates on charged electrets from spin-on organosilsesquioxane resins. *Adv. Funct. Mater.* **17,** 142–153 (2007).

41. Calhoun, M. F., Sanchez, J., Olaya, D., Gershenson, M. E. & Podzorov, V. Electronic functionalization of the surface of organic semiconductors with self-assembled monolayers. *Nat. Mater.* **7,** 84–89 (2008).

42. Hirata, I. *et al.* High-resolution spatial control of the threshold voltage of organic transistors by microcontact printing of alkyl and fluoroalkylphosphonic acid self-assembled monolayers. *Org. Electron.* **26,** 239–244 (2015).

43. Torricelli, F. *et al.* Ambipolar organic tri-gate transistor for low-power complementary electronics. *Adv. Mater.* **28,** 284–290 (2016).

44. Mariucci, L. *et al.* Current spreading effects in fully printed p-channel organic thin film transistors with Schottky source-drain contacts. *Org. Electron.* **14,** 86–93 (2013).

45. Brondijk, J. J., Spijkman, M., Torricelli, F., Blom, P. W. M. & de Leeuw, D. M. Charge transport in dual-gate organic field-effect transistors. *Appl. Phys. Lett.* **100,** 023308 (2012).

Acknowledgements

We acknowledge funding from the Dutch Technology Foundation STW, which is the Applied Science Division of the Netherlands Organisation for Scientific Research (NWO), and by the Technology Programme of the Ministry of Economic Affairs. We also thank the Polymer Vision for the transistors fabrication. F.T. and D.R. thank Holst Centre for the patent application submission. F.T. thank Matteo Ghittorelli for the useful discussions, suggestions and for his contribution to Figure 2.

Author contributions

F.T. conceived the DOFET architecture, designed the devices, performed the 2D numerical calculations, measured the transistors, analysed the data, developed the physical analysis and wrote the manuscript. L.C. contributed to the physical analysis and wrote the manuscript. D.R. contributed to the DOFET architecture idea and performed the measurements. Z.M.K.-V. contributed to the performance analysis, supported the manuscript preparation and revised the manuscript. E.C. supervised the project and revised the manuscript. All the authors discussed the results and commented on the manuscript.

Additional information

2

Subnanometre enzyme mechanics probed by single-molecule force spectroscopy

Benjamin Pelz[1], Gabriel Žoldák[1], Fabian Zeller[2], Martin Zacharias[2] & Matthias Rief[1,3]

Enzymes are molecular machines that bind substrates specifically, provide an adequate chemical environment for catalysis and exchange products rapidly, to ensure fast turnover rates. Direct information about the energetics that drive conformational changes is difficult to obtain. We used subnanometre single-molecule force spectroscopy to study the energetic drive of substrate-dependent lid closing in the enzyme adenylate kinase. Here we show that in the presence of the bisubstrate inhibitor diadenosine pentaphosphate (AP5A), closing and opening of both lids is cooperative and tightly coupled to inhibitor binding. Surprisingly, binding of the substrates ADP and ATP exhibits a much smaller energetic drive towards the fully closed state. Instead, we observe a new dominant energetic minimum with both lids half closed. Our results, combining experiment and molecular dynamics simulations, give detailed mechanical insights into how an enzyme can cope with the seemingly contradictory requirements of rapid substrate exchange and tight closing, to ensure efficient catalysis.

[1] Physik Department E22, Technische Universität München, James Franck Strasse 1, 85748 Garching, Germany. [2] Physik Department T38, Technische Universität München, 85748 Garching, Germany. [3] Munich Center for Integrated Protein Science, 81377 München, Germany. Correspondence and requests for materials should be addressed to M.R. (email: mrief@ph.tum.de).

Enzymes are machines that catalyse biochemical reactions in a highly dynamic process often involving large conformational changes. How conformational changes are coupled to catalytic turnover is still controversial[1,2]. Experimental techniques employed to answer those important questions are often limited and, although we have high-resolution crystal structures of selected states along the reaction pathways, capturing reaction dynamics and structure at the same time is difficult. As conformational changes are transitions in an energy landscape involving interatomic distance changes, enzymes may be viewed as molecular machines[3]. Single-molecule mechanical methods such as atomic force microscopy or optical tweezers have opened new ways for studying conformational transitions of molecular motors and in protein and nucleic acid folding[4–9]. In most enzymes, however, conformational motions associated with catalysis are generally very small (<1 nm), involving only few residues, and hence have so far not been directly amenable to those techniques. In this study, we have employed a high-resolution optical tweezers setup to study the mechanical coupling of substrate binding and lid closing in the enzyme adenylate kinase (AdK).

AdK catalyses the reversible conversion of ATP and AMP to two ADPs and thereby plays an important role for the energy balance of the cell. Adk consists of three domains. The ATP- and AMP-lid are responsible for binding of substrates and the CORE domain governs the overall stability of the enzyme[10,11]. During its catalytic cycle, AdK undergoes a large conformational change as both the ATP- and AMP-lid close over the CORE domain, thereby reducing unproductive active site fluctuations and minimizing non-productive hydrolysis[12–14]. Opening of AdK and subsequent product release have been suggested as the rate-limiting step for enzymatic turnover[1,15,16]. A variety of experiments and simulations have indicated that the enzyme samples closed-like states even in the absence of substrates[16–19]. Bisubstrate analogues consisting of two adenosine groups connected by a chain of phosphates, such as AP5A, have been proven strong inhibitors of AdK and bind with nanomolar affinity[20,21]. AP5A is a multi-substrate inhibitor, which induces closing of the enzyme and keeps the lids firmly shut as has been shown by multiple crystal structures[1,14]. Therefore, the AP5A-bound state has served as a model for the fully closed (fc) state of AdK. Although crystal structures convey a picture of a tightly closed substrate-bound conformation, nuclear magnetic resonance (NMR) studies indicate much more dynamics[1]. In this study, we used single-molecule optical tweezers to directly measure the substrate-dependent forces that drive AdK into a closed conformation.

acting on the protein–DNA system will vary and different forces will result in different DNA extensions that will add to the measured conformational change of the protein. To remove these linker effects, we performed a DNA elasticity correction (for details, see Methods). By changing the trap distance, the applied force bias can be varied and thus shifting the equilibrium of the conformational fluctuations of the protein.

In the presence of AP5A, we observed clear two-state transitions of AdK between an extended (lower level) and a contracted state (upper level) different by $ca.$ 1.6 nm in length. We identify the contracted conformation with the lid-closed conformation of AdK and the extended state with the lid-open conformation. This interpretation is supported by the finding that in the absence of an inhibitor, only the extended state is populated under load, while the contracted conformation is populated only in the presence of an inhibitor (see Supplementary Fig. 1). Our data directly show that closing and opening of both lids is cooperative, as the molecules always transition between the fully open and fc states. This interpretation is supported by a mutant where we attach our handles to only one lid (ATP lid) and the CORE domain (for details see Methods). We find conformational changes of only half the size ($ca.$ 0.9 nm) albeit with identical kinetics (Supplementary Fig. 4).

The conformational transitions of AdK strictly depend on the concentration of inhibitor present (see Fig. 2c and Supplementary Fig. 5a). The closing rate rises proportionally to concentration over a large investigated range covering more than three orders of magnitude (Fig. 2c). In contrast, lid opening is independent of concentration (Fig. 2d and Supplementary Fig. 5b). The kinetics can be extrapolated to zero-force conditions (see asterisks in Fig. 2c,d). The values we find for closing ($0.18 \, nM^{-1} s^{-1}$) and opening ($1 \, s^{-1}$), as well as the associated equilibrium constant (3 nM) almost exactly match published values for AP5A binding[23] and affinity[21] measured in bulk assays for *Escherichia coli* AdK. This agreement reveals a tight coupling of inhibitor binding and lid closing in the enzyme. Apparently, the complete gain in free energy on inhibitor binding is invested to keep the lids firmly shut.

Interestingly, the force dependence of the rates differs dramatically for lid closing as compared with opening. Although lid closing exhibits only a very weak dependence on force (shallow slopes in Fig. 2c), lid-opening rates couple much more strongly and change almost two orders of magnitude over the force range investigated (Fig. 2d). The independence of the closing rate on force provides direct evidence for an induced-fit mechanism of inhibitor binding. Induced-fit describes a binding mechanism where substrate binds to an open state and alters the

Results

Conformational dynamics in the presence of an inhibitor. We designed a mutant of a thermophilic variant of AdK from *Aquifex aeolicus*[1], wherein we inserted cysteine residues at positions 42 and 144 that allowed us to bind DNA handles to the enzyme and then subject the molecule to mechanical loads in an optical trap (see Figs 1 and 2a). The residues were chosen such that the distance change between the two attachment points would be maximal ($ca.$ 1.7 nm) on closing and opening of the lids as judged by the crystal structure (see Fig. 1). The variant showed wild-type activity for substrate turnover in bulk assays (see Methods). In a first set of experiments, we investigated the conformational fluctuations at constant trap positions (passive mode) in the presence of the inhibitor AP5A (see Fig. 2b). The distance between the two trap centres is held constant, imposing a constant average force bias on the fluctuating beads[4,22]. As only the distance between the two traps is held constant, the force

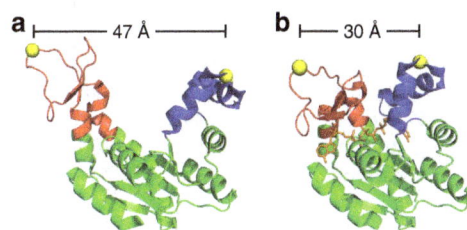

Figure 1 | Structure of AdK. The three-dimensional structure[34] of the open (**a**) and closed (**b**) conformation of AdK from *A. aeolicus* (PDB ID Code 2RH5 and 2RGX). The CORE domain is labelled in green, and ATP and AMP lid in red and blue, respectively. The closed conformation is shown in the presence of the bisubstrate inhibitor AP5A (orange). The yellow spheres indicate the position of the inserted cysteines between position 42 and 43, and 144 and 145. The distance between the cysteines in the open and closed conformation is shown above.

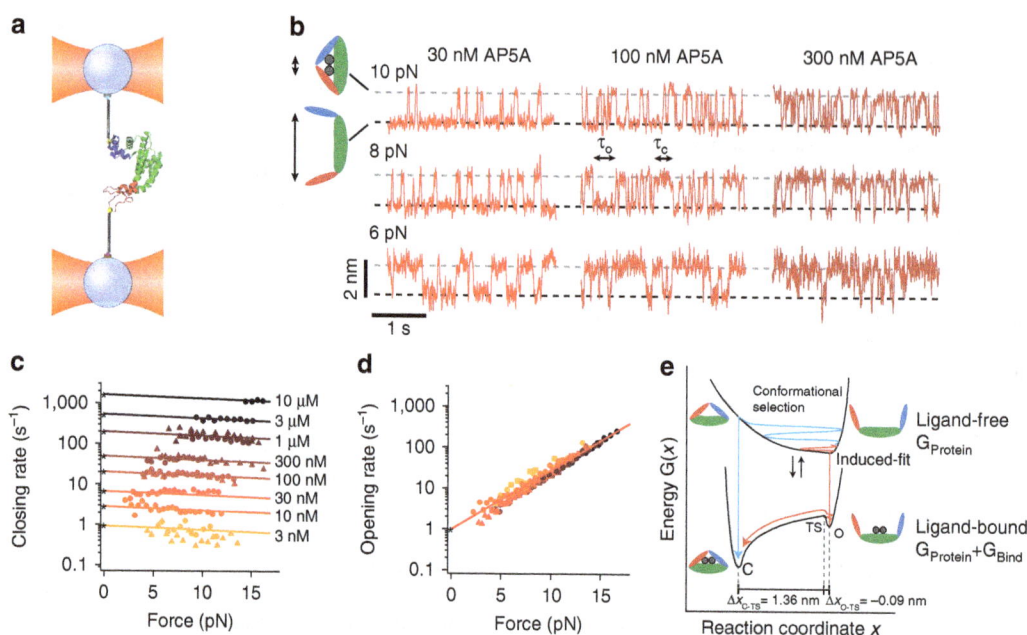

Figure 2 | Single-molecule force experiments of AdK by optical tweezers. (**a**) Optical trap assay (for details, see Methods). (**b**) Sample traces of the closing and opening fluctuations of AdK at different Mg-AP5A concentrations and force biases. The grey and black dashed lines indicate the position of the closed (contracted) and open (extended) state, respectively (see also Supplementary Fig. 2). (**c**) Closing rate as a function of force for different Mg-AP5A concentrations. Solid lines are extrapolations of the closing rates to zero force, which are indicated by the asterisks (for details, see Methods). (**d**) Opening rate as a function of force for different Mg-AP5A concentrations (colours as in **c**) (see also Supplementary Fig. 3). The solid line is a fit extrapolating the opening rate to zero force (asterisk). (**e**) Comparison of two binding and closing models for AdK and AP5A adapted from Okazaki and Takada[45]. Binding and unbinding of the inhibitor is represented as the jump between a ligand-free and ligand-bound energy landscape. Exemplary routes from open ligand-free to closed ligand-bound form are shown for the conformational selection (blue) and induced-fit model (red). For the ligand-bound energy landscape, distances from open and closed state to the transition state from optical trap experiments are shown.

energy landscape such that the protein is strongly driven into the closed state (see Fig. 2e red pathway). An alternative mechanism would involve a conformational selection where the protein fluctuates between open and closed conformations in the unbound state. Here, substrate recognizes the (rare) closed conformations in this ensemble, binds and locks the protein in this conformation (Fig. 2e blue pathway). As closing rates in our measurements are force independent, a conformational-selection mechanism can be ruled out, because force would reduce the probability for finding a closed conformation in the unbound state and hence also reduce binding and closing rates.

In a conventional view, force-dependent opening rates and force-independent closing rates indicate a transition state position close to the open state (see Fig. 2e bottom) further supporting the induced-fit mechanism. It is important to note that an induced-fit mechanism is well consistent with the unbound state being flexible and able to sample closed-like conformations as has been inferred from single-molecule fluorescence experiments and molecular dynamics (MD) simulations[16–18]. However, we can rule out that significant inhibitor binding occurs through those closed conformations, which is consistent with the crystal structure being so tightly closed that direct binding of inhibitor into the buried binding site appears unlikely[14].

Nucleotide binding and conformational fluctuations. Under catalytic conditions with substrates, AdK has to be much more dynamic, as the enzyme must ensure rapid turnover. In a next set of experiments, we therefore investigated the lid-closing mechanics in the presence of the substrate molecules ATP and ADP. In contrast to AP5A, we could not detect any conformational kinetics in the presence of either of the substrate molecules (Supplementary Figs 6 and 7). Given the expected fast

conformational fluctuations reported in NMR and in fluorescence experiments under saturating substrate conditions[1,12,16], we suspected those fluctuations to be buried in the noise in our experiments. To assess the magnitude of the conformational transition with substrate, we therefore designed a competition experiment with fixed concentrations of AP5A and varying concentrations of substrate molecules (see Fig. 3a). This assay allows us to monitor the full conformational transition induced by AP5A and, at the same time, follow any changes induced by substrates.

With increasing ATP concentrations (Fig. 3a top to bottom), the open dwell times increase significantly, whereas the closed dwells remain unaffected. Apparently, ATP binds to the open state and blocks rebinding of AP5A. As an example, the open dwells increase their lifetime from 5 ms (0 μM ATP, upper trace in Fig. 3a) to about 0.5 s at 3,000 μM (lowest trace in Fig. 3a). This means that during 99% of the open dwell time, ATP is bound to at least one binding site, therefore blocking the rebinding of AP5A from solution. This increase in dwell time provides a measure for the binding affinities of the various substrates to AdK (Fig. 3b and for details, see Methods). All affinities agree within error with published affinities from bulk experiments with no mechanical loads applied[21], proving that our tethered enzyme has fully native substrate-binding properties. Further evidence that substrates bind directly to the enzyme even without AP5A present comes from experiments where we pulled AdK to forces where part of the structure unfolds (Supplementary Fig. 6). Constant velocity experiments in the presence of nucleotides (see Supplementary Fig. 6) suggest that the observed equilibrium transition is caused by the folding and unfolding of the ATP lid. This is supported by previous studies showing that the ATP and AMP lid of the mesophilic AdK from *E. coli* have a lower thermodynamic stability than the CORE domain with the ATP

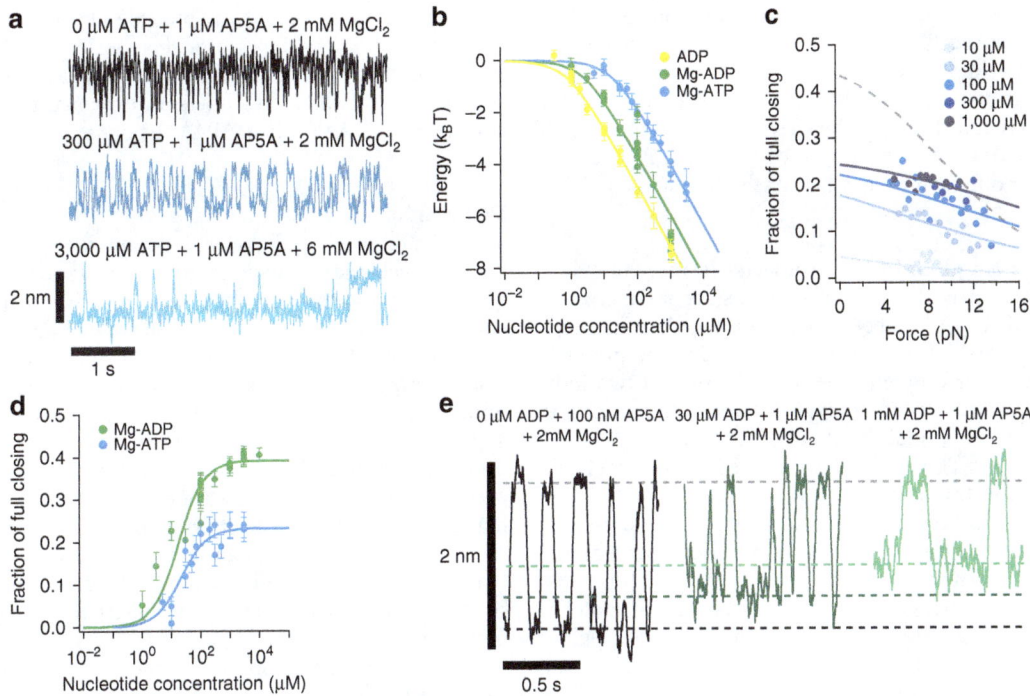

Figure 3 | Nucleotide-induced conformational changes of AdK. (**a**) Sample traces of AP5A-induced closing and opening in the single-molecule competition assay at 1 μM AP5A with varying concentrations of ATP at a force of 10.5 pN. (**b**) Decrease in apparent binding free energy of AP5A in the single-molecule competition assay for the nucleotides ADP (yellow), Mg-ADP (green) and Mg-ATP (blue). Solid lines are fits of the binding model described in Methods (equation (2)) resulting in the following dissociation constants: $K_D^{ADP} = (0.8 \pm 0.1)$ μM, $K_D^{Mg-ADP} = (2 \pm 0.1)$ μM and $K_D^{Mg-ATP} = (17 \pm 2)$ μM. (**c**) ATP-induced fraction of full closing as a function of force for different ATP concentrations from the AP5A competition assay. Solid lines are fits to equation (7) yielding a size of the conformational change of $\Delta x = 0.5$ nm. The dashed line shows the force-dependent closing expected if complete ATP lid closure is assumed. (**d**) Fraction of full closing at zero force induced by nucleotides Mg-ADP (green) and Mg-ATP (blue) as a function of nucleotide concentration extracted from the AP5A competition experiments. Solid lines are fits to the data assuming a binding model (equation (12)) with an affinity of the nucleotides to open and closed state of AdK. The resulting dissociation constants are $K_{close}^{Mg-ADP} = (19 \pm 2)$ μM, $K_{open}^{Mg-ADP} = (290 \pm 80)$ μM and $K_{close}^{Mg-ATP} = (25 \pm 5)$ μM and $K_{open}^{Mg-ATP} = (134 \pm 60)$ μM. (**e**) ADP-dependent lid closing from AP5A competition experiments. The upper dashed line indicates the position of the fc AP5A-bound state and the lower three lines indicate the position of the ADP-dependent AP5A-unbound state.

lid unfolding already at temperatures over 35 °C (refs 11,24). The higher conformational stability of the hyperthermophilic variant from *A. aeolicus* we use in our study explains why ATP lid melting is negligible at room temperature and needs substantial mechanical force to be induced. We observe that in the presence of substrates this equilibrium unfolding/folding transition is shifted to higher values demonstrating that substrate has bound and stabilized the protein structure. From these experiments we can independently extract dissociation constants from substrate binding, which lie in a very similar range as the ones obtained from the competition assay (for details, see Methods and Supplementary Fig. 8).

In contrast to AP5A (see above), the application of load apparently does not affect the binding of substrates to AdK as can be seen from the similarity of solution affinities with the ones we obtain at a pulling force of 10 pN. This provides evidence that substrate binding is not coupled to significant closing of the enzyme. Our competition assay also yields a direct comparison of the degree of lid closing between the substrate-bound and inhibitor-bound states. Even though our temporal resolution does not allow us to capture the individual conformational fluctuations induced by the rapid substrate binding/unbinding events, we can make a clear statement about the average degree of closing. Again, in the case of 3,000 μM ATP (see lower trace in Fig. 3a), the apparent closing distance between open and closed dwells induced by AP5A binding is almost identical to the one measured without ATP. This means that even though AdK carries ATP

during 99% of the time in the open dwell, on average, the enzyme is nearly open. If ATP binding induced full closing of the ATP lid, the apparent position of the open dwell level would have to be half its original value.

A closer analysis of the substrate-induced shift of the lower dwell levels is shown in Fig. 3c. For ATP, we find a substrate-dependent shift of the lower level (fraction of full closing) of *ca.* 0.2 as compared with the full closing distance. Apparently, the lid closes, but much less than full closing of one lid (0.5), let alone both lids (1.0). Two possible explanations outlined in Supplementary Fig. 10 could explain this result: (1) ATP binding occurs to a conformation with an only partially closed lid (pc) or (2) substrate binds to the open state with full affinity but drives AdK only weakly into a fc conformation resulting in a rapid equilibrium between fc and the open (o) conformation. Both of these explanations would support a conclusion that ATP binding does not strongly drive AdK into an fc ATP-lid state.

Analysis of the force dependence of the fraction of full closing (Fig. 3c) provides evidence for explanation (1) and hence the existence of a free-energy minimum in the AdK energy landscape at a pc position. We find that force only weakly affects the fraction of full closing as can be seen from the shallow slopes of the data points at all ATP concentrations. A fit to the data is consistent with a pc state (solid line fits) but not with an fc lid (dashed line; for a detailed model, see Methods). We cannot exclude transient population of the fc state; however, the major population will be the pc state. The concentration-dependent degree of ATP-induced lid closing extrapolated to zero force is shown in Fig. 3d (blue).

Similarly, in the presence of ADP, which binds to both lids, we also do not observe full closing of the enzyme. Even at the highest concentrations of ADP (1–10 mM) (see right trace in Fig. 3e), the ADP-bound level only approaches half of the fc state (Fig. 3d).

Phosphate linker length of inhibitor affects closing. How does the strong energetic drive towards the closed state of the enzyme in the presence of AP5A fit together with the population of a pc state and only a weak drive towards the fc state in the presence of substrates? A major difference between AP5A and two individual substrates is the rigid linkage between the adenosine moieties in AP5A, while individual substrates are free to move relative to each other. The rigid linkage in the inhibitor probably puts significant constraint on the conformational flexibility of AdK. Coupling of the relative distance between the substrate-binding sites to the degree of lid closing could explain why the inhibitor locks the enzyme into the closed state: if only the closed conformation offers a binding site distance that matches the length of the phosphate linker, AP5A can only bind tightly to the closed state. We tested this hypothesis by measuring lid-closing distances with the inhibitors AP4A and AP6A, where the two adenosine moieties are linked with four or six phosphate groups, respectively (see Fig. 4a), thus altering their relative distance.

We found that AP4A induces full closing of AdK indistinguishable from AP5A (see Fig. 4b). In contrast, AP6A only closes to about 0.6 nm, corresponding to a fraction of full closing of 0.4. One might argue that reduced closing due to AP6A as compared with the two other inhibitors is owed to the longer and bulkier phosphate chain that cannot be completely accommodated into an fc structure. However, the substrate-dependent closing and opening rates suggest a different explanation: we find that AP4A has a 100-fold reduced closing rate (see Fig. 4c), which is consistent with a similarly reduced affinity for AP4A measured in bulk experiments[21]. For AP6A, we measure closing/binding rates that are identical to the ones of AP5A, showing that AP6A binds with the same affinity as AP5A despite its significantly reduced closing distance (see Fig. 4d). Zero-force opening rates for AP5A, AP4A and AP6A are largely identical (see Fig. 4e). It is noteworthy that the force dependence for opening of AP6A is weaker (smaller slope), consistent with the smaller conformational change (see Methods). Our results suggest that if the distance between the two adenosine moieties is larger, full lid closing for high affinity binding is not required. Hence, under these conditions, a pc state is already an optimal configuration.

Atomistic MD free-energy simulations of AdK bound to all three inhibitors support our conclusion and provide additional structural insight (Fig. 4f and for details, see Methods). In all cases, a significant free-energy minimum at a lid closing of ~50% was observed, consistent with our interpretation that a pc state of the enzyme is a dominant feature of its energy landscape. For AP4A and AP5A binding with an added load of 15 pN, the fc state is of lower free energy than the pc state. In contrast, under the same conditions, for AP6A the pc state is lower in free energy than the fc state. Although a precise quantitative prediction of free energies cannot be expected given the known shortcomings of MD simulations, this result is at least in qualitative agreement with the experiment and in line with the experimentally observed dominant closing fractions in the inhibitor-bound states. Analysis of the simulation trajectories offers a structural explanation (see Supplementary Fig. 11). For the longer inhibitors, especially AP6A, a simultaneous binding of the two adenosine moieties to

the corresponding lid sites is possible already at partially open AdK conformations. In case of the shorter AP4A inhibitor this stable binding mode is only possible at more closed lid configurations, which also offers an explanation for the reduced closing rate of AdK when AP4A is bound.

Discussion

The possibility of studying native-state conformational energies using mechanical single-molecule methods allows us to quantitatively determine the energetic drive of conformational changes of an enzyme induced by substrate binding. Even though structural studies have suggested firmly closed states, our results now provide an important enhancement of our understanding of AdKs conformational dynamics. The two apparently contradictory requirements of firm closing ensuring catalytic activity and flexibility to ensure rapid substrate/product exchange are met by (1) a strong drive towards a pc state with high substrate affinity still allowing for substrate exchange and (2) a much weaker drive into an fc and catalytically active structure that is only transiently visited. These results show that subnanometre force spectroscopy opens up new possibilities to directly study the energetic balance of enzyme conformations and help answering the question how these conformations couple to substrate binding.

How can our results be reconciled with reports from the literature that substrates induce full closing of AdK[1,12,25]? It is important to note that our measurements always occur under a certain amount of mechanical load. Therefore, a slight bias within the enzyme towards an fc state at zero load is well consistent with our findings. For example, Wolf-Watz et al.[1] reported that under saturating substrate condition AdK rapidly fluctuates between closed and open conformations with $k_{close} = 1,370\,s^{-1}$ and $k_{open} = 40\,s^{-1}$, whereas Hanson et al.[16] measure $k_{close} = 400\,s^{-1}$ and $k_{open} = 160\,s^{-1}$ for E. coli AdK. Both these measurements indicate the open state is populated to between 3 and 30%, consistent with our finding that despite the large substrate concentrations, the binding energy does not shift the equilibrium strongly to the closed state. Our data now explain where most of the substrate-binding free energy goes: it is used for populating a pc state. This may be important for the function of AdK as an enzyme that rapidly turns over a substrate and has to allow quick substrate exchange at the same time.

The results from the experiments with the three bisubstrate inhibitors AP4A, AP5A and AP6A can give a structural explanation for the occurrence of the pc state. Although AP4A and AP5A induce conformations where the adenosine moieties are close together, AP6A-bound AdK settles at a conformation where they are further apart (Fig. 4g right). The left branch of Fig. 4g provides a picture of how the different states might blend into the catalytic cycle of AdK: when two unlinked substrate nucleotides bind, they will rapidly settle at an energetic minimum where the lids are half-closed with little energetic drive bringing them to the fc state. This will be the main substrate-bound state of the enzyme. In this state, catalysis cannot occur, because substrates are too far apart. Transiently, the fc state will then be visited in a process driven by thermal fluctuations. It is noteworthy that the fc state may even be a high-energy state and thus not energetically favoured (that is, no favourable energy bias) at least under the forces we apply. This fc state is essential for catalysis, because water molecules need to be excluded to ensure efficient phosphate transfer[12,13]. Even more than a single transition to the fc state may be necessary until catalysis occurs. After catalysis has occurred, the enzyme will switch back to the half open state and eventually release products. For rapid substrate turnover, it is important that the lids be not locked by favourable energy bias in an fc state but

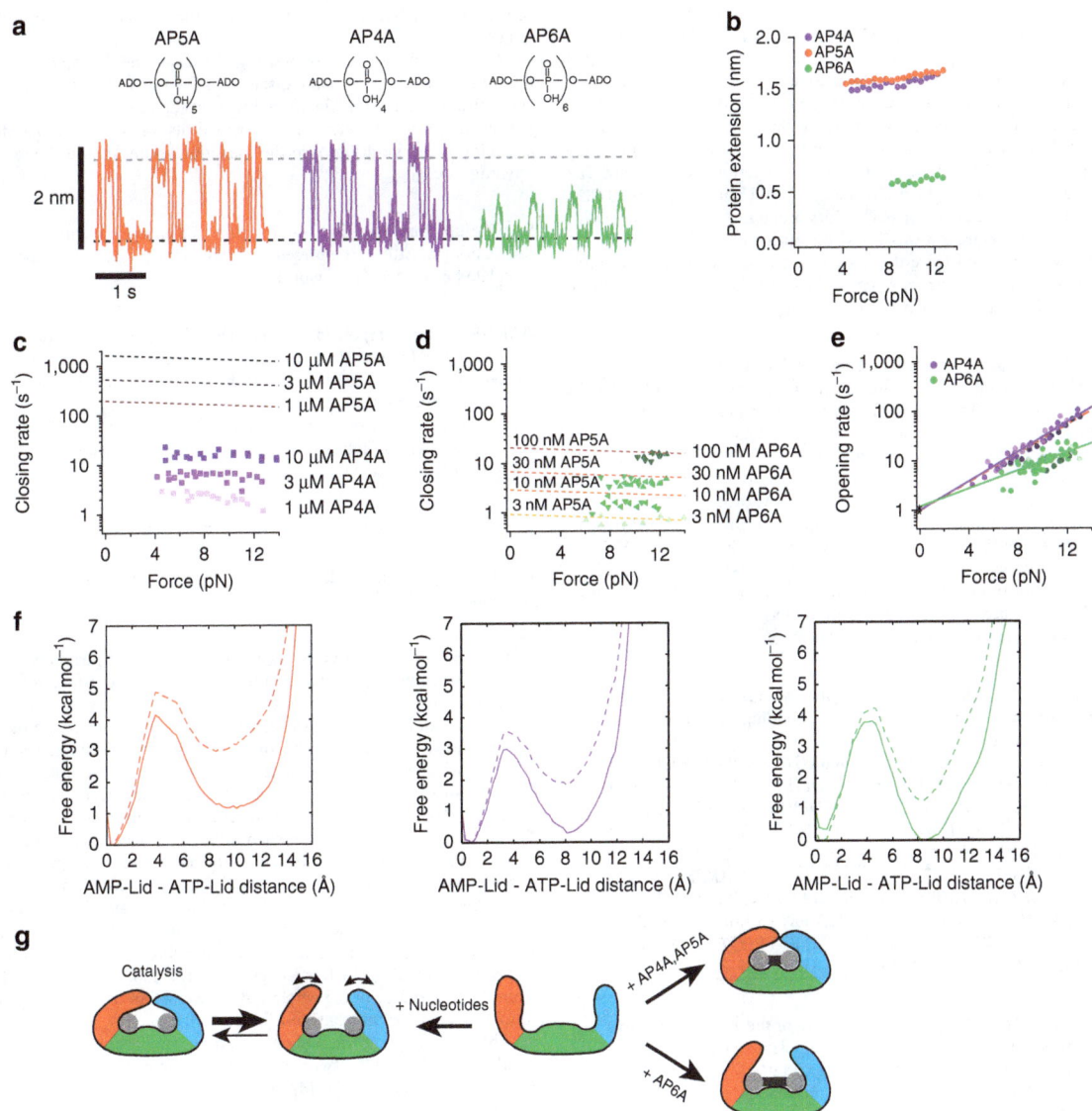

Figure 4 | Effect of the phosphate linker length between the adenosine moieties. (**a**) Equilibrium traces of the closing and opening of AdK in the presence of AP5A (left panel), AP4A (middle panel) and AP6A (right panel) at a force of 8 pN for AP5A and AP4A, and 12 pN for AP6A. The dashed lines represent the position of the fc (grey) and open (black) state of AdK. (**b**) Distance between closed and open state as a function of force in the presence of the different inhibitors AP5A (red), AP4A (purple) and AP6A (green). (**c–d**) Closing rate of AdK as a function of force for different AP4A (**c**) or AP6A (**d**) concentrations. It is noteworthy that the dashed lines are not fits to the AP4A and AP6A data, but fits to the rates obtained at the corresponding AP5A concentrations for comparison. (**e**) Opening rate of AdK as a function of force for different AP4A (purple) and AP6A (green) concentrations. Solid lines are fits to the closing rates extrapolated to zero force (asterisks). For comparison, a fit to the rates obtained at the corresponding AP5A concentrations is shown as a red dashed line. (**f**) Calculated free energy along the opening of AdK bound to AP5A (red), AP4A (purple) and AP6A (green) obtained from free-energy simulations without load (dashed lines) and with a subsequently added bias corresponding to an external load of 15 pN (solid lines). The free-energy curves are projections of free-energy landscapes (Supplementary Fig. 11) obtained from 2D H-REMD-US simulations in which the arrangements of AMP lid and ATP lid were controlled independently. (**g**) Model of the coupling between the conformational fluctuations and the binding of inhibitors or nucleotides to AdK. In the presence of nucleotides, a pc state is stably populated under force; however, AdK is still able to close fully especially in the absence of force. The binding of AP4A and AP5A induce the fc conformation of AdK, whereas the binding of AP6A stabilizes a pc state.

they open rapidly to allow product–substrate exchange. We think that the pc state ensures both rapid substrate exchange and high-affinity binding through establishing contacts between substrates, core domain and lids.

Why can we not see a kinetic time constant attributable to catalysis directly in our data? A turnover rate of $25\,\mathrm{s}^{-1}$ (that is, 40 ms) in solution lies within the limit we should be able to resolve even in our filtered traces. However, as the fc state is only transiently populated, the lids will be only fully closed during a small fraction of those 40 ms. In this case, we are not able to pick up a signal related to catalysis in the assay. In this context, it is

important to note that we only have a readout for length changes but not for chemical changes.

Moreover, in our assay, we always apply loads around 10 pN, which, in turn will speed up opening kinetics and slow down closing kinetics. If force slows transitions to the fc state also, catalysis will be slowed and may be even suppressed, offering another explanation for why we cannot observe a signal associated with catalysis. A combination of mechanical and fluorescent single-molecule measurement may, in the future, give more insight into coupling between mechanical and chemical steps.

Methods

Experimental procedures. The setup used was a custom-built, dual-beam, high-resolution optical trap setup as described previously[26,27]. Data were collected at 150 kHz and then averaged to 30 kHz before storage. Transitions between open and closed, and closed and open state were detected using Hidden Markov Model analysis on the unfiltered raw data of the difference signal of the two traps as described before[28]. The data at constant trap position shown throughout the study were low-pass filtered to 37.5 Hz. All measurements were performed in 50 mM Tris pH 7.5, 200 mM KCl, 26 U ml^{-1} glucose oxidase, 17,000 U ml^{-1} catalase, 0.65% glucose with varying concentrations of nucleotides, inhibitors, MgCl$_2$ and EDTA. The MgCl$_2$ or EDTA concentration was 2 mM for nucleotide and inhibitor concentration below 1 mM, and twice the nucleotide concentration for nucleotide concentrations above 2 mM. All errors given are statistical errors (1 s.d.) derived from the Levenberg–Marquardt fit algorithm.

Protein expression and purification. The gene for AdK of *A. aeolicus* was synthesized by Mr Gene GmbH (Regensburg). The gene was cloned into pET11a using BamHI and NdeI restriction sites. Cysteine residues were inserted between positions 42 and 43, and 144 and 145 (wt-numbering) by QuikChange mutagenesis (Agilent). To observe conformational fluctuations of the ATP-lid only, a second mutant was produced with a cysteine at the amino terminus and a cysteine residue inserted between positions 144 and 145 (AdK N-144). The proteins were expressed in *E. coli* BL21(DE3) and purified by a Ni-NTA affinity column followed by gel filtration chromatography (HPLC system, Jasco Germany, GmbH) using YMC diol-120 columns (YMC Europe, GmbH) and Superdex S200 (10/300 GL, GE Healthcare). Attachment of oligonucleotides and purification optical trap assays were performed as described previously[4,29], with an additional Ni-NTA affinity purification step for protein-oligonucleotide purification.

Protein sequences. In the following, the cysteines that are used for handle attachment are shown in bold, the sequence corresponding to AdK is shown in italics and the spacers are underlined.

AdK 42-144. <u>MAAKGELSGMI</u>*LVFLGPPGAGKGTQAKRLAKEKGFVHIS TGDILREAVQKG***C***TPLGKKAKEYMERGELVPDDLIIALIEEVFPKHGNVIFD GFPRTVKQAEALDEMLEKKGLKVDHVLLFEVPDEVVIERLSGRRINPET GEVYHVKYNPPPPG***C***VKVIQREDDKPEVIKKRLEVYREQTAPLIEYYKKKG ILRIIDASKPVE EVYRQVLEVIGDGNGTGSKALE<u>HHHHHH</u>*

AdK N-144. <u>MA</u>***C***<u>KGELSGMI</u>*LVFLGPPGAGKGTQAKRLAKEKGFVHIST GDILREAVQKGTPLGKKAKEYMERGELVPDDLIIALIEEVFPKHGNVI FDGFPRTVKQAEALDEMLEKKGLKVDHVLLFEVPDEVVIERLSGRRINPE TGEVYHVKY NPPPPG***C***VKVIQREDDKPEVIKKRLEVYREQTAPLIEYYKK KGILRIIDASKPVEEVYRQVL EVIGD*<u>GNGTGSKALEHHHHHH</u>

Enzymatic activity of AdK. The enzymatic activity of the AdK mutant was tested by a coupled spectrophotometric assay[30] (Supplementary Figs 12 and 13). The resulting Michaelis–Menten constant K_m (41 ± 11 µM for ATP, ~10 µM for AMP and 360 ± 80 µM for Mg-ADP/ADP) and the catalytic rate k_{cat} (25 ± 7 s^{-1}) are in good agreement with the values of the wild-type protein reported by Wolf-Watz *et al.*[1]

DNA elasticity correction. As our measurements occur at non-constant force, comparing the conformational changes of AdK under different experimental conditions requires corrections for trap and DNA compliances. We therefore corrected all traces obtained at constant trap distances for the non-Hookean elastic behaviour of the double-stranded DNA linker. Briefly, a force–extension curve of a DNA–protein construct was measured. Data, force versus time, were acquired, and from known distance and trap stiffness, force–extension curves were calculated. Next, the data were fitted with a worm-like chain (WLC) model to account for the DNA tether (~360 nm). From the now known WLC model parameters, the extension of the DNA handles can be calculated for each data point and subsequently subtracted. This 'constant-force' transformation removes variations in the DNA linker elasticity and trap stiffness, and allows for a better comparison of the conformational changes of AdK between different experiments. All data shown are subjected to this procedure.

Extrapolation of closing and opening rates. For determining the opening and closing rates at zero force $k_{0,i}$ and distances to the transition state Δx, the force-dependent opening and closing rates $k_i(F)$ were fitted with a Bell model[31]:

$$k_i(F) = k_{0,i} \exp\left(\frac{F \Delta x_i}{k_B T}\right), \tag{1}$$

where k_B is the Boltzmann constant and T is the absolute temperature. The distance to the transition state is a measure for the force dependence of the respective rate. A small Δx results in a small force dependence of the corresponding rate, whereas a large Δx causes the opening or closing rate to be strongly dependent on force. This behaviour can be observed for the conformational change in the presence of AP5A and AP6A. Both feature a similar zero-force opening rate; however, the force dependence of the AP5A opening rate is much bigger, as the size

of the conformational change and with it the distance to the transition state is larger.

Bell model fits were performed as global fits to the closing rate for the individual inhibitors, assuming the same distance to the transitions state Δx for all concentrations and individual zero force closing rates $k_{0,close}$. The Bell model fits to the opening rate were performed as global fits for the individual inhibitors, assuming the same distance to the transitions state Δx and the same zero force opening rate $k_{0,open}$.

Equilibrium free energies. Equilibrium free energies of the conformational fluctuations of AdK in the presence of inhibitors and in the competition assay were calculated as described previously[4].

Affinities from competition assay. The affinity of the substrates to AdK was determined as follows. First, the binding free energy of AP5A at various AP5A concentration, ΔG^0, was determined in the absence of nucleotides as described before. Second, the decrease in binding free energy due to nucleotide binding in the AP5A competition assay was determined by comparison with the AP5A-only conditions. The nucleotide concentration-dependent decrease in binding free energy can be described by

$$-\frac{\Delta G^0}{k_B T} = -\ln\left(1 + \frac{[L]}{K_D}\right), \tag{2}$$

where $[L]$ is the nucleotide concentration and K_D the dissociation constant of the nucleotides to AdK[32].

ATP lid unfolding and free energy. Stretching AdK at positions 42 and 144 shows a continuous hump-like equilibrium transition around a force of 17 pN in force extension traces (see Supplementary Fig. 6). These fast unfolding and folding transitions can be observed between 13 and 20 pN. To calculate the free energy of this transition, we integrated the force distance curves. The free energy of the whole system, including the stretching of the DNA linker, is given by the area under the force distance curve of a constant velocity experiment (for a clearer representation, Supplementary Fig. 8a shows the corresponding curves as force extension traces). To calculate the energy of the system in the unfolded state, a serial combination of DNA and polypeptide elasticity is fit to the data and the energy of the system in the unfolded state is given by the area under the fit. To reduce the influence of noise at low forces, a WLC model is also fit to the folded state. The energy of the whole system is calculated by the sum of the area under the WLC fit to the folded state up to a force of 10 pN and the area under the actual curve (orange) for forces above 10 pN. By subtracting the free energy of the system in the unfolded state (grey area) from the free energy of the whole system, we obtain the free energy of the folding/unfolding equilibrium transition (blue area).

The free energy of the folding/unfolding equilibrium transition in apo conditions is $(11.7 \pm 0.7) k_B T$. The increase in free energy of this transition in presence of the nucleotides Mg-ADP and Mg-ATP is shown in Supplementary Fig. 8b.

The free energy in the presence of nucleotides can be described similar to the competition assay as

$$-\frac{\Delta G_0}{k_B T} = -\frac{\Delta G_0^{L=0}}{k_B T} - \ln\left(1 + \frac{[L]}{K_D}\right), \tag{3}$$

where $[L]$ is the nucleotide concentration, $-\frac{\Delta G_0^{L=0}}{k_B T}$ is the free energy in apo conditions and K_D the dissociation constant of the nucleotides to AdK.

Nucleotide-induced conformational fluctuations. To determine the size of the conformational change induced by the nucleotides, the measured size of the AP5A-induced conformational fluctuations in the competition experiment is compared with the one at AP5A-only conditions. The fraction of closing induced by the nucleotides in the competition experiment is given by the difference between the size of the conformational change at AP5A-only conditions compared with the one in the presence of nucleotides in the competition experiment and normalized by the length of the conformational change in presence of AP5A only:

$$Frac(F) = \frac{L_{AP5Aonly}(F) - L_{NuclComp}(F)}{L_{AP5Aonly}(F)}, \tag{4}$$

where $L_{AP5Aonly}(F)$ is the size of the full conformational change for AP5A-only experiments (in nm) and $L_{NuclComp}(F)$ is the size of the conformational change in the nucleotide competition experiment. The system can be described by the probability to find the enzyme in either the open or closed state. The force-dependent probability to find the enzyme in the closed state is given by

$$P_{closed}(F) = \frac{k_{close}(F)}{k_{close}(F) + k_{open}(F)} = \frac{1}{1 + \frac{k_{open}(F)}{k_{close}(F)}}, \tag{5}$$

where $k_{close}(F)$ and $k_{open}(F)$ are the force-dependent closing and opening rates at a specific nucleotide concentration, respectively. Assuming the Bell model, with a zero force rate $k_{0,i}$ and a distance to the transition state Δx_i, the force dependence of the rates is given by equation (1).

With equation (1), the force dependence of the probability to find AdK in the closed state can be rewritten as:

$$P_{closed}(F) = \frac{1}{1 + \frac{k_{0,open}}{k_{0,close}} \exp\left(\frac{F\Delta x}{k_B T}\right)}. \tag{6}$$

Here, $k_{0,open}$ and $k_{0,close}$ are the zero-force opening and closing rates, respectively, and Δx is the size of the conformational change that AdK undergoes in the presence of nucleotides. To calculate the fraction of closing, equation (6) has to be extended by $A = \frac{\Delta x}{1.72\,mm}$

$$Frac(F) = \frac{A}{1 + \frac{k_{0,open}}{k_{0,close}} \exp\left(\frac{F\Delta x}{k_B T}\right)}. \tag{7}$$

A global fit of this model to the force-dependent fraction of closing at all ATP concentration was performed. One Δx for all ATP concentrations was used and an ATP concentration-dependent ratio $\frac{k_{0,open}}{k_{0,close}}$ was used as well. The global fit to the data yields the size of the conformational change induced by ATP of $\Delta x = 0.5$ nm. A similar global fit to the force-dependent fraction of closing shown in Supplementary Fig. 9 for different Mg-ADP concentrations yields the size of the conformational change induced by Mg-ADP of $\Delta x = 0.7$ nm. From the acquired fit parameters, the fraction of full closing at zero force can be determined (Fig. 3d).

To describe the concentration-dependent closing of AdK in presence of a ligand, the probability to find the enzyme in the closed conformation in the presence of a ligand is calculated

$$P_{closed} = \frac{[BC] + [UC]}{[BC] + [UC] + [BO] + [UO]}, \tag{8}$$

where $[BC]$ and $[UC]$ are the concentrations of ligand-bound and ligand-unbound closed state, respectively, and $[BO]$ and $[UO]$ are the concentrations of ligand-bound and ligand-unbound open state, respectively. The dissociation constants for the open K_{open} and closed state K_{close} are defined by

$$K_{open} = \frac{[UO]*c}{[BO]}, \tag{9}$$

and

$$K_{close} = \frac{[UO]*c}{[BC]}. \tag{10}$$

Single-molecule fluorescence experiments by Henzler-Wildman et al.[18] have shown that the equilibrium between open and closed state in the absence of nucleotides lies on the open side (opening:closing rate, $6,500\,s^{-1}$:$2,000\,s^{-1}$). Assuming a similar Δx for full closing as in the AP5A case, force will shift the equilibrium further towards the open side. At our measurement forces, the concentration $[UC]$ will be negligible compared with the other concentrations. Together with equations (9) and (10), equation (8) can be simplified to:

$$P_{closed}(c) = \frac{1}{1 + K_{close}\left(\frac{1}{c} + \frac{1}{K_{open}}\right)}. \tag{11}$$

As the size of the conformational change caused by the nucleotides is known from the fits of the force-dependent closing data, the concentration-dependent fraction of closing can be written as:

$$Frac(c) = \frac{A}{1 + K_{close}\left(\frac{1}{c} + \frac{1}{K_{open}}\right)}. \tag{12}$$

MD simulation. All-atom MD simulations were performed on AdK bound to each of the three inhibitors AP4A, AP5A and AP6A. Two-dimensional (2D) potentials of mean force (PMFs) along the opening of the two flexible lid domains were calculated by means of Hamiltonian-replica exchange umbrella sampling (H-REMD-US) simulations. The resulting 2D free-energy landscapes were projected on a one-dimensional coordinate, resembling the experimental lid-to-lid distance. A free-energy bias corresponding to a force of 15 pN was subsequently added to the free-energy projections, to enable direct comparison with the experiments.

Starting structures. The crystal structure of A. aeolicus AdK in complex with AP5A was taken from PDB:2RGX[18]. Structures of AP4A and AP6A were built manually and optimized with respect to their internal energy using Antechamber as part of Amber14 (ref. 33). Starting structures of AdK in complex with AP4A and AP6A were constructed by aligning[34] these inhibitors over AP5A in 2RGX and subsequently removing AP5A (followed by energy minimization). Using the Leap module of Amber14, a periodic octahedral box with edge length ~28 Å was constructed around each protein–inhibitor complex and filled with ~9,000 water molecules. Potassium ions were added to neutralize the overall charge of the systems.

Simulation setup. All simulations were carried out using Amber14 (ref. 33). The protein was described by the Amber ff14SB force field[35] and the water molecules by the TIP3P model[36]. The AP4A, AP5A and AP6A inhibitors were parametrized with Antechamber and the GAFF[37] force field. Specific parameters were used for the potassium ions[38]. The short-range cutoff radius was 9 Å. Long-range interactions were accounted for by the particle mesh Ewald method (Amber14 default parameters). Hydrogen bonds were constrained via the SHAKE[39] algorithm. The integration time step was 2 fs. For temperature and pressure coupling, Berendsen thermostat and barostat[40] were used with a coupling time of 1 ps.

Equilibration. The starting structures were minimized with respect to their internal energy using the steepest decent method for 10,000 steps. Several equilibration simulations of 20 ps were performed in the NPT ensemble at a pressure of 1 bar. Initially, the temperature was successively raised to 100, 200 and 298 K, while harmonic positional restraints of a force of $50.0\,kcal\,mol^{-1}\,Å^{-2}$ were applied to protein and inhibitor atoms with respect to the starting structures. In five additional simulations, the restraining potential force was successively reduced to 24.0, 12.0, 6.0, 2.0 and $0.2\,kcal\,mol^{-1}\,Å^{-2}$. A final equilibration step of 40 ps was performed without restraints. The total equilibration time of each system was 2 ns.

2D H-REMD-US simulations. The US simulations were performed in the NVT ensemble at 298 K. The reaction coordinates used for the 2D US[41] simulations were defined as the centre of mass distances between the C_α-atoms of residues 33–58 and 114–123, 154–170 (AMP-lid opening) and 125–148, and 1–28 and 75–82 (ATP-lid opening), respectively. Harmonic umbrella potentials were placed along the AMP-lid coordinate between 18.0 and 30.1 Å, and along the ATP-lid coordinate between 29.0 and 31.1 Å with an intermediate spacing of 1.1 Å. To obtain fast equilibration of the US simulation setup, two different starting configurations for each umbrella window were generated. The equilibrated structures were driven along the ATP-lid coordinate and then along the AMP-lid coordinate by applying the umbrella potentials one after another with a force constant of $4\,kcal\,mol^{-1}\,Å^{-2}$ for 100 ps. The process was then repeated in the reverse order, starting from the obtained open AdK conformation. Only during this initial process, the inhibitor atoms were restrained to their initial positions with a force of $0.025\,kcal\,mol^{-1}\,Å^{-2}$, to prevent dissociation due to the fast initial opening motion. In this way, starting configurations generated on the basis of a completely open and a completely closed structure were provided for each umbrella window position. Subsequently, two strands of 2D US simulations with a harmonic force constant of $2\,kcal\,mol^{-1}\,Å^{-2}$ were run in parallel, each based on one of the previously obtained sets of starting configurations. Via Hamiltonian replica exchanges[42], neighbouring windows within and between the two strands were allowed to swap configurations. This enables rapid diffusion of the configurations along the reaction coordinate plane and between the strands as well. The exchange scheme was realized by performing 4D H-REMD-US with adequately modified exchange groups. As starting configurations originating from two configurational extremes are provided, which are allowed to rapidly exchange towards an adequate umbrella window position, equilibration is drastically sped up in comparison with a setup based on only one starting configuration per window. For the actual sampling, the umbrella windows at the positions between 29.0 and 30.1 Å, 18.0 and 21.3 Å (top left region), and between 19.0 and 22.3 Å (bottom right region, see Supplementary Fig. 11) of the reaction coordinate plane were omitted, to obtain an advantageous number of 256 parallel simulations. Replica exchanges were attempted every 1 ps with rates of successful exchanges between 0.06 and 0.29. Each umbrella window was simulated for 6 ns (12 ns per umbrella window position), totalling up to >1.5 µs of sampling time for each system. Reaction coordinate values were saved every 0.1 ps. The 2D PMFs were calculated from the sampling data using the Weighted Histogram Analysis Method[43,44]. The first 2 ns of sampling time were omitted in the PMF calculations, to ensure adequate equilibration of the umbrella windows.

Free-energy projection and additional external force. To enable a direct comparison of the MD results with the experiments, the 2D free-energy landscapes obtained from the simulations (Supplementary Fig. 11) were projected on a coordinate roughly resembling the lid-to-lid distance as follows:

$$d_{lid-lid}(d_{AMP-lid}, d_{ATP-lid}) = \sqrt{\left(d_{AMP-lid} - d_{0,AMP-lid}\right)^2 + \left(d_{ATP-lid} - d_{0,ATP-lid}\right)^2}. \tag{13}$$

Here, $d_{AMP-lid}$ and $d_{ATP-lid}$ are the lid opening coordinates used in the H-REMD-US simulations, and $d_{0,AMP-lid}$ and $d_{0,ATP-lid}$ are the locations of the global minima in the free-energy landscapes. We hereby assumed that the lid-domain motions are mainly orthogonal. A bias corresponding to the load exerted during the experiments was subsequently added via

$$PMF_{biased}(d_{lid-lid}) = PMF(d_{lid-lid}) - F_{external} \times d_{lid-lid} - PMF_{0,biased}. \tag{14}$$

As an external force $F_{external}$ 15 pN was used, representative for the loads applied during the experiments. Finally, the resulting new minima of the biased PMFs were shifted to zero.

References

1. Wolf-Watz, M. *et al.* Linkage between dynamics and catalysis in a thermophilic-mesophilic enzyme pair. *Nat. Struct. Mol. Biol.* **11**, 945–949 (2004).
2. Pisliakov, A. V., Cao, J., Kamerlin, S. C. & Warshel, A. Enzyme millisecond conformational dynamics do not catalyze the chemical step. *Proc. Natl Acad. Sci. USA* **106**, 17359–17364 (2009).
3. Itoh, H. *et al.* Mechanically driven ATP synthesis by F1-ATPase. *Nature* **427**, 465–468 (2004).
4. Stigler, J., Ziegler, F., Gieseke, A., Gebhardt, J. C. M. & Rief, M. The complex folding network of single calmodulin molecules. *Science* **334**, 512–516 (2011).
5. Svoboda, K., Schmidt, C. F., Schnapp, B. J. & Block, S. M. Direct observation of kinesin stepping by optical trapping interferometry. *Nature* **365**, 721–727 (1993).
6. Wen, J.-D. *et al.* Following translation by single ribosomes one codon at a time. *Nature* **452**, 598–603 (2008).
7. Rief, M., Gautel, M., Oesterhelt, F., Fernandez, J. M. & Gaub, H. E. Reversible unfolding of individual titin immunoglobulin domains by AFM. *Science* **276**, 1109–1112 (1997).
8. Woodside, M. T. *et al.* Direct measurement of the full, sequence-dependent folding landscape of a nucleic acid. *Science* **314**, 1001–1004 (2006).
9. Liphardt, J., Onoa, B., Smith, S. B., Tinoco, Jr I. & Bustamante, C. Reversible unfolding of single RNA molecules by mechanical force. *Science* **292**, 733–737 (2001).
10. Bae, E. & Phillips, G. N. Roles of static and dynamic domains in stability and catalysis of adenylate kinase. *Proc. Natl Acad. Sci. USA* **103**, 2132–2137 (2006).
11. Rundqvist, L. *et al.* Noncooperative folding of subdomains in adenylate kinase. *Biochemistry* **48**, 1911–1927 (2009).
12. Kerns, S. J. *et al.* The energy landscape of adenylate kinase during catalysis. *Nat. Struct. Mol. Biol.* **22**, 124–131 (2015).
13. Berry, M. B. *et al.* The closed conformation of a highly flexible protein: the structure of *E. coli* adenylate kinase with bound AMP and AMPPNP. *Proteins* **19**, 183–198 (1994).
14. Müller, C. W. & Schulz, G. E. Structure of the complex between adenylate kinase from *Escherichia coli* and the inhibitor Ap5A refined at 1.9 A resolution. A model for a catalytic transition state. *J. Mol. Biol.* **224**, 159–177 (1992).
15. Rhoads, D. G. & Lowenstein, J. M. Initial velocity and equilibrium kinetics of myokinase. *J. Biol. Chem.* **243**, 3963–3972 (1968).
16. Hanson, J. A. *et al.* Illuminating the mechanistic roles of enzyme conformational dynamics. *Proc. Natl Acad. Sci. USA* **104**, 18055–18060 (2007).
17. Arora, K. & Brooks, C. L. Large-scale allosteric conformational transitions of adenylate kinase appear to involve a population-shift mechanism. *Proc. Natl Acad. Sci. USA* **104**, 18496–18501 (2007).
18. Henzler-Wildman, K. A. *et al.* Intrinsic motions along an enzymatic reaction trajectory. *Nature* **450**, 838–844 (2007).
19. Whitford, P. C., Gosavi, S. & Onuchic, J. N. Conformational transitions in adenylate kinase. Allosteric communication reduces misligation. *J. Biol. Chem.* **283**, 2042–2048 (2008).
20. Lienhard, G. E. & Secemski, I. I. P 1 ,P 5 -Di(adenosine-5′)pentaphosphate, a potent multisubstrate inhibitor of adenylate kinase. *J. Biol. Chem.* **248**, 1121–1123 (1973).
21. Reinstein, J. *et al.* Fluorescence and NMR investigations on the ligand binding properties of adenylate kinases. *Biochemistry* **29**, 7440–7450 (1990).
22. Elms, P. J., Chodera, J. D., Bustamante, C. J. & Marqusee, S. Limitations of constant-force-feedback experiments. *Biophys. J.* **103**, 1490–1499 (2012).
23. Kalbitzer, H. R., Marquetant, R., Rösch, P. & Schirmer, R. H. The structural isomerisation of human-muscle adenylate kinase as studied by 1H-nuclear magnetic resonance. *Eur. J. Biochem.* **126**, 531–536 (1982).
24. Schrank, T. P., Bolen, D. W. & Hilser, V. J. Rational modulation of conformational fluctuations in adenylate kinase reveals a local unfolding mechanism for allostery and functional adaptation in proteins. *Proc. Natl Acad. Sci. USA* **106**, 16984–16989 (2009).
25. Adén, J. & Wolf-Watz, M. NMR identification of transient complexes critical to adenylate kinase catalysis. *J. Am. Chem. Soc.* **129**, 14003–14012 (2007).
26. von Hansen, Y., Mehlich, A., Pelz, B., Rief, M. & Netz, R. R. Auto- and cross-power spectral analysis of dual trap optical tweezer experiments using Bayesian inference. *Rev. Sci. Instrum.* **83**, 095116 (2012).
27. Zoldák, G., Stigler, J., Pelz, B., Li, H. & Rief, M. Ultrafast folding kinetics and cooperativity of villin headpiece in single-molecule force spectroscopy. *Proc. Natl Acad. Sci. USA* **110**, 18156–18161 (2013).
28. Stigler, J. & Rief, M. Hidden markov analysis of trajectories in single-molecule experiments and the effects of missed events. *Chemphyschem* **13**, 1079–1086 (2012).
29. Cecconi, C., Shank, E. a., Bustamante, C. & Marqusee, S. Direct observation of the three-state folding of a single protein molecule. *Science* **309**, 2057–2060 (2005).
30. Saint Girons, I. *et al.* Structural and catalytic characteristics of Escherichia coli adenylate kinase. *J. Biol. Chem.* **262**, 622–629 (1987).
31. Bell, G. I. Models for the specific adhesion of cells to cells. *Science* **200**, 618–627 (1978).
32. Bodenreider, C. & Kiefhaber, T. Interpretation of protein folding psi values. *J. Mol. Biol.* **351**, 393–401 (2005).
33. Case, D. A. *et al.* AMBER 14 (Univ. California, 2014).
34. The PyMOL Molecular Graphics System, Version 1.7.4 (Schrodinger, LLC (2010).
35. Hornak, V. *et al.* Comparison of multiple Amber force fields and development of improved protein backbone parameters. *Proteins* **65**, 712–725 (2006).
36. William, L. J., Jayaraman, C., Jeffry, D. M., Roger, W. I. & Michael, L. K. Comparison of simple potential functions for simulating liquid water. *J. Chem. Phys.* **79**, 926–935 (1983).
37. Wang, J., Wolf, R. M., Caldwell, J. W., Kollman, P. A. & Case, D. A. Development and testing of a general Amber force field. *J. Comput. Chem.* **25**, 1157–1174 (2004).
38. Joung, I. S. & Cheatham, T. E. Determination of alkali and halide monovalent ion parameters for use in explicitly solvated biomolecular simulations. *J. Phys. Chem. B* **112**, 9020–9041 (2008).
39. Ryckaert, J. P., Ciccotti, G. & Berendsen, H. J. C. Numerical-integration of Cartesian equations of motion of a system with constraints - molecular-dynamics of N-alkanes. *J. Comput. Phys.* **23**, 327–341 (1977).
40. Berendsen, H. J. C., Postma, J. P. M., van Gunsteren, W. F., DiNola, A. & Haak, J. R. Molecular dynamics with coupling to an external bath. *J. Chem. Phys.* **81**, 3684–3690 (1984).
41. Torrie, G. M. & Valleau, J. P. Nonphysical sampling distributions in Monte Carlo free-energy estimation: Umbrella sampling. *J. Comput. Phys.* **23**, 187–199 (1977).
42. Fukunishi, H., Watanabe, O. & Takada, S. On the Hamiltonian replica exchange method for efficient sampling of biomolecular systems: application to protein structure prediction. *J. Chem. Phys.* **116**, 9058–9067 (2002).
43. Kumar, S., Rosenberg, J. M., Bouzida, D., Swendsen, R. H. & Kollman, P. A. The weighted histogram analysis method for free-energy calculations on biomolecules. I. The method. *J. Comput. Chem.* **13**, 1011–1021 (1992).
44. Grossfield, A. WHAM: the weighted histogram analysis method 2.0.7. http://membrane.urmc.rochester.edu/content/wham (2013).
45. Okazaki, K.-I. & Takada, S. Dynamic energy landscape view of coupled binding and protein conformational change: induced-fit versus population-shift mechanisms. *Proc. Natl Acad. Sci. USA* **105**, 11182–11187 (2008).

Acknowledgements

This work was supported by an SFB 1035 grant of Deutsche Forschungsgemeinschaft to M.R. (A05) and M.Z. (B02).

Author contributions

B.P. performed experiments. B.P., G.Z. and M.R. designed research, analysed and discussed data, and provided analytical tools and materials. F.Z. and M.Z. performed and analysed MD simulations. B.P., G.Z., F.Z., M.Z. and M.R. prepared figures, wrote and edited the manuscript.

Additional information

3

Aharonov–Bohm oscillations in Dirac semimetal Cd_3As_2 nanowires

Li-Xian Wang[1,*], Cai-Zhen Li[1,*], Da-Peng Yu[1,2,3] & Zhi-Min Liao[1,2]

Three-dimensional Dirac semimetals, three-dimensional analogues of graphene, are unusual quantum materials with massless Dirac fermions, which can be further converted to Weyl fermions by breaking time reversal or inversion symmetry. Topological surface states with Fermi arcs are predicted on the surface and have been observed by angle-resolved photoemission spectroscopy experiments. Although the exotic transport properties of the bulk Dirac cones have been demonstrated, it is still a challenge to reveal the surface states via transport measurements due to the highly conductive bulk states. Here, we show Aharonov–Bohm oscillations in individual single-crystal Cd_3As_2 nanowires with low carrier concentration and large surface-to-volume ratio, providing transport evidence of the surface state in three-dimensional Dirac semimetals. Moreover, the quantum transport can be modulated by tuning the Fermi level using a gate voltage, enabling a deeper understanding of the rich physics residing in Dirac semimetals.

[1] State Key Laboratory for Mesoscopic Physics, Department of Physics, Peking University, Beijing 100871, China. [2] Collaborative Innovation Center of Quantum Matter, Beijing, China. [3] Institute of Physics and Electronic Information, Yunnan Normal University, Kunming 650500, China. * These authors contributed equally to this work. Correspondence and requests for materials should be addressed to Z.-M.L. (email: liaozm@pku.edu.cn).

The Dirac fermions in graphene and topological insulators are confined in a two-dimensional plane, while in a three-dimensional (3D) Dirac semimetal the Dirac fermions are extended to the 3D space[1-17]. Recently, Cd_3As_2 and Na_3Bi have been predicted to be Dirac semimetals[5,6], which were further identified by angle-resolved photoemission spectroscopy experiments[7-10]. The Dirac point in 3D Dirac semimetals is constituted by two degenerated Weyl nodes and this degeneracy can be lifted by breaking time reversal or inversion symmetry[5,6]. Owing to its unique bulk Dirac cones, the 3D Dirac semimetal has become an excellent platform to investigate exotic transport properties, such as giant positive magnetoresistance (MR)[11], quantum linear MR[5,12], splitting of Landau levels[13-16] and chiral anomaly induced negative MR[17-21]. On the other hand, non-trivial surface states with arc-like Fermi surface are theoretically predicted and observed by angle-resolved photoemission spectroscopy experiments[5,6,22,23]. Nevertheless, the non-trivial surface states are not accessible by transport measurements in bulk Dirac semimetals because of the bulk-dominant conduction.

The Aharonov–Bohm (A–B) effect of individual nanowires with large surface-to-volume ratio provides an effective method to explore the transport properties of the non-trivial surface states. As the mean free path of the carriers is comparable with the perimeter of the nanowire, the electron wave is confined in the finite boundary condition along the nanowire perimeter. Consequently, the surface energy bands are enforced into discrete sub-bands because of the quantum confinement[24,25]. As sweeping the magnetic field parallel to the nanowire direction, the density of states at the Fermi level is periodically altered as each sub-band crosses the Fermi energy[24,25], and the oscillating period is described by the Φ/Φ_0, where Φ is the magnetic flux enclosed by the path of the surface carriers and $\Phi_0 = h/e$. These quantum oscillations have been observed in topological insulator nanoribbons and nanowires, which are still called as A–B oscillations in the recent literatures[26-28]. Particularly, the helical surface states in topological insulator accumulate an additional π Berry phase after travelling a closed trajectory, resulting in the conductance peaks are situated at the magnetic flux of $\left(n + \frac{1}{2}\right)h/e$ (refs 26–28), which is very distinct from ordinary h/e oscillations in normal metal rings[29].

Here we report the A–B oscillations in individual Cd_3As_2 nanowires to demonstrate the transport properties of the surface states in Dirac semimetals. A π A–B effect that the conductance oscillations peak at odd integers of $h/2e$ with a period of h/e is observed, providing transport evidence of the topological surface states of Cd_3As_2 nanowires. The A–B oscillations are tunable by altering the Fermi level of the nanowire via gate voltage, giving deeper insight of the 3D Dirac semimetal with Dirac fermion bulk states and Fermi arc surface states.

Results

Characterizations.
The Cd_3As_2 nanowires were synthesized via chemical vapour deposition method, which are with high aspect ratio and right stoichiometry (Supplementary Fig. 1). The Cd_3As_2 nanowires are of single crystalline nature and grown along the [112] direction, as shown by the high-resolution transmission electron microscopy image in Fig. 1a. The surface of individual nanowires is usually covered by a thin amorphous layer ~ 5 nm forming a core-cell structure (Supplementary Fig. 2), which may help to maintain the clean Cd_3As_2 surface immunized from the environment[30].

The schematic representation of the device and measurement configuration is shown in Fig. 1b. The temperature-dependent resistance manifests semiconducting-like behaviours

(Supplementary Fig. 3), similar to other 3D Dirac semimetal materials, such as Na_3Bi, and is attributed to the low carrier density[17,20]. Notable Shubnikov-de Haas oscillations were also observed under perpendicular magnetic field (Supplementary Fig. 4). The linear extrapolation of the Landau index plot yields the intercept ~ 0.1, which is close to the intercept of 1/8 in 3D Dirac systems with perfect π Berry phase, giving an evidence for the Dirac fermion transport in Cd_3As_2 nanowires[31].

Aharonov–Bohm oscillations. The A–B oscillations were measured as the magnetic field was applied parallel with the longitudinal axis of the nanowire. Figure 1c shows the resistance of a nanowire with diameter ~ 115 nm (Device 1) as a function of magnetic field $\mathbf{B}$ at different temperatures. In low magnetic field regime ($|\mathbf{B}| < 0.3$ T), the MR exhibits an obtuse dip because of the weak anti-localization, which is in agreement with the presence of strong spin–orbit interaction in Cd_3As_2 (ref. 5). At higher fields, the MR starts to oscillate periodically on a smooth negative MR background at 1.5 K. The negative MR has already been observed in Dirac semimetals and Weyl semimetals, which was believed to originate from the chiral anomaly induced charge pumping effect[17-21]. The magnitude of the negative MR is increased from -19% at 1.5 K to -34% at 20 K (Supplementary Fig. 5), which is consistent with previous report[20]. It should be noted that the (quasi-) ballistic transport along the perimeter of the nanowire is necessary to observe the A–B oscillations[28]. Therefore, the A–B oscillations are much more pronounced in the fine nanowire (~ 115 nm in diameter) in this work than the thick nanowire in ref. 20. As the temperature rises up to 20 K, the oscillations are nearly smeared out, indicative of its quantum nature. To focus on the quantum oscillations, the negative MR background is removed and the oscillating term $\Delta G(\mathbf{B})$ is shown in Fig. 1d. More than 10 oscillation periods (from 0 to ± 5 T) can be clearly distinguished that are distributed symmetrically around the zero-field. The fast-Fourier transform (FFT) on $\Delta G(\mathbf{B})$ shown in Fig. 1e indicates the main peak at ~ 2.55 T^{-1}, corresponding to an individual 0.38 T period, giving a remarkable signature of flux-related quantum oscillations. If the quantum interference occurs on the surface, the h/e A–B oscillations yield a cross-section area of the nanowire $\sim 0.11 \mu m^2$. The cross-section area of the Device 1 was exactly estimated to be $0.11 \pm 0.005 \mu m^2$ by direct measurements on its cut (Supplementary Fig. 6), which identifies the observed quantum oscillations are from the A–B effect.

The amplitude of oscillating term $\Delta G \sim 0.005 e^2/h$ is much smaller than the quantum conductance e^2/h. Previously, the observed amplitudes of A–B oscillations in topological insulator nanoribbons are also smaller than e^2/h (refs 26,28). The electrons circle along the nanowire perimeter in a quasi-ballistic manner, but drift along the longitudinal direction of the nanowire in a diffusive manner (mean free path $< L$). As probing the longitudinal conductance, the A–B oscillation amplitude of conductance may be reduced due to the diffusive transport in longitudinal direction. The amplitude of the A–B oscillation increases with decreasing temperature, as shown in Fig. 1f. It is found that the FFT peak amplitude obeys to T^{-1} law in the temperature range of 5–20 K, then deviates the T^{-1} dependence below ~ 5 K and increases more slowly as further decreasing the temperature. The T^{-1} temperature dependence of oscillation amplitude indicates the nature of ballistic transport in circumferential direction[28]. The deviation from T^{-1} dependence below 5 K is due to the low influence of thermal broadening on the electron wave packets[32]. The critical temperature for the deviation of T^{-1} dependence can be estimated by considering the critical thermal

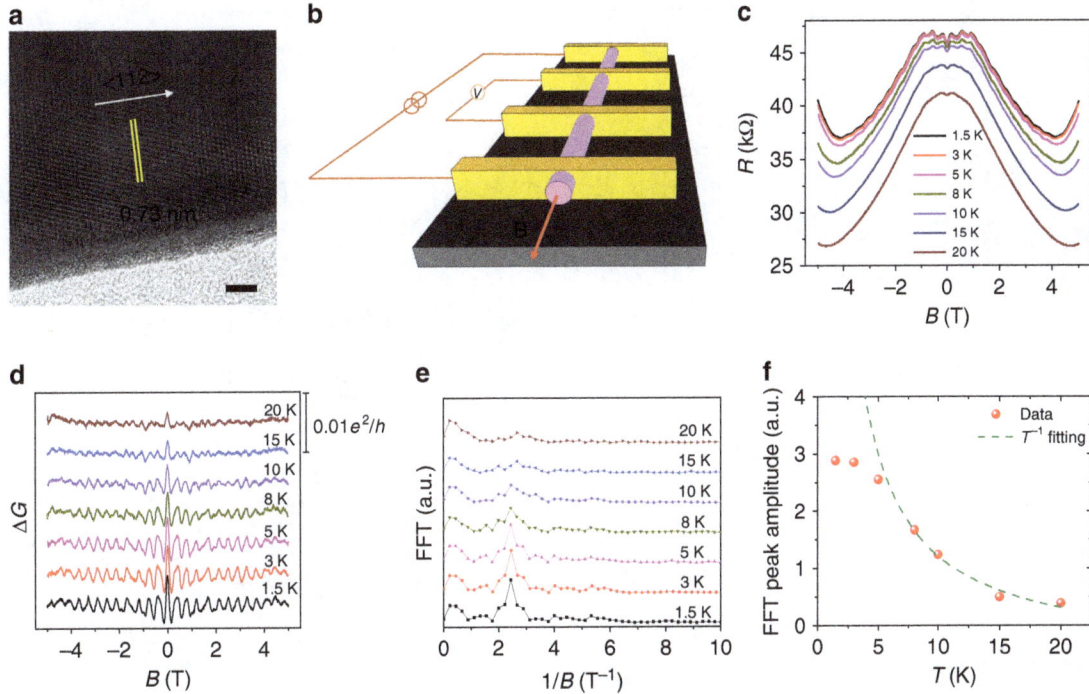

Figure 1 | Magnetotransport of a nanowire with diameter ∼115 nm (Device 1). (**a**) High-resolution transmission electron microscopy image of a typical nanowire indicates <112> growth direction with interplanar space ∼0.73 nm. Scale bar, 5 nm. (**b**) Schematic diagram of the four-terminal device with applied magnetic field aligned with the length. (**c**) Resistance as a function of magnetic field at different temperatures from 1.5 to 20 K. (**d**) The oscillations in conductance as a function of magnetic field after subtracting background. (**e**) FFT spectrums of the conductance oscillations. (**f**) Plot of the temperature dependence of the FFT peak amplitude. The dashed line indicates the fitting by the T^{-1} dependence.

energy $E_c = \frac{h}{\pi e^2 R} \Delta E$, where R is the nanowire resistance, and ΔE is the sub-band spacing due to geometric confinement[32]. According to nanowire boundary conditions, the sub-band spacing has a form of $\Delta E = \frac{h v_F}{W} = h v_F \sqrt{\frac{1}{4\pi S}}$, where W is the perimeter of nanowire, S is cross-sectional area and v_F is Fermi velocity[26]. Hence, the critical temperature is calculated to be ∼7.2 K, which is close to the experimental observation of ∼5 K.

The elastic scattering length $L_0 \sim 10$ nm of the longitudinal transport along the nanowire is estimated by the amplitude of oscillating term $\Delta G(\mathbf{B}) \sim 0.005 e^2/h$ via the Landauer–Büttiker formula of $G = \frac{e^2/h}{1 + L/L_0}$ (ref. 33), where $L \sim 2$ μm is the distance between the two voltage probes. Although the electrons circle along the nanowire perimeter in a quasi-ballistic manner, the carrier drift along the longitudinal direction of the nanowire is in a diffusive manner (mean free path $< L$). Therefore, the obtained L_0 can be underestimated due to the reduced amplitude ΔG. Nevertheless, the ballistic or quasi-ballistic transport along the nanowire perimeter must be satisfied to observe the A–B oscillations[26–28,30,34]. Therefore, it is naturally inferred that the non-trivial surface states in the Cd_3As_2 nanowire should have high carrier mobility with suppressed scatterings by topological protection. Our results are also consistent with the observations in topological insulator nanowires that are not ballistic along the nanowire length but ballistic along the nanowire perimeter[34].

The oscillations in ΔG as a function of magnetic flux Φ in unit of h/e are presented in Fig. 2a. In the two initial periods, the conductance peaks are at the integer quantum flux ($\Phi = h/e$, $2h/e$). Then, a phase-shift takes place in the range from 1 to 2 T, wherein the conductance peaks are at random flux $\Phi = 3.1h/e$ and $4.3h/e$. After the phase-shift region, the conductance peaks are stable at the half-integer quantum flux ($\Phi = (5 + 1/2)h/e$, $(6 + 1/2)h/e$,...). The conductance oscillation peaks at

half-integer of Φ_0 with a period of Φ_0 are very similar with that observed in topological insulators, which was ascribed to the spin-related π Berry phase of the topological surface states[24,25]. At $\Phi/\Phi_0 = 0$, the inter-spacing between sub-bands due to the finite boundary condition is estimated by $\Delta E = h v_F/W$ (refs 26–28). At $\Phi/\Phi_0 = n$, the surface state is fully gapped at $E = 0$. While at $\Phi/\Phi_0 = n + 1/2$, a gapless surface band appears due to the additional π Berry phase[26]. Intuitively, to observe this anomalous A–B oscillation, it requires that the Fermi-level E_F is close to $E = 0$ or the quantum confinement induced gap is large enough. Because of the high crystal quality with fewer defects, the Cd_3As_2 nanowires are of low carrier density with the order of ∼10^{17} cm^{-3}, which has been measured by field-effect gating method and found to be the crucial factor for the observation of the negative MR effect in our previous work[20]. The low carrier density also suppresses the bulk conductance with the resistance $R > 25.8$ kΩ at low temperatures, promoting the detection of the surface states of the nanowire with large surface-to-volume ratio.

To understand the phase-shift of the A–B effect in Cd_3As_2 nanowires, we would like to discuss the origin of the additional π Berry phase by considering the evolution of Dirac point under magnetic fields. In Cd_3As_2, the Dirac points are distributed equally away from the Γ point along the [001] direction at $\pm k_D$ (ref. 5). As schematically shown in Fig. 2b, without magnetic field, the Fermi arc surface states connect the two bulk Dirac nodes and each Dirac node contains two degenerate Weyl nodes. Applying an external magnetic field can break the time reversal symmetry and split the two Dirac points into two pairs of Weyl nodes along the magnetic field direction[35]. As schematically shown in Fig. 2c, under magnetic field, each surface Fermi arc connects the Weyl nodes with right-handed chirality and left-handed chirality, respectively. The chirality-non-trivial surface states are non-degenerate. The lifting of electronic state degeneracy results in an additional Berry phase π for cyclotron electrons

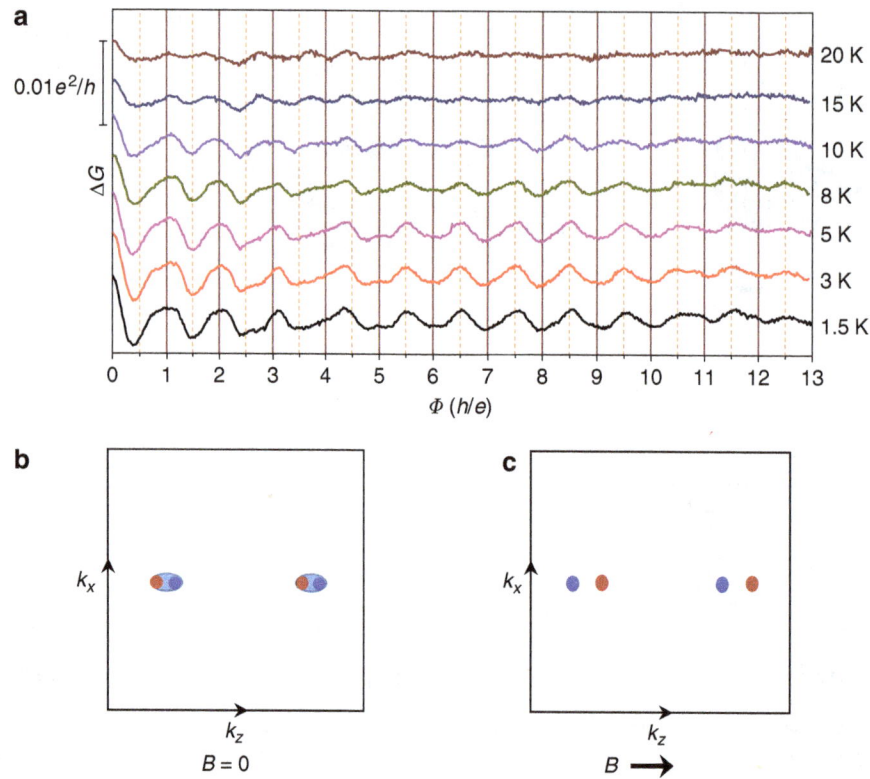

Figure 2 | Aharonov–Bohm oscillations. (**a**) Oscillating conductance as a function of magnetic flux in unit of h/e at variable temperatures. (**b,c**) Schematic diagram of the Weyl nodes projected onto the two-dimensional $k_x - k_z$ surface plane (**b**) without and (**c**) with an external magnetic field. The right-handed chirality and left-handed chirality are denoted by the red ball and blue ball, respectively. The Fermi arcs connect the Weyl nodes with opposite chirality. Without magnetic field, each Dirac node consists of two degenerate Weyl nodes. A magnetic field can separate the two degenerate Weyl nodes in the momentum space.

after cycling around one of the Weyl nodes in the low-energy band. Moreover, the analysis of Shubnikov–de Haas oscillations also suggests the presence of non-trivial π Berry phase (Supplementary Fig. 4)[16]. It is very interesting to compare our results with the A–B oscillations in topological insulator nanoribbons. The surface sub-bands have the 1D band dispersion as $E(n,k,\Phi) = \pm h v_F [\frac{k^2}{4\pi^2} + (n + 0.5 - \frac{\Phi}{\Phi_0})^2 / W^2]^{1/2}$, where the A–B oscillation in conductance is anticipated to peak at odd integers of $\Phi_0/2$ (refs 26–28). Similarly, the A–B oscillation with peaks at odd integers of $\Phi_0/2$ was also found in our Cd_3As_2 nanowires under a magnetic field larger than a critical value. This phase-shift is believed to be in close relationship to the extra π Berry phase released by the magnetic field-induced lifting of degeneracy.

Angular dependence of A–B oscillations. To verify the influence of magnetic flux on the A–B effect, the nanowire direction was tilted, forming an angle θ with the direction of the magnetic field (Fig. 3a). As shown in Fig. 3b, the magnetotransport results illustrate a negative to positive MR transition from $\theta = 0°$ (parallel case) to $\theta = 40°$. The oscillations in ΔG obtained at different tilted angles are presented in Fig. 3c. The weak anti-localization induced conductance peak at $\mathbf{B} = 0$ is not sensitive to the tilt angle, indicating its bulk origin. At low magnetic field, the periodic oscillations at high tilt angles are still discernible. By scaling the magnetic field with $B\cos\theta$, we replot the oscillations at different tilt angles in Fig. 3d. It is found that the oscillation peaks are line up in Fig. 3d, suggesting the dominant magnetic flux along the nanowire longitudinal direction. Meanwhile, the oscillation peaks split at high tilt angles, which may be due to the

deflection of the electron trajectory under the Lorentz force from the perpendicular component $B_\perp$ of the magnetic field. It is also found that the periodic oscillations vanish gradually as increasing θ under high magnetic fields. The vanishing of oscillations under high magnetic field may be due to the $B_\perp$-induced breaking of the time reversal symmetry of the surface states and a bandgap opening, which is consistent with that observed in topological insulators[26].

Gate-voltage dependence of A–B oscillations. The periodic conductance oscillation is caused by the rearrangement of the surface sub-bands in energy as varying the magnetic flux[26,28]. Therefore, the occupation of subbands can affect the pattern of oscillation in aspect of magnitude and phase. By employing a back-gate with 285 nm SiO_2, we are able to modulate the sub-band occupation by applying gate voltage (V_g), and explore how the Fermi level affects the A–B oscillations. To achieve this goal, we performed systematic gate-modulation experiments on a very thin Cd_3As_2 nanowire with diameter ~ 65 nm (Device 2), in which a large sub-band spacing $\Delta E = \frac{h v_F}{W} \sim 21$ meV can be obtained. As shown in Fig. 4a, the transfer curve indicates the ambipolar field effect of the device. The resistance saturates to a certain value at high gate voltages (> 5 V), indicating that the screening effect may hinder further gate-modulating as the Fermi-level is far away from the Dirac point. Nevertheless, as shown in Fig. 4a, the resistance can be largely tuned at small gate voltages V_g, demonstrating the effective gate-modulation as long as the screening effect is not strong enough.

We have measured the magnetotransport properties of the nanowire in the gate voltage range without obvious screening

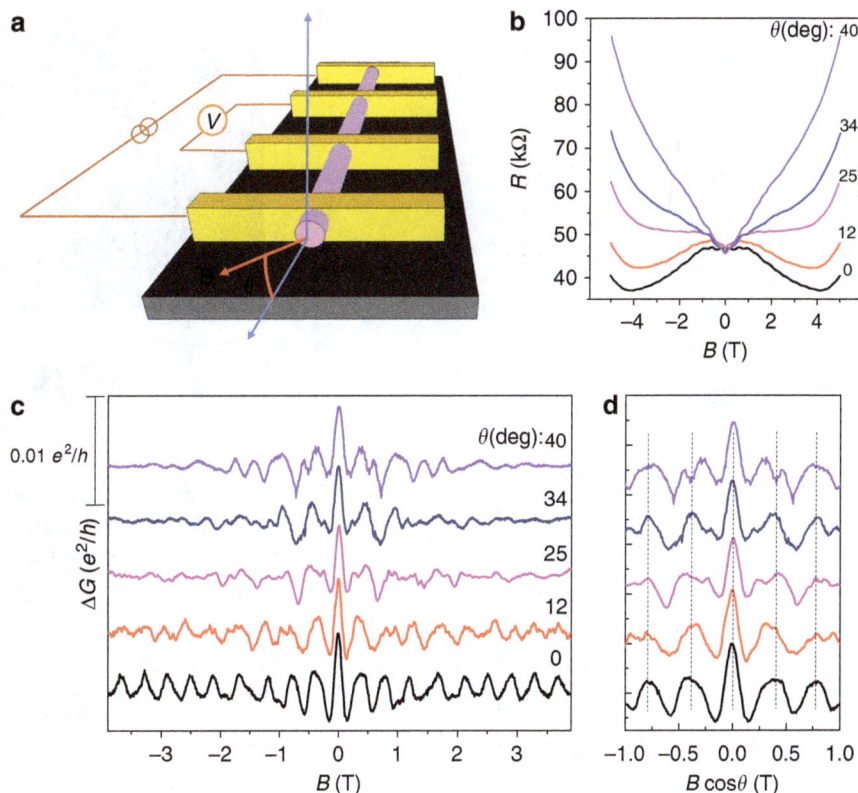

Figure 3 | Aharonov–Bohm oscillations at tilted magnetic field. (a) Schematic representation of the measurement configuration. The θ is the angle between the orientation of the magnetic field and the longitudinal direction of the nanowire. (b) Magnetoresistance at different θ angles and (c) the corresponding conductance oscillations. (d) The conductance oscillations as a function of $B\cos\theta$.

effect. As shown in Fig. 4b, the A–B oscillations in ΔG were plotted as a function of magnetic flux at different V_g of -7.5, -3, 0, 3 and 5 V. It is worth noting that recent studies on Cd_3As_2 thin films (~ 50 nm in thickness) demonstrate that a considerable bandgap > 24.9 meV may be opened at the Dirac point as a result of quantum confinement effect[36]. Because this nanowire is with diameter of ~ 65 nm, it is very likely that a bandgap is also opened at the Dirac point in the nanowire. Therefore, at $V_g = 0$ V (near the Dirac point as seen in Fig. 4a), the Fermi level is in the bandgap. As the Fermi level is inside the bandgap, the A–B oscillation thus has a normal origin, which may be responsible for the observation of the maxima of ΔG at even integers of $\Phi_0/2$ at $V_g = 0$ V, as shown in Fig. 4b. As tuning the Fermi level into the valence band, there is a phase shift of the A–B oscillations at $V_g = -3$ V. As the Fermi level further enters into the valence band with linear dispersion, the π-A–B effect is observed at $V_g = -7.5$ V, as shown in Fig. 4b, where the maxima of ΔG are at odd integers of $\Phi_0/2$.

As tuning the Fermi level into the conduction band, the A–B oscillations are still clearly observed but with more complicated features at $V_g = 3$ and 5 V, as shown in Fig. 4b. Previous studies indicate that the electron has much higher mobility than that of hole in Cd_3As_2 because of the different Fermi velocity and scattering rate with defects[9,14,37]. The magnetic field can force the bulk electrons to the surface under the classic limit of $R < L_B = \sqrt{\hbar/eB}$, that is, the nanowire radius R is smaller than the magnetic length L_B (refs 38,39). The electrons with high carrier mobility deflected from the bulk channel contribute to the surface conductance significantly, resulting in the irregular A–B oscillations. Furthermore, we have performed more systematic gate-dependent A–B oscillations on another sample with similar size (radius ~ 37 nm, Device 3), as shown in Supplementary Fig. 7. The oscillation patterns are complicated at relatively weak

magnetic fields, implying the co-contribution to the quantum oscillations from both the bulk and surface electrons. Fortunately, further increasing B may give rise to less scattering on the surface from bulk electrons as $L_B < R$, and there is a clear π-A–B effect under high magnetic field (Supplementary Fig. 7b).

In Dirac semimetals, the two bulk Dirac points are separated by $2k_D$ in momentum space. The relative magnitude of k_D and the Fermi wave vector k_F is thus crucial to describe the electronic structure in Dirac semimetals. Further increasing k_F larger than k_D, two Fermi pockets nested at $\pm k_D$ merge into one entire Fermi surface centred at Γ, called the Lifshitz transition[14]. As the system is back to degenerate system, the π Berry phase should disappear. To verify this scenario, we measured a nanowire (Device 4) with back-gate voltage up to 60 V. At $V_g = 60$ V, the oscillation peaks at relatively random positions under low magnetic fields (Supplementary Fig. 8), which may be due to the large magnetic length of the electron orbit and high carrier density. In this situation, the bulk carrier can be scattered to the surface, which makes the A–B oscillations complicated. As increasing magnetic field, the magnetic length of the bulk electron cyclotron motion decreases and the oscillation starts to peak exactly at the integer quantum flux with $\Phi = 5$, 6, 7, 8 h/e (Supplementary Fig. 8). Since the Fermi level is far from the Dirac points, the Fermi surface may enter above the Lifshitz transition, thus leading to the normal A–B oscillations.

Altshuler–Aronov–Spivak effect. For the thick nanowire device, the electron transport along the nanowire perimeter may be in the diffusive regime. Figure 5a shows a thick nanowire with diameter ~ 200 nm (Device 5). Even in a relatively narrow field range from -1 to 1 T, prominent oscillations superposed on negative MR background can be easily discerned, as shown in Fig. 5b.

a

b

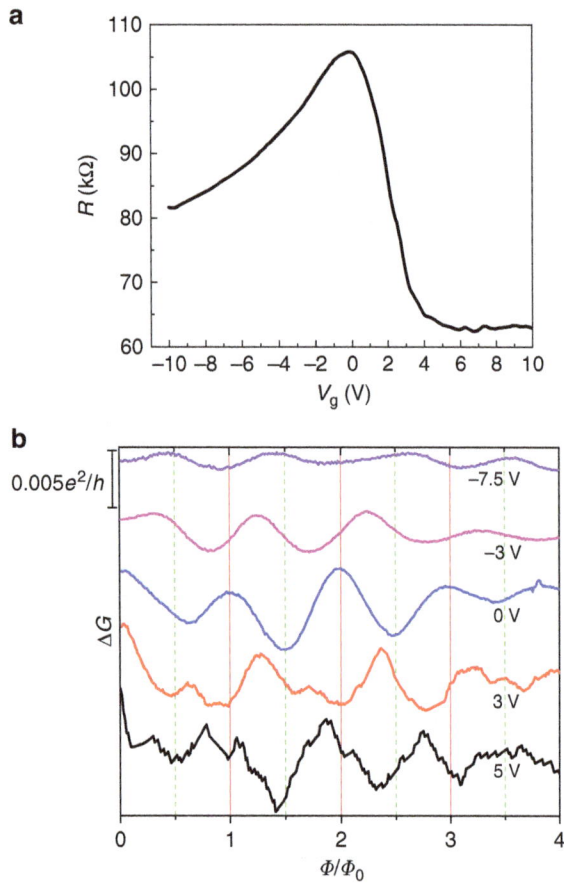

Figure 4 | Gate tuning of Aharonov–Bohm oscillations. (**a**) Gate voltage dependent resistance of a nanowire with diameter ~65 nm (Device 2) at 1.5 K. (**b**) A–B oscillations at gate voltages as denoted and at 1.5 K. The positive and negative gate voltages pull the Fermi level into the conduction band and valence band, respectively.

a

b

c

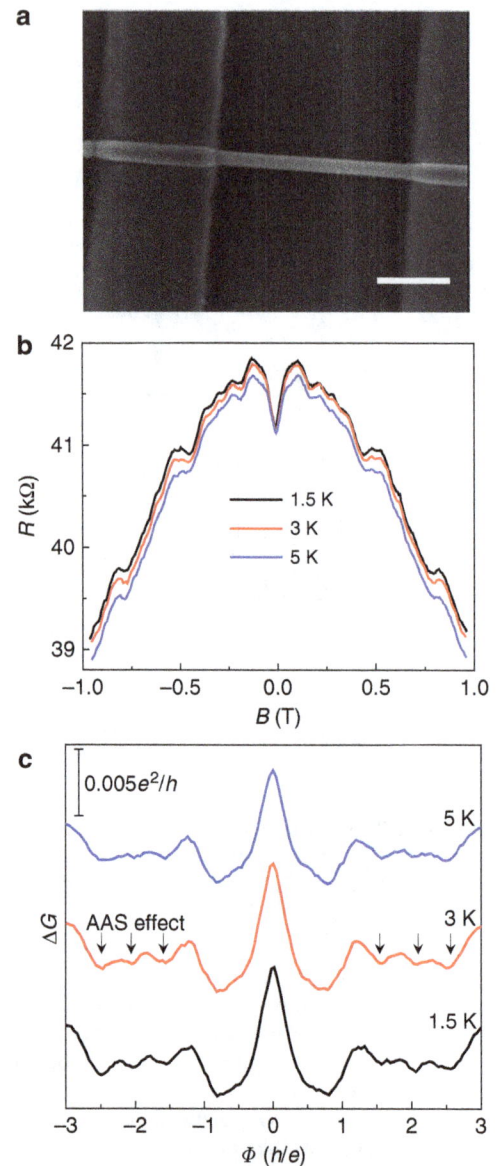

Figure 5 | Altshuler–Aronov–Spivak oscillations. (**a**) Scanning electron microscopic image of a thick nanowire with diameter about 200 nm (Device 5). Scale bar, 1 μm. (**b**) Magnetoresistance at different temperatures. (**c**) The conductance oscillations as a function of magnetic flux, showing the AAS effect.

According to the nanowire cross-sectional area (Supplementary Fig. 6), it is found that there are oscillations with $h/2e$ flux period, as noted by arrows in Fig. 5c. These $h/2e$ oscillations may be attributed to the Altshuler–Aronov–Spivak (AAS) effect in diffusive regime. Notably, the $h/2e$ oscillation peaks at $\left(n+\frac{1}{2}\right)h/2e$ ($n = 2, 3, 4$), indicating a phase shift for the $h/2e$ oscillations that needs further investigations.

Discussion

In summary, we have studied the A–B effect in individual Dirac semimetal Cd_3As_2 nanowires as the magnetic flux penetrates through the cross-section of the nanowire. As increasing magnetic field, a phase shift in A–B oscillations was observed, which is ascribed to the splitting of the Dirac node into Weyl nodes due to the magnetic field-induced time-reversal symmetry breaking. With this lifting of the Dirac node degeneracy, the conductance oscillation peaks at odd integers of $h/2e$ with period of h/e, which provides transport evidence of the topological surface states of Cd_3As_2 nanowires. As tuning the Fermi level by gate voltage, a transition between the 0 A–B effect and the π A–B effect was observed, which is self-consistent with the presence of linear dispersion relation in 1D-like quantum confined system. In thick nanowires, the A–B oscillations were largely reduced and the AAS oscillations presented due to the transition from quasi-ballistic to diffusive transport. The A–B oscillations observed in the Dirac semimetal nanowires show abundant features, offering a route to explore the exotic surface states.

Methods

Growth of the nanowires. The Cd_3As_2 nanowires were synthesized by chemical vapour deposition method in a horizontal furnace. For safety, the whole system was placed in a ventilation closet and the pumper was equipped with a cold trap to collect the dust in the exhaust. The Cd_3As_2 powders (purity > 99.99%, Alfa Aesar) placed at the centre of the tube furnace were heated from room temperature to 650 °C in 20 min after pumping and flushing the tube several times to get rid of oxygen. The Si substrates covered by a 5-nm-thick gold film were placed at the downstream away from the source ~15 cm to collect the products. The system was maintained at 650 °C for 10 min and at about 1 atm pressure with ~20 s.c.c.m Argon flow during the growth process. Then, the system cooled down to room temperature naturally.

Device fabrication. The Cd_3As_2 nanowire devices were fabricated on a Si substrate with 285 nm SiO_2 layer. To make the nanowire direction can be parallel with the direction of the magnetic field, microstrips were first prefabricated on a SiO_2/Si

square substrate as marks, which are parallel to one side of the square substrate. Then, the nanowire was transferred from the as-grown substrate to the SiO_2/Si square substrate, and the nanowire direction was aligned parallel to the prefabricated marks by a micromanipulator. The individual nanowire devices were fabricated using standard electron beam lithography techniques. To form ohmic contacts between Cd_3As_2 nanowires and electrodes, the amorphous layer of the nanowire was removed *in situ* by Ar^+ etching treatment in the metal deposition chamber. Then, $\sim$150-nm-thick Au electrodes were deposited. The linear current–voltage curve (Supplementary Fig. 9) indicates the Ohmic contacts between the nanowire and the electrodes.

Transport measurements. The SiO_2/Si square substrate with the Cd_3As_2 nanowire devices was mounted onto the sample holder in an Oxford commercial variable temperature insert to make the nanowire parallel to the direction of the magnetic field. The sample holder can be rotated continuously by an electric motor to change the angle between **E** and **B**. The electrical signals were measured using a four-terminal method by Stanford SR830 lock-in amplifiers at frequency of 17.7 Hz.

References

1. Novoselov, K. *et al.* Two-dimensional gas of massless Dirac fermions in graphene. *Nature* **438**, 197–200 (2005).
2. Zhang, Y. B., Tan, Y. W., Stormer, H. L. & Kim, P. Experimental observation of the quantum Hall effect and Berry's phase in graphene. *Nature* **438**, 201–204 (2005).
3. Xu, Y. *et al.* Observation of topological surface state quantum Hall effect in an intrinsic three-dimensional topological insulator. *Nat. Phys.* **10**, 956–963 (2014).
4. Young, S. M. *et al.* Dirac semimetal in three dimensions. *Phys. Rev. Lett.* **108**, 140405 (2012).
5. Wang, Z., Weng, H., Wu, Q., Dai, X. & Fang, Z. Three-dimensional Dirac semimetal and quantum transport in Cd_3As_2. *Phys. Rev. B* **88**, 125427 (2013).
6. Wang, Z. *et al.* Dirac semimetal and topological phase transitions in A(3)Bi (A = Na, K, Rb). *Phys. Rev. B* **85**, 195320 (2012).
7. Liu, Z. K. *et al.* Discovery of a three-dimensional topological Dirac semimetal, Na_3Bi. *Science* **343**, 864–867 (2014).
8. Liu, Z. K. *et al.* A stable three-dimensional topological Dirac semimetal Cd_3As_2. *Nat. Mater.* **13**, 677–681 (2014).
9. Neupane, M. *et al.* Observation of a three-dimensional topological Dirac semimetal phase in high-mobility Cd_3As_2. *Nat. Commun.* **5**, 3786 (2014).
10. Borisenko, S. *et al.* Experimental realization of a three-dimensional Dirac semimetal. *Phys. Rev. Lett.* **113**, 027603 (2014).
11. Liang, T. *et al.* Ultrahigh mobility and giant magnetoresistance in the Dirac semimetal Cd_3As_2. *Nat. Mater.* **14**, 280–284 (2015).
12. He, L. *et al.* Quantum transport evidence for the three-dimensional Dirac semimetal phase in Cd_3As_2. *Phys. Rev. Lett.* **113**, 246402 (2014).
13. Jeon, S. *et al.* Landau quantization and quasiparticle interference in the three-dimensional Dirac semimetal Cd_3As_2. *Nat. Mater.* **13**, 851–856 (2014).
14. Cao, J. *et al.* Landau level splitting in Cd_3As_2 under high magnetic fields. *Nat. Commun.* **6**, 6779 (2015).
15. Xiang, Z. *et al.* Angular-dependent phase factor of Shubnikov-de Haas oscillations in the Dirac semimetal Cd_3As_2. *Phys. Rev. Lett.* **115**, 226401 (2015).
16. Zhao, Y. *et al.* Anisotropic Fermi surface and quantum limit transport in high mobility three-dimensional Dirac semimetal Cd_3As_2. *Phys. Rev. X* **5**, 031037 (2015).
17. Xiong, J. *et al.* Evidence for the chiral anomaly in the Dirac semimetal Na_3Bi. *Science* **350**, 413 (2015).
18. Li, H. *et al.* Negative magnetoresistance in Dirac semimetal Cd_3As_2. *Nat. Commun.* **7**, 10301 (2016).
19. Parameswaran, S., Grover, T., Abanin, D., Pesin, D. & Vishwanath, A. Probing the chiral anomaly with nonlocal transport in three-dimensional topological semimetals. *Phys. Rev. X* **4**, 031035 (2014).
20. Li, C.-Z. *et al.* Giant negative magnetoresistance induced by the chiral anomaly in individual Cd_3As_2 nanowires. *Nat. Commun.* **6**, 10137 (2015).
21. Zhang, C. *et al.* Detection of chiral anomaly and valley transport in Dirac semimetals. Preprint at http://arxiv.org/abs/1504.07698 (2015).
22. Xu, S.-Y. *et al.* Observation of Fermi arc surface states in a topological metal. *Science* **347**, 294–298 (2015).
23. Yi, H. *et al.* Evidence of topological surface state in three-dimensional Dirac semimetal Cd_3As_2. *Sci. Rep.* **4**, 6106 (2014).
24. Zhang, Y. & Vishwanath, A. Anomalous Aharonov-Bohm conductance oscillations from topological insulator surface states. *Phys. Rev. Lett.* **105**, 206601 (2010).
25. Bardarson, J. H., Brouwer, P. & Moore, J. Aharonov-Bohm oscillations in disordered topological insulator nanowires. *Phys. Rev. Lett.* **105**, 156803 (2010).
26. Hong, S. S., Zhang, Y., Cha, J. J., Qi, X.-L. & Cui, Y. One-dimensional helical transport in topological insulator nanowire interferometers. *Nano Lett.* **14**, 2815–2821 (2014).
27. Cho, S. *et al.* Aharonov-Bohm oscillations in a quasi-ballistic three-dimensional topological insulator nanowire. *Nat. Commun.* **6**, 7634 (2015).
28. Jauregui, L. A., Pettes, M. T., Rokhinson, L. P., Shi, L. & Chen, Y. P. Magnetic field-induced helical mode and topological transitions in a topological insulator nanoribbon. *Nat. Nanotech.* Doi:10.1038/nnano.2015.293 (2016).
29. Stone, A. D. Magnetoresistance fluctuations in mesoscopic wires and rings. *Phys. Rev. Lett.* **54**, 2692 (1985).
30. Peng, H. *et al.* Aharonov-Bohm interference in topological insulator nanoribbons. *Nat. Mater.* **9**, 225–229 (2010).
31. Murakawa, H. *et al.* Detection of Berry's phase in a bulk Rashba semiconductor. *Science* **342**, 1490 (2013).
32. Washburn, S., Umbach, C., Laibowitz, R. & Webb, R. A. Temperature dependence of the normal-metal Aharonov-Bohm effect. *Phys. Rev. B* **32**, 4789 (1985).
33. Fagas, G. & Greer, J. C. Ballistic conductance in oxidized Si nanowires. *Nano Lett.* **9**, 1856–1860 (2009).
34. Xiu, F. *et al.* Manipulating surface states in topological insulator nanoribbons. *Nat. Nanotechnol.* **6**, 216–221 (2011).
35. Gorbar, E., Miransky, V. & Shovkovy, I. Engineering Weyl nodes in Dirac semimetals by a magnetic field. *Phys. Rev. B* **88**, 165105 (2013).
36. Liu, Y. *et al.* Gate-tunable quantum oscillations in ambipolar Cd_3As_2 thin films. *NPG Asia Mater.* **7**, e221 (2015).
37. Zhang, E. *et al.* Magnetotransport properties of Cd_3As_2 nanostructures. *ACS Nano* **9**, 8843 (2015).
38. Heremans, J. *et al.* Bismuth nanowire arrays: synthesis and galvanomagnetic properties. *Phys. Rev. B* **61**, 2921 (2000).
39. Nikolaeva, A. A. *et al.* Negative magnetoresistance in transverse and longitudinal magnetic fields in Bi nanowires. *J. Phys. Conf. Ser.* **150**, 022065 (2009).

Acknowledgements

We are grateful to Professors Xincheng Xie, Ji Feng, Kai Chang and Dr Haiwen Liu for inspired discussions. This work was supported by MOST (nos 2013CB934600 and 2013CB932602) and NSFC (nos 11274014, 11234001 and 11327902).

Author contributions

Z.-M.L. conceived and designed the experiments. D.-P.Y. gave scientific advice. L.-X.W. and C.-Z.L. performed the measurements. Z.-M.L. and L.-X.W. wrote the manuscript.

Additional information

4

Manipulating the interfacial structure of nanomaterials to achieve a unique combination of strength and ductility

Amirhossein Khalajhedayati[1], Zhiliang Pan[2] & Timothy J. Rupert[1,2]

The control of interfaces in engineered nanostructured materials has met limited success compared with that which has evolved in natural materials, where hierarchical structures with distinct interfacial states are often found. Such interface control could mitigate common limitations of engineering nanomaterials. For example, nanostructured metals exhibit extremely high strength, but this benefit comes at the expense of other important properties like ductility. Here, we report a technique for combining nanostructuring with recent advances capable of tuning interface structure, a complementary materials design strategy that allows for unprecedented property combinations. Copper-based alloys with both grain sizes in the nanometre range and distinct grain boundary structural features are created, using segregating dopants and a processing route that favours the formation of amorphous intergranular films. The mechanical behaviour of these alloys shows that the trade-off between strength and ductility typically observed for metallic materials is successfully avoided here.

[1] Department of Chemical Engineering and Materials Science, University of California, Irvine, California 92697, USA. [2] Department of Mechanical and Aerospace Engineering, University of California, 4200 Engineering Gateway, Irvine, California 92697, USA. Correspondence and requests for materials should be addressed to T.J.R. (email: trupert@uci.edu).

The vast majority of engineering materials are polycrystalline and tailoring the interfaces, or grain boundaries, between the many crystals within a given sample volume offers a promising approach for the discovery of new materials with improved properties. Two recent strategies for realizing such gains have been (i) increasing the number of interfaces or effectively the volume fraction of material located in the interfacial regions and (ii) tailoring the structure of grain boundaries by changing interfacial chemistry. The first strategy is characterized by a dramatic reduction in the size of the crystalline grains that comprise the material and can be termed 'nanostructuring' when this reduction brings the average grain size below ∼100 nm. The second strategy uses segregating dopant atoms to tailor the local equilibrium structure of interfaces and can be termed 'complexion engineering'[1]. Although nanostructuring and complexion engineering each offer opportunities on their own, these two material design approaches have not yet been combined in a complementary manner. Natural materials, such as dental enamel[2] and nacre[3], often have evolved to contain such hierarchical nanostructures with distinct interfacial features that serve to dramatically influence their performance. Such a strategy applied to engineering materials should allow for the realization of unique property sets, where a material has the benefits of a nanostructured material with regards to one property yet avoids the usual degradation of other properties through complexion engineering.

Nanostructuring has been used extensively in recent years to create a wide range of nanocrystalline metals and ceramics[4,5]. These materials take advantage of the fact that atoms in the grain boundaries behave differently than atoms in the grain interior, because of lower packing fractions and the lack of long-range crystalline order. For example, grain boundaries resist the transmission of plastic strain in the form of moving dislocations, as these linear defects in crystals cannot glide through the interfacial region easily. Therefore, a large reduction in grain size is equivalent to adding more obstacles to dislocation motion, leading to a dramatic increase in the material's strength following a well-known scaling law known as Hall–Petch scaling[6,7]. Wang et al.[8] reported on nanocrystalline Cu with an average grain size (d) of 30 nm and a yield strength of 760 MPa, or more than ten times higher than that of Cu with an average grain size in the micrometre range (55 MPa)[9]. Unfortunately, the addition of a large volume fraction of interfaces can have negative consequences as well. For instance, these same nanocrystalline metals are almost always brittle[9,10], with very little global plastic strain before failure, negating the value of their high strength in many cases.

Separately, it has been recognized that a grain boundary can be described as a type of phase-like structure in many cases, although these interfacial features differ from traditional bulk phases because their structure is dependent on the two abutting crystals and are therefore known as complexions[1,11]. However, similar to bulk phases, interface complexions can be analysed with equilibrium thermodynamic theories and may transform between different structures as temperature or chemical composition varies[1,12]. Complexions are difficult to observe in materials made from a single element[13], but recent research has shown that distinct interfacial structures become more accessible in alloys[14,15]. Dillon et al. created a categorization scheme based on transmission electron microscopy (TEM) observations in doped Al_2O_3 that relies on complexion thickness[11]. Six discrete complexion types were found, with increasing levels of structural disorder, which are categorized as (i) single dopant layers, (ii) nominally 'clean' grain boundaries, (iii) bilayers, (iv) multilayers, (v) intergranular films with equilibrium thicknesses in the nanometre range and (vi) wetting films with arbitrary thicknesses.

Complexion types V and VI are often disordered films without any long-range crystalline order, in which case they can both be classified as amorphous intergranular films (AIFs). An AIF is an example of a complexion that is both distinct in its local structure as well as its effect on material properties, being suggested as the cause for the previously unexplained phenomenon of activated sintering in refractory metals[14].

To date, complexions have been studied and utilized predominantly in either coarse-grained polycrystalline materials or simple bicrystal samples. Here, we report on a strategy for combining the concepts of nanostructuring and complexion engineering, to enable a new class of materials: nanomaterials with tunable grain boundary structure. We begin by developing a processing route for making such materials, with an eye on techniques that are flexible enough to create a wide variety of chemistries while also scalable so that they can eventually be used to produce bulk quantities of material. Since AIFs are a complexion type that will have atomic structure and properties which are unique compared with normal, clean grain boundaries, we focus on inducing and characterizing these features in a nanoscale grain structure. Then, the mechanical properties of the alloy are probed, to demonstrate the clear advantage of this new materials design strategy. Specifically, we find that the ductility and toughness of nanostructured metals can be greatly improved using complexion engineering without sacrificing any strength, breaking the paradigm of a direct strength-ductility trade-off that has dominated prior observations.

Results

Materials selection, processing and characterization. Cu-based alloys were selected for this investigation, as Cu is a common face-centred cubic material and our findings should be broadly applicable to many other structural metals. In addition, the deformation physics and properties of pure nanocrystalline Cu have been well-studied, giving reference points to compare against[16]. As doping of the interfacial regions is required to induce complexions, alloying elements that segregate to the grain boundaries are sought. Empirical considerations such as large atomic size mismatch and limited solubility in the bulk can give rough guidelines for dopant selection, whereas grain boundary segregation enthalpy calculations give a more direct prediction of segregation tendency[17]. As this study is concerned with mechanical behaviour, embrittlement is to be avoided and there is a need to preserve metallic bonding at the interfaces. Non-metals or poor metals such as Bi and Pb are eliminated from consideration. Finally, thermodynamic modelling tools developed by Luo and co-workers[12,18] suggest that AIFs will form if the total energy of an amorphous layer with finite thickness and two new amorphous-crystalline interfaces (that is, the entire complexion) is lower than the energy associated with the interface between two crystalline phases (that is, the original grain boundary). This means that alloy combinations that have a low volumetric free energy penalty for a liquid-like structure and can easily form amorphous solids are promising for the formation of AIFs. With all of these considerations in mind, Zr was chosen as our alloying element, as it will segregate to the grain boundaries[17], is a transition metal and binary Cu-Zr alloys have a high glass-forming ability[19].

Mechanical alloying with a high-energy ball mill produces powders with particle sizes of micrometre-scale diameter, with each particle containing many individual nanometre-scale grains. This technique can make nanostructured powders with a wide variety of chemical compositions and scale-up to bulk parts through powder metallurgy techniques is a possibility[20]. As a control, unalloyed nanocrystalline Cu with $d = 30$ nm was first produced, as shown in Fig. 1a,d. The selected area electron diffraction pattern in Fig. 1d shows that the material is a random polycrystalline sample. A Cu-3 atomic % Zr alloy was then created, as full segregation of this amount of Zr to the interfaces will put the grain boundary composition into the glass-forming range for Cu-Zr metallic glasses[21]. In the as-milled state, the alloy sample also has $d = 30$ nm (Fig. 1e). The bottom frame of Fig. 1e presents a compositional line scan through the as-milled Cu-Zr alloy, showing that Zr is found throughout the sample, mixed into both crystal interior and grain boundary regions (shown schematically in Fig. 1b). Three grain boundaries are labelled in this figure with black arrows. A small amount of grain boundary segregation is occasionally found even in the as-milled sample, as shown by the slightly elevated Zr concentration at the first boundary noted, but significant amounts of Zr were inside of the grains as well. As Zr has negligible solubility (~ 0.12 atomic % (ref. 22)) in the Cu lattice according to the bulk phase diagram, this structure is a supersaturated solid solution. To induce segregation, the powders were annealed at 950 °C for 1 h. This temperature is extremely high for Cu and Cu-Zr alloys, being $\sim 90\%$ of the melting temperature of pure Cu and $\sim 98\%$ of the solidus temperature (where the material begins to melt) of Cu-3 atomic % Zr[22]. During annealing, Zr diffuses to the grain boundaries, as shown by the compositional line scan in Fig. 1f, where two obvious examples of grain boundaries are labelled. Grain interiors are depleted of Zr after annealing, whereas the interfacial regions show intense Zr enrichment (shown schematically in Fig. 1c). Additional evidence of Zr grain boundary segregation was found by using high-angle annular

dark-field TEM imaging (see Supplementary Fig. 1 for details). The Cu-Zr alloy experiences very little grain growth, only coarsening to an average grain size of 45 nm (Fig. 1f), because of a reduction in grain boundary energy with Zr segregation[23], whereas the grain size of a pure Cu sample annealed under the same conditions coarsens to the micrometre range. Although not the explicit focus of this study, the observed thermal stability at such a high temperature suggests that this material can be consolidated into bulk pieces[20].

The high-temperature annealing treatment used to induce Zr segregation is also useful for promoting AIF formation. Shi and Luo[12] recently developed interfacial thermodynamic models and grain boundary diagrams, showing that higher temperatures usually promote thicker AIFs. A set of annealed Cu-Zr powders were quickly quenched by dropping into a large water bath in less than 1 s, freezing in any structures which are in equilibrium at 950 °C. Fresnel fringe imaging[24] was used to identify interfacial films, followed by high-resolution TEM for detailed characterization of grain boundary structure and measurement of AIF thickness. A representative example of an AIF is presented in Fig. 2a. The areas in the bottom left and top right of Fig. 2a are crystalline, as shown by the presence of lattice fringes in the image as well as sharp spots in the fast Fourier transform patterns, which denote periodic order associated with the lattice. In contrast, the region at the interface, between the two dashed lines, is amorphous and disordered with a thickness of 5.7 nm. The fast Fourier transform pattern shows no sign of long-range crystalline order in this case and is completely featureless. An estimation of the diffusion length for this system shows that these

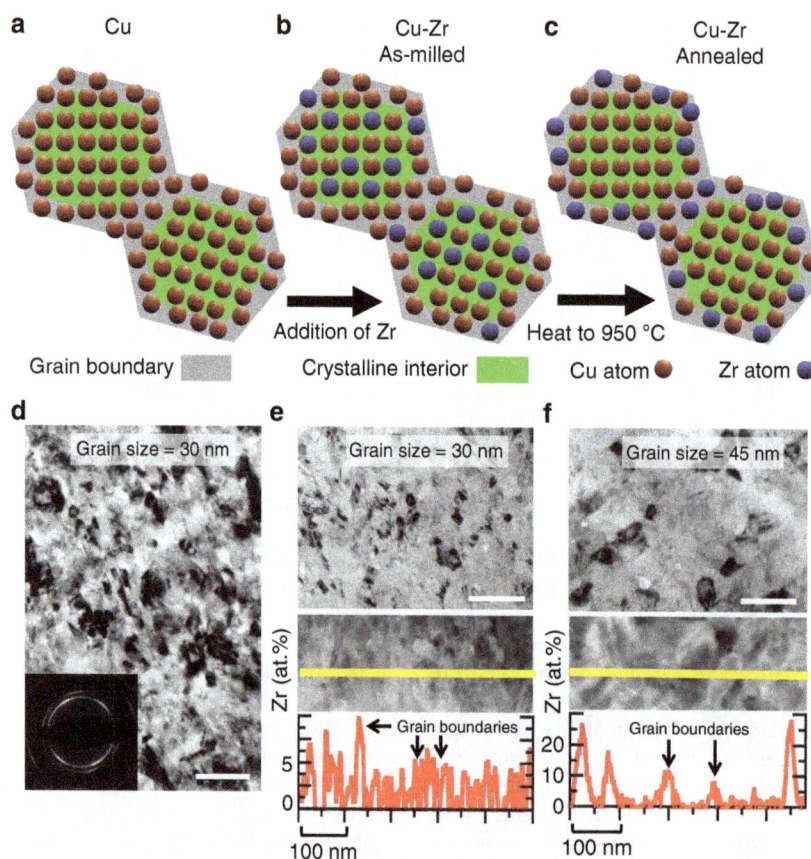

Figure 1 | Alloy design strategy for adding segregating dopants. Pure Cu (**a**) can be converted to an alloy by adding Zr during ball milling (**b**). Although the Zr is mixed throughout the grain structure, annealing treatments can then be used to induce preferential Zr segregation to the grain boundaries (**c**). Both the (**d**) pure Cu and (**e**) Cu-Zr as-milled samples have an average grain size of 30 nm (scale bars, 100 nm). (**f**) Annealing the Cu-Zr sample at 950 °C for 1 h allows for segregation and only causes coarsening to a grain size of 45 nm (scale bar, 100 nm). The same level of Zr segregation was found in both the quenched and air-cooled samples.

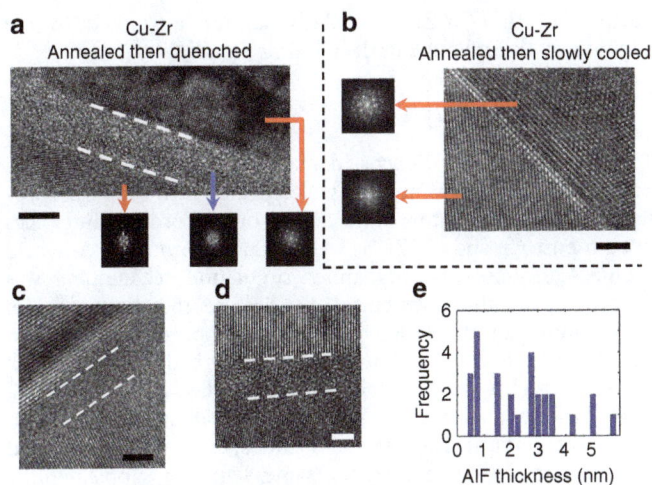

Figure 2 | High-resolution TEM images of grain boundary structure in nanocrystalline Cu-Zr alloys. (**a**) An amorphous intergranular film with thickness of 5.7 nm was observed at a grain boundary after quickly quenching from 950 °C (scale bar, 5 nm). (**b**) In contrast, grain boundaries in a slowly cooled sample, with structures that are in equilibrium near ambient temperatures, are all ordered interfaces (scale bar, 2 nm). Insets are fast Fourier transform patterns, highlighting the disordered nature of the interface in **a**. (**c,d**) Additional examples of amorphous complexions in the quenched sample (scale bars, 2 nm), whereas **e** summarizes the measurements from the 28 interfacial films found here.

AIFs cannot be a metastable phase, as the high temperature used for annealing provides more than enough kinetic driving force for equilibrium chemical distributions to form within the nanocrystalline grain structure (see Supplementary Note 1 for detailed calculations).

To ensure that the observed AIFs were truly complexions, driven by alloy thermodynamics with a combination of grain boundary dopants and elevated temperatures required, and did not form through another process such as solid-state amorphization, which is purely driven by local composition, another set of powders was annealed at 950 °C and then slowly cooled back to room temperature over a period of ~5 min, or ~300 times longer than the quenched sample. Such a cooling schedule gives a sample with interfaces that are in equilibrium at ambient temperatures. No AIFs were observed in this specimen and all interfaces studied were ordered, with a representative example shown in Fig. 2b. Crystalline order is present all of the way up to the boundary. Zr dopants were still found to have segregated to the grain boundaries, with compositional line scans appearing nominally identical to the one shown in Fig. 1f. Although the boundaries were still heavily doped with Zr, the lack of AIFs proves that the amorphous structure shown in Fig. 2a is only in equilibrium at high temperature, in line with thermodynamic theories of disordered complexions.

Additional examples of AIFs from the quenched Cu-Zr sample are shown in Fig. 2c,d with thicknesses of 2.2 and 2.7 nm, respectively. Twenty-eight AIFs were studied with high-resolution TEM, with a distribution of measured AIF thicknesses shown in Fig. 2e. In addition, a number of ordered grain boundaries were observed in this sample as well. As a whole, this means that not all grain boundaries have the same structure in the quenched Cu-Zr alloy, but rather there is a distribution of complexion types. Possible explanations for these variations include differences in grain boundary energy, local segregation state or slight differences in local cooling rates during quenching. The existence of amorphous structures below the solidus temperature of the alloy

suggests that these features are stable AIFs with thicknesses determined by a combination of local grain boundary chemical composition and high temperature (type V), although some of the thicker AIFs could possibly be wetting films (type VI). Even though the annealing temperature was below the solidus temperature for Cu-3 atomic % Zr, it was above the eutectic temperature, the lowest possible melting temperature over all mixing ratios, which occurs at higher Zr concentrations. It is possible that some boundaries have higher levels of Zr segregation and move into a different region of the phase diagram where wetting films can form at 950 °C. However, previous observations of complexions in Al_2O_3 showed that film thickness changes significantly along individual wetting films[25], whereas each individual boundary studied here showed no measureable variation in thickness. In addition, wetting films are often much thicker (at least 10 nm thick) than the films observed here. Even if there is some uncertainty in the complexion type of the thickest AIFs, the existence of many nanometre-thick amorphous interfacial structures is clear.

Mechanical behaviour and the effect of complexions. To demonstrate the potential utility of the material described in this report, a nanostructured metal with disordered amorphous grain boundaries, we turn our attention to mechanical behaviour. Nanocrystalline metals have very high strengths, but are generally also very brittle and exhibit little ductility, with elongations or strain-to-failure values of only a few percent typically reported[9,10]. This brittle nature often restricts the usage of nanocrystalline materials in technological applications. Prior attempts to increase the ductility of nanocrystalline metals, for example, by encouraging mechanically driven grain coarsening[26] or by inducing superplasticity due to sulfur contamination[27,28] or inclusion of a phase with low melting temperature[29], have come with dramatic reductions in strength. The high strength of a nanocrystalline material comes from the fact that traditional dislocation plasticity mechanisms are shut off at these extremely fine crystallite sizes, requiring dislocations to be nucleated at an interface, propagate across the nanograin and then be absorbed at the opposite grain boundary. The repeated dislocation absorption required for such a physical mechanism may be the cause of the brittle nature of nanocrystalline metals, as atomistic simulations show that a single absorption event leads to high local stresses at the grain boundary[30], whereas multiple absorption events result in crack nucleation[31]. Such a hypothesis is supported by in situ TEM deformation studies, which show that nanometre size cracks nucleate at grain boundary sites during plastic deformation[32]. As a result, damage tolerant grain boundaries should be the key to designing ductile nanocrystalline metals, making control of grain boundary structure a priority.

Small cylindrical specimens were created in the powder particles using a focused ion beam microscope, then tested in compression to quantify strength and in bending to quantify strain-to-failure. Representative examples from each test for pure nanocrystalline Cu are shown in Fig. 3a–c, demonstrating the typical response of a nanocrystalline metal to which we can then compare our Cu-Zr results. Figure 3a shows that pure Cu has a high-yield strength of 740 MPa but the pillar cracks and fails during compression, resembling the failure of a brittle ceramic. Compression experiments are not able to quantify the ductility of a material, which is defined as the ability to deform under tensile stress, so pillar bending experiments were used to measure plastic strain-to-failure (see Supplementary Methods for detailed experimental techniques). Videos of representative beam bending experiments for pure Cu and Cu-Zr with AIFs are shown in Supplementary Movies 1 and 2, respectively. Figure 3b,c shows

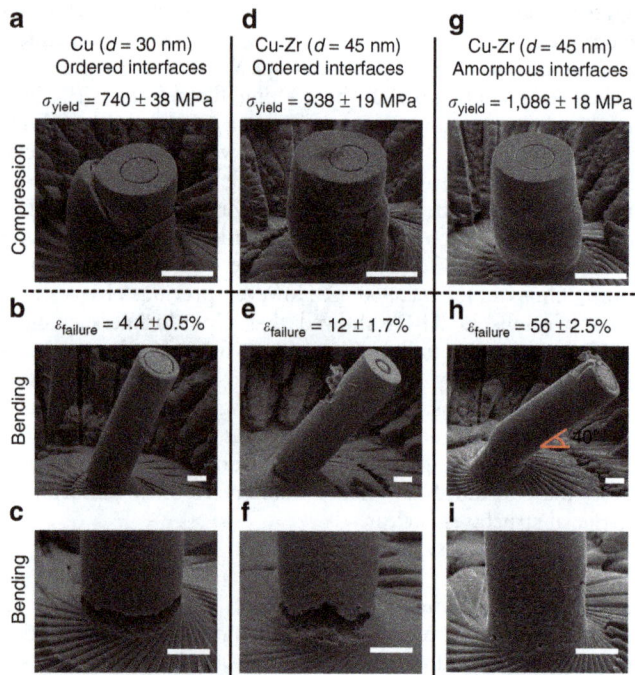

Figure 3 | Mechanical testing results from nanocrystalline Cu and Cu-Zr samples. Failed micropillars after compression, as well as yield strength measurements, are shown for (**a**) pure Cu with ordered interfaces, (**d**) Cu-Zr with ordered interfaces and (**g**) Cu-Zr with amorphous intergranular films (scale bars, 5 μm). Failed micropillars after bending, as well as strain-to-failure measurements, are shown for (**b,c**) pure Cu with ordered interfaces, (**e,f**) Cu-Zr with ordered interfaces and (**h,i**) Cu-Zr with amorphous intergranular films (scale bars, 2 μm). A nanocrystalline alloy with AIFs can be both stronger and more ductile than its traditional, pure metal counterpart.

this same brittle response during bending, where a crack forms early during the experiment and there is little plastic strain built up at the base of the pillar, from which strain-to-failure can be measured as 4.4%. This low value is nearly identical to a previous report for nanocrystalline Cu, produced by a different method and tested in uniaxial tension, that had the same average grain size (30 nm) and a similar yield strength (760 MPa)[8]. Representative examples for each test from our Cu-Zr alloy with AIFs are shown in Fig. 3g–i, providing evidence of a strong yet ductile response. Figure 3g shows that the Cu-Zr with AIFs is even stronger than the pure Cu, having a yield strength of 1,086 MPa, but more importantly the pillar compresses in a stable, homogeneous manner. Figure 3h shows that the pillar is ductile during the bending experiment as well and a strain-to-failure of 56% was measured for these alloys, more than ten times larger than the pure Cu value. A magnified view of the pillar base is presented in Fig. 3i, showing that extensive plasticity has occurred and that, even at this large plastic strain, there are only small, unconnected cracks instead of a crack completely across the surface, suggesting that this strain-to-failure value may even be slightly conservative for this alloy. To highlight the importance of AIFs for the observed behaviour, mechanical tests were also performed on the Cu-Zr sample that was slowly cooled in order to have ordered grain boundaries. These results, presented in Fig. 3d–f, demonstrate that this alloy is brittle as well, mimicking the pure Cu sample by crumbling in compression and failing at a small strain in bending. As the only difference between the Cu-Zr samples is the grain boundary structure, with grain size and Zr segregation state identical for both samples, the increased

ductility of the Cu-Zr with AIFs can be attributed to the addition of amorphous interfacial films alone.

Discussion

To understand why AIFs act as damage tolerant grain boundaries, molecular dynamics simulations of repeated dislocation absorption at room temperature were carried out on ordered interfaces in both pure Cu and Cu-Zr, as well as Cu-Zr with AIFs. A hybrid Monte Carlo/molecular dynamics simulation technique[33] was used to create these structures and ensure that they were in equilibrium. For the ordered interfaces in pure Cu, shown in Fig. 4a, very few dislocations can be absorbed before a crack nucleates, leading to cracking at a low applied shear strain (γ). The crack then grows quickly after a small additional strain increment until it runs through the specimen. The ordered interfaces in Cu-Zr behaved the same way (see Supplementary Fig. 5 for details). On the other hand, the AIFs diffuse the strain concentration brought by dislocation absorption into a wider region within the boundary (Fig. 4b), delaying the crack nucleation event until a much larger applied shear strain is reached. Pan and Rupert studied the mechanics of this problem systematically in a recent paper, finding that crack nucleation as well as crack growth rate is suppressed as AIF thickness increases[34]. Brandl et al. even showed that an AIF can exert attractive forces on nearby dislocations as it is elastically softer than the crystalline lattice, pulling dislocations towards the interface where they can then be absorbed[35]. Experiments on crystalline-amorphous nanocomposite films support this concept as well, as the addition of a thin amorphous layer was found to improve strain-to-failure when compared with films with only a crystalline phase[36–38]. Although the nanocomposites outperform the crystalline films, the strain-to-failure values were only measured to be in the range of 4–14% (much lower than the values reported here), likely because the AIFs were only added in one direction of the film. Donohue et al.[39] reported on another nanolaminate comprised of 90-nm-thick layers of crystalline Cu separated by amorphous Pd-Si layers of 10 nm thickness. Although the monolithic amorphous films had negligible ductility by themselves, the nanolaminate structure had a strain-to-failure of 3%. In contrast, our materials have a fully three-dimensional grain boundary network decorated with AIFs. In our case, by adding amorphous interfacial features, it is possible to have the high strength characteristic of a nanocrystalline metal without the typical brittle failure (Fig. 4c). In fact, the strain-to-failure and plastic flow of a nanocrystalline metal with AIFs (Fig. 4d) resembles that of a coarse-grained or single crystalline metal (Fig. 4e).

The observation of high strength with large ductility fits a recent narrative that nanocrystalline materials are intrinsically ductile on the nanoscale[40], but suffer from premature failure due to strain localization or cracking[41]. If this premature failure can be suppressed, then the traditional trade-off between strength and ductility can be avoided. One recently developed path for suppression of this premature failure is to incorporate nanocrystalline metals as the outer layer in a gradient architecture, where grain size changes continuously from nanometres to micrometres to avoid early plastic necking[42,43]. The results presented here for Cu-Zr alloys created by a unique processing route can be interpreted as another example of this design strategy, with an explicit focus on resisting crack nucleation and growth through the introduction of grain boundary complexions. A strength-ductility synergy is achieved and our Cu-Zr alloys outperform traditional metals. A compilation of data from the literature for Cu and Cu alloys is presented in Fig. 5, where strain-to-failure is plotted as a function of yield strength. The properties

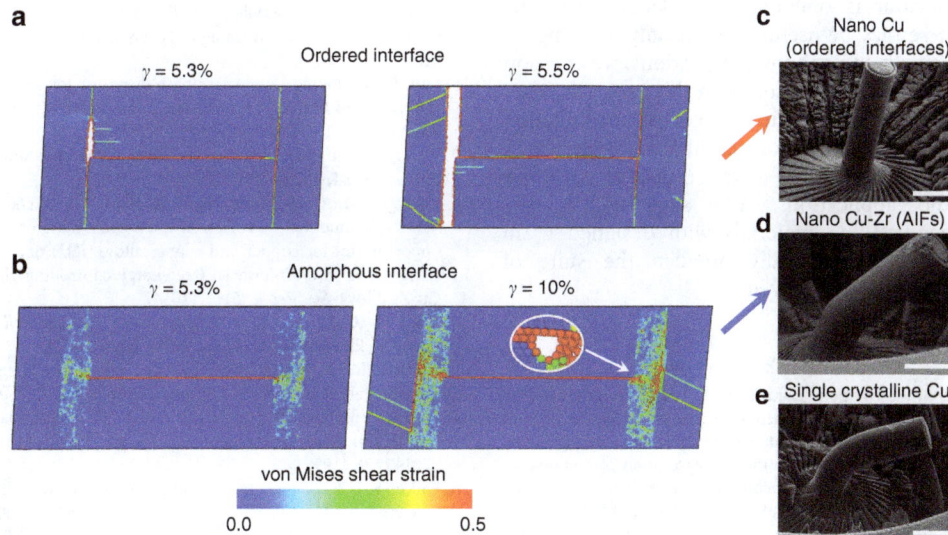

Figure 4 | Connection between interfacial structure and damage tolerance. Molecular dynamics simulations of dislocation absorption at (**a**) an ordered grain boundary and (**b**) an amorphous intergranular film show the formation and propagation of crack damage. The ordered interface quickly fractures, whereas the amorphous interface diffuses the strain concentration brought by dislocation absorption and fracture is delayed. The delay of failure explains why (**d**) nanocrystalline Cu-Zr with AIFs has ductility reminiscent of (**e**) coarse-grained Cu while retaining the high strength of (**c**) nanocrystalline Cu (scale bars, 5 μm).

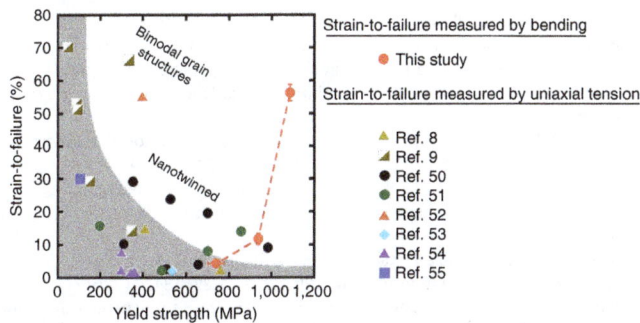

Figure 5 | Strain-to-failure and yield strength for Cu and Cu-based alloys. The vast majority of data fall within the grey envelope, with recent advanced alloys pushing slightly outside this limit. Our Cu-Zr alloy breaks the expected trend, with both higher strength and ductility than the pure Cu sample. All strain-to-failure data comes from material that failed under tensile stresses, either under uniaxial tension (literature data) or from the tensile side of a bending experiment (this study). Literature data are taken from refs 8,9,50–55.

of traditional Cu, both microcrystalline and nanocrystalline, fall within the grey envelope, demonstrating a clear trade-off between strength and ductility. Advanced alloys, such as Cu with a bimodal grain structure (micrometre-scale grains embedded in a matrix of nanocrystalline grains) and nanotwinned Cu, push beyond these limits and demonstrate improved combinations of these properties. The Cu and Cu-Zr alloys described in this paper are shown as red points on the plot. It is important to note that our strength measurements come from microcompression tests and strain-to-failure is taken from bending experiments, whereas the data from the literature comes from tensile experiments. The pure Cu sample exhibits classical behaviour and falls within the grey envelope. The strength and strain-to-failure measurements for our pure nanocrystalline Cu are remarkably similar to those of Wang et al.[8], taken from nanocrystalline Cu with an identical grain size created by surface mechanical attrition and appearing as a yellow triangle in Fig. 5. The close agreement between these values demonstrates that it is reasonable to compare our

micropillar experiments to the uniaxial tension results available in the literature. Our Cu-Zr with AIFs breaks the traditional paradigm, being both stronger and more ductile than the pure Cu sample. Although the improved ductility has already been discussed, the increased strength comes from the segregation of Zr to the interfaces. Vo et al.[44] used atomistic simulation techniques to study how interfacial segregation altered the grain boundary energy and strength of nanocrystalline Cu. These authors found that boundary energy and strength were intimately related, with strength increasing monotonically as grain boundary energy was reduced by doping. Physically, this means reducing grain boundary energy makes it harder for dislocations to nucleate and then propagate through the grain (the deformation mechanisms, which have been tied to yield strength[45,46]). In our case, Zr segregation to grain boundary sites reduces grain boundary energy, which is the inherent driving force for segregation in general but is also supported by the increased thermal stability of Cu-Zr compared with the pure Cu sample. There is even a small increase in strength when going from Cu-Zr with ordered boundaries to Cu-Zr with AIFs. Complexion formation is fundamentally driven by the material seeking a reduction in total interface energy, so this should provide a strengthening contribution. Dislocation nucleation may also be affected by the shift from a crystalline–crystalline interface in the ordered grain boundary to an amorphous–crystalline interface in the case of the AIF. As a whole, the concurrent increase in both strength and ductility shows that control of grain boundary structure is a promising pathway for the design of optimized structural materials.

Our study shows that the previously separate techniques of nanostructuring and complexion engineering can be combined in a complementary manner, introducing a new feature into the design toolbox of materials scientists: grain boundary structure. Unique Cu-Zr alloys are reported, created through powder metallurgy techniques, with AIFs 'frozen' into the interfaces as a result of Zr segregation and rapid quenching from high-temperature ageing treatments. These new materials demonstrate a combination of strength and ductility that is not found in traditional Cu alloys, behaviour that can be attributed directly to the fracture resistance of AIFs. However, it is important to note that mechanical behaviour is one of many places where the materials described here may be useful. Even if only focusing on nanocrystalline metals with amorphous complexions, we envision materials that can be rapidly consolidated because of faster diffusion along AIFs or that are more tolerant to radiation damage with AIFs acting as sinks for residual point defects[47]. Amorphous interfaces could also act as fast transport paths for energy or electronic applications. We suggest that the introduction of nanocrystalline materials with designed grain boundary structures will dramatically broaden the suite of material properties that can be achieved.

Methods

Powder processing and characterization. Nanostructured metals were created by ball milling high-purity powders of the constituent elements with a hardened steel vial and milling media in an argon environment. Powders were then encapsulated in quartz tubes under vacuum and annealed to induce segregation of dopants, either followed by slow air cooling or fast quenching in a water bath. Milling in argon and annealing under vacuum was done to avoid oxidation of the powders. The phase content of the powders was measured using X-ray diffraction, whereas grain size and detailed grain boundary structure were characterized with TEM. The yield strength of each alloy was measured by making micrometre-sized pillars in each powder, followed by compression with an Agilent G200 nanoindenter using a flat triangular diamond tip and an engineering strain rate of $3.1 \times 10^{-4} \, \text{s}^{-1}$. Engineering stress–strain curves were calculated and are presented in Supplementary Fig. 2. Yield strengths were calculated based on a 0.7% plastic strain offset, following the work of Brandstetter et al.[48] Strain-to-failure was calculated from another set of micropillars that were bent with a micro-manipulator inside of a scanning electron microscope. The process for estimating strain-to-failure is illustrated in detail in Supplementary Fig. 3.

Atomistic modelling. Classical molecular dynamics simulations were performed using the Large-scale Atomic/Molecular Massively Parallel Simulator (LAMMPS) package[49]. The starting configurations for the atomistic simulations were created using a hybrid Monte Carlo/molecular dynamics method capable of finding equilibrium chemical composition and structural configuration. Further details of the processing, mechanical testing and simulation methods can be found in the Supplementary Methods section.

References

1. Cantwell, P. R. et al. Grain boundary complexions. Acta Mater. **62**, 1–48 (2014).
2. Gordon, L. M. et al. Amorphous intergranular phases control the properties of rodent tooth enamel. Science **347**, 746–750 (2015).
3. Sun, J. & Bhushan, B. Hierarchical structure and mechanical properties of nacre: a review. RSC Adv **2**, 7617–7632 (2012).
4. Gleiter, H. Nanocrystalline Materials. Prog. Mater. Sci. **33**, 223–315 (1989).
5. Gleiter, H. Nanostructured materials: basic concepts and microstructure. Acta Mater. **48**, 1–29 (2000).
6. Hall, E. O. The deformation and ageing of mild steel: III. Discussion of results. Proc. Phys. Soc. Lond. B **64**, 747–753 (1951).
7. Petch, N. J. The cleavage strength of polycrystals. J. Iron Steel Inst **174**, 25–28 (1953).
8. Wang, Y. M. et al. Microsample tensile testing of nanocrystalline copper. Scr. Mater **48**, 1581–1586 (2003).
9. Wang, Y., Chen, M., Zhou, F. & Ma, E. High tensile ductility in a nanostructured metal. Nature **419**, 912–915 (2002).
10. Meyers, M. A., Mishra, A. & Benson, D. J. Mechanical properties of nanocrystalline materials. Prog. Mater. Sci. **51**, 427–556 (2006).
11. Dillon, S. J., Tang, M., Carter, W. C. & Harmer, M. P. Complexion: a new concept for kinetic engineering in materials science. Acta Mater. **55**, 6208–6218 (2007).
12. Shi, X. M. & Luo, J. Developing grain boundary diagrams as a materials science tool: a case study of nickel-doped molybdenum. Phys. Rev. B **84**, 014105 (2011).
13. Hsieh, T. E. & Balluffi, R. W. Experimental study of grain boundary melting in aluminum. Acta Metall. **37**, 1637–1644 (1989).
14. Gupta, V. K., Yoon, D. H., Meyer, H. M. & Luo, J. Thin intergranular films and solid-state activated sintering in nickel-doped tungsten. Acta Mater. **55**, 3131–3142 (2007).
15. Shi, X. M. & Luo, J. Grain boundary wetting and prewetting in Ni-doped Mo. Appl. Phys. Lett. **94**, 251908 (2009).
16. Schiotz, J. & Jacobsen, K. W. A maximum in the strength of nanocrystalline copper. Science **301**, 1357–1359 (2003).
17. Murdoch, H. A. & Schuh, C. A. Estimation of grain boundary segregation enthalpy and its role in stable nanocrystalline alloy design. J. Mater. Res. **28**, 2154–2163 (2013).
18. Luo, J. Developing interfacial phase diagrams for applications in activated sintering and beyond: current status and future directions. J. Am. Ceram. Soc. **95**, 2358–2371 (2012).
19. Wang, D. et al. Bulk metallic glass formation in the binary Cu–Zr system. Appl. Phys. Lett. **84**, 4029–4031 (2004).
20. Tschopp, M. A., Murdoch, H. A., Kecskes, L. J. & Darling, K. A. 'Bulk' nanocrystalline metals: review of the current state of the art and future opportunities for copper and copper alloys. JOM **66**, 1000–1019 (2014).
21. Kang, D.-H. et al. Interfacial free energy controlling glass-forming ability of Cu-Zr alloys. Sci. Rep **4**, 5167 (2014).
22. Smithells, C. J., Gale, W. F. & Totemeier, T. C. Smithells Metals Reference Book 8th edn (Elsevier Butterworth-Heinemann, 2004).
23. Atwater, M. A., Scattergood, R. O. & Koch, C. C. The stabilization of nanocrystalline copper by zirconium. Mater. Sci. Eng. A **559**, 250–256 (2013).
24. Jin, Q., Wilkinson, D. S. & Weatherly, G. C. Determination of grain-boundary film thickness by the fresnel fringe imaging technique. J. Eur. Ceram. Soc. **18**, 2281–2286 (1998).
25. Dillon, S. J. & Harmer, M. P. Multiple grain boundary transitions in ceramics: a case study of alumina. Acta Mater. **55**, 5247–5254 (2007).
26. Gianola, D. S., Mendis, B. G., Cheng, X. M. & Hemker, K. J. Grain-size stabilization by impurities and effect on stress-coupled grain growth in nanocrystalline Al thin films. Mater. Sci. Eng. A **483**, 637–640 (2008).
27. McFadden, S. X. & Mukherjee, A. K. Sulfur and superplasticity in electrodeposited ultrafine-grained Ni. Mater. Sci. Eng. A **395**, 265–268 (2005).
28. McFadden, S. X., Mishra, R. S., Valiev, R. Z., Zhilyaev, A. P. & Mukherjee, A. K. Low-temperature superplasticity in nanostructured nickel and metal alloys. Nature **398**, 684–686 (1999).
29. Valiev, R. et al. Unusual super-ductility at room temperature in an ultrafine-grained aluminum alloy. J. Mater. Sci. **45**, 4718–4724 (2010).
30. Bitzek, E., Brandl, C., Weygand, D., Derlet, P. M. & Van Swygenhoven, H. Atomistic simulation of a dislocation shear loop interacting with grain boundaries in nanocrystalline aluminium. Modell. Simul. Mater. Sci. Eng **17**, 055008 (2009).
31. Pan, Z. & Rupert, T. J. Damage nucleation from repeated dislocation absorption at a grain boundary. Comput. Mater. Sci. **93**, 206–209 (2014).
32. Kumar, K. S., Suresh, S., Chisholm, M. F., Horton, J. A. & Wang, P. Deformation of electrodeposited nanocrystalline nickel. Acta Mater. **51**, 387–405 (2003).
33. Sadigh, B. et al. Scalable parallel Monte Carlo algorithm for atomistic simulations of precipitation in alloys. Phys. Rev. B **85**, 184203 (2012).
34. Pan, Z. & Rupert, T. J. Amorphous intergranular films as toughening structural features. Acta Mater. **89**, 205–214 (2015).
35. Brandl, C., Germann, T. C. & Misra, A. Structure and shear deformation of metallic crystalline–amorphous interfaces. Acta Mater. **61**, 3600–3611 (2013).
36. Wang, Y. M., Li, J., Hamza, A. V. & Barbee, T. W. Ductile crystalline–amorphous nanolaminates. Proc. Natl Acad. Sci. USA **104**, 11155–11160 (2007).
37. Nieh, T. G. & Wadsworth, J. Bypassing shear band nucleation and ductilization of an amorphous–crystalline nanolaminate in tension. Intermetallics **16**, 1156–1159 (2008).
38. Nieh, T. G., Barbee, T. W. & Wadsworth, J. Tensile properties of a free-standing Cu/Zr nanolaminate (or compositionally-modulated thin film). Scr. Mater **41**, 929–935 (1999).
39. Donohue, A., Spaepen, F., Hoagland, R. G. & Misra, A. Suppression of the shear band instability during plastic flow of nanometer-scale confined metallic glasses. Appl. Phys. Lett. **91**, 241905 (2007).
40. Hasnaoui, A., Van Swygenhoven, H. & Derlet, P. M. Dimples on nanocrystalline fracture surfaces as evidence for shear plane formation. Science **300**, 1550–1552 (2003).
41. Lu, K. Making strong nanomaterials ductile with gradients. Science **345**, 1455–1456 (2014).
42. Fang, T. H., Li, W. L., Tao, N. R. & Lu, K. Revealing extraordinary intrinsic tensile plasticity in gradient nano-grained copper. Science **331**, 1587–1590 (2011).
43. Wu, X., Jiang, P., Chen, L., Yuan, F. & Zhu, Y. T. Extraordinary strain hardening by gradient structure. Proc. Natl Acad. Sci. USA **111**, 7197–7201 (2014).
44. Vo, N. Q. et al. Reaching theoretical strengths in nanocrystalline Cu by grain boundary doping. Scr. Mater **65**, 660–663 (2011).
45. Van Swygenhoven, H., Derlet, P. M. & Froseth, A. G. Nucleation and propagation of dislocations in nanocrystalline fcc metals. Acta Mater. **54**, 1975–1983 (2006).
46. Rupert, T. J., Trenkle, J. C. & Schuh, C. A. Enhanced solid solution effects on the strength of nanocrystalline alloys. Acta Mater. **59**, 1619–1631 (2011).
47. Ludy, J. E. & Rupert, T. J. Amorphous intergranular films act as ultra-efficient point defect sinks during collision cascades. Scr. Mater **110**, 37–40 (2016).

48. Brandstetter, S. *et al.* From micro- to macroplasticity. *Adv. Mater.* **18,** 1545–1548 (2006).
49. Plimpton, S. Fast parallel algorithms for short-range molecular-dynamics *J. Comput. Phys.* **117,** 1–19 (1995).
50. Lu, L., Chen, X., Huang, X. & Lu, K. Revealing the maximum strength in nanotwinned copper. *Science* **323,** 607–610 (2009).
51. Dao, M., Lu, L., Shen, Y. F. & Suresh, S. Strength, strain-rate sensitivity and ductility of copper with nanoscale twins. *Acta Mater.* **54,** 5421–5432 (2006).
52. Valiev, R. Z., Alexandrov, I. V., Zhu, Y. T. & Lowe, T. C. Paradox of strength and ductility in metals processed by severe plastic deformation. *J. Mater. Res.* **17,** 5–8 (2002).
53. Legros, M., Elliott, B. R., Rittner, M. N., Weertman, J. R. & Hemker, K. J. Microsample tensile testing of nanocrystalline metals. *Philos. Mag. A* **80,** 1017–1026 (2000).
54. Sanders, P. G., Eastman, J. A. & Weertman, J. R. Elastic and tensile behavior of nanocrystalline copper and palladium. *Acta Mater.* **45,** 4019–4025 (1997).
55. Lu, L., Wang, L. B., Ding, B. Z. & Lu, K. High-tensile ductility in nanocrystalline copper. *J. Mater. Res.* **15,** 270–273 (2000).

Acknowledgements

This study was supported by the US Army Research Office under Grant W911NF-12-1-0511. Materials characterization was performed at the Laboratory for Electron and X-ray Instrumentation (LEXI) at UC Irvine, using instrumentation funded in part by the National Science Foundation Center for Chemistry at the Space-Time Limit (CHE-082913). We thank Professor Daniel S. Gianola (University of Pennsylvania), Professor Kevin J. Hemker (Johns Hopkins University) and Professor Christopher A. Schuh (Massachusetts Institute of Technology) for valuable discussions and feedback on this manuscript.

Author contributions

A.K. and T.J.R. designed the processing route and experiments. A.K. implemented materials synthesis and mechanical testing experiments. Z.P. performed the atomistic simulations. T.J.R. wrote the initial manuscript with input from all authors. All authors contributed to discussion of the results, provided input and approved the final manuscript.

Additional information

5

Ferroelasticity and domain physics in two-dimensional transition metal dichalcogenide monolayers

Wenbin Li[1] & Ju Li[2]

Monolayers of transition metal dichalcogenides can exist in several structural polymorphs, including 2H, 1T and 1T'. The low-symmetry 1T' phase has three orientation variants, resulting from the three equivalent directions of Peierls distortion in the parental 1T phase. Using first-principles calculations, we predict that mechanical strain can switch the relative thermodynamic stability between the orientation variants of the 1T' phase. We find that such strain-induced variant switching only requires a few percent elastic strain, which is eminently achievable experimentally with transition metal dichalcogenide monolayers. Calculations indicate that the transformation barrier associated with such variant switching is small (<0.2 eV per chemical formula unit), suggesting that strain-induced variant switching can happen under laboratory conditions. Monolayers of transition metal dichalcogenides with 1T' structure therefore have the potential to be ferroelastic and shape memory materials with interesting domain physics.

[1] Research Laboratory of Electronics, Massachusetts Institute of Technology, Cambridge, Massachusetts 02139, USA. [2] Department of Nuclear Science and Engineering and Department of Materials Science and Engineering, Massachusetts Institute of Technology, Cambridge, Massachusetts 02139, USA. Correspondence and requests for materials should be addressed to J.L. (email: liju@mit.edu).

The discovery of two-dimensional (2D) atomic crystals[1] has fuelled intensive research efforts on this new class of materials, revealing fundamentally new physics and properties[2-6] that could be essential for next-generation nanoscale devices. Monolayers of group VI transition metal dichalcogenides (TMDs) with chemical formula MX_2, where M is Mo or W and X stands for S, Se or Te, have in particular attracted much recent attention due to their semiconducting, optical and valleytronic properties[4,7-9]. Owning to their atomic thickness, the TMD monolayers have extraordinary mechanical flexibility and strength, capable of sustaining up to 10% of elastic strain before failure[10,11], which enables significant dynamical tuning of their properties by strain engineering[12] and makes them attractive for application in ultrathin flexible electronics[13-15].

MX_2 monolayers can exist in several polytypic structures, including 2H, 1T and 1T'[16-18]. In the semiconducting 2H phase, the atomic stacking sequence within a single XMX monolayer is Bernal (ABA) and the M–X coordination is trigonal prismatic. In contrast, in the 1T phase, the XMX stacking sequence is rhombohedral (ABC), and the M and X atoms form octahedral coordination. The 1T phase is metallic, but was found to be unstable to Peierls distortion[19,20], where two adjacent lines of metal atoms along the highest symmetry directions can dimerize and form parallel chains of M atoms. This leads to the formation of 1T' phase[17,18], in which the octahedral coordination between M and X atoms becomes distorted, and the symmetry of the crystal structure is reduced. While the thermodynamically stable phase of most group VI MX_2 monolayers under ambient conditions is 2H, the ground-state phase of WTe_2 has 1T' structure[16,21]. For other MX_2 monolayers, the 1T' phase is usually metastable, but large transition barriers of order 1 eV per formula unit exist between 1T' and 2H (ref. 22), suggesting that the 1T' phase can be stabilized under appropriate thermal or chemical conditions. In particular, the energetic difference between the 2H and 1T' phase of $MoTe_2$ is rather small[19], suggesting that the 1T' phase can be stabilized relatively easily. Indeed, single crystals and few-layer films of $MoTe_2$ in 1T' phase have been synthesized on a large scale recently[20,23,24]. It has also been theoretically proposed that the 2H to 1T' transition in $MoTe_2$ monolayers can be induced by experimentally accessible tensile strain[19].

The low-symmetry 1T' phase of TMD monolayers harbours extraordinary properties that have only started to be revealed, which, for example, includes enhanced catalytic activities[25], large, non-saturating magnetoresistance[21] and quantum spin Hall effect[22].

A ferroelastic material is defined by the existence of two or more equally stable orientation variants, which can be switched from one variant to another without diffusion by the application of external stress[26,27]. A ferroelastic phase usually forms through a structural phase transition (or a hypothetical one) that reduces the symmetry of a prototype phase. The low-symmetry ferroelastic phase possesses several orientation states (domain variants) with different spontaneous strain[28], that is, the distortion of the unit cell relative to that in the prototype phase. The difference in spontaneous strain between different variants enables external stress to couple energetically with the strain state of the system and drive orientation switch, analogous to the switching of spontaneous polarization by external electric field in a ferroelectric material. In a ferroelastic crystal, domains of different orientations can coexist and form twin boundaries. On activation by appropriate external stress, those twin boundaries can move in a glissile fashion, resulting in the growth of one orientation state at the expense of another, as well as hysteretic stress–strain response[27].

In this article, we focus on the possibility of ferroelastic behaviours in 1T'–MX_2 monolayers. A notable feature associated

with the 1T' phase that has hitherto been overlooked is that it has three distinct orientation variants, resulting from the three equivalent directions of structural distortion in the parental 1T phase. Our density functional theory (DFT) calculations indicate that ferroelastic switching can occur between the different orientation variants of the 1T' phase with a few percent of elastic strain, which is experimentally achievable for MX_2 monolayers.

Results

Crystal structures and transformation strains. We use WTe_2 monolayers as a representative of 1T'–MX_2 to illustrate the possibility of 2D ferroelasticity. Figure 1 shows the atomistic structures of 1T–WTe_2 and 1T'–WTe_2 monolayers. In the 1T phase, the W atoms arrange in 2D triangular lattice, which is sandwiched between two Te atomic layers. The 2D primitive cell of the 1T phase is a 120° rhombus with side length t_0. Due to Fermi surface nesting induced Peierls distortion[20], adjacent parallel lines of W atoms along the high-symmetry [100], [010] or [$\bar{1}$10] directions in the 1T structure can spontaneously dimerize and result in the formation of 1T' phase, with distorted octahedral coordination. The 2D primitive cell of the 1T' phase is a rectangle with dimensions $a \times b$, which corresponds to the $1 \times \sqrt{3}$ supercell of the 1T phase. Because of the $P\bar{3}m2$ space group symmetry of the 1T phase, there are three symmetry-equivalent directions of structural distortion in the 1T phase. These directions are labelled on Fig. 2a as direction 1, 2 and 3 on the 2D triangular lattice formed by W atoms. The atomistic structures of the three orientation variants formed by structural distortion in the 1T phase along the three directions are shown in Fig. 2b–d. Hereafter, we refer to the three orientation states as the O1, O2 and O3 variant, respectively.

The spontaneous transformation strains associated with the 1T to 1T' transformation can be compared between the three variants based on the $2 \times 2\sqrt{3}$ supercell of the prototype 1T phase. All the three orientation variants of the 1T' phase, namely, O1, O2 and O3, can be derived through the Peierls distortion of this supercell and the atoms within the supercell along the corresponding orientation direction. Namely, the $2 \times 2\sqrt{3}$

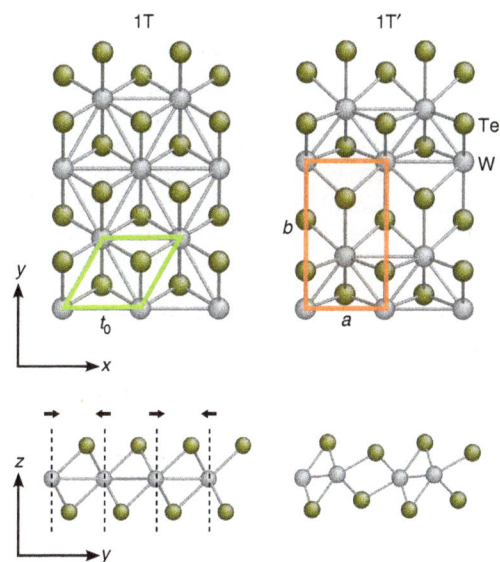

Figure 1 | Atomistic structure of of 1T–WTe_2 and 1T'–WTe_2 monolayers. The 2D primitive cells of 1T and 1T' are highlighted in green and red, respectively. The primitive cell of 1T' corresponds to the $1 \times \sqrt{3}$ supercell of 1T. The 1T' phase can be derived via the structural distortion of the 1T phase, which is schematically illustrated in the side views. These features are generic to all other group VI MX_2 monolayers.

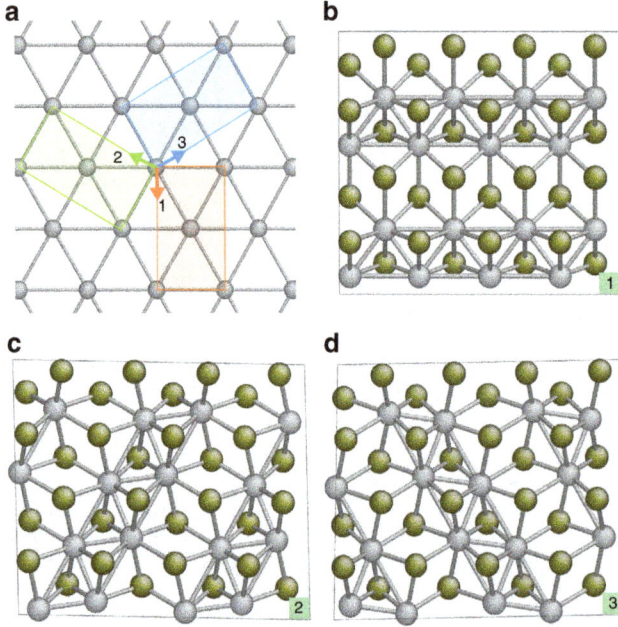

Figure 2 | Three orientation variants of 1T′-MX₂ monolayers. (a) The three symmetry-equivalent directions of structural distortion in the 1T structure are indicated by arrows and numerical labels on the triangular lattice formed by M atoms. The corresponding primitive cells of the 1T′ phase after structural distortion are represented as shaded rectangles. **(b–d)** Relaxed atomistic structure of the 1T′ phase after structural distortion along the three different directions, which are referred to as orientation variants O1, O2 and O3, respectively.

supercell of 1T can transform to become the supercells of all three variants of the 1T′ phase. In Cartesian coordinates, the 2D basis vectors $\mathbf{h}_1$ and $\mathbf{h}_2$ of the 1T supercell can be written as $\mathbf{h}_1 = 2t_0\hat{\mathbf{x}}$, $\mathbf{h}_2 = 2\sqrt{3}t_0\hat{\mathbf{y}}$, where $\hat{\mathbf{x}}$ and $\hat{\mathbf{y}}$ are the unit vectors along the x and y directions labelled on Fig. 1. A supercell matrix $\mathbf{H}_0 = \{\mathbf{h}_1,\mathbf{h}_2\}$ can be constructed, where $\mathbf{h}_1$ and $\mathbf{h}_2$ are treated as column vectors, that is,

$$\mathbf{H}_0 = \begin{pmatrix} 2t_0 & 0 \\ 0 & 2\sqrt{3}t_0 \end{pmatrix}. \tag{1}$$

After transforming to the 1T′ phase, the distorted supercell matrix corresponding to the O1, O2 and O3 variants will be denoted by $\mathbf{H}_1$, $\mathbf{H}_2$ and $\mathbf{H}_3$, respectively. These new supercell matrices can be related to the original supercell matrix by transformation matrices $\mathbf{J}_i$, which map the undistorted supercell to the distorted supercells. Namely, $\mathbf{H}_i = \mathbf{J}_i\mathbf{H}_0$, where the subscript i stands for the i-th orientation variant. The transformation strain matrices $\boldsymbol{\eta}_i$ associated with different variants can then be calculated from $\mathbf{J}_i$ based on the definition of Green-Lagrange strain tensor:

$$\boldsymbol{\eta}_i = \frac{1}{2}\left(\mathbf{J}_i^{\mathrm{T}}\mathbf{J}_i - \mathbf{I}\right) = \frac{1}{2}\left[\left(\mathbf{H}_0^{-1}\right)^{\mathrm{T}}\mathbf{H}_i^{\mathrm{T}}\mathbf{H}_i\mathbf{H}_0^{-1} - \mathbf{I}\right]. \tag{2}$$

Here, the superscripts -1 and T denote matrix inversion and transposition, respectively. $\mathbf{I}$ is a 2×2 identity matrix. The 2D transformation strain tensor $\boldsymbol{\eta}_i$ has the following symmetric form:

$$\eta = \begin{pmatrix} \varepsilon_{xx} & \varepsilon_{xy} \\ \varepsilon_{xy} & \varepsilon_{yy} \end{pmatrix}, \tag{3}$$

where ε_{xx} and ε_{yy} are the tensile/compressive strain along x or y direction, and ε_{xy} is the shear strain component.

We have employed DFT calculations to obtain the equilibrium supercell vectors and the relaxed atomic coordinates of the O1, O2 and O3 variants, resulting from the distortion of $2\times2\sqrt{3}$

supercell in the 1T prototype phase. The supercell matrices $\mathbf{H}_i$ for different MX₂ monolayers are tabulated in Supplementary Table 1. From the supercell matrices, the spontaneous transformation strain tensors $\boldsymbol{\eta}_i$ can be evaluated, which are listed in Supplementary Table 2. For WTe₂ monolayers, the transformation strain matrices form 1T to 1T′ are

$$\eta_1 = \begin{pmatrix} -0.005 & 0.0 \\ 0.0 & 0.039 \end{pmatrix},$$

$$\eta_2 = \begin{pmatrix} 0.029 & -0.019 \\ -0.019 & 0.006 \end{pmatrix}, \tag{4}$$

$$\eta_3 = \begin{pmatrix} 0.029 & 0.019 \\ 0.019 & 0.007 \end{pmatrix}.$$

The difference in transformation strain between the three variants of 1T′ suggests that one may switch the relative thermodynamic stability between different variants by applying suitable external mechanical stress. Since the equilibrium structure of WTe₂ is 1T′, it is informative to directly compare the distorted supercell of the three variants by computing the relative supercell strain associated with the transformation from one variant to another. This can be carried out again using the supercells of the three variants derived from the common $2\times2\sqrt{3}$ supercell in the prototype 1T phase. The reference configuration for computing the supercell strain is now chosen to be the O1 variant of 1T′ phase, and we use ε_i^j to denote the transformation strain tensor from variant i to j. Calculations based on the same definition of strain tensor as in equation (2) give the transformation strain associated with O1–O2 and O1–O3 switching to be

$$\varepsilon_1^2 = \begin{pmatrix} 0.034 & -0.019 \\ -0.019 & -0.030 \end{pmatrix},$$

$$\varepsilon_1^3 = \begin{pmatrix} 0.033 & 0.019 \\ 0.019 & -0.030 \end{pmatrix}. \tag{5}$$

It then follows that, starting with the O1 variant of 1T′ in a strain-free state, after imposing an external strain of magnitude ε_1^2 on the monolayer, the system would be in a thermodynamically more favourable state by transforming to the O2 variant, since both O1 and O2 belong to the same 1T′ structure, but O1 will have higher strain energy than O2. The same argument applies to any other two variants. Hence, the relative energetic stability between the different orientation variants of 1T′ phase can be controlled by external stress or strain.

Variant energetics under biaxial and shear strain. To study in detail the relative thermodynamic stability of different variants when external mechanical deformation is imposed on a 1T′–MX₂ monolayer, we have used DFT to calculate the potential energy surfaces of the three variants of 1T′ as a function of 2D supercell dimensions. We first investigate the possibility of mechanically switching the O1 variant to O2 or O3 variant by applying biaxial strain to the system, again using WTe₂ monolayer as an example. The strain-free 2×2 supercell of the O1 variant, derived from the distortion of the aforementioned $2\times2\sqrt{3}$ supercell in the parental 1T phase, is chosen to be the reference system. The 2×2 supercell of the O1 variant has dimensions $2a \times 2b$ within the x–y plane of 2D monolayer. We adjust the dimensions of the supercell along x and y directions independently, with the values of a and b range from -10 to 10% of engineering strain at an equal step of 2%. At each pair of (a,b), the atomic coordinates within the supercell are relaxed. We also compute the energies of O2 and O3 variants when their supercell dimensions are fixed to be the same as O1. The energies U of all three variants are computed on a

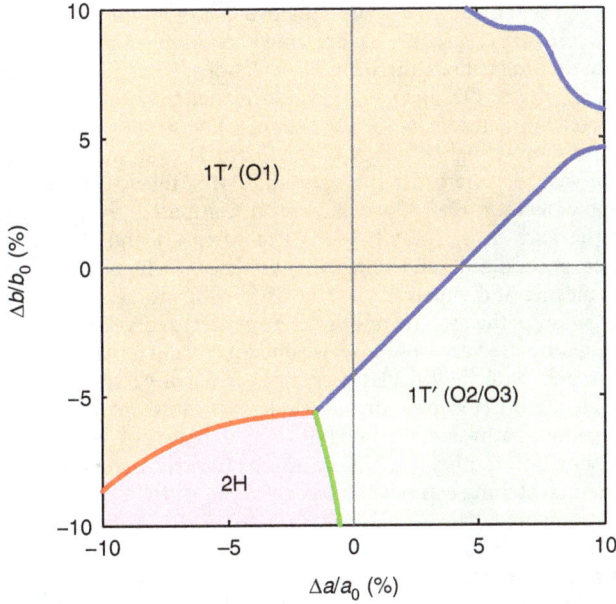

Figure 3 | Intersection contours of the energy surfaces between the different orientation variants of 1T′-MoTe$_2$ and between 2H and 1T′. The lattice constants a and b, corresponding to the dimensions of the rectangular primitive cell of the O1 variant in the 1T′ phase, are represented as percent engineering strain with respect to the equilibrium lattice constants a_0 and b_0. The regions of lower-energy phase/variant are labelled and shaded in different colours.

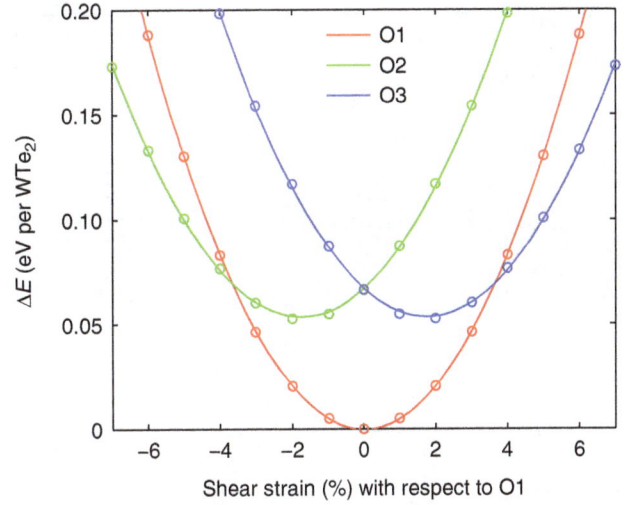

Figure 4 | Potential energies of the three variants of 1T′-WTe$_2$ monolayers as a function of shear strain with respect to the equilibrium supercell of the O1 variant. The O2 or O3 variant becomes energetically more favourable when negative- or positive-shear strain of a few percent is imposed on the O1 variant.

11×11 grid in the (a,b) space, giving a total number of 121 data points distributed evenly around the equilibrium lattice constants of the O1 variant. Smooth potential energy surfaces are then constructed by approximating the intermediate values of $U(a,b)$ using 2D spline interpolation, which allows us to directly compare the relative energetic stability of the O1, O2 and O3 variants in the full (a,b) space. In addition, the $U(a,b)$ for the 2H phase is computed for comparison, as a previous study indicates that strain-induced phase transformation between the 2H and 1T′ phases can happen in MX$_2$ monolayers[19].

After obtaining the potential energy surfaces for all the three variants of 1T′ as well as the 2H phase, the lowest-energy variants/phases in the (a,b) space are determined. The result is shown in Fig. 3, where we label the lowest-energy variant/phase in each region of phase space and plot the intersection boundaries between two neighbouring variants/phases. An important feature of Fig. 3 is that the potential energy surfaces of O1 and O2/O3 variants intersect at a few percent of biaxial supercell strain, which is experimentally achievable in MX$_2$ monolayers[10,11]. The O2 and O3 variants are grouped together in Fig. 3 because their potential energies in the (a,b) space are essentially the same. This can be rationalized by the fact that the supercells of both variants can be derived from the distortion of the $2 \times 2\sqrt{3}$ supercell in the 1T phase, and their distortion directions are related by mirror symmetry along the y direction in the 1T phase, as can be seen from Fig. 2. Since biaxial strain does not break the mirror symmetry of 1T phase along the y axis, the O2 and O3 variants are still mirror images of each other and have the same energy. We however expect that shear strain, which breaks the mirror symmetry, can distinguish the energies of all the three variants of 1T′. Indeed, Fig. 4 shows that, when shear strain ε_{xy} of magnitude $> 3.5\%$ is imposed on the O1 variant, O3 becomes the lowest-energy variant within the strained supercell. If the sign of ε_{xy} is reversed, then the O2 variant has lower energy than both O1 and O3.

Figure 3 indicates that the 2H phase of WTe$_2$ monolayer only takes a small region in the (a,b) space as the lowest-energy phase. This result is different from the study by Duerloo *et al.*[19] of strain-induced phase transformation between the 2H and 1T′ phases of MX$_2$ monolayers, as the authors did not take into account the existence of orientation variant degrees of freedom in the 1T′ phase.

Figure 3 also shows that the fastest route to switching the energetic order between O1 and O2/O3 in the (a,b) space is by applying tensile strain along the a axis of the O1 variant, which is the direction of dimerized metal-atom chains, while simultaneously applying compressive strain along the b axis. It is however known that 2D MX$_2$ monolayers usually cannot sustain large compressive strain due to compression-induced buckling response and formation of ripplocations[29]. On the contrary, experiments have demonstrated that 2H–MX$_2$ monolayers can withstand tensile elastic strain as large as 10% before mechanical failure[10,11]. Hence, it may be experimentally more convenient to realize variant switching in 1T′-WTe$_2$ by uniaxially stretching it along the a axis, which is the direction of dimerized tungsten atoms. This axis can be identified by mechanical cleavage or by the anisotropic response to external fields that is expected for the low-symmetry 1T′ structure[21].

In Supplementary Fig. 1, we have also computed the intersection contours of the potential energy surfaces between the O1 and O2/O3 variants for other 1T′-MX$_2$ monolayers, including MoS$_2$, MoSe$_2$, MoTe$_2$, WS$_2$ and WSe$_2$. The results are very similar to WTe$_2$, indicating that strain-induced switching of thermodynamic stability between different orientation variants is generic to MX$_2$ monolayers with 1T′ structure.

Variant energetics under uniaxial tension. We emphasize that the strain at which the potential energy surfaces of different variants intersects is not the same as the strain at which variant switching becomes thermodynamically favourable. The system can minimize its free energy by choosing a state where different variants (or phases) coexist, akin to the two-phase region in chemical-composition phase diagrams. Under constant temperature and fixed external strain (supercell dimensions), the thermodynamic potential that determines the relative variant/

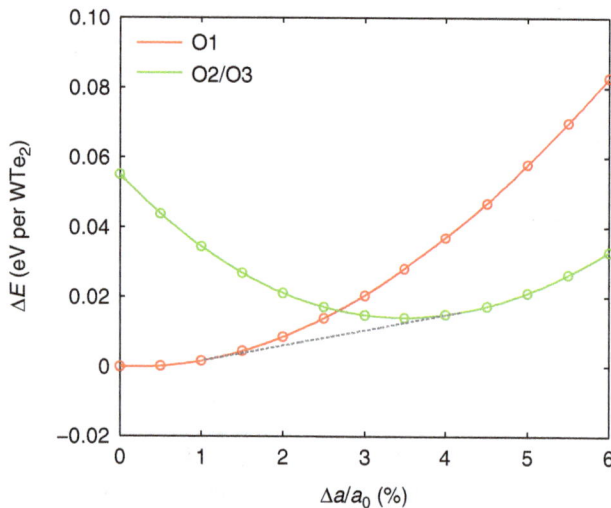

Figure 5 | Potential energy curves of the different variants of 1T′-WTe$_2$ uniaxially strained along the *a* axis. The dashed line is the common tangent between the two curves. The uniaxial strain is calculated with respect to the equilibrium supercell dimension of the O1 variant along the *x* direction.

phase stability is the Helmholtz free energy $F = E - TS$, where E is internal energy that includes both potential energy U and kinetic energy, T is temperature and S is entropy. Because all the three variants O1, O2 and O3 belong to the same 1T′ structure, and because the entropy of solids (mainly vibrational) is relatively insensitive to small deformation, we can use the potential energy U of different variants, computed by DFT at zero temperature, to compare the free energies of different variants at ambient conditions. In Fig. 5, we plot the potential energy curve of the O1 variant of 1T′-WTe$_2$ when it is uniaxially stretched along the *a* axis. Consistent with typical experimental set-ups for uniaxial deformation, the stress of the supercell along the *b* axis is relaxed to zero. This corresponds to free boundary, or zero stress ($\sigma_y = 0$) condition along the *b* axis. In Fig. 5, we also plot the potential energy of the O2/O3 variant in a rectangular supercell with the same dimension along the *a* axis and the same boundary condition along the *b* axis. A common tangent can be constructed between the energy curves of O1 and O2/O3, which intersects the two curves at uniaxial strains equal to 1% and 4%, respectively. Between these two values, the system can lower its energy by existing in a state where both O1 and O2/O3 variants coexist. This indicates that the formation of O2/O3 variants becomes thermodynamically favourable when the uniaxial strain along the *a* axis of O1 is as low as 1%.

Kinetic aspects of variant switching. Up to now, we have only considered the thermodynamic aspects of variant switching in the 1T′-MX$_2$ monolayers. Our results suggest that it becomes thermodynamically favourable for the O1 variant of 1T′-WTe$_2$ monolayers to switch to the other two variants when applying uniaxial strain around 1% along the direction of dimerized tungsten atom chains. However, if the kinetic barrier associated with variant switching is too high, such variant switching may not occur under normal experimental conditions and timescale, and the materials would still not be ferroelastic. We have therefore computed the transition barrier associated with the switching between the O1, O2 and O3 variants using climbing image nudged elastic band (NEB) method[30]. The result of our calculation for the variant switching between the O1 and O2 variants of WTe$_2$ monolayer is shown in Fig. 6. We find the transition barrier of variant switching is only 0.22 eV per formula

unit. Very similar results are obtained for orientation switching between other variants, as presented in Supplementary Fig. 2. Note that to facilitate these NEB calculations, we impose supercell strains on the O2 or O3 variants such that they have the same supercell dimensions of the O1 variants. The strain energy results in the slightly higher energy of the O2 or O3 variant that would otherwise be energetically degenerate with the O1 variant. In Supplementary Fig. 3, we have also computed the transition barrier and the pathway between stress-free O1 and O2 variants using generalized solid-state NEB method[31], which allows both the atomic and supercell degrees of freedom to relax along the transition pathway. The results of the generalized solid-state NEB calculation are very close to those obtained using a fixed-supercell approach, with the calculated energy barrier of variant switching equals to 0.19 eV per formula unit. We note that, while the transition pathway illustrated in Fig. 6 may not be the only possible one, if other pathways exist, the transformation barrier of variant switching can only be smaller or equal than the values we have obtained.

In Supplementary Fig. 4, we have also computed the transition barriers for other MX$_2$ monolayers, and the barriers obtained are even lower than 1T′-WTe$_2$ monolayers. The transition barriers of variant switching are significantly lower than the barriers of phase transition between the 2H and 1T′ phase[19], which we computed to be 0.8 eV per formula unit for 1T′-WTe$_2$ monolayers at the equilibrium lattice constants of the 2H phase. The much smaller transition barriers associated with the variant switching within the 1T′ phase as compared with the 1T′ to 2H phase transition has an intuitive geometric explanation. Variant switching between the orientation variants of 1T′ phase only involves the distortion of M–X octahedral coordination, while the 1T′ to 2H phase transition requires the complete change of M–X coordination pattern from octahedral to trigonal prismatic.

According to transition-state theory, assuming a characteristic attempt frequency of 10 THz, which is the typical frequency of optical phonons in 1T′-MX$_2$ monolayers[22], a 0.2 eV barrier is associated with a timescale of around 0.2 ns. Although the actual barrier of forming a critical nucleus of new variant may involve multiple formula units, and other factors such as interfaces and pre-existing defects may also affect the transformation kinetics, the much smaller barrier associated with variant switching within the 1T′ phase as compared with 2H to 1T′ phase transition[19,20] suggests that ferroelastic variant switching in 1T′-MX$_2$ monolayers is very likely to happen under normal laboratory experimental conditions.

Ferroelastic domain boundaries. A direct consequence of strain-induced variant switching in 1T′-MX$_2$ monolayers is the formation of domain boundaries between different orientation variants. Strain-induced ferroelastic switching between the O1, O2 and O3 variants can lead to the formation of three possible types of coherent twin boundaries, between O1 and O2, O1 and O3, and between O2 and O3, which we refer to as O1–O2, O1–O3 and O2–O3, respectively. The DFT-relaxed atomistic structures of the three different types of twinning domain boundaries in 1T′-WTe$_2$ monolayers under zero external stress are shown in Fig. 7. The three domain boundaries are energetically degenerate, and they are related to each other by 120° rotational symmetry operation. Unlike their three-dimensional (3D) counterparts, where the domain boundaries are 2D, the boundaries formed between the domains of 2D MX$_2$ monolayers are quasi-one dimensional (1D) in nature, which may impart them unique properties. We have calculated the domain boundary energies associated with the three types of the 1D domain boundaries and found they have small formation energies. Our DFT calculations give the domain boundary energies of MoS$_2$,

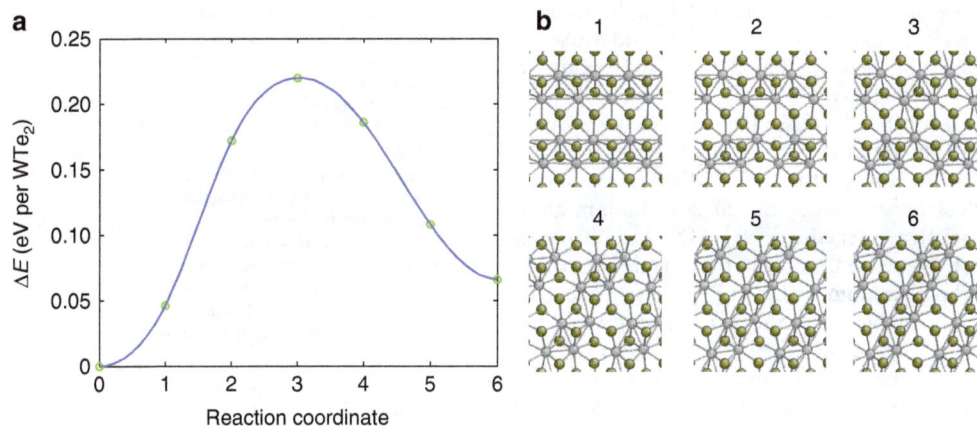

Figure 6 | NEB calculation of transformation barrier and the pathway for orientation switching. The initial configuration of the NEB calculation is the O1 variant of 1T′–WTe$_2$ monolayers at ground state (zero external stress), while the final state is the strained O2 variant with the same supercell dimensions as those of the O1 variant. (**a**) Change of the system energy per chemical formula unit as a function of reaction coordinate. (**b**) The corresponding atomistic structure of the system along the reaction coordinate. Four different supercells were used to carry out the NEB calculations: 2 × 2, 2 × 4, 4 × 2 and 4 × 4 supercell of the 1T′ phase, which all give identical results.

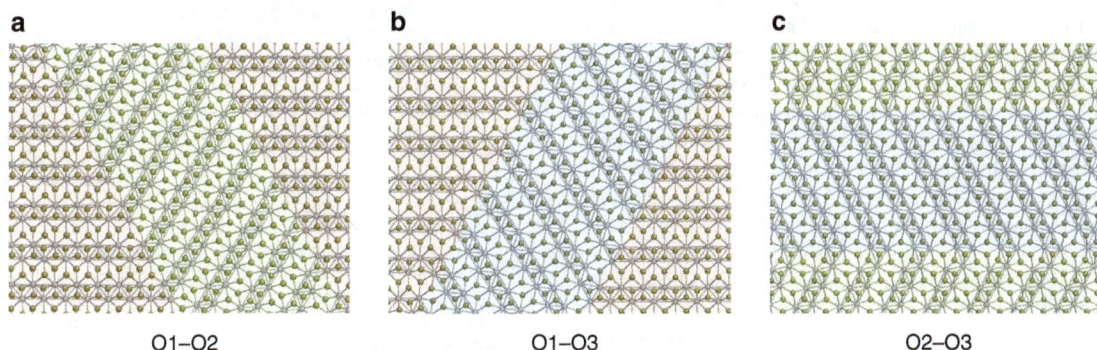

O1–O2   O1–O3   O2–O3

Figure 7 | Ferroelastic domain boundaries between different orientation variants. (**a–c**) Illustrate the DFT-relaxed atomistic structures of domain boundaries formed between the O1 and O2, O1 and O3, and O2 and O3 variants of 1T′–WTe$_2$ monolayers, respectively. To help guide the eyes, domains of different variants are shaded in distinct colours: O1 variant in orange, O2 variant in green and O3 variant in blue.

MoSe$_2$, MoTe$_2$, WS$_2$, WSe$_2$ and WTe$_2$ monolayers to be 27, 46, 40, 22, 51 and 52 meV · Å^{-1}, respectively. In comparison, the formation energy of another type of 1D defects in 3D crystals, dislocations, is in the order of several hundred meV per Angstrom. Such small-domain boundary energies will facilitate the ferroelastic switching between different orientation variants.

Discussion

The thermodynamic and kinetic analysis above have provided strong evidence that strain-induced ferroelastic switching of orientation variants can occur in 1T′–MX$_2$ monolayers, with a few percent of local strain. Our calculations indicate that variant switching can most easily happen when stretching the 1T′–MX$_2$ monolayers along the direction of dimerized metal chains. This prediction, if experimentally realized, will render 1T′–MX$_2$ monolayers as the first class of 2D ferroelastic materials[32]. Signatures of such ferroelastic switching in experiments include hysteresis in stress–strain curves[27], and the existence of a force plateau when the externally applied strain is beyond a critical value that corresponds to the onset of variant coexistence. Direct experimental proof of ferroelastic domain switching may be realized by carrying out *in situ* transmission electron microscopy experiments of mechanical deformation of 1T′–MX$_2$ monolayers. As the domains of different variants have different crystallo-

graphic orientations, the migration of domain walls during strain-induced variant switching can be observed by dark-field transmission electron microscopy, which has been demonstrated for domain imagining in graphene and MoS$_2$ monolayers[33,34]. Selective area electron diffraction could also reveal the formation of twinning domains, as variant switching results in the rotation of the underlying Bravais lattice of the 1T′ structure, which will manifest in selective area electron diffraction as the rotation of diffraction patterns.

Our prediction of ferroelasticity in the TMD monolayers can be readily tested experimentally in 1T′–WTe$_2$ and 1T′–MoTe$_2$, for which bulk single crystals have been synthesized on a large scale and exfoliated down to the monolayer or few-layer regime[20,21]. Recently, large-area and high-quality MoTe$_2$ few layers in 1T′ phase have been grown via chemical vapour deposition[23,24]. Local and controlled phase transformation of MoTe$_2$ from the 2H to 1T′ phase can also be realized using laser ablation[35]. In principle, ferroelastic domain switching can be observed not only in monolayers but also in few-layer samples, since dimerized metal chains within different layers of the 1T′ phase orient along the same direction in naturally grown crystals[36].

For other group VI MX$_2$ that include MoS$_2$, MoSe$_2$, WS$_2$ and WSe$_2$, as the 2H phase is energetically more stable than the 1T′ phase under normal conditions, the 1T′ phase can be realized using a phase engineering approach[37,38]. The 1T or 1T′ phase of these materials are now actively being explored for applications in energy and electronics[38]. Monolayers of WS_2, MoS_2 and $MoSe_2$ in 1T′ phase have been obtained via liquid-phase exfoliation of the bulk crystals intercalated by alkaline metals[18,25,39]. The transformation from the 2H phase to 1T′ phase by alkaline metal intercalation is attributed to charge transfer from the intercalated alkali atoms to the TMDs[37]. We have performed DFT calculations to study the effect of lithium atom adsorption on the relative energetics of $2H-MX_2$ and $1T′-MX_2$ monolayers, and it is indeed shown that, with increased amount of adsorbed lithium, the 1T′ phase becomes energetically more favourable than the 2H phase for all the MX_2 monolayers, as illustrated in Supplementary Fig. 5. It has also been proposed that the substitutional doping of MX_2 with elements having more valence electrons (for example, Re) than the transition metal ions can be another effective way to stabilize the 1T′ phase[40].

From an application perspective, ferroelastic behaviours have close connection to the shape memory effect, which has been exploited to make actuators in a wide range of industries. In 3D, shape memory alloys (SMAs) is a well known and technologically important class of ferroelastic materials. SMAs can undergo a diffusionless martensitic phase transformation below a critical temperature, from the high-temperature austenite to the low-temperature martensite phase[41]. The martensite phase of SMAs is ferroelastic: it has several equivalent orientations or variants, that can be switched from one to another by an appropriate uniaxial or shear stress. The martensite phase in SMAs can undergo large inelastic deformation through stress-induced migration of twin boundaries between different variants. On heating the deformed crystal above the martensitic phase transformation temperature, the martensite phase can revert to the austenite phase and recover its original shape before deformation. For the ferroelastic $1T′-MX_2$ monolayers, if the 1T′ phase can be reversibly transformed to the 1T or 2H phase under external stimuli (which do not have to be thermal but could be other fields), then MX_2 monolayers could be 2D shape memory materials, with operating principles similar to SMAs. As such, $1T′-MX_2$ monolayers can be used to make ultrathin actuators for applications in nanoscale-integrated electromechanical systems.

In closing, we would like to make a few additional comments. First, since the $1T′-MX_2$ monolayers were predicted to be quantum spin Hall insulators[22], topological effect plays an important role in determining their electronic properties. The twinning domain boundaries formed through ferroelastic switching in the 1T′ phase, which are 1D defects in 2D quantum materials, may possess exotic physics and provide a rich playground for domain boundary engineering[42]. Second, the possibility of ferroelastic switching in 2D materials may not be limited to group VI MX_2 monolayers, considering the rich family of 2D materials[43]. Our initial studies indicate that several other TMD monolayers, including ReS_2, $NbTe_2$ and $TaTe_2$, which have low-symmetry distorted crystal structures similar to 1T′[16], also possess distinct orientation variants and could be ferroelastic as well. Indeed, experimental evidence of local strain-induced orientational switching in ReS_2 and $ReSe_2$ monolayers has recently been reported[44]. Our finding of potential 2D ferroelastic behaviours in monolayer materials could therefore open doors to many exciting discoveries in 2D materials with low-symmetry distorted crystal structures, which may also include ferroelectric, ferromagnetic and multiferroic behaviours in the future.

Methods

First-principles calculations. DFT calculations were performed using the Vienna Ab initio Simulation Package with a plane-wave basis set[45,46] and the projector-augmented wave[47] pseudopotentials. Exchange-correlation effects were treated using the generalized gradient approximation[48] in the Perdew–Burke–Ernzerhof form[49]. The kinetic energy cutoff for wavefunction expansion was fixed to be 350 eV. The TMD monolayers were modelled in supercells with a vacuum region in the direction perpendicular to the 2D planes of the monolayers (the z direction). The length of the supercells along the z direction was chosen to be 20 Å. Brillouin zone integration employed a Gamma point centred $m \times n \times 1$ Monkhorst-Pack[50] k-point grid and a Gaussian smearing of 50 meV, where the numbers m and n were chosen such that the k-point sampling spacing is < 0.1 Å^{-1} along the supercell reciprocal vectors in the x–y plane. The energy convergence thresholds for electronic and ionic relaxations were 10^{-6} and 10^{-5} eV, respectively. The maximum residual forces resulted from these convergence criteria are smaller than 5×10^{-3} eV Å^{-1}. We provide the DFT-relaxed atomistic structures of the O1, O2 and O3 variants of $1T′-MX_2$ monolayers in the POSCAR format of Vienna Ab initio Simulation Package, as listed in Supplementary Tables 4–9.

References

1. Novoselov, K. S. *et al.* Two-dimensional atomic crystals. *Proc. Natl Acad. Sci. USA* **102**, 10451–10453 (2005).
2. Castro Neto, A. H., Guinea, F., Peres, N. M. R., Novoselov, K. S. & Geim, A. K. The electronic properties of graphene. *Rev. Mod. Phys.* **81**, 109–162 (2009).
3. Lee, C., Wei, X. D., Kysar, J. W. & Hone, J. Measurement of the elastic properties and intrinsic strength of monolayer graphene. *Science* **321**, 385–388 (2008).
4. Wang, Q. H., Kalantar-Zadeh, K., Kis, A., Coleman, J. N. & Strano, M. S. Electronics and optoelectronics of two-dimensional transition metal dichalcogenides. *Nat. Nanotechnol.* **7**, 699–712 (2012).
5. Xu, X., Yao, W., Xiao, D. & Heinz, T. F. Spin and pseudospins in layered transition metal dichalcogenides. *Nat. Phys.* **10**, 343–350 (2014).
6. Xia, F., Wang, H., Xiao, D., Dubey, M. & Ramasubramaniam, A. Two-dimensional material nanophotonics. *Nat. Photon.* **8**, 899–907 (2014).
7. Mak, K. F., Lee, C., Hone, J., Shan, J. & Heinz, T. F. Atomically thin MoS_2: A new direct-gap semiconductor. *Phys. Rev. Lett.* **105**, 136805 (2010).
8. Mak, K. F., He, K., Shan, J. & Heinz, T. F. Control of valley polarization in monolayer MoS_2 by optical helicity. *Nat. Nanotechnol.* **7**, 494–498 (2012).
9. Zeng, H., Dai, J., Yao, W., Xiao, D. & Cui, X. Valley polarization in MoS_2 monolayers by optical pumping. *Nat. Nanotechnol.* **7**, 490–493 (2012).
10. Bertolazzi, S., Brivio, J. & Kis, A. Stretching and breaking of ultrathin MoS_2. *ACS Nano* **5**, 9703–9709 (2011).
11. Cooper, R. C. *et al.* Nonlinear elastic behavior of two-dimensional molybdenum disulfide. *Phys. Rev. B* **87**, 035423 (2013).
12. Feng, J., Qian, X. F., Huang, C. W. & Li, J. Strain-engineered artificial atom as a broad-spectrum solar energy funnel. *Nat. Photon.* **6**, 866–872 (2012).
13. Radisavljevic, B., Radenovic, A., Brivio, J., Giacometti, V. & Kis, A. Single-layer MoS_2 transistors. *Nat. Nanotechnol.* **6**, 147–150 (2011).
14. Lee, G.-H. *et al.* Flexible and transparent MoS_2 field-effect transistors on hexagonal boron nitride-graphene heterostructures. *ACS Nano* **7**, 7931–7936 (2013).
15. Fiori, G. *et al.* Electronics based on two-dimensional materials. *Nat. Nanotechnol.* **9**, 768–779 (2014).
16. Wilson, J. A. & Yoffe, A. D. Transition metal dichalcogenides discussion and interpretation of observed optical, electrical and structural properties. *Adv. Phys.* **18**, 193–335 (1969).
17. Heising, J. & Kanatzidis, M. G. Structure of restacked MoS_2 and WS_2 elucidated by electron crystallography. *J. Am. Chem. Soc.* **121**, 638–643 (1999).
18. Eda, G. *et al.* Coherent atomic and electronic heterostructures of single-layer MoS_2. *ACS Nano* **6**, 7311–7317 (2012).
19. Duerloo, K. A. N., Li, Y. & Reed, E. J. Structural phase transitions in two-dimensional Mo- and W-dichalcogenide monolayers. *Nat. Commun.* **5**, 4214 (2014).
20. Keum, D. H. *et al.* Bandgap opening in few-layered monoclinic $MoTe_2$. *Nat. Phys.* **11**, 482–486 (2015).
21. Ali, M. N. *et al.* Large, non-saturating magne-toresistance in WTe_2. *Nature* **514**, 205–208 (2014).
22. Qian, X.-F., Liu, J.-W., Fu, L. & Li, J. Quantum spin Hall effect in two-dimensional transition metal dichalcogenides. *Science* **346**, 1344–1347 (2014).
23. Park, J. C. *et al.* Phase-engineered synthesis of centimeter-scale 1T′-and 2H-molybdenum ditelluride thin films. *ACS Nano* **9**, 6548–6554 (2015).
24. Zhou, L. *et al.* Large-area synthesis of high-quality uniform few-layer $MoTe_2$. *J. Am. Chem. Soc.* **137**, 11892–11895 (2015).
25. Voiry, D. *et al.* Enhanced catalytic activity in strained chemically exfoliated WS_2 nanosheets for hydrogen evolution. *Nat. Mater.* **12**, 850–855 (2013).
26. Aizu, K. Possible species of ferroelastic crystals and of simultaneously ferroelectric and fer-roelastic crystals. *J. Phys. Soc. Jpn* **27**, 387–396 (1969).
27. Salje, E. K. H. Ferroelastic materials. *Annu. Rev. Mater. Res.* **42**, 265–283 (2012).
28. Salje, E. K. H. *Phase Transitions in Ferroelastic and Co-Elastic Crystals* (Cambridge Univ. Press, 1990).

29. Kushima, A., Qian, X.-F., Zhao, P., Zhang, S.-L. & Li, J. Ripplocations in van der Waals layers. *Nano Lett.* **15,** 1302–1308 (2015).

30. Henkelman, G., Uberuaga, B. P. & Jonsson, H. A climbing image nudged elastic band method for finding saddle points and minimum energy paths. *J. Chem. Phys.* **113,** 9901–9904 (2000).

31. Sheppard, D., Xiao, P., Chemelewski, W., Johnson, D. D. & Henkelman, G. A generalized solid-state nudged elastic band method. *J. Chem. Phys.* **136,** 074103 (2012).

32. Tang, S., Mahanti, S. D. & Kalia, R. K. Ferroelastic phase transition in two-dimensional molecular solids. *Phys. Rev. Lett.* **56,** 484–487 (1986).

33. Huang, P. Y. *et al.* Grains and grain boundaries in single-layer graphene atomic patchwork quilts. *Nature* **469,** 389–392 (2011).

34. van der Zande, A. M. *et al.* Grains and grain boundaries in highly crystalline monolayer molybdenum disulphide. *Nat. Mater.* **12,** 554–561 (2013).

35. Cho, S. *et al.* Phase patterning for ohmic homojunction contact in MoTe$_2$. *Science* **349,** 625–628 (2015).

36. Brown, B. E. The crystal structures of WTe$_2$ and high-temperature MoTe$_2$. *Acta Crystallogr.* **20,** 268–274 (1966).

37. Voiry, D., Mohite, A. & Chhowalla, M. Phase engineering of transition metal dichalco-genides. *Chem. Soc. Rev.* **44,** 2702–2712 (2015).

38. Chhowalla, M., Voiry, D., Yang, J., Shin, H. S. & Loh, K. P. Phase-engineered transition-metal dichalcogenides for energy and electronics. *MRS Bull.* **40,** 585–591 (2015).

39. Gordon, R. A., Yang, D., Crozier, E. D., Jiang, D. T. & Frindt, R. F. Structures of exfoliated single layers of WS$_2$, MoS$_2$, and MoSe$_2$ in aqueous suspension. *Phys. Rev. B* **65,** 125407 (2002).

40. Enyashin, A. N. *et al.* New route for stabilization of 1T-WS$_2$ and MoS$_2$ phases. *J. Phys. Chem. C* **115,** 24586–24591 (2011).

41. Bhattacharya, K. *Microstructure of Martensite: Why It Forms and How IT Gives Rise to the Shape-Memory Effect* (Oxford Univ. Press, 2003).

42. Salje, E. & Zhang, H. Domain boundary engineering. *Phase Transit.* **82,** 452–469 (2009).

43. Lebegue, S., Bjorkman, T., Klintenberg, M., Nieminen, R. M. & Eriksson, O. Two-dimensional materials from data filtering and Ab Initio calculations. *Phys. Rev. X* **3,** 031002 (2013).

44. Lin, Y.-C. *et al.* Single-layer ReS$_2$: two-dimensional semiconductor with tunable in-plane anisotropy. *ACS Nano* **9,** 11249–11257 (2015).

45. Kresse, G. & Furthmiiller, J. Efficiency of ab-initio total energy calculations for metals and semiconductors using a plane-wave basis set. *Comput. Mater. Sci.* **6,** 15–50 (1996).

46. Kresse, G. & Furthmiiller, J. Efficient iterative schemes for ab initio total-energy calculations using a plane-wave basis set. *Phys. Rev. B* **54,** 11169–11186 (1996).

47. Blochl, P. E. Projector augmented-wave method. *Phys. Rev. B* **50,** 17953–17979 (1994).

48. Perdew, J. P. *et al.* Atoms, molecules, solids, and surfaces: Applications of the generalized gradient approximation for exchange and correlation. *Phys. Rev. B* **46,** 6671–6687 (1992).

49. Perdew, J. P., Burke, K. & Ernzerhof, M. Generalized gradient approximation made simple. *Phys. Rev. Lett.* **77,** 3865–3868 (1996).

50. Monkhorst, H. J. & Pack, J. D. Special points for Brillouin-zone integrations. *Phys. Rev. B* **13,** 5188–5192 (1976).

Acknowledgements

This work was primarily supported by the Center for Excitonics, an Energy Frontier Research Center funded by the US Department of Energy, Office of Science, Basic Energy Sciences under award no. DE-SC0001088. Our work was also supported by National Science Foundation under grant no. DMR-1410636. Computational time on the Extreme Science and Engineering Discovery Environment (XSEDE) under the grant no. TG-DMR130038 is gratefully acknowledged.

Author contributions

J.L. and W.L. designed the research. W.L. carried out the first-principles calculations. W.L. and J.L. wrote the paper.

Additional information

6

Directed block copolymer self-assembly implemented via surface-embedded electrets

Mei-Ling Wu[1,2,3], Dong Wang[1,2] & Li-Jun Wan[1,2]

Block copolymer (BCP) nanolithography is widely recognized as a promising complementary approach to circumvent the feature size limits of conventional photolithography. The directed self-assembly of BCP thin film to form ordered nanostructures with controlled orientation and localized pattern has been the key challenge for practical nanolithography applications. Here we show that BCP nanopatterns can be directed on localized surface electrets defined by electron-beam irradiation to realize diverse features in a simple, effective and non-destructive manner. Charged electrets can generate a built-in electric field in BCP thin film and induce the formation of perpendicularly oriented microdomain of BCP film. The electret-directed orientation control of BCP film can be either integrated with mask-based patterning technique or realized by electron-beam direct-writing method to fabricate microscale arbitrary lateral patterns down to single BCP cylinder nanopattern. The electret-directed BCP self-assembly could provide an alternative means for BCP-based nanolithography, with high resolution.

[1] Key Laboratory of Molecular Nanostructure and Nanotechnology, Institute of Chemistry, Chinese Academy of Sciences (CAS), Beijing 100190, China. [2] Beijing National Laboratory for Molecular Sciences, Beijing 100190, China. [3] University of CAS, Beijing 100049, China. Correspondence and requests for materials should be addressed to D.W. (email: wangd@iccas.ac.cn).

Block copolymer (BCP) self-assembly can spontaneously generate periodic arrays of microdomains with versatile morphology and nanoscale feature size in the range of ca. 10–100 nm (ref. 1). The well-defined nanostructures of BCP have been utilized as templates to control spatial order of other functional materials for a variety of applications[2,3] such as microelectronic devices[4,5], photovoltaic devices[6], magnetic storage devices[7], microreactors[8] and porous filtration membranes[9]. In particular, BCP-based nanolithography[10,11], which utilizes the self-assembled BCP thin film to pattern the substrates, has been widely recognized as a viable alternative or complementary approach to conventional photolithography[12,13]. It is promised to address the conflict between the continuous demanding to shrink the feature size of electronic devices and the technology and cost limits of photolithography. Owing to the favourable advantages such as high throughput, low cost and most notably, compatibility with current nanolithography streamline, BCP-based nanolithography has been targeted as one of the most important next-generation lithography techniques in the International Technology Roadmap for Semiconductors[12].

For a typical BCP-based nanolithography, orientation of domains normal to the substrates is essential to facilitate robust pattern transfer[13,14]. The most common way to control the orientation of BCP microdomains is to fabricate a neutral surface through surface energy modification to balance its interfacial energy with both BCP components[15,16]. Taking advantage of the dielectric constant difference of each block, orientation control of the BCP self-assembly can also be achieved by applying an external electric field, although it is generally limited to BCP films with micrometre-scale thickness[17,18]. In addition, surface topology modulation[19,20], solvent annealing[21,22] and other methods[23,24] have also been proposed to control BCP microdomain orientation. Besides the perpendicular orientation, controlled BCP self-assembly with lateral order and localized nanostructures[25–28] is also necessary in practical lithography processes to realize pattern registration and addressable device-oriented patterns with low defects[29,30]. Directed self-assembly, which integrates the BCP self-assembly with the traditional lithography processes, has been developed to achieve oriented and lateral ordering of BCP films[31,32] utilizing chemical epitaxy and graphoepitaxy methods[33–36]. In addition, electrohydrodynamic jet printing was utilized to fabricate complex and hierarchical patterns of BCP films[37,38]. While these methods have improved the ordering of BCP films and advanced the process of registration and addressing, they are generally subject to complex lithographic steps, limited substrates and/or physical destruction of the substrates. It would be rewarding to develop a simple, effective and non-destructive method to realize device-oriented features with versatile patterns and high nanopatterning resolution.

Herein, we introduce a novel method to achieve simultaneously orientation control and localization of BCP self-assembly implemented by surface-embedded electrets, which is fabricated by electron-beam (e-beam) irradiation. Electrets are materials that retain trapped charges or polarization, forming a quasi-permanent, macroscopic electric field around the perimeter[39–41]. Similar to external electric filed, which can align the BCP blocks, the electric field around the electrets is also found to be able to control the orientation of BCP film. The surface electrets-mediated BCP self-assembly provides a facile approach to BCP nanopatterning, and can yield customized nanopatterns with arbitrary geometry and high fidelity by employing an e-beam direct-writing technique. More attractively, by tuning the parameters of e-beam irradiation carefully, we can realize the formation of individual cylinder nanopattern with extremely high

accuracy, which represents the utmost resolution of BCP-based nanolithography and is critical for the formation of contact and via holes in integrated circuit.

Results

Fabrication of nanopatterns. The overall fabricating process of electrets-induced polystyrene-b-poly(methyl methacrylate) (PS-b-PMMA) self-assembly is shown in Fig. 1a. Briefly, a SiO$_2$/Si wafer was first irradiated by e-beam, with proper dose and then spin-coated with BCP film. After thermal annealing, the BCP film underwent microphase separation and self-assembled into stable cylindrical or lamellar nanostructures, depending on the molecular weight ratio of the blocks. Figure 1b shows the plain view scanning electron microscopy (SEM) image of a cylinder-forming BCP thin film on the SiO$_2$/Si wafer, of which the central region has been irradiated by the e-beam. The as-obtained hexagonal arrays (Fig. 1b and Supplementary Fig. 1) in the irradiated region prove that microdomains of PS-b-PMMAs orient normal to the substrate surface. Similar results are found in the lamellar BCP film. Fingerprint arrays are formed in the irradiated area, with the lamella microdomains oriented normal to the substrate (Fig. 1c and Supplementary Fig. 2). In contrast, in the area where SiO$_2$/Si wafers have not been treated by e-beam irradiation, parallel orientations are observed (Supplementary Fig. 3). To probe the internal perpendicular orientation of the entire thin films, we

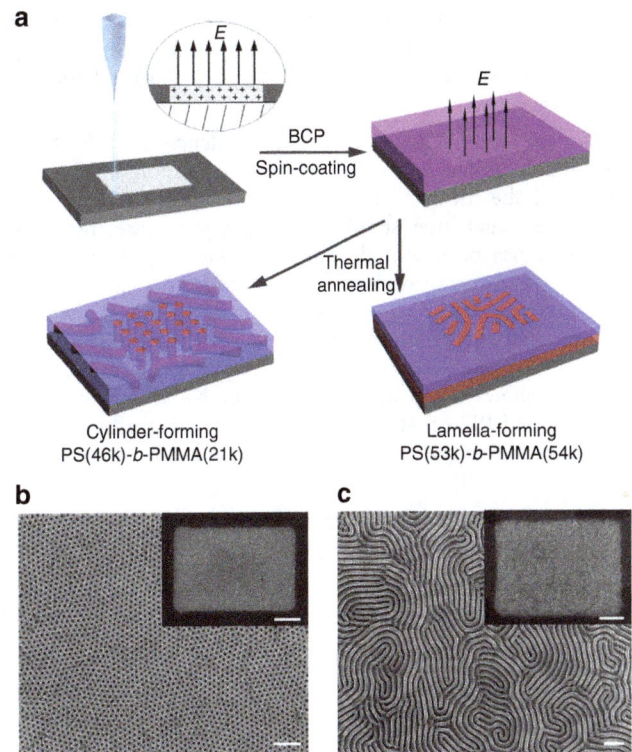

Figure 1 | PS-b-PMMA self-assembly on e-beam-irradiated SiO$_2$/Si substrate. (a) Schematic description for fabricating PS-b-PMMA films with perpendicular orientation on e-beam-irradiated SiO$_2$/Si substrates. SEM images of PS-b-PMMA films self-assembled on the SiO$_2$/Si wafers whose central regions were irradiated by e-beam: **(b)** Cylinder-forming PS-b-PMMA (46k-b-21k) with thickness of 37 nm and **(c)** lamella-forming PS-b-PMMA (53k-b-54k) with thickness of 56 nm (scale bars, 200 nm). The insets in **b** and **c** are the corresponding low-magnification images (scale bars, 1 μm), in which the bright contrast indicates a uniform formation of the perpendicular orientation. The samples were treated by Ar plasma to selectively remove PMMA to improve the SEM image contrast.

performed grazing-incidence small-angle X-ray scattering (GISAXS) experiment. In the two-dimensional GISAXS pattern of cylinder PS-*b*-PMMA film self-assembled on the irradiated SiO$_2$/Si of $\sim 1\,cm^2$ (Supplementary Fig. 4a), the symmetric scattering peaks are confined in the Q_y direction and the first-order scattering peaks at $Q_y = 0.196\,nm^{-1}$ reflect the perpendicularly oriented microdomains of the entire film with a period (domain space) $L_0 = 32.1\,nm$ ($2\pi/Q_y$). Similarly, perpendicular orientation of lamellae is supported by the GISAXS pattern in Supplementary Fig. 4b ($Q_y = 0.119\,nm^{-1}$ and $L_0 = 52.8\,nm$). In addition, the e-beam-induced BCP self-assembly is found widely feasible; all the tested PS-*b*-PMMAs of different molecular weights and block ratios show vertically oriented patterns (Supplementary Fig. 5). The above results indicate that with the aid of e-beam irradiation, we can readily achieve perpendicularly oriented microdomain of PS-*b*-PMMA films.

The e-beam irradiation is considered critical to induce BCP orientation. To ascertain this point, we studied the influence of e-beam parameters on the self-assembly of BCP film. Here we use the area ratio of perpendicularly oriented cylinder microdomains ($C_\perp$) in the whole irradiated region, that is, coverage of $C_\perp$, to represent the degree of orientation. As shown in Fig. 2, the coverage of $C_\perp$ increases gradually with e-beam dose (dose = beam current $\times$ irradiation time/area) and then stabilizes to $\sim 100\%$ at an optimized e-beam dose ($\sim 200\,mC\,cm^{-2}$). A threshold e-beam dose is required to achieve 100% coverage of $C_\perp$, which is affected by the e-beam parameters, such as beam current and accelerating voltage (Supplementary Fig. 6). The dependence of the BCP orientation on the e-beam parameters indicates that the BCP self-assembly by this method hinges largely on the surface charges induced by e-beam irradiation.

Mechanism study. The orientation dependence of PS-*b*-PMMA self-assembly on irradiation dosage provides an important cue to understand the mechanism. In this work, native layer of SiO$_2$ exists on Si wafer after standard cleaning procedure (thickness: 2.3 nm). It has been well documented that SiO$_2$ is an electret material that can trap charges[41,42]. We preformed Kelvin probe force microscope (KPFM) to trace the charges of the SiO$_2$ surface, as it can measure relative surface potential and is an effective technique to uncover the charge trapping of substrates[43]. Figure 3a shows a schematic diagram of KPFM measurement and a typical KPFM mapping image of a SiO$_2$/Si wafer, of which the central region has been irradiated by e-beam. According to

the KPFM image, the contact potential difference (CPD, between the substrate and the tip) in the irradiated region increases significantly compared with the pristine region. The positive CPD reflects that positive charges are trapped in the SiO$_2$ layer after e-beam irradiation, forming a surface-embedded electret. The CPD increment (ΔCPD) increases gradually with the e-beam dose and finally stabilizes to 300–350 mV (Fig. 3b). The evolution tendency of ΔCPD is similar to that of nanopattern coverage (Fig. 2), indicating that orientation of BCP film is closely correlated to the charge state of the surface.

In addition to KPFM measurement, SEM and X-ray photoelectronic spectroscopy (XPS) can also reflect the charging behaviour of the SiO$_2$ layer. The SEM image of electret irradiated by e-beam shows higher contrast difference compared with the non-irradiated region (Supplementary Fig. 7a). The voltage contrast image is attributed to the charging of the electret[44]. XPS was performed to examine the surface constitution and chemical states of the SiO$_2$/Si wafers before and after irradiation (Supplementary Fig. 7b). Binding energies of Si–Si and Si–O bonds increase gradually with the e-beam irradiation, and 0.3 eV increment of the Si–Si binding energy is found in the SiO$_2$/Si substrate after irradiation (3.9 mC cm^{-2}). The small positive shift of binding energy indicates the accumulation of positive charges (holes). The positive charges result from the net outcome of electron injection and secondary electron emission during irradiation. As the secondary electron emitting from the wafer occurs mainly on the surface and cannot be counteracted by the electron injection, accumulated holes are formed on the SiO$_2$ layer, thereby charging the SiO$_2$ electret[42].

Earlier reports have revealed that external electric field can align the blocks of PS-*b*-PMMA along the field and control its orientation due to the dielectric permittivity difference of blocks[18,45]. The e-beam irradiation introduces positive charges into the surface-embedded electret and generates a static electric field E at the SiO$_2$/Si surface (inset in Fig. 1a). For a planar electret, whose lateral dimension is much larger than its thickness D and the film thickness L, one has

$$E = \frac{\varphi}{L + D\varepsilon_{AB}/\varepsilon_s} \qquad (1)$$

where φ is the surface potential at the electret (determined by the charges trapped in the electret), ε_{AB} is the mean (arithmetic) dielectric constant of diblock copolymer A–B (Supplementary Discussion) and ε_s is the dielectric constant of the electret substrate[41]. Therefore, the electric field strength is 1.5–9.5 V μm^{-1} through BCP films with thicknesses of 30–200 nm (Supplementary Discussion). These values are comparable with typical external fields (1.0–40 V μm^{-1}, Supplementary Table 1) used previously in the electric field alignment of the BCP films. As the electric field is strong enough to compensate the energetic penalty from the orientation change[46–48], it would induce PS-*b*-PMMA films to orient normal to the electret substrate.

It is noted that the surface energy of substrate could also be changed after e-beam irradiation. To clarify whether electric field or the surface-wetting property underlies the orientation control, a series of control experiments was conducted. A thin self-assembled monolayer of phenyltrichlorosilane was grafted onto the e-beam-irradiated SiO$_2$/Si wafer, which would alter the surface energy but should not shield the electric field. As a result, on the irradiated region, PS-*b*-PMMA microdomains can still form perpendicular orientation similar to the uncoated samples, whereas parallel orientation is obtained in the non-irradiated area (Supplementary Fig. 8). Thus, the electric field is considered the driving force for BCP alignment; if the orientation control were caused by the surface energy change during e-beam irradiation,

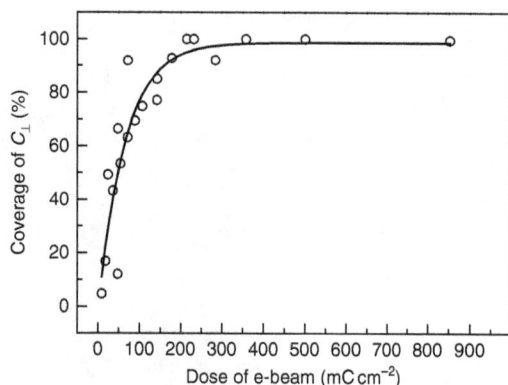

Figure 2 | Evolution of coverage of $C_\perp$ with e-beam dose. E-beam dose is controlled by changing irradiation area and time. Coverage of $C_\perp$ is calculated by the area ratio of $C_\perp$ in the whole irradiated region. The line is a fitted curve based on the scattered points. All experiments were conducted using a 16.9-pA beam current, a 5-kV accelerating voltage and PS(46k)-*b*-PMMA(21k) films with thickness of 37 nm.

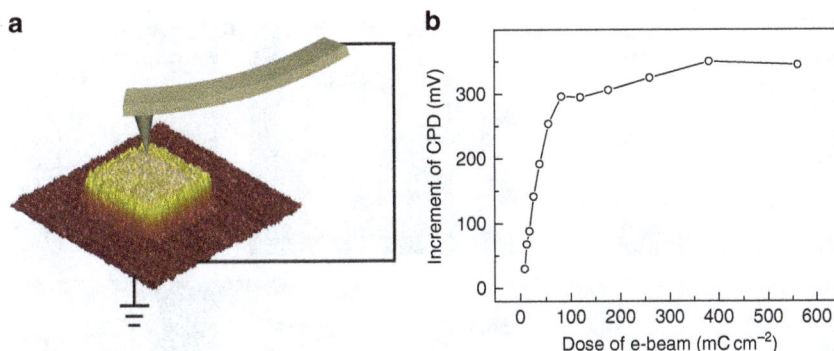

Figure 3 | Surface potential analysis of electrets. (**a**) Schematic diagram of KPFM measurement and a representative surface potential distribution of a SiO$_2$/Si surface with central region irradiated by e-beam. (**b**) Evolution of corresponding incremental CPD (ΔCPD) with e-beam dose.

the self-assembled monolayer modification should have removed the surface energy difference at the irradiated/non-irradiated regions and the morphology of BCP film should be the same. Furthermore, we removed the charges in the SiO$_2$/Si electrets deliberately by thermal annealing (>500 °C; electrets are relatively stable at ~250 °C) or by piranha solution cleaning before PS-b-PMMA self-assembly. Parallel orientation is obtained on the SiO$_2$/Si without the electric field alignment (Supplementary Fig. 9). Third, we have extended the present method to achieve perpendicular orientation of BCP film on other electret substrates. Uniform vertical cylindrical PS-b-PMMA films are attained on various e-beam-irradiated electrets, including SiO$_2$/Si, PI and Si$_3$N$_4$/SiO$_2$/Si (nitride–oxide–silicon (NOS); Supplementary Fig. 10). Intriguingly, wettability experiment shows that charged SiO$_2$/Si is preferential to PMMA block, charged PI is relatively preferential to PS block and charged NOS is non-preferential for both blocks (see Supplementary Table 2 and Supplementary Discussion for more discussion). Perpendicular orientation of PS-b-PMMA film is successfully achieved regardless of the preferential wetting properties (SiO$_2$/Si, PI). These results further prove that the electric field from the charged surface electret is the critical driving force for the perpendicular orientation of PS-b-PMMA films.

Features and merits. As the directed self-assembly of PS-b-PMMA film is implemented with surface-embedded electrets, this method exhibits several pronounced advantages. First, this method is feasible, requiring virtually no chemical pretreatments, and is applicable to many electret materials. As discussed above, other electret substrates besides SiO$_2$ are suitable for this method. In addition, this method is applicable to SiO$_2$ layer with a wide thickness range (Supplementary Fig. 11).

In the conventional approaches to control the BCP orientation by surface energy modification, the BCP thickness is generally limited to below ~3 L_0 in PS-b-PMMA. For thicker films, driving force from neutral interfaces declines and the vertical microdomains would not penetrate through the entire film[10,49]. External electric field-based alignment method is applicable to film thickness of micrometres, but it is difficult to be applied to BCP with nanometre scale, as BCP film with sufficient thickness is required for electrode separation[31]. In contrast, the built-in electric field originated from the surface electret eliminates the limitation of external electric field to apply to film thickness down to tens of nanometres. As shown in Supplementary Fig. 12, the PS-b-PMMA films of different film thicknesses up to 161 nm can orient normal to surface of the electret and the degree of long-range lateral order slightly increases with the film thickness. We can further fabricate PS(53k)-b-PMMA(54k) film with thickness

of 208 nm (~3.9 L_0), with vertical orientation penetrating the entire film thickness (Supplementary Fig. 13a). Moreover, in the GISAXS pattern (Supplementary Fig. 13b), the pronounced first-order peak at $Q_y = 0.12$ nm^{-1} and the high-order peaks with relative Q_y ratios of 1:2:3:4 also evidence that the lamella BCP microdomains orient normal to the substrate throughout the entire film.

The electrets-mediated BCP orientation can yield nanopatterns in millimetre scale (mm scale) and produce customized patterns. By increasing e-beam diameter and beam current, the process of e-beam irradiation can be significantly accelerated. For example, when e-beam of 20 μm in diameter (see Mode 3 in the Method) is employed, the irradiation speed increases from 0.4 to 2.5×10^3 μm^2 s^{-1} and it takes ~30 min to irradiate a region of 3.0×2.2 mm^2. As shown in Supplementary Fig. 14, perpendicularly oriented cylinder morphology is realized in several square millimetres with high uniformity. The orientation quality of the mm-scale BCP film is comparable to those fabricated by surface energy modification, surface topology modulation and external electric alignment[15,18,50]. In addition, the electret implantation can be guided by customized masks to realize patterned BCP orientation in mm scale (see Supplementary Fig. 15 for the process and pattern images).

Finally, the electret charging can be realized by the technique of e-beam direct writing (EBD) for patterns with precise size and customized geometry. EBD technique has been extensively applied in fabricating patterned templates as a mask-free method. By exploiting EBD to manipulate the e-beam moving path, we can fabricate e-beam-irradiated region with arbitrary geometries and sizes. The fine and arbitrary features of EBD endow this method with splendid abilities to produce versatile user-defined layouts for localized and self-registered BCP nanopatterns. We elucidate this combined technique with a pattern of Tai Chi diagram. The Tai Chi-patterned electret was first fabricated by EBD, which was demonstrated by the KPFM image (Fig. 4a). After BCP self-assembly on the customized electret pattern, Tai Chi patterns can be easily built with perpendicularly oriented lamella microdomains in high fidelity (Fig. 4b,c and Supplementary Fig. 16). Using BCP with different molecular weight ratios, we can form nanopattern of fingerprints and dots in Yin and Yang deliberately (see Supplementary Fig. 16 for more nanopatterns with Tai Chi diagram). Furthermore, patterns of stars and word 'NANO' with perpendicular orientation of cylinder-forming BCP can be also fabricated successfully by a similar method (Fig. 4d–f).

Intriguingly, we further demonstrate orientation control over individual cylinders, which represents the ultimate resolution limit for directed BCP self-assembly, as a most attractive feature by controlling the size of the surface-embedded electrets with a

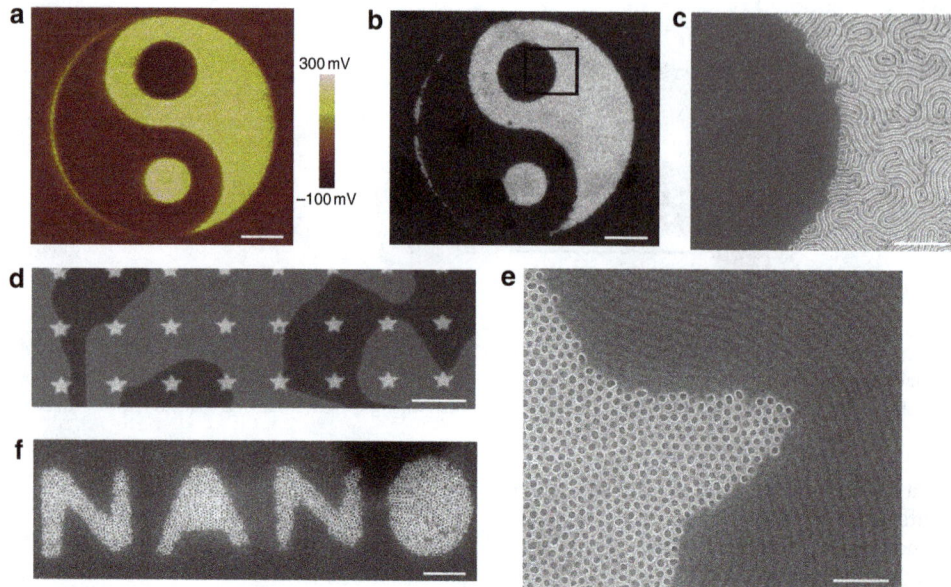

Figure 4 | BCP nanopatterns with user-defined diagrams. (a) Surface potential image of a Tai Chi diagram fabricated by e-beam irradiation with EBD. **(b,c)** SEM images of the corresponding self-assembled morphology of PS(53k)-b-PMMA(54k) film (thickness: 56 nm). Spatially oriented PS(46k)-b-PMMA(21k) film with various geometries: **(d)** patterned array of stars; **(e)** magnified view of **d**; **(f)** pattern of letters NANO. The film thickness in **d,e** and **f** is 37 nm. Scale bars, 2 μm **(a,b)**; 500 nm **(c)**; 5 μm **(d)**; 200 nm **(e)**; 500 nm **(f)**.

highly focused e-beam. Patterns of isolated hole or several holes can be formed with an irradiated circle region by careful e-beam manipulation (Fig. 5). Highly focused e-beam can precisely control the spatial distribution of the electret, at surface of which local electric field is strong enough to control the orientation of PS-b-PMMA for nucleation in a minimum volume[51]. As shown in Fig. 5b, individual cylinders can be realized on the SiO$_2$ surface electret obtained from superfine e-beam irradiation. The control over individual microdomain of BCP self-assembly, as the resolution limit of the BCP nanopattern, is of paramount importance for application as contact holes in integrated circuit[12]. Moreover, the high-resolution nanopattern implemented on the localized electrets is free from small guiding templates or physical destruction of the substrate, and thus remarkably facilitates the process of subsequent high-resolution nanopattern transfer to other substrates[35]. Increasing the irradiated size or e-beam dosage, several cylinders are achieved in a controllable manner (Fig. 5c–e). An array composed of ordered vertical cylinder can also be successfully fabricated on an irradiated linear region (Supplementary Fig. 17a). When applying a lamella-forming polymer to such linear electret, vertical lamellae orient normal to the domain boundary (Supplementary Fig. 17b) for minimizing the free energy[52]. We can envision that the approach of surface electret-directed BCP self-assembly holds feasibility and opportunity to form accurate registering patterns provided with ultra-high-resolution e-beam[53].

Discussion

We have experimentally demonstrated the mechanism and appealing merits of the electret-mediated BCP self-assembly. The charged electret introduces an electric field that aligns the BCP film. In the electric field E over the electret, diblock copolymer A–B consisting of blocks with different dielectric constants (ε_A and ε_B) is aligned by the driving force[47,54]

$$f_e \propto \frac{(\Delta\varepsilon)^2}{\varepsilon_{AB}} E^2 \qquad (2)$$

where $\Delta\varepsilon = |\varepsilon_A - \varepsilon_B|$ is the dielectric constant contrast of blocks A

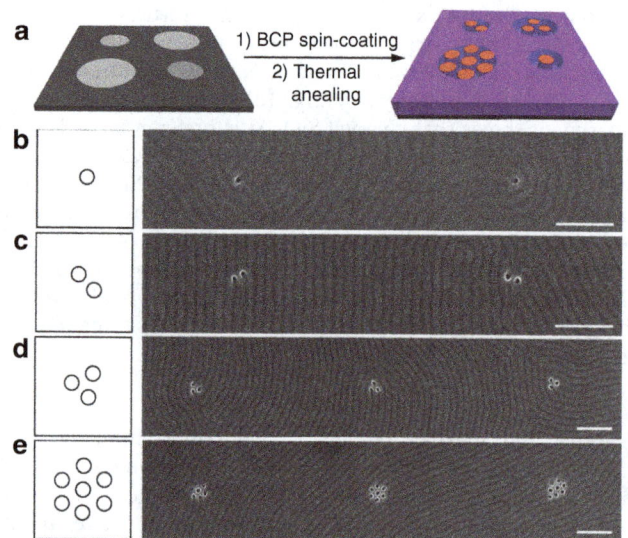

Figure 5 | High-resolution BCP patterns realized by e-beam control. (a) Schematic process of high-resolution perpendicular cylindrical BCP nanopattern self-assembled on charged electrets with controlled circular size and e-beam dose. **(b–e)** SEM images of individual hole and patterned holes of PS-b-PMMA film after PMMA-block removal (scale bars, 200 nm). Each hole corresponds to the position of a vertically oriented PMMA microdomain of PS-b-PMMA film. The film thickness is 37 nm.

and B, and ε_{AB} is the dielectric constant of A–B. The microdomains would be aligned parallel to the electric field (that is, perpendicular to the substrate) as long as the electrostatic energy (F_e), which favours the perpendicular orientation of BCP, overcomes the energy penalty from the BCP orientation transformation (F_p, mainly the elastic free energy and interfacial energy). Under a critical condition, the change of free energy (ΔF_e and ΔF_p) of BCP film from parallel to perpendicular orientation equals 0, that is,

$$\Delta F = \Delta F_e + \Delta F_p = 0 \qquad (3)$$

The critical surface potential (φ_c) can be derived from equation (3), which is the threshold of surface potential to compensate the surface energy penalty and provides a theoretical assessment of possibility to align the BCP via the charged electret. Key parameters influencing φ_c include dielectric constants (ε_A and ε_B), BCP film thickness L and interfacial energy mismatch δ. On the basis of the lamellar diblock copolymer A–B (in a strong-segregation regime) and a planar electret model, the dependence of φ_c on these parameters is theoretically calculated and plotted in Supplementary Fig. 18 (see Supplementary Discussion for details). According the theoretical estimation, the influence of dielectric constant contrast $\Delta\varepsilon$ (electrically relative) is significant, whereas the effect of interfacial energy is minor. The interfacial energy influences the BCP self-assembly by changing the threshold of the surface potential. The dependence of the electret-directed self-assembly on the film thickness is also insignificant.

In this study, the surface electrets were charged via e-beam irradiation to produce BCP nanopatterns with sizes ranging from nanoscale to mm scale. In fact, electrets can also be charged via corona discharge or a metal-coated polydimethylsiloxane stamp, to yield inch-scale built-in electric field, which is promising to make the electrets-aided patterning practically scalable. It is noteworthy that though this BCP self-assembly method applying local electric field can effectively achieve spatial orientation control of PS-*b*-PMMA film, further work is necessary to control the defects and achieve highly ordered BCP assembly.

In summary, we have presented a novel approach to achieve orientation control of the self-assembled PS-*b*-PMMA film with the aid of surface electret. The surface-embedded electret, which is charged by e-beam irradiation, generates static electric field in polymer film and control the orientation of BCP film perpendicular on the surface. This method is marked with simplicity and compatibility. Various customized patterns are achievable with size ranging from millimetre with a mask-based patterning technique, to nanometre or micrometre integrated with EBD. By precisely tuning the e-beam parameters, we demonstrate the formation of individual cylinder nanopattern, which represents the resolution limit of BCP nanopattern and provides promising prospect in achieving accurate registering.

Methods

Cleaning of substrates. Silicon wafers were cleaned in piranha solution at boiling water bath for 30 min and then rinsed with super-pure water. Polyamide film was cleaned in ethanol and acetone solution with ultrasonic cleaner. Si substrate with Si$_3$N$_4$/SiO$_2$ (2.3 nm/2.3 nm) double layer (NOS) was fabricated by magnetron sputtering and was directly used. All other substrates were blow-dried with nitrogen flow before use.

E-beam irradiation and post treatment. Three e-beam irradiation modes were used in this work for different experimental purposes.

Mode 1: E-beam irradiated in repeated raster scan mode, producing numerous frames. The scan time for each frame was 472 ms. Each scan consisted of 768 × 512 pixels, and the irradiation time was 1 μs for each pixel. Diameter of the e-beam was ~5 nm. E-beam dose was determined by beam current, irradiation time and irradiation area.

Mode 2: E-beam irradiated in single-frame mode (the same as EBD technique). E-beam dose was user-defined. The point-to-point spacing was 4.6 nm. Diameter of the e-beam was ~5 nm.

Mode 3: E-beam irradiation is the same as that in Mode 1, except that the e-beam diameter was ~20 μm.

Irradiation process in Modes 1 and 2 were performed on Helios NanoLab 600i (FEI company, Germany). The chamber pressure was ~2.0×10^{-4} Pa and the working distance was 4 mm. Irradiation process in Mode 3 was performed on the electron microprobe (Shimadzu Corp., Japan). The chamber pressure was ~2.0×10^{-4} Pa and the working distance was 5 mm.

Modes 1 and 2 were used to fabricate small-size or patterned irradiation region. Beam current and accelerating voltage can be adjusted as required. Unless otherwise stated, the beam current was 16.9 pA and the accelerating voltage was 5 kV. Beam current was measured accurately by a Faraday cup. To form the isolated or superfine patterns, e-beam dose was ~50 mC cm^{-2} and the irradiated region was set as a circle with a diameter of 50–100 nm.

Mode 3 was used to form large-area irradiation region (mm scale). The beam current was 100 nA and the accelerating voltage was 15 kV. The SiO$_2$/Si wafer was irradiated for 30 min to form an irradiated region of 3 × 2.25 mm^2. To achieve patterned irradiation, Cu grid was used as a mask to block partial e-beam irradiation.

To study the post-treatment effects on the BCP self-assembly, SiO$_2$/Si substrates irradiated by Mode 1 were annealed at 250 or 500 °C for 2 h in Ar flow or immersed in piranha solution (H$_2$SO$_4$/H$_2$O$_2$ mixture) at 100 °C for 0.5 h. To alter the surface energy of SiO$_2$/Si, SiO$_2$/Si substrate irradiated by Mode 1 was immersed in phenylteichlorosilane/toluene mixture (1 vol%) for 0.5 h and was then washed in toluene to remove the unreacted silane.

Formation of PS-*b*-PMMA film. Cylinder-forming PS-*b*-PMMAs (46k-*b*-21k, 68k-*b*-33.5k and 35k-*b*-12.5k g mol^{-1}) and lamella-forming PS-*b*-PMMAs (53k-*b*-54k and 23k-*b*-22k g mol^{-1}) diblock copolymers were purchased from Polymer Source. The BCPs were dissolved in toluene and spin-coated onto the irradiated SiO$_2$/Si wafers. Unless otherwise stated, the film thickness of PS-*b*-PMMA (46k-*b*-21k) and PS-*b*-PMMA (53k-*b*-54k) was 37 and 56 nm, respectively, which was determined by a spectroscopic ellipsometer (SE 850 DUV, SENTECH Instruments). The BCP films were annealed at 250 °C for 2 h in Ar flow for microphase separation.

KPFM measurement. The KPFM measurement was operated on Multimode 8 (Bruker Corp.). The samples were prepared with the Helios 600i system, and a marker was added near the irradiated zone to locate the irradiated zone quickly under the charge-coupled device of KPFM system. We used Si probe with 1.5 N m^{-1} force constant and 250 kHz resonance frequency. Scan velocity was 0.6 Hz and number of scan lines was 512. All KPFM images were analysed using Nanoscope Analysis software.

Characterization. The surface morphology of the BCP films was observed by Helios 600i SEM system with an accelerating voltage of 5 kV. To increase the contrast of SEM image, all the annealed samples were treated with Ar plasma (power, 50 W; flow rate, 20 s.c.c.m.) to remove the PMMA domains selectively. GISAXS experiments were conducted at beamline 16B of the Shanghai Synchrotron Radiation Facility. The synchrotron X-ray energy was 10 keV and sample-to-detector distance was 1,820 mm. The incident angle of X-ray beam was 0.185°. A vacuum guide tube in which the scattered beam passed through was used to minimize air scattering. The two-dimensional GISAXS patterns were recorded on a Mar 165 charge-coupled device detector (2,048 × 2,048 pixels, 80 μm per pixel). The surface elemental information was analysed by the X-ray photoelectron spectroscopy on the Thermo Scientific ESCALab 250Xi using 200 W Al Kα radiation. The irradiated SiO$_2$/Si wafers for GISAXS and XPS measurement were prepared by Mode 3.

References

1. Tseng, Y. C. & Darling, S. B. Block copolymer nanostructures for technology. *Polymers* **2**, 470–489 (2010).
2. Chai, J., Wang, D., Fan, X. & Buriak, J. M. Assembly of aligned linear metallic patterns on silicon. *Nat. Nanotechnol.* **2**, 500–506 (2007).
3. Jeong, S.-J. *et al.* Universal block copolymer lithography for metals, semiconductors, ceramics, and polymers. *Adv. Mater.* **20**, 1898–1904 (2008).
4. Black, C. T. *et al.* Polymer self assembly in semiconductor microelectronics. *IBM J. Res. Dev.* **51**, 605–633 (2007).
5. Black, C. T. Self-aligned self assembly of multi-nanowire silicon field effect transistors. *Appl. Phys. Lett.* **87**, 163116 (2005).
6. Hua, B., Lin, Q., Zhang, Q. & Fan, Z. Efficient photon management with nanostructures for photovoltaics. *Nanoscale* **5**, 6627–6640 (2013).
7. Naito, K., Hieda, H., Sakurai, M., Kamata, Y. & Asakawa, K. 2.5-inch disk patterned media prepared by an artificially assisted self-assembling method. *IEEE Trans. Magn.* **38**, 1949–1951 (2002).
8. Deng, X., Buriak, J. M., Dai, P. X., Wan, L. J. & Wang, D. Block copolymer-templated chemical nanopatterning on pyrolyzed photoresist carbon films. *Chem. Commun.* **48**, 9741–9743 (2012).
9. Yang, S. Y. *et al.* Nanoporous membranes with ultrahigh selectivity and flux for the filtration of viruses. *Adv. Mater.* **18**, 709–712 (2006).
10. Bang, J., Jeong, U., Ryu du, Y., Russell, T. P. & Hawker, C. J. Block copolymer nanolithography: translation of molecular level control to nanoscale patterns. *Adv. Mater.* **21**, 4769–4792 (2009).
11. Bates, C. M., Maher, M. J., Janes, D. W., Ellison, C. J. & Willson, C. G. Block copolymer lithography. *Macromolecules* **47**, 2–12 (2014).

12. The international technology roadmap for semiconductors: lithography. http://www.itrs2.net/itrs-reports.html (2015).

13. Kim, S. Y., Gwyther, J., Manners, I., Chaikin, P. M. & Register, R. A. Metal-containing block copolymer thin films yield wire grid polarizers with high aspect ratio. *Adv. Mater.* **26**, 791–795 (2014).

14. Park, O. H. *et al.* High aspect-ratio cylindrical nanopore arrays and their use for templating titania nanoposts. *Adv. Mater.* **20**, 738–742 (2008).

15. Mansky, P., Liu, Y., Huang, E., Russell, T. P. & Hawker, C. J. Controlling polymer-surface interactions with random copolymer brushes. *Science* **275**, 1458–1460 (1997).

16. Bates, C. M. *et al.* Polarity-switching top coats enable orientation of sub-10-nm block copolymer domains. *Science* **338**, 775–779 (2012).

17. Violetta, O. *et al.* Electric field alignment of a block copolymer nanopattern: direct observation of the microscopic mechanism. *ACS Nano* **3**, 1091–1096 (2009).

18. Thurn-Albrecht, T. *et al.* Ultrahigh-density nanowire arrays grown in self-assembled diblock copolymer templates. *Science* **290**, 2126–2129 (2000).

19. Man, X., Tang, J., Zhou, P., Yan, D. & Andelman, D. Lamellar diblock copolymers on rough substrates: self-consistent field theory studies. *Macromolecules* **48**, 7689–7697 (2015).

20. Hong, S. W. *et al.* Unidirectionally aligned line patterns driven by entropic effects on faceted surfaces. *Proc. Natl Acad. Sci. USA* **109**, 1402–1406 (2012).

21. Gu, X., Gunkel, I., Hexemer, A., Gu, W. & Russell, T. P. An in situ grazing incidence X-ray scattering study of block copolymer thin films during solvent vapor annealing. *Adv. Mater.* **26**, 273–281 (2014).

22. Lo, T. Y. *et al.* Phase transitions of polystyrene-b-poly(dimethylsiloxane) in solvents of varying selectivity. *Macromolecules* **46**, 7513–7524 (2013).

23. Thebault, P. *et al.* Tailoring nanostructures using copolymer nanoimprint lithography. *Adv. Mater.* **24**, 1952–1955 (2012).

24. Yu, B. *et al.* Confinement-induced novel morphologies of block copolymers. *Phys. Rev. Lett.* **96**, 138306 (2006).

25. Park, S. *et al.* Macroscopic 10-terabit-per-square-inch arrays from block copolymers with lateral order. *Science* **323**, 1030–1033 (2009).

26. Tang, C., Lennon, E. M., Fredrickson, G. H., Kramer, E. J. & Hawker, C. J. Evolution of block copolymer lithography to highly ordered square arrays. *Science* **322**, 429–432 (2008).

27. Majewski, P. W., Rahman, A., Black, C. T. & Yager, K. G. Arbitrary lattice symmetries via block copolymer nanomeshes. *Nat. Commun.* **6**, 7448 (2015).

28. Maher, M. J. *et al.* Photopatternable interfaces for block copolymer lithography. *ACS Macro Lett.* **3**, 824–828 (2014).

29. Tsai, H. *et al.* Two-dimensional pattern formation using graphoepitaxy of PS-b-PMMA block copolymers for advanced FinFET device and circuit fabrication. *ACS Nano* **8**, 5227–5232 (2014).

30. Doerk, G. S. *et al.* Pattern placement accuracy in block copolymer directed self-assembly based on chemical epitaxy. *ACS Nano* **7**, 276–285 (2013).

31. Hu, H., Gopinadhan, M. & Osuji, C. O. Directed self-assembly of block copolymers: a tutorial review of strategies for enabling nanotechnology with soft matter. *Soft Matter* **10**, 3867–3889 (2014).

32. Cheng, J. Y., Ross, C. A., Smith, H. I. & Thomas, E. L. Templated self-assembly of block copolymers: top-down helps bottom-up. *Adv. Mater.* **18**, 2505–2521 (2006).

33. Mark, P. S. *et al.* Directed assembly of block copolymer blends into nonregular device-oriented structures. *Science* **308**, 1442–1446 (2005).

34. Ion, B. *et al.* Graphoepitaxy of self-assembled block copolymers on two-dimensional periodic patterned templates. *Science* **321**, 939–943 (2008).

35. Yi, H. *et al.* Flexible control of block copolymer directed self-assembly using small, topographical templates: potential lithography solution for integrated circuit contact hole patterning. *Adv. Mater.* **24**, 3107–3114 (2012).

36. Doerk, G. S. *et al.* Enabling complex nanoscale pattern customization using directed self-assembly. *Nat. Commun.* **5**, 5805 (2014).

37. Onses, M. S. *et al.* Hierarchical patterns of three-dimensional block-copolymer films formed by electrohydrodynamic jet printing and self-assembly. *Nat. Nanotechnol.* **8**, 667–675 (2013).

38. Onses, M. S. *et al.* Block copolymer assembly on nanoscale patterns of polymer brushes formed by electrohydrodynamic jet printing. *ACS Nano* **8**, 6606–6613 (2014).

39. Jacobs, H. O. & Whitesides, G. M. Submicrometer patterning of charge in thin-film electrets. *Science* **291**, 1763–1766 (2001).

40. Zhao, D. *et al.* Self-organization of thin polymer films guided by electrostatic charges on the substrate. *Small* **7**, 2326–2333 (2011).

41. Sessler, G. M. *Electrets* 2nd edn (Springer, 1987).

42. Li, W.-Q. & Zhang, H.-B. The positive charging effect of dielectric films irradiated by a focused electron beam. *Appl. Surf. Sci.* **256**, 3482–3492 (2010).

43. Tsui, B. Y., Hsieh, C. M., Su, P. C., Tzeng, S. D. & Gwo, S. Two-dimensional carrier profiling by Kelvin-probe force microscopy. *Jpn J. Appl. Phys.* **47**, 4448–4453 (2008).

44. Nakasugi, T., Ando, A., Sugihara, K., Miyoshi, M. & Okumura, K. New registration technique using voltage-contrast images for low-energy electron-beam lithography. *Proc. SPIE* **4343**, 334–341, (2001).

45. Liedel, C., Pester, C. W., Ruppel, M., Urban, V. S. & Boker, A. Beyond orientation: the impact of electric fields on block copolymers. *Macromol. Chem. Phys.* **213**, 259–269 (2012).

46. Thurn-Albrecht, T., DeRouchey, J., Russell, T. P. & Jaeger, H. M. Overcoming interfacial interactions with electric fields. *Macromolecules* **33**, 3250–3253 (2000).

47. Kyrylyuk, A. V. & Fraaije, J. G. Electric field versus surface alignment in confined films of a diblock copolymer melt. *J. Chem. Phys.* **125**, 164716 (2006).

48. Kyrylyuk, A. V., Sevink, G. J. A., Zvelindovsky, A. V. & Fraaije, J. G. E. M. Simulations of electric field induced lamellar alignment in block copolymers in the presence of selective electrodes. *Macromol. Theory Simul.* **12**, 508–511 (2003).

49. Maher, M. J. *et al.* Interfacial design for block copolymer thin films. *Chem. Mater.* **26**, 1471–1479 (2014).

50. Kulkarni, M. M., Yager, K. G., Sharma, A. & Karim, A. Combinatorial block copolymer ordering on tunable rough substrates. *Macromolecules* **45**, 4303–4314 (2012).

51. Gopinadhan, M., Majewski, P. W. & Osuji, C. O. Facile alignment of amorphous poly(ethylene oxide) microdomains in a liquid crystalline block copolymer using magnetic fields: toward odered electrolyte membranes. *Macromolecules* **43**, 3286–3293 (2010).

52. Shin, D. O. *et al.* One-dimensional nanoassembly of block copolymers tailored by chemically patterned surfaces. *Macromolecules* **42**, 1189–1193 (2009).

53. van Dorp, W. F. *et al.* Molecule-by-molecule writing using a focused electron beam. *ACS Nano* **6**, 10076–10081 (2012).

54. Ashok, B., Muthukumar, M. & Russell, T. P. Confined thin film diblock copolymer in the presence of an electric field. *J. Chem. Phys.* **115**, 1559–1564 (2001).

Acknowledgements

This work was supported by National Natural Science Foundation of China (21433011, 21127901 and 91527303) and the Strategic Priority Research Program of the Chinese Academy of Sciences (Grant No. XDB12020100). We thank Shanghai Synchrotron Radiation Facility and Dr Yuzhu Wang for the help in the GISAXS experiments and Prof. Shien-Der Tzeng in National Tsing Hua University for providing the NOS samples.

Author contributions

D.W. and M.-L.W. conceived the idea; M.-L.W. conducted the experiments; all the authors participated in discussing the results; M.-L.W. and D.W. drafted the manuscript.

Additional information

7

Exciton localization in solution-processed organolead trihalide perovskites

Haiping He[1], Qianqian Yu[1], Hui Li[1], Jing Li[1], Junjie Si[1], Yizheng Jin[2], Nana Wang[3], Jianpu Wang[3], Jingwen He[4], Xinke Wang[4], Yan Zhang[4] & Zhizhen Ye[1]

Organolead trihalide perovskites have attracted great attention due to the stunning advances in both photovoltaic and light-emitting devices. However, the photophysical properties, especially the recombination dynamics of photogenerated carriers, of this class of materials are controversial. Here we report that under an excitation level close to the working regime of solar cells, the recombination of photogenerated carriers in solution-processed methylammonium–lead–halide films is dominated by excitons weakly localized in band tail states. This scenario is evidenced by experiments of spectral-dependent luminescence decay, excitation density-dependent luminescence and frequency-dependent terahertz photo-conductivity. The exciton localization effect is found to be general for several solution-processed hybrid perovskite films prepared by different methods. Our results provide insights into the charge transport and recombination mechanism in perovskite films and help to unravel their potential for high-performance optoelectronic devices.

[1] State Key Laboratory of Silicon Materials, School of Materials Science and Engineering, Zhejiang University, Hangzhou 310027, China. [2] Center for Chemistry of High-Performance and Novel Materials and State Key Laboratory of Silicon Materials, Department of Chemistry, Zhejiang University, Hangzhou 310027, China. [3] Key Laboratory of Flexible Electronics (KLOFE) & Institute of Advanced Materials (IAM), Jiangsu National Synergetic Innovation Center for Advanced Materials (SICAM), Nanjing Tech University (NanjingTech), 30 South Puzhu Road, Nanjing 211816, China. [4] Department of Physics, Capital Normal University, Beijing Key Lab for Metamaterials and Devices, and Key Laboratory of Terahertz Optoelectronics, Ministry of Education, Beijing 100048, China. Correspondence and requests for materials should be addressed to H.H. (email: hphe@zju.edu.cn) or to Y.J. (email: yizhengjin@zju.edu.cn) or to Z.Y. (email: yezz@zju.edu.cn).

Recently, organolead trihalide perovskites have been utilized in low-temperature solution-processed photovoltaics[1–6] and light-emitting devices[7–10]. Certified power conversion efficiency approaching 20.1% has been realized[6]. The impressive photovoltaic performance is believed to originate from the long-distance and balanced diffusion of charge carriers[11–12]. Remarkably, the solution-processed perovskite films also exhibit superior luminescence properties. Optically pumped lasing with low thresholds and tunable wavelengths[7], and bright light-emitting diodes[9–10] have been demonstrated.

Despite these remarkable advances, knowledge of the photophysical properties of the perovskites is still lacking. One of the key questions concerns the recombination dynamics of photogenerated charges: whether exciton or free carrier (FC) is the dominant recombination channel in organolead trihalide perovskites. The answer will help to interpret the seemingly counterintuitive facts that organolead trihalide perovskites can act both as extraordinary photovoltaic materials and superior gain mediums for lasing[13]. In general, photovoltaic materials require efficient separation of photocarriers, and lasing materials require high recombination rates. The reported exciton binding energy of the perovskites[14,15] is comparable to the thermal energy at room temperature (RT), which arouses arguments that in such a case excited states will tend to dissociate into FCs rather than recombine radiatively.

Several groups have used photoluminescence (PL) to study the competition between exciton and FC in organolead trihalide perovskites. In a few steady-state PL studies, the RT PL was attributed to exciton recombination,[16–18] but the conclusions lacked solid evidence. Recently, several groups[8,14,19,20] attributed the RT PL to FC recombination. For example, D'Innocenzo et al.[14] argued that excitons generated by low-density excitation are almost fully ionized at RT when the exciton binding energy is moderately larger than the RT thermal energy. The band filling effect[8] and quadratic dependence of the PL intensity on the excitation intensity[19], the two characteristic features of FC recombination, were observed at relatively high excitation levels. We note that the observation of FC recombination at

relatively high excitation levels is not surprising because the reduced exciton binding energy originated from the screening effect of FCs, a phenomenon that has been well established in many semiconductors[21,22].

Here we show that under an excitation level close to the working regime of solar cells, the radiative recombination of photogenerated carriers in solution-processed $CH_3NH_3PbX_3$ perovskites is dominated by excitons localized in band tail states. The excitonic nature of the emission is evidenced by the excellent power-law dependence of the PL intensity on the excitation intensity expected for bound excitons, and is supported by the PL lineshape analysis. The localization effect is supported by the spectral dependence of the PL lifetime and frequency-dependent THz photoconductivity results. We also show that the exciton localization effect is general in several solution-process perovskite films.

Results

Evidence for exciton localization in $CH_3NH_3PbBr_3$ films. We use solution-processed $CH_3NH_3PbBr_3$ films for the PL studies. The films show good crystalline and optical quality (Supplementary Fig. 1). To avoid degradation induced by air exposure, all samples were prepared in a nitrogen-filled glove box, coated with a polymethyl methacrylate (PMMA) layer and were measured in vacuum (10^{-1} Pa). The $CH_3NH_3PbBr_3$ film shows emission at 2.35 eV, in agreement with the reported values[23,24].

Near-band-edge emission in semiconductors may have several origins, including exciton recombination, FC recombination (also known as band-to-band transition), free-to-bound recombination and donor–acceptor pair recombination. To determine which process is dominant in our samples, we measured the PL spectra under various excitation densities close to or lower than the photovoltaic working regime ($\sim 5 \times 10^{14}$ cm^{-3}) (refs 14,25). The PL lineshapes in Fig. 1a are almost identical in the whole range of excitation intensity. The PL intensity shows excellent power-law dependence on the excitation power, with a power-law exponent of 1.179. In direct bandgap semiconductors and

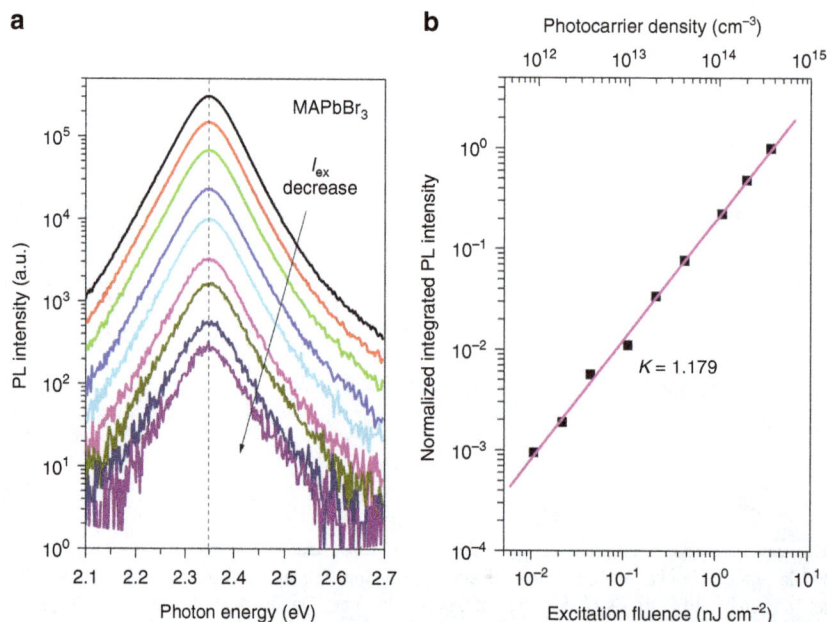

Figure 1 | Excitation density-dependent PL of solution-processed $CH_3NH_3PbBr_3$ films. (a) Steady-state PL spectra recorded with excitation density from 0.01 to 4 nJ cm^{-2}. All spectra are measured in vacuum at RT. In all spectra, the peak energy (indicated by the dashed line), lineshape and linewidth are identical within the experimental error. **(b)** Logarithm plot of the integrated PL intensity versus excitation density. The data show a power-law dependence with $k = 1.179$.

under non-resonant excitation conditions, the integrated PL intensity (I_{PL}) is a power-law function of the excitation density[26],

$$I_{PL} \sim I_{ex}^k \qquad (1)$$

with $k = 2$ for FC recombination, $1 < k < 2$ for recombination of excitons (including free excitons and bound excitons) and $k < 1$ for free-to-bound recombination and donor–acceptor pair. The model was further refined by Shibata et al[27], who provided an analytical formula to confirm that $1 < k < 2$, even for free excitons. The physics behind the process is the photo-neutralization of the donors/acceptors, which are present in all semiconductors and result in competitive recombination channels. Our k value agrees well with those reported for excitons in semiconductors[26,27].

Figure 2a shows the RT PL decay curves monitored at different excitation energies. On the low-energy side, the PL lifetime is almost constant for all emission energies. On the high-energy side, the PL lifetime decreases with increasing emission energy. The PL decay curves can be well fitted by the thermalized stretching exponential line shape[28,29]

$$I(t) = I_1 \exp\left(-\frac{t}{\tau_1}\right) + I_2 \exp\left[-\left(\frac{t}{\tau_2}\right)^\beta\right], \qquad (2)$$

where τ_i is the decay time and I_i is the weight factor of each decay channel. A typical fitting result is plotted in Fig. 2b (for all fitting results, see Supplementary Fig. 2). The fitting curves do not match normal mono-exponential or simple stretched-exponential decay (Supplementary Fig. 2c). Stretched-exponential decay is regarded as evidence of the exciton localization, in which the parameter β is related to the dimensionality of the localizing centres. The former exponential term in equation (2) represents the relaxation of free or extended states towards localized states, whereas the latter stretched-exponential term accounts for the communication between the localized states. We found that all the decay curves can be well fitted with a constant β of 0.43 ± 0.03. Both lifetimes τ_2 and τ_1 show clear spectral dependence (Fig. 2c; Supplementary Fig. 2d). It markedly decreases on the high-energy side, while remaining constant on the low-energy side.

The spectral dependence of τ_2 can be described by a well-established model[30] for excitons localized in the tail states

$$\tau(E) = \frac{\tau_{LE}}{1 + \exp\left(\frac{E - E_{me}}{E_0}\right)}, \qquad (3)$$

where τ_{LE} is the lifetime of localized excitons, E_{me} can be regarded as the mobility edge and E_0 is a characteristic energy of the density of band tail states, which can be a measure of the localization energy. The best fit of the data gives $\tau_{LE} = 61$ ns, $E_{me} = 2.419$ eV and $E_0 = 41$ meV. The localization energy, 41 meV, is higher than the RT thermal energy, which is consistent with localized excitons being observed even at RT.

We provide evidence to show that the PL in our perovskite films is not due to FC recombination but to a localization effect. The first evidence comes from the PL spectra under high excitation. As shown in Supplementary Fig. 3, increasing the excitation density leads to the occurrence and increase of a higher energy tail at ~2.41 eV. The result suggests that the main peak at 2.32 eV does not originate from FC recombination because FC emission has the highest energy among all radiative recombination. With increasing excitation density, the 2.32 eV peak is still dominant over the 2.41 eV FC emission and is almost unshifted. The unshifted luminescence agrees with the reported[31] feature of localized excitons, which means that the many-particle effect of these localized excitons is weak. The second evidence comes from the frequency-dependent THz measurements (the experimental details can be found in Supplementary methods). Supplementary Figure 4 plots the photoconductivity induced by 400-nm pump pulses. The real part of the induced photoconductivity $\Delta\sigma_1(\omega)$ decreases with decreasing frequency, and the imaginary part $\Delta\sigma_2(\omega)$ has a negative value at low frequency. The results do not support the Drude model for FCs[32], which predicts increasing $\Delta\sigma_1(\omega)$ with decreasing frequency and positive $\Delta\sigma_2(\omega)$. The real part increasing with frequency and the negative imaginary part are typical signatures of carrier localization[33,34]. The low conductivity at low energy is consistent with the insulating nature of the charge-neutral excitons[32]. Such features build up immediately after excitation ($\Delta t = 3$ ps) and are maintained within the entire timescale ($\Delta t = 300$ ps).

The above results indicate that the PL decay is dominated by recombination of localized excitons rather than FCs at RT. We also conducted excitation density-dependent PL and spectral-dependent lifetime measurements on $CH_3NH_3PbBr_3$ films at 237 K. These experiments were designed to check whether the assignment of localized excitons is valid for the perovskite films at low temperatures, which have higher quantum efficiency than that of the films at RT. The temperature of 237 K was chosen based on the cubic-to-tetragonal phase transition[35] of $CH_3NH_3PbBr_3$. As shown in Fig. 3, the relative internal quantum efficiency (IQE) at 237 K is estimated to be ~80% based on the temperature-dependent PL intensity (Supplementary Note 1), a method that has been widely used in inorganic compound

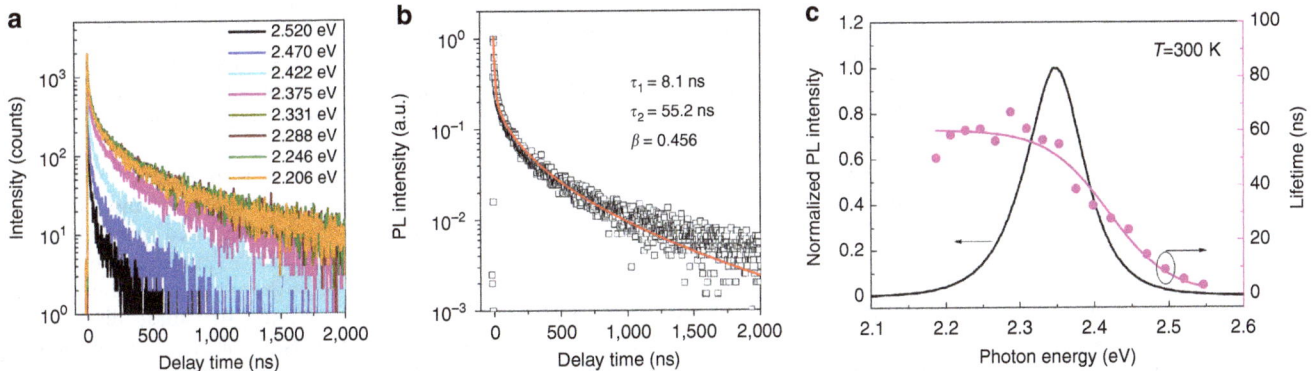

Figure 2 | Spectral-dependent PL decay of solution-processed $CH_3NH_3PbBr_3$ films. (a) PL decay curves monitored at various emission energies. The lifetime decreases markedly on the high-energy side of the emission. (b) Typical fitting of a decay curve by the thermalized stretching exponential model described by equation (2). (c) The lifetime of localized excitons τ_2 (circles) as a function of emission energy. The data are fitted with equation (3) (magenta line), with the lifetime of localized exciton $\tau_{LE} = 60.5$ ns, mobility edge of 2.419 eV and the localization energy $E_0 = 40.9$ meV.

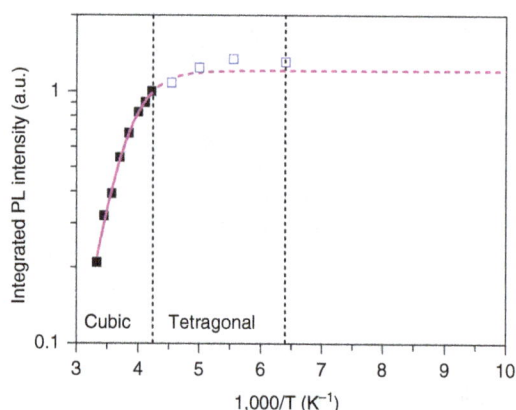

Figure 3 | PL thermal quenching behaviour of solution-processed CH₃NH₃PbBr₃ film. Integrated PL intensity as a function of the reciprocal temperature under low excitation (photocarrier density $\sim 3.7 \times 10^{14}$ cm^{-3}). There is a cubic (RT) to tetragonal phase transition at 236 K, and tetragonal-to-tetragonal phase transition at 155 K. The magenta solid curve represents the fitting result for the cubic data (Supplementary Note 1). The magenta dashed curve is the extrapolation result. The tetragonal data (open square) are plotted for comparison.

semiconductors[36]. The results (Supplementary Fig. 5) show that the k value and PL decay features at 237 K are similar to those at RT, indicating that localized excitons still dominate the PL under the condition of high quantum yield.

Recently, there have been arguments that at RT excitons cannot exist in perovskites because the dielectric constant of perovskite materials is very large[37–40], which results in dielectric screening and consequent dissociation of excitons. However, our evidence for exciton localization suggests that the role of the dielectric screening effect might be more complicated than expected. In fact, excitons do exist in materials with large dielectric constants. In ferroelectric oxides, such as SrTiO₃ and FeBiO₃, it is accepted that self-trapped excitons[41] and charge transfer vibronic excitons may exist[42]. The mechanism for these excitons can be quite different from the normal one. For example, charge transfer vibronic excitons is correlated electronic and hole polarons induced by the Jahn–Teller type lattice distortion, which are localized and can be quite stable at RT[42]. In view of these facts, we suggest that excitons can also exist in trihalide perovskites, despite the possible large dielectric constant under illumination. As additional experimental evidence, the absorption of exciton resonance is observed in the optical absorption spectra of CH₃NH₃PbBr₃ film (Supplementary Fig. 1b). Recent analysis of the absorption spectra suggested that bound exciton states also exist in CH₃NH₃PbI₃ films[15].

The properties of perovskite films may depend on the preparation methods and/or processing conditions. To test whether the exciton localization is a general mechanism, we prepared CH₃NH₃PbBr₃ films by the two-step method and measured the excitation-dependent PL and spectral-dependent lifetime (Supplementary Fig. 6). The samples also exhibit features similar to that shown in Fig. 2 and Fig. 3, indicating the localized exciton nature.

Exciton localization in CH₃NH₃PbI(Cl)₃ films.

The emissions from solution-processed CH₃NH₃PbI₃ and CH₃NH₃PbI₃ $_{-x}$Cl$_x$ films are also dominated by localized excitons. The power-law dependence of the PL intensity on the excitation density reveals k values of ~ 1.5 (Fig. 4a), and the PL decay shows spectral dependence (Supplementary Fig. 7) similar to that observed in CH₃NH₃PbBr₃. The k values are larger than the value obtained in

CH₃NH₃PbBr₃. According to the theory developed by Schmidt et al.[26] and Shibata et al.[27], such a difference mainly represents the different material properties such as the probabilities of radiative recombination and competitive nonradiative recombination. For example, the decrease of crystal perfection is expected to increase the k value. Moreover, the contribution of FC recombination may also lead to a change of k value. The PL lineshape analysis (Supplementary Fig. 8; Supplementary Note 2) indicates a small fraction ($\sim 9.5\%$) of FC recombination in the emission of CH₃NH₃PbI₃ $_{-x}$Cl$_x$ film, which is reasonable because the reported exciton binding energy[43] of CH₃NH₃PbI₃ and CH₃NH₃PbI₃ $_{-x}$Cl$_x$ is lower than CH₃NH₃PbBr₃, so the thermal dissociation of excitons is easier. The coexistence of exciton and FC recombination in the PL spectra has been observed in other materials[44] with exciton binding energy comparable to the RT thermal energy. Excitation-dependent PL at low temperature reveals smaller k values (Supplementary Fig. 9), which can be interpreted by reduced thermal dissociation of excitons at low temperature.

It is of interest to determine whether the conclusion of localized excitons in perovskite materials is valid for the same material in a photovoltaic device structure. We construct such a structure using CH₃NH₃PbI₃ $_{-x}$Cl$_x$, as shown in the inset of Fig. 4a. The PL intensity still shows power-law dependence on the excitation density, with a k value of 1.547, very close to the result of the bare film. Moreover, the PL decay spectra (Supplementary Fig. 10) show dependence on the emission energy and the lifetime can be well described by equation (3) with $E_0 \sim 17$ meV, as seen in Fig. 4b. The results indicate that the PL of CH₃NH₃PbI₃ $_{-x}$Cl$_x$ film in a typical photovoltaic structure is also dominated by the localized excitons rather than FC recombination.

Discussion

The physical picture of the recombination of localized excitons is illustrated in Fig. 5. The density of localized states is approximated by an exponential tail with the form of $\sim \exp(-E/E_0)$. The excitons can be either partly localized (one carrier is localized with another carrier bound to it by Coulomb attraction) or wholly localized[21]. With increasing energy, the localized excitons may transit to the extended exciton states (approaching free excitons) at the transition region known as the mobility edge. Under low excitation, most of the photocarriers occupy the tail states. The picture can also be understood in the space coordinates as shown in Fig. 5b. In this case, the tail states are represented by the local potential minima in the conduction and/or valence bands. The photogenerated carriers transfer to these potential minima to form localized excitons, which have much longer lifetime than free excitons due to the transfer between localized states[28,29]. The long lifetime of the localized excitons accounts for the observed long PL lifetime. This phenomenon is also observed in inorganic semiconductors. For example, localized exciton lifetime as long as 65 ns at RT has been reported[29] in InGaN.

Tail states are very common in semiconductors and can be induced by doping, compositional changes and structural deformation[21,45]. Although solution-processed perovskites are materials with reasonable crystal quality, structure imperfections are inevitable. For example, the large rotational freedom of the polar CH₃NH₃⁺ cation can produce structural disorder[37,46]. Unintentional/intentional doping is possible. The weak bonding between lead and halogens may also produce local disorder, especially in the surface and crystal boundary region. Recent studies[47] revealed that the grain boundaries exhibited faster nonradiative decay. Other results[48] suggest that perovskites with larger grains exhibit better photovoltaic performance. Given these

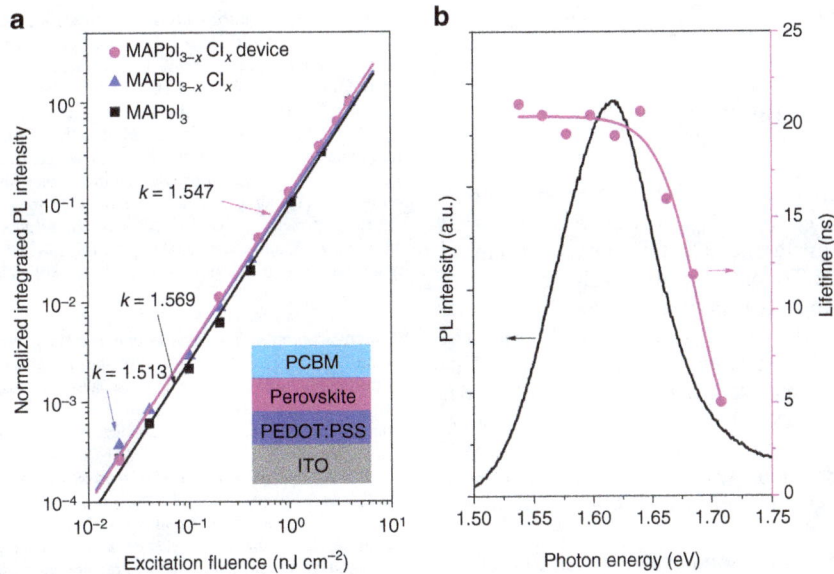

Figure 4 | Steady-state PL spectra and transient PL decay of CH₃NH₃PbI₃ and CH₃NH₃PbI₃₋ₓClₓ. (a) Logarithm plot of the integrated PL intensity versus excitation density for $CH_3NH_3PbI_3$ and $CH_3NH_3PbI_{3-x}Cl_x$ films. Insert is a typical photovoltaic structure with a $CH_3NH_3PbI_{3-x}Cl_x$ film sandwiched between two charge-transporting interlayers. The data show good power-law dependence with k values of 1.569, 1.513 and 1.547. **(b)** PL lifetime (solid circles) of ITO/PEDOT:PSS/ $CH_3NH_3PbI_{3-x}Cl_x$/PCBM photovoltaic structure as a function of the emission energy. The data are fitted with equation (3) (magenta line), with the lifetime of localized exciton $\tau_{LE} = 20.5$ ns, mobility edge 1.689 eV and localization energy $E_0 = 17.3$ meV. The PL lifetime is greatly reduced due to the quenching effects of the adjacent layers.

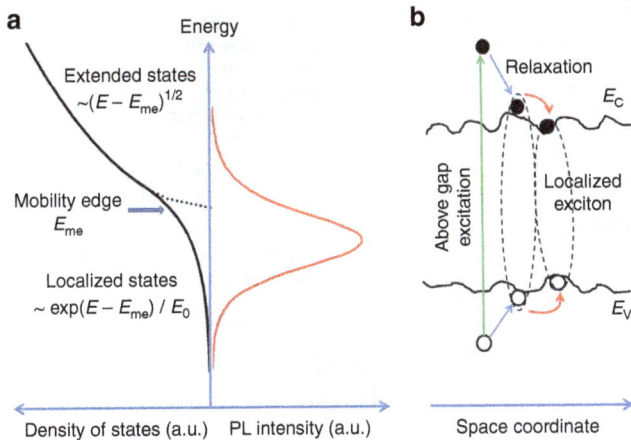

Figure 5 | Physical picture of exciton localization. (a) The density of the extended and localized states, and emission in perovskites (schematic). The density of the localized states is approximated by an exponential tail with the form of $\sim \exp(-E/E_0)$. The localized and extended states are divided by the mobility edge. Under low excitation, the excitons mainly occupy the localized states. Under high excitation, the localized states can be filled and the photocarriers also occupy the extended states, leading to emissions of free carriers. **(b)** Schematic drawing of exciton localization in space coordinates. With the presence of structural disorder, the tails of the localized states form local potential fluctuation in the energy bands. These potential minima can localize electrons and holes to form localized excitons. The carriers can transfer between the local potential minima, leading to long PL lifetime.

facts, it is reasonable to suggest the existence of tail states in solution-processed perovskite films.

The energy and intensity of localized state-related emission depend strongly on the nature (for example, energy level) and density of the localized states. In the case of strong localization, such as deep-level centres, the emission energy may be

substantially reduced. The localized states may emit weakly provided their density is low. In the present case, strong near-band-edge emission can be realized, indicating that excitons are weakly localized in the band tail states. The weak localization is evidenced by the relatively small localization energy reflected by the E_0 values in equation (3) and by the small Stokes shift of the luminescence (Supplementary Fig. 1b). The localization effect can even be beneficial to light-emitting devices. Localized exciton may contribute to the optical gain because the localized states can be easily filled, provided their density is not too high[31,49,50]. We performed temperature-dependent PL measurements under moderate excitation to check whether nonradiative channels become dominant when the excitation density increases. Supplementary Fig. 11 shows that the relative IQE of moderate excitation is much higher than that of low excitation, in agreement with the result reported by Deschler *et al*[8]. This result implies that other decay channels such as Auger recombination do not become dominant, which will be of great benefit to low threshold lasing. We conducted PL experiments at high excitation (photocarrier concentration up to $\sim 10^{19}$ cm^{-3}). Amplified spontaneous emission was observed on the low-energy side (at 2.26 eV) of the localized exciton emission with a threshold of $\sim 300 \mu J$ cm^{-2} (Supplementary Fig. 3b). It is noteworthy that FC emission (~ 2.41 eV) emerges with increasing excitation density. These results suggest that the amplified spontaneous emission is likely from localized excitons, which indicates that the optical gain can come from the filled localized states, and it is possible to achieve a low threshold by controlling the density of localized states. Moreover, exciton localization may also increase the luminescence efficiency because the oscillator strength of the transitions in the localized states is greatly enhanced[49,51].

We emphasize that the generation of localized excitons at excitation levels close to the working regime of solar cells does not conflict the fact that the hybrid perovskites are superior photovoltaic materials. The localized excitons can diffuse by a thermally activated multiple trapping–escaping process[52,53]. This process leads to an exponential dependence of the diffusion coefficient on the exciton energy (E_{exc}), as well as on the temperature[53]

$$D = D_0 \exp(\frac{E_{exc} - E_{me}}{k_B T}), \quad (4)$$

where D_0 is the diffusion coefficient of the extended excitons. Taking the energetic distance $E_{exc} - E_{me} \sim 70$ meV (as extracted from Figs 2c,4b), we estimate the RT diffusion coefficient of localized excitons D as $0.067 D_0$, only approximately one order of magnitude lower than the free exciton. If we invoke the reported[11,12], D value of $0.011 - 0.054$ cm^2 s^{-1} for the perovskites, a D_0 of $\sim 0.16 - 0.80$ cm^2 s^{-1} is obtained. This value is comparable to the reported values[53,54] for CdS$_{1-x}$Se$_x$ (0.3 cm^2 s^{-1}) and In$_{1-x}$Ga$_x$N (0.5–1.1 cm^2 s^{-1}), two alloy semiconductors known as the typical examples for the study of localized excitons. In combination with the very long lifetime, a long carrier diffusion length in perovskites is expected in the scenario of exciton localization.

In summary, we have presented solid evidence that the RT PL in organolead trihalide perovskites is dominated by weakly localized excitons. The evidence includes: (i) the excellent power-law dependence of the PL intensity on the excitation intensity with $1 < k < 2$, (ii) the localization effect indicated by the spectral dependence of the PL lifetime and frequency-dependent THz photoconductivity and (iii) the coexistence of exciton and FC recombination under high excitation. Exciton localization is suggested as the origin of the long PL lifetime in this class of materials. We find that the localization effect is general in solution-process perovskite films due to the presence of crystal imperfections. The localization of excitons strongly influences the transport and recombination properties of perovskite materials. The dominance of the localized exciton in the recombination channels as well as its higher IQE under moderate excitation, strongly suggests that it is possible to utilize these benefits to realize low threshold lasing in perovskites, as has been demonstrated in III–V and II–VI semiconductors and devices. The elaborate tailoring of the localization effect in perovskites is thus highly attractive in designing future high-performance optoelectronic devices.

Methods

Synthesis of perovskite films. All the indium-tin oxide (ITO)-coated glass substrates were cleaned sequentially in deionized water, ethanol, acetone and oxygen plasma before spin-coating. The perovskite CH$_3$NH$_3$PbI$_{3-x}$Cl$_x$ was prepared according to the reported procedure[11]. Methylamine iodide was prepared by reacting 33 wt % methylamine in ethanol (Sigma-Aldrich), with 57 wt % hydroiodic acid in water (Sigma-Aldrich), at RT. Hydroiodic acid was added dropwise while stirring. After drying at 100 °C, the resultant white powder was dried overnight in a vacuum oven and was recrystallized from ethanol before use. To form the CH$_3$NH$_3$PbI$_{3-x}$Cl$_x$ precursor solution, methylammonium iodide and lead (II) chloride (Sigma-Aldrich) were dissolved in anhydrous N,N-dimethylformamide (DMF) in a 3:1 molar ratio of methylamine iodide to PbCl$_2$, with final concentrations 0.88 M lead chloride and 2.64 M methylammonium iodide. The precursor was filtered through a 220-nm polytetrafluoroethylene (PTFE) filter head, then spin-coated at 3,000 r.p.m. for 30 s on ITO-coated glass; finally, it was annealed at 95 °C for ~ 10 min. CH$_3$NH$_3$PbI$_3$ was prepared via the sequential deposition route[3]. A PbI$_2$ (Sigma-Aldrich) solution in DMF (462 mg ml^{-1}) was spin-coated on glass substrate and then kept at 70 °C. After drying, the films were dipped in a solution of CH$_3$NH$_3$I in 2-propanol (10 mg ml^{-1}) for ~ 60 s and rinsed with 2-propanol, and then spin-coated to form uniform CH$_3$NH$_3$PbI$_3$ thin films. For CH$_3$NH$_3$PbBr$_3$ preparation,[24] CH$_3$NH$_3$Br was first prepared by mixing methylamine with hydrobromic acid (48% in water; CAUTION: exothermic reaction) in 1:1 molar ratio in a 100-ml flask under continuous stirring at 0 °C for 2 h. CH$_3$NH$_3$Br was then crystallized by removing the solvent in an evaporator, washing three times in diethyl ether for 30 min and filtering the precipitate. The material, in the form of white crystals, was then dried in vacuum at 60 °C for 24 h and was then kept in a dark, dry environment until further use. A 20-wt % solution of CH$_3$NH$_3$PbBr$_3$ was prepared by mixing PbBr$_2$ and CH$_3$NH$_3$Br in a 1:3 molar ratio in DMF. The precursor was spin-coated at 4,000 r.p.m. for 30 s on ITO-coated glass and was annealed at 60 °C for ~ 10 min.

Fabrication of the photovoltaic device structure. The structure of ITO/PEDOT:PSS/perovskite/PCBM was fabricated on patterned ITO-coated glass substrates (sheet resistance: 15 Ω sq^{-1}). The substrates were cleaned sequentially in acetone, ethanol, deionized water and ethanol for 10 min each, followed by oxygen plasma treatment for 15 min. A poly(3,4-ethylenedioxythiophene):poly(styrenesulfonate) (PEDOT:PSS) layer was spin-coated onto the substrates at 4,000 r.p.m. for 60 s and then annealed in air at 140 °C. The sample was transferred into a glove box. Next, the CH$_3$NH$_3$PbI$_{3-x}$Cl$_x$ precursor solution was spin-coated at 3,000 r.p.m. for 45 s, followed by annealing on a hot plate at 100 °C for ~ 60 min. The [6,6]-phenyl-C61-butyric acid methyl ester (PCBM) layers were deposited from a 30 mg ml^{-1} chlorobenzene solution at 2,000 r.p.m. for 45 s. To avoid the degradation induced by air exposure, the devices were packaged with glass.

PL measurements. The steady-state and time-resolved PL measurements were performed on a FLS920 fluorescence spectrophotometer (Edinburgh Instruments). To minimize the effect of air exposure, all the measurements were taken in vacuum with pressure of < 0.01 torr. Pulsed laser diodes (EPL405 and EPL635) with tunable repeating frequency of 20 kHz to 20 MHz were used as the excitation source. For CH$_3$NH$_3$PbI$_3$ and CH$_3$NH$_3$PbI$_{3-x}$Cl$_x$ films, EPL635 with a wavelength of 638.8 nm and pulse width of 86.4 ps was used. For CH$_3$NH$_3$PbBr$_3$, we used EPL405 with a wavelength of 404.2 nm and pulse width of 58.6 ps. The excitation fluence for both wavelengths was ~ 4 nJ cm^{-2}. For excitation density-dependent measurements at the low level, the lasers operated at 20 MHz and the light fluence was tuned by a neutral attenuator. For moderate excitation experiments, a 355 nm frequency-tripled Nd:YAG laser (FTSS 355-50, CryLaS GmbH) with a pulse width of 1 ns and a repetition rate of 100 Hz was used. For lifetime measurements by time-correlated single-photon counting, the lasers operated at 200 kHz. The temperature-dependent measurements were performed with a closed-cycle helium cryostat.

Calculation of the photocarrier density. The corresponding photocarrier density can be calculated as

ρ_{exc} = light fluence density of a single pulse/(photon energy × optical penetration depth) = 4 nJ cm^{-2}/(3.07 eV × 1.6 × 10^{-19} J × 220 nm) $\sim 3.7 \times 10^{14}$ cm^{-3}

Here the optical penetration depth of CH$_3$NH$_3$PbBr$_3$ is taken as ~ 220 nm (ref. 55). In this calculation, we assume a constant excitation because the effective excited volume is remarkably larger than the directly excited one due to the carrier diffusion during the long carrier lifetime[21]. For CH$_3$NH$_3$PbI$_3$ and CH$_3$NH$_3$PbI$_{3-x}$Cl$_x$, the photon energy of the excitation laser was 1.94 eV, and the optical penetration depth was 250 nm (ref. 1); hence, the photocarrier density was $\sim 5.2 \times 10^{14}$ cm^{-3}.

References

1. Kojima, A., Teshima, K., Shirai, Y. & Miyasaka, T. Organometal halide perovskites as visible-light sensitizers for photovoltaic cells. *J. Am. Chem. Soc.* **131**, 6050–6051 (2009).
2. Lee, M. M., Teuscher, J., Miyasaka, T., Murakami, T. N. & Snaith, H. J. Efficient hybrid solar cells based on meso-superstructured organometal halide perovskites. *Science* **338**, 643–647 (2012).
3. Burschka, J. *et al.* Sequential deposition as a route to high-performance perovskite-sensitized solar cells. *Nature* **499**, 316–319 (2013).
4. Jeon, N. J. *et al.* Compositional engineering of perovskite materials for high-performance solar cells. *Nature* **517**, 476–480 (2015).
5. Bai, S. *et al.* High-performance planar heterojunction perovskite solar cells: preserving long charge carrier diffusion lengths and interfacial engineering. *Nano Res.* **7**, 1749–1758 (2014).
6. Research Cell Efficiency Records. Available at http://www.nrel.gov/ncpv/ (2015).
7. Xing, G. *et al.* Low-temperature solution-processed wavelength-tunable perovskites for lasing. *Nat. Mater.* **13**, 476–480 (2014).
8. Deschler, F. *et al.* High photoluminescence efficiency and optically pumped lasing in solution-processed mixed halide perovskite semiconductors. *J. Phys. Chem. Lett.* **5**, 1421–1426 (2014).
9. Tan, Z. K. *et al.* Bright light-emitting diodes based on organometal halide perovskite. *Nat. Nanotechnol* **9**, 687–692 (2014).
10. Wang, J. P. *et al.* Interfacial control toward efficient and low-voltage perovskite light-emitting diodes. *Adv. Mater.* **27**, 2311–2316 (2015).
11. Stranks, S. D. *et al.* Electron-hole diffusion lengths exceeding 1 micrometer in an organometal trihalide perovskite absorber. *Science* **342**, 341–344 (2013).
12. Xing, G. C. *et al.* Long-range balanced electron and hole-transport lengths in organic-inorganic CH$_3$NH$_3$PbI$_3$. *Science* **342**, 344–347 (2013).
13. Laquai, F. All-round perovskites. *Nat. Mater.* **13**, 429–430 (2014).
14. D'Innocenzo, V. *et al.* Excitons versus free charges in organo-lead tri-halide perovskites. *Nat. Commun.* **5**, 3586–3591 (2014).
15. Saba, M. *et al.* Correlated electron-hole plasma in organometal perovskites. *Nat. Commun.* **5**, 5049–5058 (2014).

16. Sun, S. *et al.* The origin of high efficiency in low-temperature solution-processable bilayer organometal halide hybrid solar cells. *Energy Environ. Sci.* **7**, 399–407 (2014).

17. Savenije, T. J. *et al.* Thermally activated exciton dissociation and recombination control the organometal halide perovskite carrier dynamics. *J. Phys. Chem. Lett.* **5**, 2189–2194 (2014).

18. Zhang, W. *et al.* Enhancement of perovskite-based solar cells employing core-shell metal nanoparticles. *Nano Lett.* **13**, 4505–4510 (2013).

19. Yamada, Y., Nakamura, T., Endo, M., Wakamiya, A. & Kanemitsu, Y. Photocarrier recombination dynamics in perovskite $CH_3NH_3PbI_3$ for solar cell applications. *J. Am. Chem. Soc.* **136**, 11610–11613 (2014).

20. Lin, Q. Q., Armin, A., Nagiri, R. C. R., Burn, P. L. & Meredith, P. Electro-optics of perovskite solar cells. *Nat. Photon.* **9**, 106–112 (2015).

21. Klingshirn, C. F. *Semiconductor Optics* (Springer-Verlag, 1997).

22. Cingolani, R. *et al.* Radiative recombination processes in wide-band-gap II–VI quantum wells: the interplay between excitons and free carriers. *J. Opt. Soc. Am. B* **13**, 1268–1277 (1996).

23. Wehrenfennig, C., Eperon, G. E., Johnston, M. B., Snaith, H. J. & Herz, L. M. High charge carrier mobilities and lifetimes in organolead trihalide perovskites. *Adv. Mater.* **26**, 1584–1589 (2014).

24. Edri, E., Kirmayer, S., Cahen, D. & Hodes, G. High open-circuit voltage solar cells based on organic-inorganic lead bromide perovskite. *J. Phys. Chem. Lett.* **4**, 897–902 (2013).

25. Stranks, S. D. *et al.* Recombination kinetics in organic-inorganic perovskites: excitons, free charge, and subgap states. *Phys. Rev. Appl.* **2**, 034007 (2014).

26. Schmidt, T., Lischka, K. & Zulehner, W. Excitation-power dependence of the near-band-edge photoluminescence of semiconductors. *Phys. Rev. B* **45**, 8989–8994 (1992).

27. Shibata, H. *et al.* Excitation-power dependence of free exciton photoluminescence of semiconductors. *Jpn J. Appl. Phys.* **44**, 6113–6114 (2005).

28. Sun, Y. J. *et al.* Nonpolar $In_xGa_{1-x}N$/GaN (1-100) multiple quantum wells grown on γ-$LiAlO_2$ (100) by plasma-assisted molecular-beam epitaxy. *Phys. Rev. B* **67**, 041306 (2003).

29. Onuma, T. *et al.* Localized exciton dynamics in nonpolar (11-20) $In_xGa_{1-x}N$ multiple quantum wells grown on GaN templates prepared by lateral epitaxial overgrowth. *Appl. Phys. Lett.* **86**, 151918 (2005).

30. Gourdon, C. & Lavallard, P. Exciton transfer between localized states in $CdS_{1-x}Se_x$ alloys. *Phys.Stat. Sol. (b)* **153**, 641–652 (1989).

31. Majumder, F. A., Shevel, S., Lyssenko, V. G., Swoboda, H. E. & Klingshirn, C. Luminescence and gain spectroscopy of disordered $CdS_{1-x}Se_x$ under high excitation. *Z. Phys. B* **66**, 409–418 (1987).

32. Kaindl, R. A., Carnahan, M. A., Hagele, D., Lovenich, R. & Chemla, D. S. Ultrafast terahertz probes of transient conducting and insulating phases in an electron-hole gas. *Nature* **423**, 734–738 (2003).

33. La-o-vorakiat, C. *et al.* Elucidating the role of disorder and free-carrier recombination kinetics in $CH_3NH_3PbI_3$ perovskite films. *Nat. Commun.* **6**, 7903–7909 (2015).

34. Nemec, H., Kuzel, P. & Sundström, V. Charge transport in nanostructured materials for solar energy conversion studied by time-resolved terahertz spectroscopy. *J. Photochem. Photobiol. A.* **215**, 123–139 (2010).

35. Mashiyama, H., Kurihara, Y. & Azetsu, T. Disordered cubic perovskite structure of $CH_3NH_3PbX_3$ (X = Cl, Br, I). *J. Korean Phys. Soc.* **32**, S156–S158 (1998).

36. Okamoto, K. *et al.* Surface-plasmon-enhanced light emitters based on InGaN quantum wells. *Nat. Mater.* **3**, 601–605 (2004).

37. Juarez-Perez, E. J. *et al.* Photoinduced giant dielectric constant in lead halide perovskite solar cells. *J. Phys. Chem. Lett.* **5**, 2390–2394 (2014).

38. Frost, J. M. *et al.* Atomistic origins of high-performance in hybrid halide perovskite solar cells. *Nano Lett.* **14**, 2584–2590 (2014).

39. Menendez-Proupin, E., Palacios, P., Wahnon, P. & Conesa, J. C. Self-consistent relativistic band structure of the $CH_3NH_3PbI_3$ perovskite. *Phys. Rev. B* **90**, 045207 (2014).

40. Even, J., Pedesseau, L. & Katan, C. Analysis of multivalley and multibandgap absorption and enhancement of free carriers related to exciton screening in hybrid perovskites. *J. Phys. Chem. C* **118**, 11566–11572 (2014).

41. Kan, D. *et al.* Blue-light emission at room temperature from Ar^+-irradiated $SrTiO_3$. *Nat. Mater.* **4**, 816–819 (2005).

42. Vikhnin, V. S., Eglitis, R. I., Kapphan, S. E., Borstel, G. & Kotomin, E. A. Polaronic-type excitons in ferroelectric oxides: microscopic calculations and experimental manifestation. *Phys. Rev. B.* **65**, 104304 (2002).

43. Tanaka, K. *et al.* Comparative study on the excitons in lead-halide-based perovskite type crystals $CH_3NH_3PbBr_3$ $CH_3NH_3PbI_3$. *Solid State Commun.* **127**, 619–623 (2003).

44. Wang, H., Wong, K. S. & Wong, G. K. L. in *Proceedings of SPIE* **3624**, 13–24 (International Society for Optical Engineering, San Jose, CA, USA, 1999).

45. Permogorov, S. & Reznitsky, A. Effect of disorder on the optical-spectra of wide-gap II-VI semiconductor solid-solutions. *J. Lumin.* **52**, 201–223 (1992).

46. Wasylishen, R. E., Knop, O. & Macdonald, J. B. *Solid State Commun.* **56**, 581 (1985).

47. deQuilettes, D. W. *et al.* Impact of microstructure on local carrier lifetime in perovskite solar cells. *Science* **348**, 683–686 (2015).

48. Nie, W. Y. *et al.* High-efficiency solution-processed perovskite solar cells with millimeter-scale grains. *Science* **347**, 522–525 (2015).

49. Satake, A. *et al.* Localized exciton and its stimulated emission in surface mode from single-layer $In_xGa_{1-x}N$. *Phys. Rev. B* **57**, R2041–R2044 (1998).

50. Chen, R. *et al.* Exciton localization and optical properties improvement in nanocrystal-embedded ZnO core-shell nanowires. *Nano Lett.* **13**, 734–739 (2013).

51. O'Donnell, K. P., Martin, R. W. & Middleton, P. G. Origin of luminescence from InGaN diodes. *Phys. Rev. Lett.* **82**, 237–240 (1999).

52. Chichibu, S. F. *et al.* Recombination dynamics of localized excitons in cubic $In_xGa_{1-x}N$/GaN multiple quantum wells grown by radio frequency molecular beam epitaxy on 3C–SiC substrate. *J. Vac. Sci. Technol. B* **21**, 1856–1862 (2003).

53. Schwab, H., Pantke, K. H., Hvam, J. M. & Klingshirn, C. Measurements of exciton diffusion by degenerate four-wave mixing in $CdS_{1-x}Se_x$. *Phys. Rev. B* **46**, 7528–7532 (1992).

54. Okamoto, K. *et al.* in *Proceedings of SPIE (International Society for Optical Engineering)* **4278**, 157–150 (Bellingham, WA, USA, 2001).

55. Kumawat, N. K., Dey, A., Narasimhan, K. L. & Kabra, D. Near infrared to visible electroluminescent diodes based on organometallic halide perovskites: structural and optical investigation. *ACS Photon.* **2**, 349–354 (2015).

Acknowledgements

This work was supported by the Natural Science Foundation of China (nos 51372223, 91333203, 51522209, 11474164, 61405091, 11474249 and 91433204), the Program for Innovative Research Team in University of Ministry of Education of China (no. IRT13037), the National Basic Research Program of China- Fundamental Studies of Perovskite Solar Cells (2015CB932200), the Natural Science Foundation of Jiangsu Province, China (BK20131413, BK20140952), the National 973 Program of China (2015CB654901), the Synergetic Innovation Center for Organic Electronics and Information Displays and the Fundamental Research Funds for the Central Universities (nos. 2014FZA4008 and 2015FZA3005). We thank Mr Yunzhou Deng (Zhejiang University) for his help in the numerical fitting of the PL spectra.

Author contributions

H.H. and Z.Y. supervised the study. H.H., Q.Y., H.L. and J.L. contributed to the PL measurements and analysis. Q.Y., J.S. and N.W. contributed to the synthesis and characterization of the materials. J.H., X.W. and Y.Z. contributed to the THz measurements and analysis. Y.J. and J.W. provided input to the data analysis and discussed the results. H.H. and Y.J. wrote the manuscript. All authors assisted in manuscript preparation.

Additional information

8

Graphene oxide/metal nanocrystal multilaminates as the atomic limit for safe and selective hydrogen storage

Eun Seon Cho[1], Anne M. Ruminski[1], Shaul Aloni[1], Yi-Sheng Liu[2], Jinghua Guo[2] & Jeffrey J. Urban[1]

Interest in hydrogen fuel is growing for automotive applications; however, safe, dense, solid-state hydrogen storage remains a formidable scientific challenge. Metal hydrides offer ample storage capacity and do not require cryogens or exceedingly high pressures for operation. However, hydrides have largely been abandoned because of oxidative instability and sluggish kinetics. We report a new, environmentally stable hydrogen storage material constructed of Mg nanocrystals encapsulated by atomically thin and gas-selective reduced graphene oxide (rGO) sheets. This material, protected from oxygen and moisture by the rGO layers, exhibits exceptionally dense hydrogen storage (6.5 wt% and 0.105 kg H_2 per litre in the total composite). As rGO is atomically thin, this approach minimizes inactive mass in the composite, while also providing a kinetic enhancement to hydrogen sorption performance. These multilaminates of rGO-Mg are able to deliver exceptionally dense hydrogen storage and provide a material platform for harnessing the attributes of sensitive nanomaterials in demanding environments.

[1] The Molecular Foundry, Materials Sciences Division, Lawrence Berkeley National Laboratory, 1 Cyclotron Road, Berkeley, California 94720, USA. [2] The Advanced Light Source, Lawrence Berkeley National Laboratory, Berkeley, California 94720, USA. Correspondence and requests for materials should be addressed to J.J.U. (email: jjurban@lbl.gov).

The established environmental impacts resulting from fossil fuels have stimulated urgent efforts to decarbonize our fuel sources. Hydrogen is the ultimate carbon-free energy carrier—it possesses the highest energy density among chemical fuels and water is the sole combustion product. Although major car manufacturers have made commitments to hydrogen as a 'fuel of the future'[1], hydrogen storage for FCEVs (fuel cell electric vehicles) currently relies on compressed gas tanks[2]. These are unable to meet long-term storage targets and severely compromise on-board occupancy. Metal hydrides for solid-state hydrogen storage are one of the few materials capable of providing sufficient storage density required to meet these long-term targets. However, simultaneously meeting gravimetric, volumetric, thermodynamic and kinetic requirements has proven challenging, owing to the strong binding enthalpies for the metal hydride bonds, long diffusion path lengths and oxidative instability of zero-valent metals. Although nano-structuring has been shown to optimize binding enthalpies[3], synthesis and oxidative stabilization of metallic nanocrystals remains a challenge[4]. Protection strategies against oxidization and sintering of nanocrystals often involve embedding these crystals in dense matrices, which add considerable 'dead' mass to the composite, in turn decreasing gravimetric and volumetric density. Thus, although metal hydrides show the most promise for non-cryogenic applications, it remains true that no single material has met all of these essential criteria[5,6].

Here we demonstrate mixed dimensional reduced graphene oxide (GO)/Mg nanocrystal hybrids as a novel high-performance materials platform for solid-state hydrogen storage. After the first report of the preparation of individual graphene sheets in 2004 (refs 7,8), its unique optoelectronic properties attracted great attention. GO, formerly considered just a precursor for the synthesis of graphene, has begun to find independent applications in water purification and gas separations due to its hydrophilicity, chemical structure and atomistic pore size diameters[9,10]. For example, GO membranes have recently been explored as materials for crucial gas separation challenges. Interestingly, these studies have shown extreme permeability for H_2 relative to other atmospheric gases such as O_2 and N_2, thus providing a potential avenue for use as an atomically thin, selective barrier layer for sensitive hydrogen storage materials. Furthermore, related studies have shown that reduction of GO to form reduced GO (rGO) further results in a dramatic decrease in water permeance, while maintaining desirable gas permeability characteristics[11]. In this study, we have prepared mixed dimensional laminates of two-dimensional (2D) rGO filled with Mg nanocrystals for hydrogen storage applications (Fig. 1a). In this composite, rGO serves as the atomic limit for barrier layer materials in functional composites, providing maximum environmental protection for the least possible amount of inactive mass—theoretically, for single sheet rGO protection, up to 98 wt% of the composite mass can be active Mg—thus, this rGO composite approach yields the greatest performance in selective hydrogen permeability and kinetic enhancement in hydrogen storage. As illustrated in Fig. 1a, rGO sheets function as a protective layer against Mg nanocrystal oxidation by preventing the permeation of O_2 and H_2O, while still allowing hydrogen to easily penetrate, diffuse along the layers and be released. Moreover, beyond the crucial gas barrier behaviour, we demonstrate that the rGO layers add functionality to the laminates by reducing the activation energies associated with hydrogen absorption and desorption, key kinetically limiting steps for traditional metal hydride systems. Several studies have shown that carbon-based materials such as carbon fibres, nanotubes and graphite exhibit a beneficial catalytic effect on the kinetics and cyclability of hydrogen absorption and

desorption of metal hydrides[12–14]. Although there are other reports using graphitic materials in composites for Li-ion battery applications, to our knowledge there have been no reports that take advantage of both the unique catalytic properties and high variability in gas permeability of rGO to synergistically yield new functionality. For the nanolaminate system presented, rGO layers are ideal encapsulating materials: they provide atomically thin structure to minimize added mass, catalytically enhanced rate-limiting hydrogen absorption/desorption events and protective barriers to prevent degradation of Mg nanocrystals.

Results

Synthesis and characterization of rGO-Mg multilaminates. The majority of reported composites consisting of metals and carbon materials are prepared via ball milling or solidification with either polymers or carbon frameworks. However, ball-milled materials are notoriously polydisperse, which introduces corresponding inhomogeneity in properties. Moreover, such energy-intensive processes can intrinsically introduce unwanted morphological disruptions and chemical inhomogeneities, all of which detract from performance. By contrast, we have developed a direct, one-pot co-reduction, thus simultaneously forming both pristine, monodisperse nanocrystals and the desired rGO without energy-intensive processing or ligand chemistries. Observing that current approaches for reduction of GO and reduction of metal precursors to form Mg nanocrystals both rely on similar methodologies[4,15–17], we synthesized rGO-Mg nanocomposites via a facile solution-based co-reduction method. In this process, the Mg^{2+} precursor is stabilized by GO and both of them are reduced by lithium naphthalenide. The obtained rGO-Mg was characterized via transmission electron microscopy (TEM) and X-ray diffraction (XRD), as shown in Fig. 1b,c. Mg nanocrystals were 3.26 nm diameter (± 0.87 nm) based on TEM images (Supplementary Figs 1 and 2), presenting fine monodisperse nanocrystals, compared with other metal hydrides prepared by conventional method such as ball milling. Although Scherrer analysis of the XRD peak width indicates larger crystallite size of ~ 15 nm (Supplementary Table 1), hundreds of TEM images over dozens of samples consistently show 3.26-nm-sized Mg nanocrystals. The observation of individual nanocrystals in TEM was challenging due to the low electron density of Mg, although in our materials Mg nanocrystals of diameter 3–4 nm are present and confirmed by multiple characterization methods. However, it is probable that there exist some clusters and networks of Mg particles in the sample as the XRD results imply the possible existence of a minor fraction comprising agglomerations of nanocrystals; as we discuss later, the H_2 storage properties are most consistent with ~ few nanometre-sized crystallites. The area of the peaks in electron energy loss spectroscopy (EELS) measurements (Mg >>> C > O) indicated a high density of Mg within the composite (Fig. 1d). In addition, despite containing a highly dense packing of Mg nanocrystals, the nanolaminates were observed to be remarkably environmentally stable. To investigate the limits of stability, rGO-Mg samples were exposed to air and characterized over time by XRD and TEM (Fig. 1c and Supplementary Fig. 1); remarkably, even after 3 months of air exposure, the nanocrystals remained almost entirely zero-valent crystalline Mg, while showing invasion of only a low-intensity $Mg(OH)_2$ peak (Supplementary Fig. 3 and an additional corrosion test result in Supplementary Fig. 4 and Supplementary Discussion). Moreover, to show the reliability offered by this approach, we completely exposed the sample to air and then demonstrated hydrogen cycling. This is not possible with any other hydride technology with comparable storage density. Although none of the techniques used for

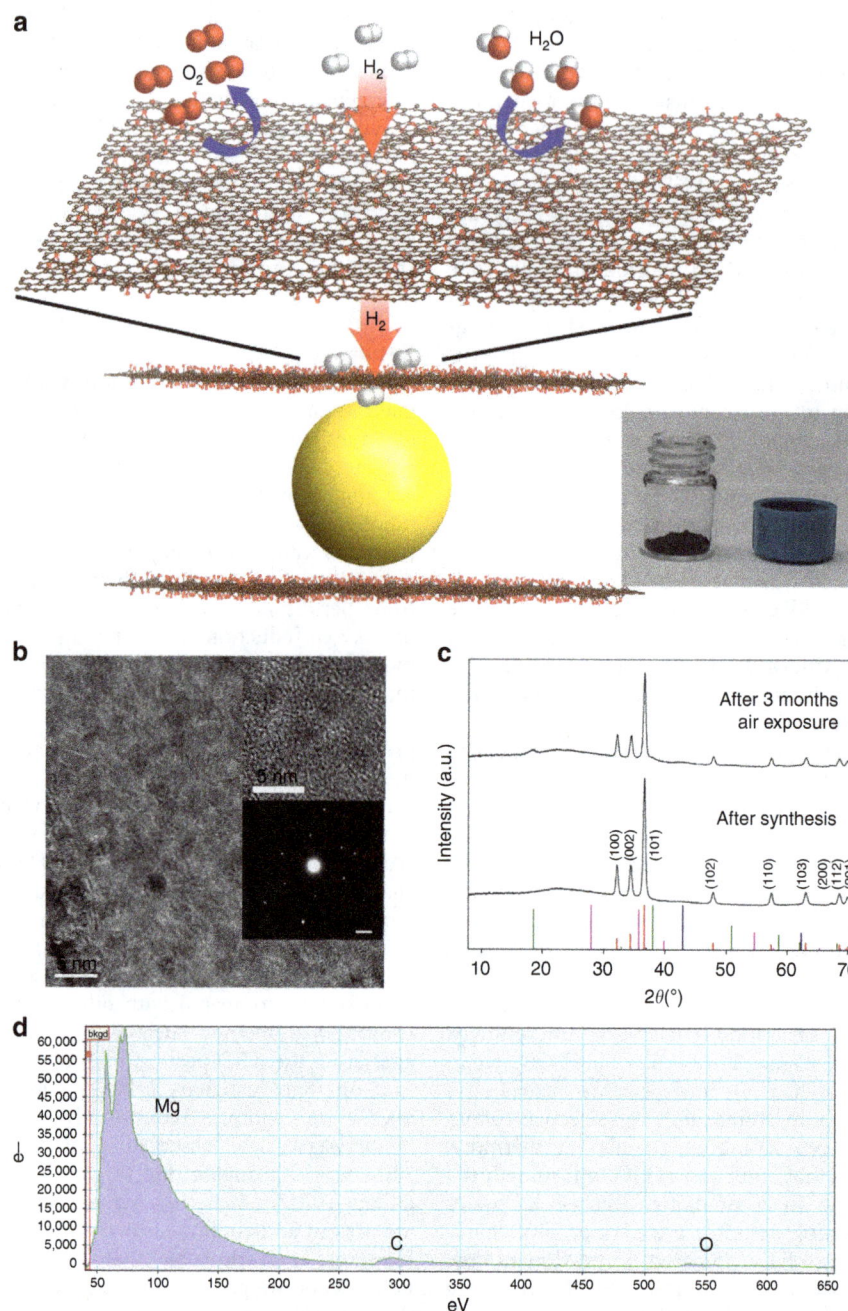

Figure 1 | Mixed dimensional rGO-Mg laminates for stable and energetically-dense hydrogen storage. (**a**) Illustrations depicting the structure of the rGO-Mg nanolaminates where rGO layers prevent O_2 and H_2O from penetrating, while allowing the diffusion of H_2 (inset: photograph of rGO-Mg in air). (**b**) TEM images of rGO-Mg showing the high density of Mg nanocrystals with no visible aggregates. The upper inset is a high-resolution image and the lower inset is diffraction pattern where the hexagonal dots are matched to Mg (100), corresponding to a d-spacing of 2.778 Å (JCPDS 04-0770). (More TEM images and analysis are provided in Supplementary Figs 1 and 2.) (**c**) XRD spectra of rGO-Mg after synthesis and after 3 months in air with indices of peaks (JCPDS 04-0770; the bottom bars represent an XRD pattern of hexagonal Mg (hcp) (red), tetragonal MgH_2 (pink), hexagonal $Mg(OH)_2$ (green) and cubic MgO (blue)). (**d**) EELS spectrum of a representative rGO-Mg composite flake suspended over a hole in the TEM support grid. The spectrum shows a dominant Mg L-edge peak and a carbon K-edge peak, indicating a large quantity of Mg crystals within the rGO support.

characterization show evidence for extensive formation of oxides of Mg (MgO or $Mg(OH)_2$), it is probable that a thin, self-terminating layer of oxide or sub-oxide may form on the nanocrystals; however, this oxide does not measurably have an impact on performance (Fig. 1d and Supplementary Fig. 5).

Hydrogen capacity and kinetic analysis. Hydrogen absorption and desorption characteristics of the rGO-Mg composite were tested using a Sieverts PCT-Pro instrument at 15 bar H_2 and

0 bar, respectively, as shown in Fig. 2a. Hydrogen uptake was immediate and the formation of MgH_2 was confirmed by XRD (Fig. 2c) and electron diffraction (Supplementary Fig. 1). The hydrogen absorption capacity of the composite was 6.5 wt% and 0.105 kg H_2 per litre in the total composite, which is the highest capacity reported (calculated on a full composite mass basis) using metal hydrides under the comparable conditions. This corresponds to 7.56 wt% H_2 in Mg nanocrystals, which is 99.5% of the theoretical value (7.6 wt%)[5]. Given the atomically thin

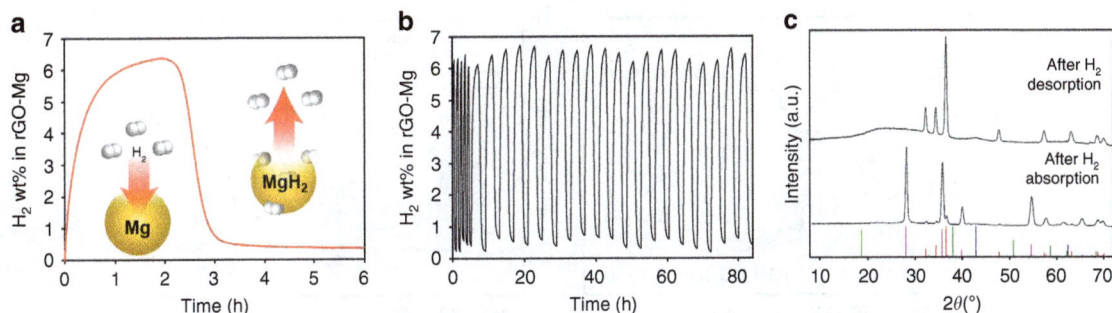

Figure 2 | Hydrogen absorption/desorption characterization of rGO-Mg multilaminates. (**a**) Hydrogen absorption/desorption (at 200 °C and 15 bar H_2/300 °C and 0 bar) for the prepared rGO-Mg multilaminates. (**b**) Hydrogen absorption/desorption cycling of rGO-Mg multilaminates that were first exposed to air overnight. The first 5 cycles were performed at 250 °C and 15 bar H_2/350 °C and 0 bar, and the additional 20 cycles at 200 °C and 15 bar H_2/300 °C and 0 bar. (**c**) XRD spectra of rGO-Mg after absorption/desorption (the bottom bars represent the XRD patterns of Mg (red), MgH_2 (pink), $Mg(OH)_2$ (green) and MgO (blue)).

nature of the encapsulation, these nanocomposites achieve denser packing of metal nanocrystals than is possible by any competing approach, leading to optimized storage density. Furthermore, hydrogen also readily desorbed up to 6.12 wt% in the composite, thus demonstrating excellent reversibility. To verify that the hydrogen absorption was not caused by the presence of GO in the composite, control studies using only GO were conducted and exhibited minimal (<0.2 wt% in GO) absorption at 200 °C and 250 °C (Supplementary Fig. 6). This is a negligible contribution, given that the amount of GO in the multilaminates is <2 wt% overall. To analyse the kinetics, the activation energy (E_a) for hydrogen absorption/desorption was determined from measurements at three different temperatures, fitting the result with the Johnson–Mehl–Avrami model (details are described in Supplementary Discussion, Supplementary Figs 7 and 8, and Supplementary Table 2)[18]. The E_a values were 60.8 and 92.9 kJ mol^{-1} for absorption and desorption, respectively, consistent with one-dimensional nucleation and growth as shown previously[4,19]. Remarkably, these kinetics are comparable to transition metal-catalysed bulk metal-hydride systems, and the overall capacity and kinetics greatly surpass the best environmentally robust samples made up to date[4,17]. We ascribe the performance of our composite to the unique features of this multilaminate: the nanoscale size of the Mg crystals is comparable to molecular diffusion lengths, which enables near-complete conversion to the metal hydride (99.5% of the theoretical value) and the interaction of the Mg nanocrystals with the rGO layers protects against invasion of oxygen, while enabling rapid surface diffusion of hydrogen, which enhances kinetics[20]. Indeed, rGO-Mg hydrogen absorption/desorption is faster than Mg-polymer composites containing nanocrystals of similar size (Supplementary Fig. 9 and Supplementary Discussion). Consistent with previous studies, the diffusion of hydrogen atoms was facilitated by the interaction between Mg and carbon layers, enhancing both the hydrogen capacity and kinetics of Mg (additional data in Supplementary Fig. 10 and Supplementary Discussion)[21,22]. Cycle tests were performed at 250 °C/350 °C for 5 cycles and at 200 °C/300 °C for additional 20 cycles (Fig. 2b). Importantly, each cycle test was conducted successively to better mimic real-world fueling/re-fueling conditions without manual evacuation in between each cycle (additional experimental results in Supplementary Fig. 11). The capacity and kinetics were well preserved during further cycles, although a slight decrease in the capacity was observed. Noticeably, the Mg nanocrystal size and size distributions were well preserved after several absorption/desorption cycles without sintering or grain growth (Supplementary Fig. 1). Although bulk metal hydrides are

susceptible to mechanical fracture and cracking due to the large volume expansion on hydriding (ca. 33% from Mg to MgH_2), the high Young's modulus of rGO enables it to robustly encase the Mg nanocrystals during expansion/contraction events without fracture and prevent macroscale sintering.

Structural analysis of rGO-Mg composite. X-ray absorption near-edge structure (XANES) measurements were performed to probe the interactions between rGO and Mg nanocrystals (Fig. 3a). Compared with GO, increased intensity of carbon K-edge at 288.4 and 290.3 eV were observed, corresponding to carbon atoms attached to oxygen or other oxygen-containing chemical species. From this, we infer that the multilaminates are uniquely stabilized by the formation of interfacial Mg–O–C bonds forged during synthesis[23]. We believe this to be the basis of the exceptional stability of these multilaminates. In addition, the zero-valent state of Mg nanocrystals was verified at Mg L-edge measurement as both total electron yield and total fluorescence yield scans display an explicit Mg metal peak (49.8 eV) without distinct MgO peaks (Fig. 3b and additional analysis in Supplementary Fig. 5 and Supplementary Discussion), consistent with XRD and EELS spectra. The structural evolution of GO during synthesis and hydrogen cycling was studied using Raman spectroscopy (Fig. 3c,d). The intensity ratio of D and G peaks ($I(D)/I(G)$) increased after rGO-Mg synthesis, indicating that the average domain size of sp^2 hybridized regions was decreased as GO was reduced[24]. The 2D peak, whose position and shape depends on the number of graphene layers, shifted to lower frequency (2,701 cm^{-1} to 2,685 cm^{-1}) and its full width at half maximum also decreased on the formation of rGO-Mg (Fig. 3d). This suggests that few, if any, isolated multilayers of rGO exist in the composite, and that most rGO layers are actively wrapping Mg nanocrystals[25,26] (additional analysis in Supplementary Fig. 12 and Supplementary Discussion). No change was observed in the Raman spectra of freshly synthesized rGO-Mg in comparison with samples studied after hydrogen cycling. Importantly, $I(D)/I(G)$ ratios remained consistent as well (1.370 after synthesis and 1.337 after cycling), indicating that the defect density, a key attribute of rGO responsible for selective hydrogen transport, was well maintained even after several hydrogen absorption and desorption cycles[25]. In addition, the chemical environment of GO and rGO-Mg were investigated via X-ray photoelectron spectroscopy (Fig. 4). Peaks associated with oxygen-containing functional groups in the GO diminished after the formation of rGO-Mg, confirming reduction of GO[27]. The rGO-Mg composite contained an additional peak at 282.5 eV, which is attributed to the interaction between carbon species and

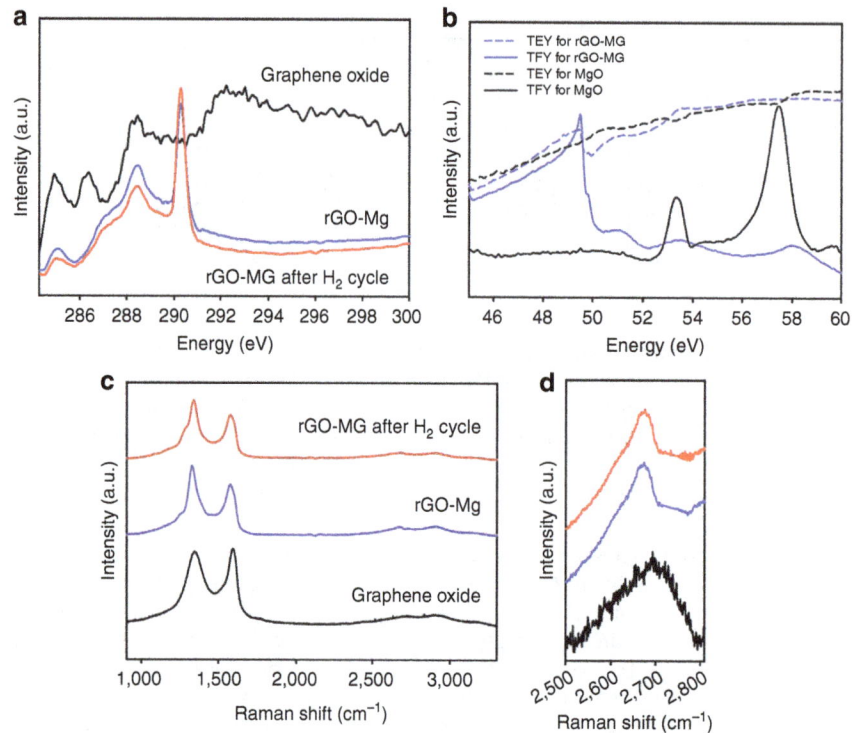

Figure 3 | XANES and Raman spectral analysis of GO and rGO-Mg multilaminates before and after hydrogen cycling. XANES spectra of (**a**) GO, rGO-Mg after synthesis and after cycling at carbon K-edge, and (**b**) rGO-Mg after synthesis and MgO at Mg L-edge. Raman spectra of (**c**) GO, rGO-Mg after synthesis and after H_2 cycling, and (**d**) the 2D peak region.

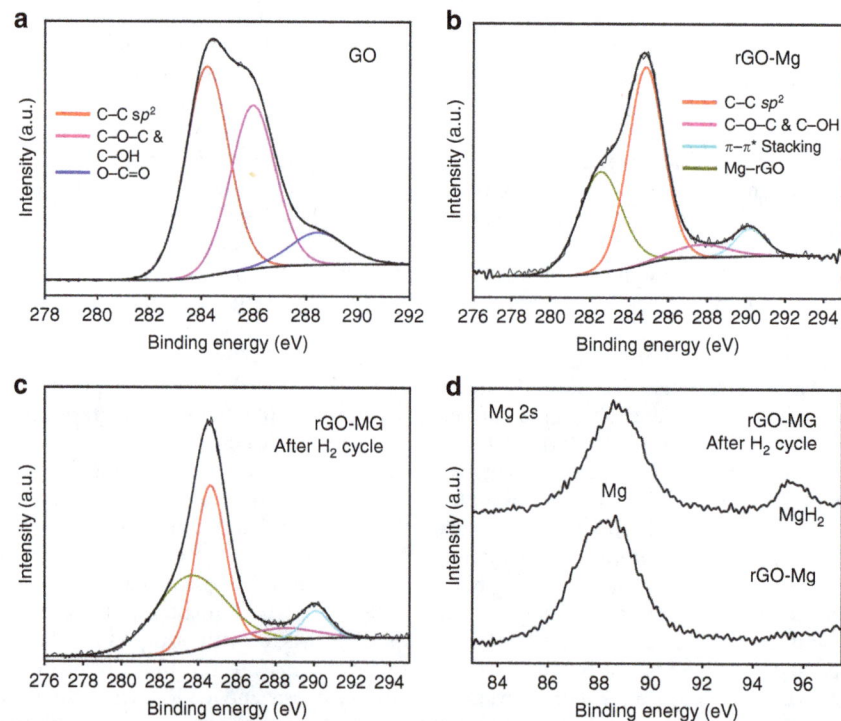

Figure 4 | X-ray photoelectron spectra (XPS) of rGO-Mg after synthesis and after hydrogen cycling. XPS spectra (C, 1s) of (**a**) GO, (**b**) rGO-Mg after synthesis and (**c**) rGO-Mg after H_2 cycling. (**d**) XPS pattern (Mg, 2s) for rGO-Mg after synthesis and after H_2 cycling.

metal particles[28,29], corresponding to the interaction of rGO and Mg nanocrystals. Furthermore, a prominent π–π^* stacking peak was observed at 290.1 eV, resulting from Mg nanocrystal wrapping, which was also observed by TEM (Supplementary

Fig. 1). In the Mg 2s spectrum, one additional peak appears in the higher energy region after hydrogen cycling, implying a new chemical state, consistent with MgH_2 (ref. 30). The Mg and MgH_2 contents were 91.45% and 8.55%, respectively, after completing cycle test based on Mg 2s spectrum, explaining a slight amount of the remaining hydride phase in the end of cycling.

Discussion

We have developed a facile method of preparing the densest possible loading of reactive nanocrystals safely into a composite material, a crucial step forward for enhancing the energy density of nanomaterials. As a result, our rGO-Mg multilaminates offer exceptional environmental stability and unsurpassed hydrogen storage capability, exceeding that offered by any other non-cryogenic reversible material. We believe that these results suggest the possibility of practical solid-state hydrogen storage and use in the near future. Furthermore, this work shows that atomically thin 2D materials can be used to simultaneously protect nanocrystals from ambient conditions, while also imparting new functionality. Such stable mixed dimensional laminates of zero-valent nanocrystalline metals can be extended to a variety of additional applications, including batteries, catalysis, encapsulants and energetic materials.

Methods

Synthesis of rGO-Mg nanocomposite. The composites of rGO-Mg were synthesized in an argon glove box. GO was ball milled for 10 min before use, to break it down to GO platelets so that it can effectively make a complex with bis(cyclopentadienyl) magnesium (Cp_2Mg). To prepare the lithium naphthalenide solution, naphthalene (2.40 g, 0.0187 mol) was dissolved in 120 ml of tetrahydrofuran (THF), followed by the immediate addition of Li metal (0.36 g, 0.0253 mol), leading to a dark green solution. GO (6.25 mg) was dispersed in 12.5 ml of THF under Ar, sealed in a container and sonicated for 1.5 h. A Cp_2Mg solution (2.31 g, 0.015 mol, in 22.5 ml of THF) was added to GO solution afterwards, stirring for 30 min. The resulting GO/Cp_2Mg solution was added to the lithium naphthalenide solution and magnetically stirred for 2 h. The product was centrifuged (10,000 r.p.m., 20 min) and washed with THF (10,000 r.p.m., 20 min) twice, followed by vacuum drying overnight.

Characterization and instrumentation. High-resolution TEM was performed using JEOL 2100-F Field-Emission Analytical Transmission Electron operated at 120 kV and equipped with Oxford INCA energy-dispersive electron X-ray spectrometer and Tridiem Gatan imaging Filter and spectrometer. The powder samples were dispersed on lacey carbon grids from THF solutions. Elemental analysis of the EELS and energy-dispersive spectra was performed using Digital Micrograph software (Gatan Inc.). XRD patterns were acquired with a Bruker AXS D8 Discover GADDS X-Ray Diffractometer, using Cu Kα radiation ($\lambda = 0.154$ nm). Hydrogen absorption/desorption measurement was performed, using a HyEnergy PCT Pro-2000 at 15/0 bar of H_2 at different temperatures. XANES spectroscopy was performed on Beamline 8.0.1.3 and 4.0.3 at the Advanced Light Source. The energy resolution at carbon K-edge and Mg L-edge was set to 0.1 eV and the experimental chamber had a base pressure of at most 1×10^{-8} torr. An highly ordered pyrolytic graphite (HOPG) reference sample was measured before and after all XANES experiments for energy calibration. The XANES spectra were recorded using total electron yield and total fluorescence yield detection modes. The Raman spectra of GO and rGO-Mg samples were collected using Horiba Jobin Yvon LabRAM ARAMIS automated scanning confocal Raman microscope with a 532-nm excitation source and X-ray photoelectron spectra were obtained via PHI 5400 X-ray Photoelectron Spectroscopy System with Al Kα. The Mg content in the composite was determined by inductively coupled plasma-optical emission spectroscopy at the Advanced Light Source Life Sciences Division and Environmental.

References

1. Tollefson, J. Fuel of the future? *Nature* **464,** 1262–1264 (2010).
2. Schlapbach, L. & Zuttel, A. Hydrogen-storage materials for mobile applications. *Nature* **414,** 353–358 (2001).
3. Shao, H. Y., Xin, G. B., Zheng, J., Li, X. G. & Akiba, E. Nanotechnology in Mg-based materials for hydrogen storage. *Nano Energy* **1,** 590–601 (2012).
4. Jeon, K. J. et al. Air-stable magnesium nanocomposites provide rapid and high-capacity hydrogen storage without using heavy-metal catalysts. *Nat. Mater.* **10,** 286–290 (2011).
5. Aguey-Zinsou, K. F. & Ares-Fernandez, J. R. Hydrogen in magnesium: new perspectives toward functional stores. *Energy Environ. Sci.* **3,** 526–543 (2010).
6. Yang, J., Sudik, A., Wolverton, C. & Siegel, D. J. High capacity hydrogen storage materials: attributes for automotive applications and techniques for materials discovery. *Chem. Soc. Rev.* **39,** 656–675 (2010).
7. Novoselov, K. S. et al. Electric field effect in atomically thin carbon films. *Science* **306,** 666–669 (2004).
8. Geim, A. K. Graphene status and prospects. *Science* **324,** 1530–1534 (2009).
9. Joshi, R. K. et al. Precise and ultrafast molecular sieving through graphene oxide membranes. *Science* **343,** 752–754 (2014).
10. Kim, H. W. et al. Selective gas transport through few-layered graphene and graphene oxide membranes. *Science* **342,** 91–95 (2013).
11. Li, H. et al. Ultrathin, molecular-sieving graphene oxide membranes for selective hydrogen separation. *Science* **342,** 95–98 (2013).
12. Adelhelm, P. & de Jongh, P. E. The impact of carbon materials on the hydrogen storage properties of light metal hydrides. *J. Mater. Chem.* **21,** 2417–2427 (2011).
13. Yao, X. D. et al. Metallic and carbon nanotube-catalyzed coupling of hydrogenation in magnesium. *J. Am. Chem. Soc.* **129,** 15650–15654 (2007).
14. Liu, X. F., Peaslee, D. & Majzoub, E. H. Tailoring the hydrogen storage properties of $Li_4BN_3H_{10}$ by confinement into highly ordered nanoporous carbon. *J. Mater. Chem. A* **1,** 3926–3931 (2013).
15. Jung, H., Yang, S. J., Kim, T., Kang, J. H. & Park, C. R. Ultrafast room-temperature reduction of graphene oxide to graphene with excellent dispersibility by lithium naphthalenide. *Carbon* **63,** 165–174 (2013).
16. Rieke, R. D. Preparation of organometallic compounds from highly reactive metal powders. *Science* **246,** 1260–1264 (1989).
17. Ruminski, A. M., Bardhan, R., Brand, A., Aloni, S. & Urban, J. J. Synergistic enhancement of hydrogen storage and air stability via Mg nanocrystal-polymer interfacial interactions. *Energy Environ. Sci.* **6,** 3267–3271 (2013).
18. Fernandez, J. F. & Sanchez, C. R. Rate determining step in the absorption and desorption of hydrogen by magnesium. *J. Alloy Compd.* **340,** 189–198 (2002).
19. Norberg, N. S., Arthur, T. S., Fredrick, S. J. & Prieto, A. L. Size-dependent hydrogen storage properties of Mg nanocrystals prepared from solution. *J. Am. Chem. Soc.* **133,** 10679–10681 (2011).
20. Denis, A., Sellier, E., Aymonier, C. & Bobet, J. L. Hydrogen sorption properties of magnesium particles decorated with metallic nanoparticles as catalyst. *J. Alloy Compd.* **476,** 152–159 (2009).
21. Du, A. J., Smith, S. C., Yao, X. D. & Lu, G. Q. Catalytic effects of subsurface carbon in the chemisorption of hydrogen on a Mg(0001) surface: an ab-initio study. *J. Phys. Chem. B* **110,** 1814–1819 (2006).
22. Yao, X. D. et al. Mg-based nanocomposites with high capacity and fast kinetics for hydrogen storage. *J. Phys. Chem. B* **110,** 11697–11703 (2006).
23. Wang, H. L. et al. An ultrafast nickel-iron battery from strongly coupled inorganic nanoparticle/nanocarbon hybrid materials. *Nat. Commun.* **3,** 917 (2012).
24. Ha, H. W., Kim, I. Y., Hwang, S. J. & Ruoff, R. S. One-pot synthesis of platinum nanoparticles embedded on reduced graphene oxide for oxygen reduction in methanol fuel cells. *Electrochem. Solid State Lett.* **14,** B70–B73 (2011).
25. Ferrari, A. C. & Basko, D. M. Raman spectroscopy as a versatile tool for studying the properties of graphene. *Nat. Nanotechnol.* **8,** 235–246 (2013).
26. Ferrari, A. C. et al. Raman spectrum of graphene and graphene layers. *Phys. Rev. Lett.* **97,** 187401 (2006).
27. Ganguly, A., Sharma, S., Papakonstantinou, P. & Hamilton, J. Probing the thermal deoxygenation of graphene oxide using high-resolution *in situ* X-ray-based spectroscopies. *J. Phys. Chem. C* **115,** 17009–17019 (2011).
28. Jia, L. S. et al. Visible-light-induced photocatalyst based on C-doped LaCoO3 synthesized by novel microorganism chelate method. *Catal. Commun.* **10,** 1230–1234 (2009).
29. Wang, H. Q., Wu, Z. B. & Liu, Y. A simple two-step template approach for preparing carbon-doped mesoporous TiO_2 hollow microspheres. *J. Phys. Chem. C* **113,** 13317–13324 (2009).
30. Friedrichs, O., Kolodziejczyk, L., Sanchez-Lopez, J. C., Lopez-Cartes, C. & Fernandez, A. Synthesis of nanocrystalline MgH_2 powder by gas-phase condensation and in situ hydridation: TEM, XPS and XRD study. *J. Alloy Compd.* **434,** 721–724 (2007).

Acknowledgements

Work at the Molecular Foundry and the Advanced Light Source was supported by the Office of Science, Office of Basic Energy Sciences, of the U.S. Department of Energy under Contract Number DE-AC02-05CH11231. We thank Yi-De Chuang for XANES experimental support. This material is based on work supported by the Department of Energy (DOE) through the Bay Area Photovoltaic Consortium (BAPVC) under Award Number DE-EE0004946 and also in part under the US-India Partnership to Advance Clean Energy-Research (PACE-R) for the Solar Energy Research Institute for India and the United States (SERIIUS), funded jointly by the U.S. Department of Energy (Office of Science, Office of Basic Energy Sciences, and Energy Efficiency and Renewable Energy, Solar Energy Technology Program, under Subcontract DE-AC36-08GO28308 to the National Renewable Energy Laboratory, Golden, Colorado) and the Government of India, through the Department of Science and Technology under Subcontract IUSSTF/JCERDC-SERIIUS/2012 dated 22 November 2012. We sincerely appreciate Jeong Yun Kim and Jayoung Kim for assisting graphic work.

Author contributions

J.J.U. and E.S.C. conceived and designed the experiments. E.S.C. synthesized and characterized the composite and performed the experiments. S.A. contributed towards TEM and EELS acquisition and analysis. Y.-S.L. and J.-H.G. performed XANES experiments. E.S.C., A.M.R. and J.J.U. analysed the data and wrote the manuscript. All authors discussed the results and commented on the manuscript.

Additional information

Competing financial interests: The authors declare no competing financial interests.

Elemental superdoping of graphene and carbon nanotubes

9

Yuan Liu[1,2], Yuting Shen[3], Litao Sun[3], Jincheng Li[4], Chang Liu[4], Wencai Ren[4], Feng Li[4], Libo Gao[1,2], Jie Chen[1,2], Fuchi Liu[1,2], Yuanyuan Sun[1,2], Nujiang Tang[1,2], Hui-Ming Cheng[4] & Youwei Du[1,2]

Doping of low-dimensional graphitic materials, including graphene, graphene quantum dots and single-wall carbon nanotubes with nitrogen, sulfur or boron can significantly change their properties. We report that simple fluorination followed by annealing in a dopant source can superdope low-dimensional graphitic materials with a high level of N, S or B. The superdoping results in the following doping levels: (i) for graphene, 29.82, 17.55 and 10.79 at% for N-, S- and B-doping, respectively; (ii) for graphene quantum dots, 36.38 at% for N-doping; and (iii) for single-wall carbon nanotubes, 7.79 and 10.66 at% for N- and S-doping, respectively. As an example, the N-superdoping of graphene can greatly increase the capacitive energy storage, increase the efficiency of the oxygen reduction reaction and induce ferromagnetism. Furthermore, by changing the degree of fluorination, the doping level can be tuned over a wide range, which is important for optimizing the performance of doped low-dimensional graphitic materials.

[1] Nanjing National Laboratory of Microstructures, Nanjing University, Nanjing 210093, China. [2] Collaborative Innovation Center of Advanced Microstructures, Nanjing University, Nanjing 210093, China. [3] FEI Nano-Pico Center, Key Laboratory of MEMS of Ministry of Education, Collaborative Innovation Center for Micro/Nano Fabrication, Device and System, Southeast University, Nanjing 210096, China. [4] Advanced Carbon Division, Shenyang National Laboratory for Materials Science, Institute of Metal Research, Chinese Academy of Sciences, 72 Wenhua Road, Shenyang, Liaoning Province 110016, China. Correspondence and requests for materials should be addressed to N.T. (email: tangnujiang@nju.edu.cn) or to L.S. (email: slt@seu.edu.cn) or to H.-M.C. (email: cheng@imr.ac.cn).

Low-dimensional graphitic materials (LDGMs) such as two-dimensional graphene sheets, one-dimensional single-wall carbon nanotubes (SWCNTs) and zero-dimensional graphene quantum dots (GQDs) show many properties that are quite different from those of bulk graphite and these allow them to be used in a wide range of applications[1–6]. It has been demonstrated that doping the graphitic lattice of LDGMs with heteroatoms such as N, S or B can change their physical and chemical properties, and significantly increase their number of applications[7–35].

Numerous studies have shown that a high doping level is often needed to obtain LDGMs with the desired properties[23–32]. To date, various methods have been developed for the doping of LDGMs with N, S or B. In general, there are two different doping processes: (i) doping during their synthesis and (ii) doping post synthesis. Among these, thermal annealing is a universal postsynthesis approach, whereby doping can be realized by annealing the LDGMs in the appropriate dopant sources. Compared with other methods, this has some distinct advantages such as its high doping efficiency (a higher doping level and a wider level of adjustment than by using other methods), its wide applicability to LDGMs of any dimensions (graphene, GQDs or SWCNTs) and forms (powders, thin films or single nanodevices), and its ability to recover the sp^2 network by thermal annealing[29] and so on.

As these heteroatoms are mainly doped by entering vacancies in the graphitic lattice, their doping level is often limited by the low concentration of vacancies in LDGMs[32]. For high-level doping, creating vacancies in the graphitic lattice is key and this has been achieved by oxidation and subsequent de-oxidation[8], NH_3 etching[31] or ion irradiation[36]. Even so, the maximum doping levels (defined as N/C, S/C and B/C × 100 at% for N-, S- and B-doping, respectively) reported by the thermal annealing of LDGMs are quite low[29–31]. For graphene, the maximum values for N, S and B doping are 14.68, 1.7 and 3.64 at%, respectively. For GQDs, the maximum N-doping level is 5.0 at%. For SWCNTs, the N-doping level of 5.0 at% has been realized via *in situ* doping[16]. In contrast, almost no N- or S-doping is possible in SWCNTs by postsynthesis doping. Therefore, achieving a high doping level in LDGMs remains a critical challenge[29–32].

Graphene doped with a high-level of N is highly desirable for high-performance supercapacitors, fuel cells, energy-storage devices, ferromagnetic materials[6,29–31] and so on. For example, the specific capacitance of pristine graphene can be increased from 69 to 280 F g^{-1} by doping with 2.51 at% N[37] and to 326 F g^{-1} by doping with 10.13 at% N[38]. Pristine graphene lacks catalytic activity in the oxygen reduction reaction (ORR)[39] and it is believed that N-doping can transform pristine graphene to an effective metal-free electrocatalyst for the ORR, and that a higher N level results in a higher ORR activity[40]. N-doping of graphene to 8.8 at% induces ferromagnetism with a high magnetization of 1.66 emu g^{-1} and a Curie temperature (T_C) of 100.2 K (ref. 26).

In this study, we show that by fluorination followed by thermal annealing, it is possible to dope all three types of LDGMs with very high levels of N, S or B. The maximum doping levels obtained for graphene are 29.82 at% for N-doping (NG), 17.55 at% for S-doping (SG) and 10.79 at% for B-doping (BG). These are respectively 2, 10 and 3 times higher than the maximum levels previously reported. For GQDs the maximum N doping level is 36.38 at% (N-GQDs), 7 times higher than the reported maximum level, and for SWCNTs the maximum N and S doping levels are respectively 7.79 (N-SWCNTs) and 10.66 at% (S-SWCNTs), which are significantly higher than the near-zero levels currently reported. These doping amounts can be easily and precisely tuned over a very wide range by changing the degree of fluorination. We therefore call this highly efficient

fluorination-assisted doping method 'superdoping'. In addition, we report some excellent properties of superdoped NG, such as its high capacitance (390 F g^{-1}), superior ORR properties (onset potential of − 0.05 eV and a current density of − 4.99 mA cm^{-2}) and near room-temperature (RT) ferromagnetism (a high magnetization of 2.3 emu g^{-1} and a high T_C of 250 K).

Results

Superdoping of LDGMs with N. We doped LDGMs with N by fluorination followed by annealing in ammonia. As N atoms mainly enter the graphite lattice through vacancies, creating vacancies in the lattice is necessary to achieve high levels of N-doping. It is known that carbon monofluoride is unstable at high temperatures[41], and that thermal defluorination of fluorinated LDGMs (F-LDGMs) by annealing consumes C atoms. Thus, it is reasonable to speculate that fluorination followed by thermal defluorination will (i) generate a high concentration of vacancies in the graphitic lattice, which will stimulate high-level N-doping and (ii) allow one to tune the vacancy concentration by changing the degree of fluorination, which may result in an ability to tune the N-doping level. In other words, one can dope LDGMs with a very high level of N by fluorination followed by thermal defluorination, and also can tune the N-doping level over a very wide range by changing the degree of fluorination. We explored this phenomenon using graphene, GQDs and SWCNTs (Supplementary Figs 1–3).

To study the structural transformation of LDGMs during fluorination and annealing in ammonia vapour, we carefully characterized pristine, fluorinated and N-doped LDGMs (N-LDGMs) by transmission electron microscopy (TEM), X-ray photoelectron spectroscopy (XPS) and Raman spectroscopy. It was found that, for the original materials, (i) the graphene consists of a few sheets of micrometre lateral size (Fig. 1a) and has a low oxygen content of 4.31 at% (Fig. 1d); (ii) GQDs are mainly distributed in the size range 1–7 nm and have a high crystallinity (Fig. 1b and Supplementary Fig. 1a,b) and an oxygen content of 5.24 at% (Fig. 1e); and (iii) the SWCNTs have diameters of 1–2 nm (Fig. 1c and Supplementary Fig. 2). For fluorination, the LDGMs were heated in XeF$_2$ (ref. 42). Based on the XPS results of the F-LDGMs (Fig. 1d–f), we can calculate the fluorination degrees (defined as F/C × 100 at%). Both fluorinated graphene (FG) and fluorinated GQDs (F-GQDs) have high fluorination degrees of 103.52 and 102.04 at%, respectively. In contrast, fluorinated SWCNTs (F-SWCNTs) have a relatively low fluorination degree of 34.24 at%.

We found that the optimum annealing temperature to obtain high-level doping of FG in ammonia is 500 °C (Supplementary Table 1) and this is therefore used as the annealing temperature for F-GQDs and F-SWCNTs. After annealing in ammonia, the disappearance of the F peak is accompanied by the appearance of a clear N peak at ∼400 eV in all three N-LDGMs, indicating complete defluorination and effective N-doping (Fig. 1d–f). This can also be confirmed by fine scans of the C 1s XPS spectra (Supplementary Fig. 4). Based on the XPS results (Fig. 1d–f), we calculated the N-doping levels to be 29.82, 36.38 and 7.79 at% for NG, N-GQDs and N-SWCNTs, respectively. The N level of N-GQDs is higher than that of NG, which may be attributed to the larger number of edge sites for N-doping in the GDQs. The N contents of NG and N-GQDs are two and seven times the corresponding maximum values reported[29–31]. In contrast, the N level of N-SWCNTs is relatively low, which may result from the relatively low fluorination degree obtained. However, one should note that the N-doping level obtained by the annealing of SWCNTs in ammonia is almost undetectable in the present studies. It is reasonable to assume that the N-doping

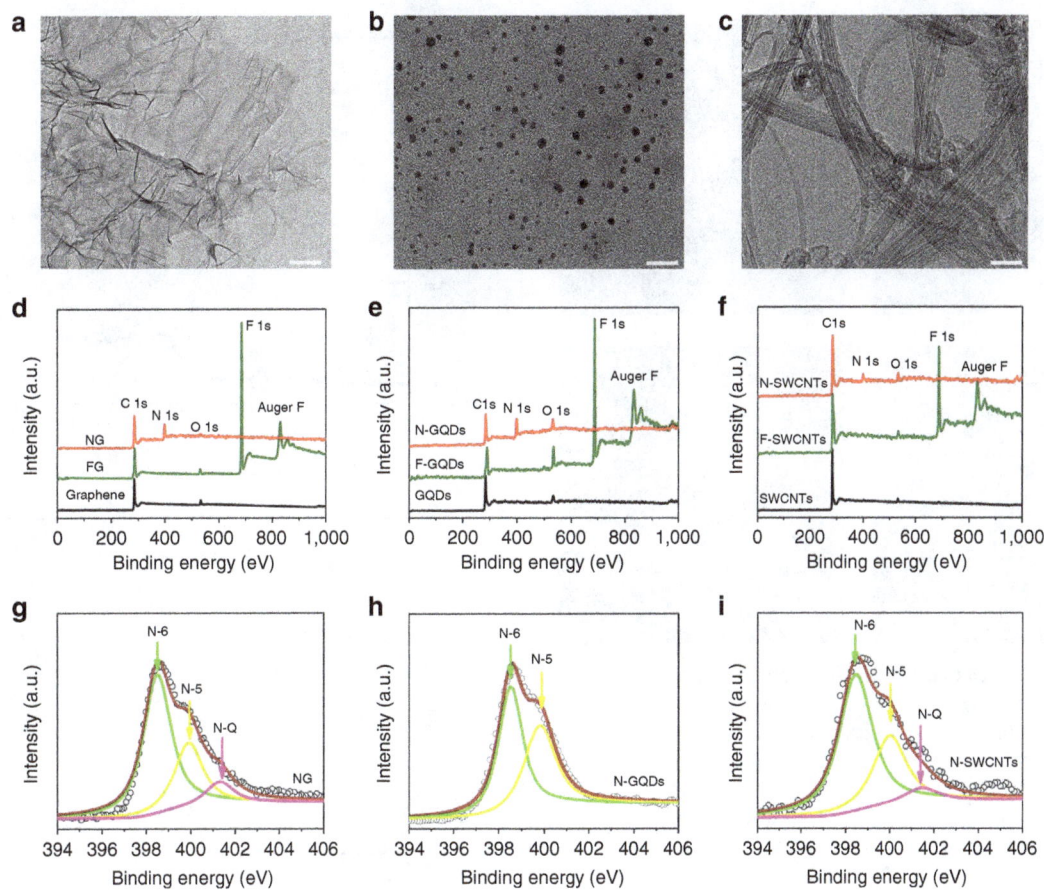

Figure 1 | Superdoping of LDGMs with N. TEM images of (**a**) graphene, (**b**) GQDs and (**c**) SWCNTs. Scale bars, 500 nm (**a**); 20 nm (**b**); 50 nm (**c**). XPS spectra over a wide range of binding energies of (**d**) graphene, FG and NG; (**e**) GQDs, F-GQDs and N-GQDs; and (**f**) SWCNTs, F-SWCNTs and N-SWCNTs. The fine-scanned N 1s XPS spectra of (**g**) NG, (**h**) N-GQDs and (**i**) N-SWCNTs. The dots are measured data and the solid lines are fitted curves.

level in SWCNTs can be further increased if one can increase the fluorination degree.

Fine scans of the N 1s XPS spectra of the three N-LDGMs were performed and deconvoluted into three sub-peaks located at ~398.3, 400.0 and 401.4 eV (Fig. 1g–i), which may be respectively attributed to pyridinic N (N-6), pyrrolic N (N-5) and graphitic-like N (N-Q) types[29–32]. It is found that N-6 and N-5 dominate the N-doping in all three N-LDGMs (Supplementary Table 2), similar to the reported results from N-LDGMs produced by annealing in ammonia. Based on the fact that both N-6 and N-5 are mainly doped in vacancies in the graphitic lattice[29–31], their high levels indicate a high concentration of vacancies in N-LDGMs. We directly annealed the pristine LDGMs in ammonia and found that the N levels are low, with values of 4.91, 10.18 and 0.67 at% for NG', N-GQDs' and N-SWCNTs' (Supplementary Fig. 5), respectively. This implies that the vacancy concentrations of the three original LDGMs are low, and that the fluorination/defluorination process generates a high concentration of vacancies, which facilitate N-doping.

Demonstration of uniformly distributed N atoms in NG sheets. A uniform distribution of N atoms in the graphite lattice is important to achieve the properties of N-doped LDGMs needed for new applications. We carefully investigated the distribution of N atoms in NG by energy-filtered TEM in an image aberration-corrected TEM. As shown in Fig. 2a–c, it is found that the N atoms are uniformly distributed in the NG sheet, indicating that uniformly distributed vacancies were generated after the thermal defluorination of FG. It is known that, as carbon monofluoride is unstable at a high temperature, thermal defluorination of FG can

consume C and F atoms instead of only F atoms, and thus introduce vacancies in the graphite lattice[41]. It is considered that a mono-vacancy can contribute three N-6 atoms (left of Fig. 2d) and a di-vacancy can introduce two N-6 atoms and one N-5 atom (right of Fig. 2d)[31]. Considering the fact that the levels of N-6 and N-5 are very high and the ratio of N-5/N-6 is 0.5 for NG, one can suggest that thermal defluorination of a fluorinated graphite lattice in ammonia generates a high-concentration di-vacancies, which significantly stimulate N-doping.

An *in situ* investigation of the simultaneous steps of vacancy generation by thermal defluorination and N-doping is difficult. We divided the process into two separate steps: the first is defluorination of FG in Ar, to obtain defluorinated graphene, and the second is N-doping of defluorinated graphene by annealing in ammonia, to obtain NG''. It is surprising that the N levels of NG'' are as high as 21.03, 13.23 and 6.61 at% for total N, N-6 and N-5 (Supplementary Fig. 6 and Supplementary Table 3), respectively. This indicates that thermal defluorination of FG in Ar generates a high concentration of vacancies and thus facilitates N-doping. However, the total N level of NG'' is only ~66.5% of the value of NG by fluorination followed by annealing, indicating that *in situ* N-doping during thermal defluorination is important in achieving the high N-doping level. The lower amount of N in NG'' may be attributed to the fact that *in situ* N-doping restrains the reconfiguration of defluorination-generated di-vacancies.

Superdoping of graphene and SWCNTs with S or B. There is no doubt that both S- and B-doping are less effective than N-doping of graphene[9,43]. To investigate the scalability of our superdoping method, we doped graphene with S (SG) and with B (BG) by

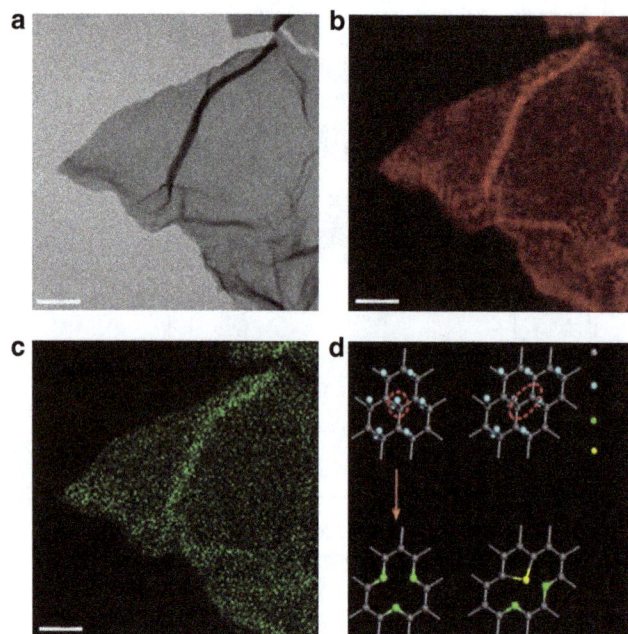

Figure 2 | Demonstration of uniformly distributed N atoms in NG sheets.
(**a**) Energy-filtered TEM (EFTEM) image (zero loss) of the NG sheet.
(**b,c**) EFTEM elemental mapping images obtained by using electron energy loss (EEL) edges of C (284 eV, depicted in red in **b**), and N (401 eV, depicted in green in **c**). Scale bars, 100 nm. (**d**) Schematic of defluorination and N-doping of a fluorinated graphite lattice. The red ring and elliptical ring in the fluorinated graphite lattice indicate that a mono- (i) and di-vacancy (ii) are generated after thermal defluorination. Three N-6 atoms (iii), and two N-6 and one N-5 atoms (iv) are generated by substituting three C atoms at the mono- and di-vacancy, respectively.

this method (see Methods for details of synthesis methods). For comparison we directly doped graphene with S (SG') by annealing it in sulfur vapour and with B (BG') by annealing it in BCl$_3$ vapour (see Methods for details of synthesis methods). It was found that, compared with direct annealing the combination of fluorination and annealing resulted in significant increases in the doping levels. These are from 4.66 to 17.55 at% for S and from 3.2 to 10.79 at% for B. It was also found that SG contains sulfur atoms that are doped into the graphite lattice, forming N-6- and N-5-like structures with neighbouring carbon atoms (Supplementary Fig. 7). We also doped SWCNTs with S by fluorination followed by annealing in sulfur vapour (S-SWCNTs) and by directly annealing in sulfur vapour (S-SWCNTs'), and again observed a drastic increase in the S level (from 3.49 to 10.66 at%, calculations based on Fig. 3c). It is clear that both the S- and B-doping levels of LDGMs by our method are much higher than the maximum values previously reported[17,19–23,29–32]. All these results demonstrate the wide suitability of this fluorination-followed-by-annealing method for the superdoping of LDGMs. In view of the fact that both S- and B-doping occur at vacant lattice sites in the graphite lattice of LDGMs, the high doping levels of both S and B further confirm that the thermal defluorination of F-LDGMs can generate a high-concentration of vacancies and facilitate doping.

Adjusting doping levels by varying the fluorination degree. Both the N- and S-doping levels of LDGMs are generally so low that it is virtually impossible to control the doping level. To synthesize FG, F-GQDs and F-SWCNTs with different fluorination degrees, we changed the mass ratio of LDGMs to XeF$_2$. From the XPS data on the F-LDGMs and N-LDGMs

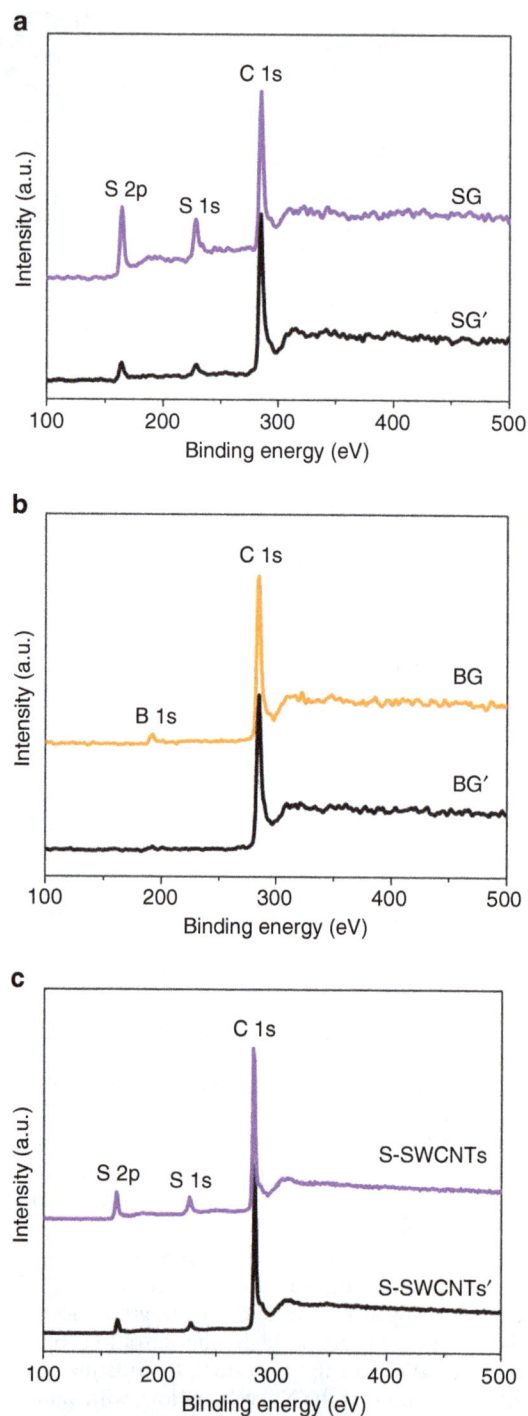

Figure 3 | Superdoping of graphene and SWCNTs with S or B. XPS spectra over a wide range of binding energies of (**a**) SG obtained by fluorination followed by annealing in sulfur vapour and SG' by annealing in sulfur vapour; (**b**) BG obtained by fluorination followed by annealing in BCl$_3$ vapour and BG' obtained by annealing in BCl$_3$ vapour; and (**c**) S-SWCNTs obtained by fluorination followed by annealing in sulfur and S-SWCNTs by annealing in sulfur vapour.

(Supplementary Fig. 8 and Supplementary Tables 4–6), we obtained the dependence of the N-doping level on the fluorination degree of the LDGMs. The results show that by varying the fluorination degree the corresponding N level can be adjusted over a wide range. For graphene this was from 4.91 to 29.82, whereas for GQDs it was from 10.18 to 36.38, and for SWCNTs from 0.67 to 7.79 at% (Fig. 4a). Similar experiments to obtain the dependence of the S-doping level on the fluorination degree of graphene (see Methods for details about synthesis methods) showed that it could be adjusted from 4.66 to 17.55 at% (Fig. 4b). For both N-LDGMs and SG, the doping level increases almost linearly with the increase of the fluorination degree. These results indicate that the doping levels of N, S or B can be precisely tuned over a very wide range by varying the fluorination degree of LDGMs, and further demonstrate that the thermal defluorination of F-LDGMs generates a high concentration of vacancies and facilitates superdoping.

Performance of N-superdoped graphene. As reported, N-, S- or B-doping can change the physical and chemical properties of LDGMs and thus greatly widen their potential applications[7-35]. The successful superdoping of LDGMs with N, S or B offers an effective platform to study the fundamental properties and explore the applications of superdoped graphene, GQDs and SWCNTs. Graphene doped with high levels of N-6 and/or N-5 is highly desirable for supercapacitors, fuel cells and energy-storage devices, and as a ferromagnetic material[26,29-31] and so on. As examples, we have investigated only supercapacitors, the ORR, magnetism and photoluminescence (PL).

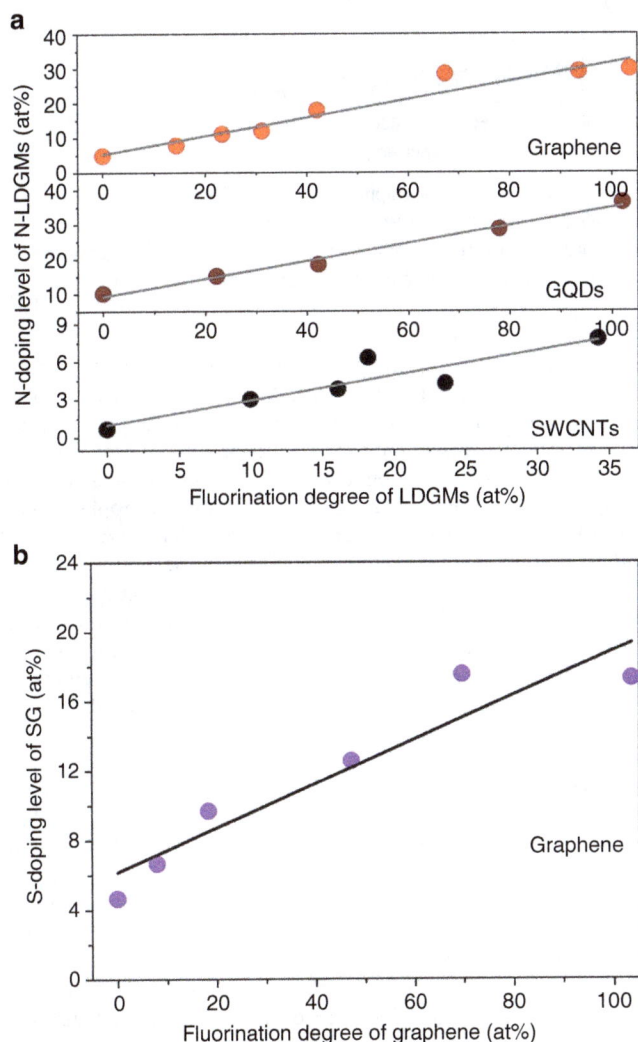

Figure 4 | Adjustability of the N- and S-doping levels by varying the fluorination degrees of LDGMs. (**a**) Dependence of the N-doping level on the fluorination degree of LDGMs. (**b**) Dependence of the S-doping level on the fluorination degree of graphene.

Graphene materials have recently shown better performance in supercapacitor devices than conventional carbon electrodes, and the specific capacity of pristine graphene can be increased four times by N-doping with N-5 and/or N-6 (ref. 37). The reason for this is considered to be that N-doping can increase the binding energy of electrolyte ions at the graphene surface, resulting in N-doped surfaces being able to accommodate more ions and thus have a higher charge-storage capacity. The performance of our graphene and NG supercapacitors was characterized through galvanostatic charge-discharge and cyclic voltammetry (CV) measurements. From the CV curves (Fig. 5a), the corresponding specific capacitance (C_s) was calculated of 353.8 F g^{-1}, which is 66% higher than that of graphene (213.2 F g^{-1}). The high C_s of NG can also be confirmed by the galvanostatic charge/discharge curves and the specific capacitances of the electrode at 5 and 10 A g^{-1} were calculated to be about 390 and 354 F g^{-1}, respectively (Supplementary Fig. 9a). Our NG exhibits the best performance among all graphene-based supercapacitors ever reported[29-31], demonstrating that a high N-doping level can greatly improve the specific capacitance of graphene. Furthermore, one finds that the CV curves are in the form of a triangle with little distortion, implying excellent capacitance behaviour and charge/discharge properties. CV curves of the NG electrode were also measured at different scan rates, ranging from 10 to 100 mV s^{-1} (Supplementary Fig. 9b). Near-rectangular CV curves without obvious redox peaks were observed, indicating the ideal capacitive behaviour of the supercapacitor. It is noteworthy that even at a high scan rate of 100 mV s^{-1}, the shape of the CV curve remained close to rectangular and the specific capacitance retains 67.75% of its initial specific capacitance measured at 5 mV s^{-1}. Moreover, it was found that the capacitance deterioration was less than 5.0% after 5,000 cycles (Supplementary Fig. 9c). All these results show that NG has a good rate capability and high capacity even under very fast charge transfer, suggesting its potential practical use in supercapacitors.

Pristine graphene lacks catalytic activity in the ORR and is not efficient in facilitating electron transfer[39]. However, as the doping of nitrogen in the graphite lattice can increase the electron density of states near the Fermi level, N-doping can improve the ORR activity of graphene, making it a promising metal-free electrocatalyst for the ORR[21]. It has been demonstrated that the high conductivity is favourable to the ORR activity of NG, and this can be achieved by annealing at 800 °C (refs 44,45). We annealed FG in ammonia at temperatures from 400 to 800 °C and also found that 800 °C is the optimum temperature for ORR performance (Supplementary Fig. 10). Figure 5b shows the linear sweep voltammetry curves obtained for graphene and NG-800 for the same mass deposited on a glassy-carbon rotating disk electrode in an O$_2$-saturated 0.1 M KOH aqueous solution at a scan rate of 5 mV s^{-1}. Graphene shows an ORR onset potential at −0.25 V and has a cathodic current density of *ca.* −3.03 mA cm^{-2}. After N-doping, the onset potential and the current density increase to −0.05 V and *ca.* −4.99 mA cm^{-2}, respectively, close to those for Pt/C (−0.02 V and *ca.* −5.38 mA cm^{-2}). It is clear that the onset potential of NG-800 is more positive than that of other N-doped carbon ORR catalysts reported[10,21,29-31,43,46,47], indicating a more efficient ORR process on NG.

We have demonstrated that N-doping can increase the magnetization of graphene and induce ferromagnetism with a T_C at 100.2 K (ref. 26). We carried out magnetic measurements on graphene and NG, and found that at 300 K only pure diamagnetism can be observed in both materials (Supplementary Fig. 11a). At 2 K, graphene has pure spin 1/2 paramagnetism with a low magnetization of 0.14 emu g^{-1} (Fig. 5c and Supplementary Fig. 11b,c). Interestingly, NG shows clear ferromagnetism with an

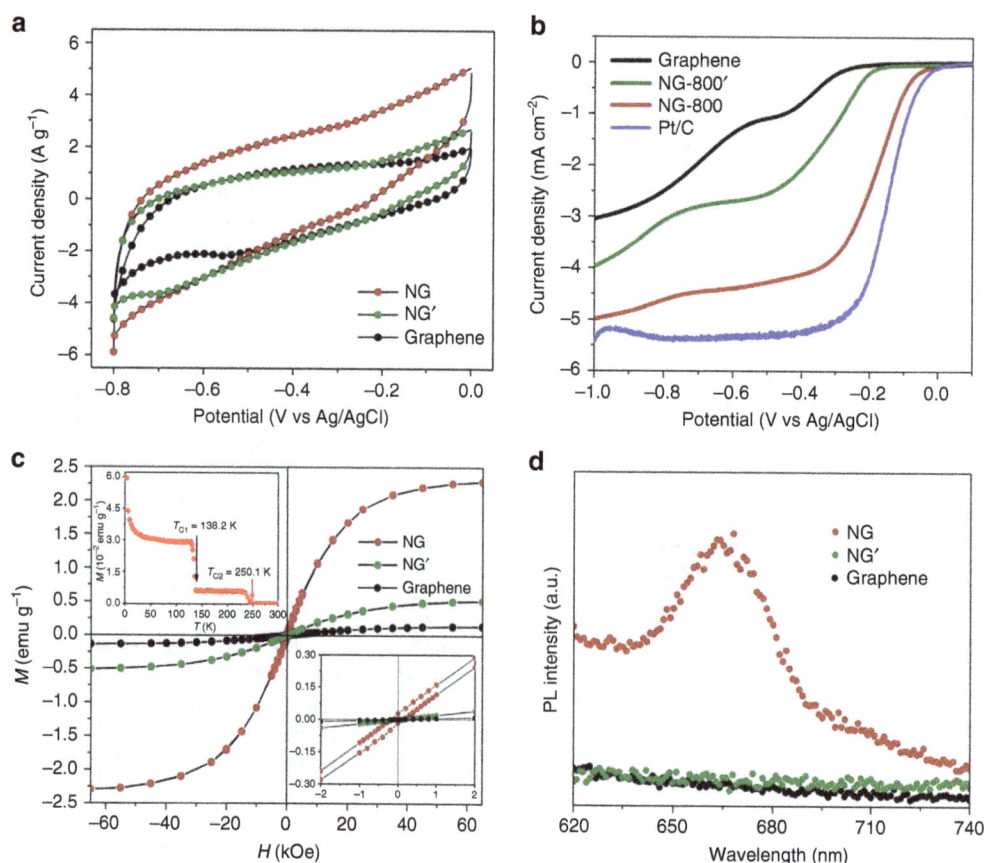

Figure 5 | Performance of superdoped graphene with a high level of N. (a) Representative CV curves of graphene, NG' and NG in 6 M KOH solution at a scan rate of 5 mV s^{-1}. **(b)** Linear sweep voltammetry (LSV) curves for graphene, NG-800', NG-800 and Pt/C on a rotating disk electrode (RDE; 1,600 r.p.m.) in 0.1 M KOH solution with a scan rate of 5 mV s^{-1}. **(c)** Typical M–H curves of graphene, NG' and NG measured at 2 K. Insets are the M–T curve of NG measured from 2 to 300 K under an applied field H = 500 Oe (top left of panel) and part of the magnetization curves (bottom right of panel). **(d)** PL spectra of graphene, NG' and NG excited at 500 nm.

obvious coercive field of 190 Oe and a remanent magnetization of 0.028 emu g^{-1} (bottom right inset of Fig. 5c). Moreover, the magnetization of NG is high, up to 2.3 emu g^{-1} (Supplementary Fig. 11d), which is the highest recorded intrinsic magnetization for graphene derivatives. We performed an M–T measurement and found two clear T_C's at 138.2 and 250.1 K (top left inset of Fig. 5c). This can also be confirmed by the M–H curves measured at 100, 200 and 300 K (Supplementary Fig. 11e and Supplementary Note 1) and the zero-field cooling M–T curve (Supplementary Fig. 11f). Apparently, high-level N-doping results in a significant increase in the magnetization of graphene and the generation of near RT ferromagnetism. In addition, creating a PL peak in visible region in graphene is important for various applications. It is very interesting to find an obvious PL at 677 nm (Fig. 5d), indicating the clear creation of a PL peak in visible region by high level N-doping.

We have shown four examples of the excellent performance of NG such as the highest capacitance, superior ORR properties close to those of Pt/C, near-RT ferromagnetism with the highest magnetization and the clear creation of a PL peak in visible region. These properties are far superior to those reported for graphene or any other graphene derivatives. For comparison, we have investigated supercapacitors, magnetism and the PL of NG', and ORR of NG-800' obtained by direct annealing pristine graphene in ammonia. The conventional N-doped graphene materials show much lower performance regarding capacitance (Fig. 5a), ORR (Fig. 5b), magnetism (Fig. 5c) and PL (Fig. 5d) than those with very high N contents produced by superdoping. Clearly, all these results demonstrate that a high level of N-doping

can significantly change the properties of graphene. Furthermore, the superdoping of graphene with N over an adjustable wide range will allow one to search for the optimum N-doping level for each property. Consequently, we examined the performance of NG with different N-doping levels and found that in capacitance (Supplementary Fig. 12a,b), magnetism (Supplementary Figs 13–16, Supplementary Table 7 and Supplementary Note 2) and PL (Supplementary Fig. 17), the higher the N-doping level, the better the performance. As a high annealing temperature is good for ORR activity, we prepared NG-800 samples with different N-doping levels by annealing FG samples with different fluorination degrees at 800 °C in ammonia (Supplementary Table 8). The results (Supplementary Fig. 18) showed that the higher the N-doping level, the better the performance.

Discussion

We have reported a fluorination-followed-by-annealing method for the efficient doping of graphene, GQDs and SWCNTs with very high levels of N, S or B. The excellent performance of NG with a very high N-doping level in supercapacity, the ORR, magnetism and PL have proved the importance of the high doping level of heteroatoms in LDGMs. Moreover, we have illustrated that the doping level of N or S can be precisely adjusted over a very wide range, which demonstrates the high efficiency of this doping method and allows one to search for the optimum doping level for the performance optimization of graphene, GQDs and SWCNTs in many applications. We believe that the method may also be applicable to the superdoping of LDGMs with other heteroatoms such as P, Si and so on, in which vacancies are necessary for the doping, and to superdope LDGMs in other forms, such as thin films and single nanodevices, with high levels of N, S or B.

Methods

Graphene. Graphene was synthesized by annealing of graphene oxide (GO) in Ar at 700 °C (ref. 26). Herein, GO was prepared by Hummer's method. In a typical experiment, the mixture of 8 g graphite, 8 g NaNO$_3$, 48 g KMnO$_4$ and 384 ml condensed H$_2$SO$_4$ was stirred for 1.5 h at 0 °C, and then followed by another 2 h stirring at 35 °C. Thereafter, 320 ml H$_2$O was added into the mixture within 15 min at a steady flow and the premixture of 800 ml H$_2$O and 40 ml H$_2$O$_2$ was added within 6 min at a steady flow. During the process, the temperature of the solution was fixed at 35 °C. The obtained solution was washed for 13 times with deionized (DI) water by repeated 30-min centrifugation at 13,000 r.p.m. Next, the sediment was re-dispersed into DI water and centrifuged at 6,000 r.p.m. for 10 min and only the supernatant was left to get GO sheets with high few-layer ratio. To obtain highly pure GO sample without contamination, GO was washed with hydrochloric acid for 15 times and then with DI water for 10 times by repeated 30-min centrifugation at 13,000 r.p.m. Thereafter, GO powder was obtained after freeze drying. Finally, graphene was obtained followed by Ar annealing at 700 °C.

Graphene quantum dots. GO quantum dots (GOQDs) were prepared from Vulcan XC-72 carbon black (Cabot Corporation, USA) by a modified Dong's method[48]. In a typical procedure, 1.6 g XC-72 carbon black was put into a 1,000-ml round bottom flask and 300 ml HNO$_3$ (15 mol l^{-1}) was added. Then, the mixture was stirred by magnetic stirrer and was heated in an oil bath at 135 °C for 24 h. After the mixture was heated in the oil bath at 180 °C for 10 h in Ar (60 ml min^{-1}) to evaporate the concentrated nitric acid, a light black solid was obtained. The light black solid was redissolved in 2 l DI water followed by centrifuging (13,000 r.p.m.) for 30 min to remove the unoxidized carbon black or big particles and a supernatant was obtained. Subsequently, the obtained supernatant was diluted to 3.6 l solution with DI water and then the obtained solution was vacuum filtered through 220-nm microporous membrane to further remove the unoxidized carbon black or big particles. Next, the obtained solution was further filtered through 25-nm microporous membrane to remove big particles. Thereafter, the reddish-brown GOQDs solution was obtained. After rotary evaporation at 80 °C to remove the water of the solution, 150 ml concentrated GOQDs solution was obtained. The concentrated solution was dried by vacuum freeze drying and ∼1.1 g brown GOQDs powder was obtained. After annealing GOQDs powder in Ar at 700 °C for 1 h, GQDs were obtained.

Single-wall carbon nanotubes. SWCNTs (XFS02, purity: >90%; outer diameter: <2 nm, length: 5–30 µm) were purchased from Nanjing XFNANO Materials Tech Co., Ltd (Nanjing, China).

Fluorination of graphene. The FG samples (FG-1, FG-2, FG-3, FG-4, FG-5, FG-6 and FG) were obtained by annealing the mixture of graphene and XeF$_2$ with different mass ratios in a Teflon container at 200 °C for 30 h in Ar.

Fluorination of GQDs. The F-GQDs samples (F-GQDs-1, F-GQDs-2 and F-GQDs-3) were obtained by annealing the mixture of GQDs and XeF$_2$ (Alfa Aesar, China) with different mass ratios in a Teflon container at 200 °C for 30 h in Ar[42].

Fluorination of SWCNTs. F-SWCNT-1, F-SWCNT-2 and F-SWCNT-3 were obtained by annealing the mixture of SWCNT and XeF$_2$ with different mass ratios in a Teflon container at 200 °C for 30 h in Ar. F-SWCNT-4 and F-SWCNT were obtained by annealing the mixture of F-SWCNT-3 and XeF$_2$ for 30 h in Teflon container in Ar at 220 °C and 240 °C, respectively.

N-doping of FG and graphene. The NG samples (NG-1, NG-2, NG-3, NG-4, NG-5, NG-6 and NG) or NG' were prepared by annealing the FG samples (FG-1, FG-2, FG-3, FG-4, FG-5, FG-6 and FG) or pristine graphene in ammonia at 500 °C for 1 h under ambient pressure. Herein, the numeric number is the serial number of the FG samples with different fluorination degree. Using the serial number of the FG samples, which correspond to NG samples (NG-1, NG-2, NG-3, NG-4, NG-5 and NG-6) with different N-doping level, are capable of being put into a one-to-one relationship.

The NG samples (NG-400, NG-500 (viz. NG), NG-600, NG-700 and NG-800, numeric numbers denote the annealing temperature) were prepared by annealing FG in ammonia at different temperatures for 1 h under ambient pressure. In short, the FG samples were spread on a quartz boat that was placed in a tubular furnace (0.4 m in length). The quartz tube reactor (1.0 m in length) in a tubular furnace (0.4 m in length). The quartz tube reactor is movable with protective atmosphere and thus the quartz boat can be introduced into and pulled out from the tubular furnace by simply moving the quartz tube reactor. Next, ammonia (99.99%) at a rate of 100 s.c.c.m. was

maintained for 30 min, to get rid of air. After heating the furnace from RT to 500 °C, the quartz boat was introduced into the furnace, while ammonia was kept at a rate of 20 s.c.c.m. After 1 h, the tube was rapidly moved out and the NG samples were obtained.

The NG-800 samples (NG-800-1, NG-800-2, NG-800-3 and NG-800) or NG-800' were prepared by annealing the FG samples (FG-1, FG-3, FG-5 and FG) or graphene in ammonia at 800 °C for 1 h under ambient pressure.

NG" was prepared by two steps. First, FG was annealed in Ar at 500 °C for 0.5 h, after that, it was annealed in ammonia at 500 °C for 1 h. In short, Ar (99.99%) at a rate of 100 s.c.c.m. was maintained for 20 min, to get rid of air; after annealing the furnace from RT to 500 °C, FG were introduced into the reaction tube with Ar rate of 20 s.c.c.m. for 0.5 h at atmospheric pressure. Second, the ammonia was introduced into the reaction tube with the rate of 20 s.c.c.m. for 1 h. After the furnace cooled to RT, NG" was obtained.

N-doping of F-GQDs and GQDs. The N-GQDs samples (N-GQDs-1, N-GQDs-2, N-GQDs-3 and N-GQDs) or N-GQDs' were prepared by annealing the F-GQDs samples (F-GQDs-1, F-GQDs-2, F-GQDs-3 and F-GQDs) or GQDs in ammonia at 500 °C for 1 h under ambient pressure. In short, the F-GQDs samples or GQDs were was spread on a quartz boat, which was placed in a quartz reaction tube in a chemical vapour deposition furnace, and then ammonia (99.99%) at a rate of 100 s.c.c.m. was maintained for 30 min, to get rid of air. After heating the furnace from RT to 500 °C, the quartz boat was introduced into the furnace, while ammonia was kept at a rate of 20 s.c.c.m. After 1 h, the tube was rapidly moved out and the N-GQDs samples were obtained.

N-doping of F-SWCNTs and SWCNTs. The N-SWCNTs samples (N-SWCNTs-1, N-SWCNTs-2, N-SWCNTs-3, N-SWCNTs-4 and N-SWCNTs) or N-SWCNTs' were prepared by annealing the F-SWCNTs samples (F-SWCNTs-1, F-SWCNTs-2, F-SWCNTs-3, F-SWCNTs-4 and F-SWCNTs) or SWCNTs in ammonia at 500 °C for 1 h under ambient pressure.

S-doping of F-LDGMs and LDGMs. S-doped LDGMs (S-LDGMs) were prepared by annealing F-LDGMs or LDGMs in sulfur gas carried by Ar. The sulfur powder (Mat-cn, China) was placed in a quartz boat that was placed in a quartz reaction tube in a tubular furnace at Ar inlet and F-LDGMs or LDGMs were put in another quartz boat at the outlet. Before heating, the system was flushed by Ar at a rate of 100 s.c.c.m. for 30 min. Then, under a constant Ar flow rate of (10 s.c.c.m.), the furnace was heated to 400 °C and held at this temperature for 10 min. After that, the furnace was cooled to RT and S-LDGMs were obtained.

B-doping of FG and graphene. BG and BG' were respectively prepared by annealing FG and graphene in a gas mixture of BCl$_3$ (∼99.99%) and Ar (1:4 v/v) with a total flow rate of 100 ml min^{-1} at 800 °C for 1 h under ambient pressure[9]. FG or graphene was spread on a quartz boat that was placed in a quartz reaction tube in a chemical vapour deposition furnace and then Ar at a flow rate of 100 s.c.c.m. was maintained for 30 min, to get rid of the air. After heating the furnace from RT to 800 °C, the quartz boat was introduced into the furnace, while the gas mixture flow of BCl$_3$ and Ar was maintained at 100 s.c.c.m. One hour later, the tube was rapidly moved out from the furnace in Ar and BG or BG' were obtained.

For details about the masses of pristine, fluorinated and N-LDGMs, S-LDGMs or B-doped LDGMs used in the synthesis, see Supplementary Tables 9–15.

Microstructure characterization. The structure was investigated by a TEM (model JEM–2100, Japan). The sizes of GQDs were calculated by Adobe Photoshop software (version CS6). XPS measurements were performed on a PHI5000 VersaProbe (ULVAC-PHI, Japan). XPS was performed using 200 W mono-chromated Al Kα radiation. A 500-µm X-ray spot was used for XPS analysis. Typically, the hydrocarbon C 1s line at 284.8 eV from adventitious carbon was used as an energy reference. The XPS peak fitting programme XPSPEAK 4.1 was used for the spectra processing. Raman spectra were obtained by an Raman system (Renishaw, England) using a 514.5-nm laser as the light source. Elemental mapping images were recorded in an image aberration-corrected TEM (FEI TITAN 80-300 operating at 80 kV) using the energy-filtered TEM mode.

Electrochemical measurements. The electrochemical properties of the as-obtained products were investigated using a three-electrode cell configuration at RT. The working electrodes were fabricated by mixing the prepared powders with 15 wt% acetylene black and 15 wt% polytetrafluorene-ethylene binder. A small amount of N-methyl-2-pyrrolidone was added to the mixture, to produce a homogeneous paste. Next, the mixture was pressed onto nickel foam current collectors (1.0 cm × 1.0 cm) to make electrodes. Platinum foil and Ag/AgCl were used as the counter and reference electrodes, respectively. Before the electro-chemical test, the prepared electrode was soaked overnight in a 6-M KOH solution. Electrochemical characterization was carried out in a conventional three-electrode cell with a 6-M KOH aqueous solution as the electrolyte. CV and galvanostatic charge–discharge measurements were conducted on an electrochemical workstation (CHI 660D, CH Instrument, USA). From the CV curves, the corresponding specific capacitance (C_s) can be calculated based on the following equation

$$C_s = \frac{\int i dv}{2 \times m \times \Delta V \times S} \quad (1)$$

where $\int i dv$ is the integrated area under the CV curve, m is the mass of the electrode active material in grams, ΔV is the scanned potential window in volts and S is the scan rate in volts per second. From the galvanostatic charge–discharge curve, the specific capacitance can be calculated from $C_s = I \Delta t / \Delta V m$, where I is the discharge current, Δt is the total discharge time, m is the mass of active material and ΔV is the potential difference in the discharge process[49].

ORR measurements. ORR measurements were performed on a computer-controlled potentiostat (CHI 760E, CH Instrument, USA) with a three-electrode cell equipped with gas flow system. A Pt wire and a Ag/AgCl electrode filled with saturated KCl aqueous solution were used as the counter electrode and reference electrode, respectively. A 0.1-M KOH solution was prepared as the electrolyte and saturated with oxygen by bubbling oxygen gas through it for 30 min before measuring ORR activities. To prepare the NG-loaded working electrode, NG was dispersed in a mixture of water and isopropanol (4:1 v/v) containing 0.05 wt.% Nafion. The electrocatalytic activities of NG for the ORR were measured using a rotating disk electrode (Pine Instruments, MSR analytical rotator, USA) with a scan rate of 5 mV s^{-1}. NG dispersion (10 µl of 2 mg ml^{-1}) was transferred onto the GC electrode (5 mm diameter, 0.196 cm^2 geometric area).

Magnetic measurements. The magnetic properties of the powdered samples were measured using a superconducting quantum interference device magnetometer with a sensitivity better than 10^{-8} emu (Quantum Design MPMS–XL, USA) and all data were corrected for the diamagnetic contribution by subtracting the corresponding linear diamagnetic background measured at 300 K. The concentrations of magnetic impurity elements (such as Fe, Co, Ni or Mn) of all the samples are below 40 p.p.m. (Supplementary Table 16) measured by ICP spectrometry (Jarrell–Ash, USA). The magnetization at 2 K is composed of two parts: paramagnetization (PM) and ferromagnetization, which can be expressed as $M_{total} = M_{para} + M_{ferro}$. Considering the fact that ferromagnetic mass magnetization can saturate at a high applied field approximately, one can fit the PM at a high applied field. The PM is fitted to the Brillouin function

$$M = M_s \left[\frac{2J+1}{2J} \text{Coth} \left(\frac{2J+1}{2J} x \right) - \frac{1}{2J} \text{Coth} \left(\frac{x}{2J} \right) \right] \quad (2)$$

where $x = gJ\mu_B H/(k_B T)$, $M_s = NgJ\mu_B$, k_B the Boltzmann constant, N the number of magnetic moments present, J the angular momentum number and g is the Landau factor. By subtracting the paramagnetic signal from the observed data, one can obtain the remaining ferromagnetic moment. From the saturated paramagnetic magnetization added to the saturated ferromagnetic magnetization, the M_s of the sample can be calculated.

PL measurements. The PL spectra were measured at ambient conditions by a spectrofluorophotometer (Shimadzu RF-5301PC, Japan) using a Xe lamp as the light source. For PL spectra investigation, ~ 1 mg of powdered sample was ultrasonically dispersed in 20 ml ethanol for 1 h. After that, the solution was used.

References

1. Kim, K. S. et al. Large-scale pattern growth of graphene films for stretchable transparent electrodes. Nature 457, 706–710 (2009).
2. Novoselov, K. S. et al. Two-dimensional gas of massless Dirac fermions in graphene. Nature 438, 197–200 (2005).
3. Wang, L. et al. Gram-scale synthesis of single-crystalline graphene quantum dots with superior optical properties. Nat. Commun. 5, 5357 (2014).
4. Lu, J., Yeo, P. S. E., Gan, C. K., Wu, P. & Loh, K. P. Transforming C-60 molecules into graphene quantum dots. Nat. Nanotechnol. 6, 247–252 (2011).
5. Cao, Q. et al. Medium-scale carbon nanotube thin-film integrated circuits on flexible plastic substrates. Nature 454, 495–500 (2008).
6. Zhang, M. et al. Strong, transparent, multifunctional, carbon nanotube sheets. Science 309, 1215–1219 (2005).
7. Li, M. et al. Synthesis and upconversion luminescence of N-doped graphene quantum dots. Appl. Phys. Lett. 101, 103107 (2012).
8. Li, Y. et al. Nitrogen-doped graphene quantum dots with oxygen-rich functional groups. J. Am. Chem. Soc. 134, 15–18 (2012).
9. Wu, Z., Ren, W., Xu, L., Li, F. & Cheng, H. Doped graphene sheets as anode materials with superhigh rate and large capacity for lithium ion batteries. ACS Nano 5, 5463–5471 (2011).
10. Li, Q. Q., Zhang, S., Dai, L. M. & Li, L. S. Nitrogen-doped colloidal graphene quantum dots and their size-dependent electrocatalytic activity for the oxygen reduction reaction. J. Am. Chem. Soc. 134, 18932–18935 (2012).
11. Yang, L. J. et al. Boron-doped carbon nanotubes as metal-free electrocatalysts for the oxygen reduction reaction. Angew. Chem. Int. Ed. 50, 7132–7135 (2011).
12. Gong, K. P., Du, F., Xia, Z. H., Durstock, M. & Dai, L. M. Nitrogen-doped carbon nanotube arrays with high electrocatalytic activity for oxygen reduction. Science 323, 760–764 (2009).
13. Stephan, O. et al. Doping graphitic and carbon nanotube structures with boron and nitrogen. Science 266, 1683–1685 (1994).
14. Wang, X. R. et al. N-doping of graphene through electrothermal reactions with ammonia. Science 324, 768–771 (2009).
15. Lee, J. M. et al. Selective electron- or hole-transport enhancement in bulk-heterojunction organic solar cells with N- or B-doped carbon nanotubes. Adv. Mater. 23, 629–633 (2011).
16. Yu, D. S., Zhang, Q. & Dai, L. M. Highly efficient metal-free growth of nitrogen-doped single-walled carbon nanotubes on plasma-etched substrates for oxygen reduction. J. Am. Chem. Soc. 132, 15127–15129 (2010).
17. Sheng, Z. H., Gao, H. L., Bao, W. J., Wang, F. B. & Xia, X. H. Synthesis of boron doped graphene for oxygen reduction reaction in fuel cells. J. Mater. Chem. 22, 390–395 (2012).
18. Park, S. et al. Chemical structures of hydrazine-treated graphene oxide and generation of aromatic nitrogen doping. Nat. Commun. 3, 638 (2012).
19. Yang, Z. et al. Sulfur-doped graphene as an efficient metal-free cathode catalyst for oxygen reduction. ACS Nano 6, 205–211 (2012).
20. Ma, Z. L. et al. Sulfur-doped graphene derived from cycled lithium-sulfur batteries as a metal-free electrocatalyst for the oxygen reduction reaction. Angew. Chem. Int. Ed. 54, 1888–1892 (2015).
21. Liang, J., Jiao, Y., Jaroniec, M. & Qiao, S. Z. Sulfur and nitrogen dual-doped mesoporous graphene electrocatalyst for oxygen reduction with synergistically enhanced performance. Angew. Chem. Int. Ed. 51, 11496–11500 (2012).
22. Wang, S. Y. et al. BCN graphene as efficient metal-free electrocatalyst for the oxygen reduction reaction. Angew. Chem. Int. Ed. 51, 4209–4212 (2012).
23. Panchokarla, L. S. et al. Synthesis, structure, and properties of boron- and nitrogen-doped graphene. Adv. Mater. 21, 4726–4730 (2009).
24. Nevidomskyy, A. H., Csanyi, G. & Payne, M. C. Chemically active substitutional nitrogen impurity in carbon nanotubes. Phys. Rev. Lett. 91, 105502 (2003).
25. Zhao, L. Y. et al. Visualizing individual nitrogen dopants in monolayer graphene. Science 333, 999–1003 (2011).
26. Liu, Y. et al. Realization of ferromagnetic graphene oxide with high magnetization by doping graphene oxide with nitrogen. Sci. Rep. 3, 2566 (2013).
27. Tang, Y. B. et al. Tunable band gaps and p-type transport properties of boron-doped graphenes by controllable ion doping using reactive microwave plasma. ACS Nano 6, 1970–1978 (2012).
28. Terrones, H., Lv, R. T., Terrones, M. & Dresselhaus, M. S. The role of defects and doping in 2D graphene sheets and 1D nanoribbons. Rep. Prog. Phys. 75, 062501 (2012).
29. Wang, X. W. et al. Heteroatom-doped graphene materials: syntheses, properties and applications. Chem. Soc. Rev. 43, 7067–7098 (2014).
30. Rao, C. N. R., Gopalakrishnan, K. & Govindaraj, A. Synthesis, properties and applications of graphene doped with boron, nitrogen and other elements. Nano Today 9, 324–343 (2014).
31. Maiti, U. N. et al. 25th anniversary article: chemically modified/doped carbon nanotubes & graphene for optimized nanostructures & nanodevices. Adv. Mater. 26, 40–67 (2014).
32. Zhang, Y. X., Zhang, J. & Su, D. S. Substitutional doping of carbon nanotubes with heteroatoms and their chemical applications. Chemsuschem 7, 1240–1250 (2014).
33. Dong, S., Chen, X., Zhang, X. & Cui, G. Nanostructured transition metal nitrides for energy storage and fuel cells. Coord. Chem. Rev. 257, 1946–1956 (2013).
34. Wang, H. et al. Nitrogen-doped graphene nanosheets with excellent lithium storage properties. J. Mater. Chem. 21, 5430–5434 (2011).
35. Li, X. L. et al. Simultaneous nitrogen doping and reduction of graphene oxide. J. Am. Chem. Soc. 131, 15939–15944 (2009).
36. Guo, B. D. et al. Controllable N-doping of graphene. Nano Lett. 10, 4975–4980 (2010).
37. Jeong, H. M. et al. Nitrogen-doped graphene for high-performance ultracapacitors and the importance of nitrogen-doped sites at basal planes. Nano Lett. 11, 2472–2477 (2011).
38. Sun, L. et al. Nitrogen-doped graphene with high nitrogen level via a one-step hydrothermal reaction of graphene oxide with urea for superior capacitive energy storage. RSC Adv. 2, 4498–4506 (2012).
39. Zheng, Y., Jiao, Y., Ge, L., Jaroniec, M. & Qiao, S. Z. Two-step boron and nitrogen doping in graphene for enhanced synergistic catalysis. Angew. Chem. Int. Ed. 52, 3110–3116 (2013).
40. Wang, D.-W. & Su, D. Heterogeneous nanocarbon materials for oxygen reduction reaction. Energy Environ. Sci. 7, 576–591 (2014).
41. Kuriakose, A. K. & Margrave, J. L. Mass spectrometric studies of the thermal decomposition of poly(carbon monofluoride). Inorg. Chem. 4, 1639–1641 (1965).
42. Feng, Q. et al. Obtaining high localized spin magnetic moments by fluorination of reduced graphene oxide. ACS Nano 7, 6729–6734 (2013).

43. Yang, S. B. *et al.* Efficient synthesis of heteroatom (N or S)-doped graphene based on ultrathin graphene oxide-porous silica sheets for oxygen reduction reactions. *Adv. Funct. Mater.* **22,** 3634–3640 (2012).

44. Lai, L. F. *et al.* Exploration of the active center structure of nitrogen-doped graphene-based catalysts for oxygen reduction reaction. *Energy Environ. Sci.* **5,** 7936–7942 (2012).

45. Dai, L. M., Xue, Y. H., Qu, L. T., Choi, H. J. & Baek, J. B. Metal-free catalysts for oxygen reduction reaction. *Chem. Rev.* **115,** 4823–4892 (2015).

46. Wang, H. B., Maiyalagan, T. & Wang, X. Review on recent progress in nitrogen-doped graphene: synthesis, characterization, and its potential applications. *ACS Catal.* **2,** 781–794 (2012).

47. Qu, L. T., Liu, Y., Baek, J. B. & Dai, L. M. Nitrogen-doped graphene as efficient metal-free electrocatalyst for oxygen reduction in fuel cells. *ACS Nano* **4,** 1321–1326 (2010).

48. Dong, Y. Q. *et al.* One-step and high yield simultaneous preparation of single- and multi-layer graphene quantum dots from CX-72 carbon black. *J. Mater. Chem.* **22,** 8764–8766 (2012).

49. Wu, Z. S. *et al.* Three-dimensional nitrogen and boron co-doped graphene for high-performance all-solid-state supercapacitors. *Adv. Mater.* **24,** 5130–5135 (2012).

Acknowledgements

This work was supported by the State Key Program for Basic Research (Grant Nos. 2014CB921102 and 2012CB932304) of China, National Natural Science Foundation of China (Grant Nos. U1232210, 51420105003, 51221264, 51325205, 51290273, 51521091, 51172240 and 11504142) and Chinese Academy of Sciences (No. KGZD-EW-T06).

Author contributions

N.J.T. and H.M.C. conceived the project and designed the experiments with Y.L. Y.L. contributed to sample fabrication. F.C.L. and Y.Y.S. synthesized GQDs. Y.L., Y.T.S., L.T.S. and N.J.T. performed the preliminary TEM characterization of the samples. Y.L. carried out PL and magnetic measurements. Y.L., J.C.L., C.L. and N.J.T. designed and carried out the electrochemical and electrocatalysts measurement. N.J.T. and Y.L. analyzed the data. N.J.T. and Y.L. wrote the paper. All authors discussed the results and commented on the manuscript.

Additional information

10

Flow invariant droplet formation for stable parallel microreactors

Carson T. Riche[1], Emily J. Roberts[2], Malancha Gupta[1,2], Richard L. Brutchey[2] & Noah Malmstadt[1,2]

The translation of batch chemistries onto continuous flow platforms requires addressing the issues of consistent fluidic behaviour, channel fouling and high-throughput processing. Droplet microfluidic technologies reduce channel fouling and provide an improved level of control over heat and mass transfer to control reaction kinetics. However, in conventional geometries, the droplet size is sensitive to changes in flow rates. Here we report a three-dimensional droplet generating device that exhibits flow invariant behaviour and is robust to fluctuations in flow rate. In addition, the droplet generator is capable of producing droplet volumes spanning four orders of magnitude. We apply this device in a parallel network to synthesize platinum nanoparticles using an ionic liquid solvent, demonstrate reproducible synthesis after recycling the ionic liquid, and double the reaction yield compared with an analogous batch synthesis.

[1] Mork Family Department of Chemical Engineering and Materials Science, University of Southern California, 925 Bloom Walk, HED 216, Los Angeles, California 90089, USA. [2] Department of Chemistry, University of Southern California, Los Angeles, California 90089, USA. Correspondence and requests for materials should be addressed to R.L.B. (email: brutchey@usc.edu) or to N.M. (email: malmstad@usc.edu).

Continuous flow microfluidic reactors are powerful tools for synthesizing chemicals and materials[1–6]. Microreactors allow for efficient heat transfer, excellent control of local mixing conditions and provide an ideal format for studying reaction kinetics on a small scale[7]. Microfluidic systems also offer a clear and appealing route to scale-up via massively parallel operation. Scaling by parallelization has clear advantages over traditional scale-up approaches. In contrast to a scale-up approach that relies on increasing the size of a single batch reactor, scale-up by parallelization does not change the local reaction conditions in terms of mixing uniformity and temperature distribution, which are critically sensitive variables for certain chemistries. For example, the scale-up of colloidal inorganic nanoparticle syntheses to yield kg quantities is difficult to execute in conventional batch reactors because higher reagent concentrations or increased reaction volumes affect mass and thermal transport, which in turn affect nucleation and growth, leading to loss of particle quality and poor process reproducibility[8]. Microfluidic parallelization can circumvent these issues and enable a simplified and more predictable scale-up route, as demonstrated in a parallel network of planar droplet formation devices[9].

One major challenge in implementing parallel microreactor systems is developing control and design strategies that guarantee uniform fluidic behaviour across an ensemble of reactors[10,11]. Here, we address this issue by presenting a fluidic design based on a three-dimensional (3D)-printed channel junction that allows for geometrically controlled two-phase liquid-in-liquid droplet formation that is robust to fluctuations in driving pressure and flow rate. Microfluidic liquid-in-liquid droplet flows facilitate rapid homogenization of reactants[12]. They are, therefore, ideal for reactions that are sensitive to concentration gradients and local mixing conditions. Droplets are isolated from each other and the channel walls to eliminate dispersion effects and prevent device fouling[13]. The design presented here enables an ensemble of parallel reactors with consistent droplet formation behaviour regardless of inconsistencies in the feed pressure or flow rate across the reactor bank. In contrast, prior attempts at nanoparticle synthesis scale-up using continuous flow have focused on modifying single-channel devices to employ larger droplets and increased operating flow rates[14,15]. Although these approaches produce good quality nanoparticles, this strategy cannot be scaled indefinitely.

An additional advantage of the droplet formation geometry we present here is that it can be rapidly reconfigured to produce a variety of droplet volumes spanning four orders of magnitude. This range of droplet sizes is used in many applications, including biomimetic vesicle formation, cell encapsulation and millifluidic reactor platforms[16,17]. In traditional T-junction and flow-focusing droplet formation devices, the operating parameters (that is, flow rates) allow for a relatively narrow range of droplet sizes to be accessed by a single device geometry[18]. Switching to a different droplet size regime requires redesigning (and refabricating) the device. In the device geometry we present here, droplet size is set by the diameter of an easily interchangeable outlet component. Different droplet sizes are accessible by swapping out this modular component. This control mechanism is in contrast to upstream geometrical control exhibited in planar droplet formation devices where the inlet geometry governs droplet size. Coupled with the relative insensitivity of droplet formation to flow rates, this design represents an important innovation in microfluidic droplet formation.

In addition to demonstrating the robust operation of this droplet formation geometry in a parallel system, we show it operating as the key element of a droplet microreactor applied to the synthesis of metal nanoparticles. In this paper, we demonstrate the first platinum nanoparticle (PtNP) synthesis using a continuous flow droplet microreactor. A key aspect of our continuous flow synthesis is the use of ionic liquid droplets as the dispersed phase and reaction medium. Ionic liquids are gaining interest as solvents for precious metal nanoparticle synthesis because of their ability to colloidally stabilize nanoparticles and induce high nucleation rates resulting in more monodisperse nanoparticle ensembles[19]. This, coupled with their environmental health, safety and sustainability advantages over volatile and flammable organic solvents[20], makes ionic liquids promising solvents for large-scale nanofabrication reactions. Herein, the synthesis of PtNPs is performed in ionic liquid droplets that are successfully recycled and reused to produce PtNPs over multiple runs in high fidelity.

Results

3D-printed microfluidic droplet generators. This microfluidic droplet generator is designed to be an easy-to-operate device that forms consistent droplet sizes across a broad range of inlet pressures or flow rates. The 3D geometry is manufactured using stereolithographic (SLA) printing technologies. This geometry is designed to interface with commercially available tubing (outer diameter (OD) = 1/16 inch) to create a droplet generating chip that does not require any fabrication steps in a clean room facility and can passively form droplet volumes spanning four orders of magnitude. Tubing connections form an elastomeric seal that resists leaking for water flow rates beyond $8 \, l \, h^{-1}$. Although the 3D printing technology is quickly advancing to create higher resolution features, a major concern is that the smallest channel feature that has been produced in the commonly used Watershed material is $400 \, \mu m$. Herein, we overcome this barrier by (i) fabricating a device with a 250-μm feature and (ii) using the printed device as a fluidic manifold while relying on higher resolution extruded tubing to control droplet size. The channels deliver the laminated dispersed and continuous phases to the outlet in a perpendicular orientation (Fig. 1a,b). Droplets pinch off as the fluids enter the vertical outlet tubing (Fig. 1c,d). The entire droplet formation process can be seen in the Supplementary Movie 1.

The basic 3D-printed device shown here can be used to form a broad range of droplets sizes by interfacing it with outlet tubing of various inner diameters. The size of the outlet tubing determines the droplet size such that the droplet sizes are similar to the inner diameter of the outlet tubing (Fig. 2a). In comparison, a planar T-junction geometry produces droplets with a size governed by the geometry of the dispersed phase inlet[21]. Regardless of the outlet tubing size, the droplet population is monodisperse, which allows for the collected droplets to self-assemble into hexagonally close packed arrays (Supplementary Figs 1–3).

We also present a droplet generator where the vertical cavity is integrated into the SLA manufactured device rather than being provided by external tubing (Fig. 1b,d). The breakup process is imaged in fully printed devices with cylindrical sizes of inner diameter (ID) = 250 and 500 μm (Supplementary Movie 2). We observe droplets forming at the point where the horizontal flow turns vertical, as in the droplet generators with externally connected tubing acting as the vertical cavity. As the dispersed and continuous phases enter the vertically oriented cylinder, the dispersed phase segments into droplets. This 250 μm cylinder is the smallest channel reported using the SLA process with transparent Watershed resin (FineLine Prototyping)[22]. Post printing, FineLine Prototyping clears unreacted resin from the vertical cavity and the devices are used as received.

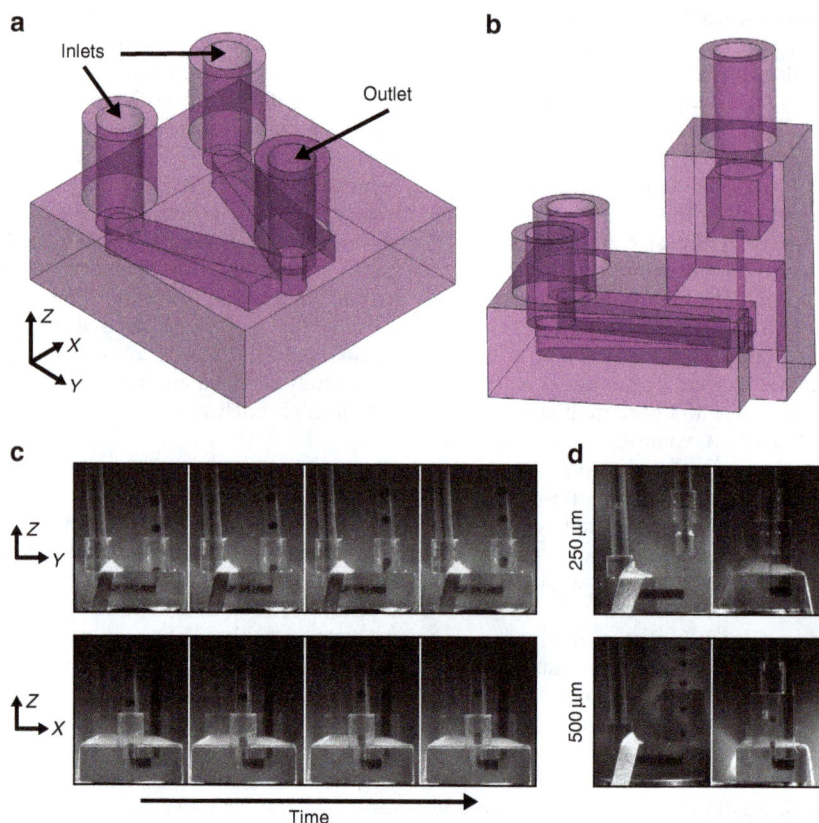

Figure 1 | Renderings of 3D droplet generators and images of droplet formation process. (**a**) Computer-aided design (CAD) rendering of a droplet generator with two inlets for the dispersed and continuous phases and a single outlet that accepts tubing (OD = 1/16 inch) with various IDs to control the droplet size. (**b**) CAD rendering of a droplet generator in which the vertical segment is fully constructed by stereolithography (SLA) rather than being formed by external tubing. (**c**) Micrographs depicting different views of the device during the droplet breakup process. (**d**) Micrographs of the droplet breakup process in full SLA droplet generators with an outlet size of 250 or 500 μm.

Flow invariant droplet formation. The defining characteristic of this droplet-forming device is that uniform droplets can be formed while varying the flow rate ratio. We demonstrate flow invariant droplet formation for six different commercially available sizes of outlet tubing. The smallest and largest inner diameters are 25 and 762 μm, respectively. For each tubing size, flow invariant droplet formation is observed up to an upper limit flow rate (Fig. 2b). By analysing the sizes of collected droplets, we determine the upper limit of the flow invariant regime for each tubing size. The data presented are collected at this upper limit for the continuous phase flow rate of 0.2, 5, 10, 20, 80 and 180 ml h^{-1} corresponding to the outlet tubing with an inner diameter of 25, 127, 178, 254, 508 and 762 μm, respectively. Below this upper limit, the droplet size is approximately the same as the diameter of the outlet tubing (Fig. 2b).

The upper limit of the flow invariant regime can be expressed in terms of the capillary number (*Ca*) of the system. For all outlet tubing sizes, this value (calculated using the inner diameter of the outlet as the characteristic length and the continuous phase flow rate as characteristic velocity) is about 10^{-3}. Below this capillary number, the droplet size is independent of the flow rate ratio for values of 1:20 and 1:2 (Fig. 2c). The droplet diameters are plotted versus the fractional droplet number. These data show the entire population of droplets. The total number of droplets is normalized to one and represented as a fractional droplet number. We also observe a consistent droplet size for intermediate flow rate ratios (Supplementary Fig. 4). At higher capillary numbers, in the flow-dependent regime, the droplet formation process transitions to a jetting mechanism (Supplementary Movie 3). The same flow invariant behaviour is

observed for a more viscous dispersed phase of 70 wt% glycerol in water (Supplementary Fig. 5). However, the threshold for invariance shifted to a lower capillary number. In the flow invariant regime, the median droplet sizes have a coefficient of variation <3%.

The output of the droplet generator is dependent on the geometry of the outlet tubing and not dependent on the surface chemistry of the channel before the outlet. We modify the internal channel surfaces by depositing a poly(ethylene glycol diacylate) or poly(1*H*,1*H*,2*H*,2*H* perfluorodecylarcylate-co-ethylene glycol diacrylate) polymeric film via initiated chemical vapour deposition (iCVD). The two coatings have water contact angles of 60° and 120°, respectively. The native (uncoated) material has a contact angle of 100°. With the same outlet tubing (ID = 254 μm), the coated devices produce the same size droplets as the unmodified device (Fig. 2d). The roughness of the channel surfaces was the same before and after coating by scanning electron microscopy analysis (Supplementary Fig. 6).

Device parallelization. A single-droplet formation device can be used as a single unit in a highly parallelized system of *n* units to linearly create *n*-fold droplets and *n*-fold throughput. Ideally, a parallelized system of single-droplet generators is designed to have an equal pressure drop and resistance over each channel to create identical flow conditions in each droplet generator. However, feedback between channels arises because of unequal numbers of droplets flowing in each channel[23]. This imbalance leads to flow rate fluctuations that alter the final droplet size, but the fluctuations are irrelevant in the 3D droplet generators

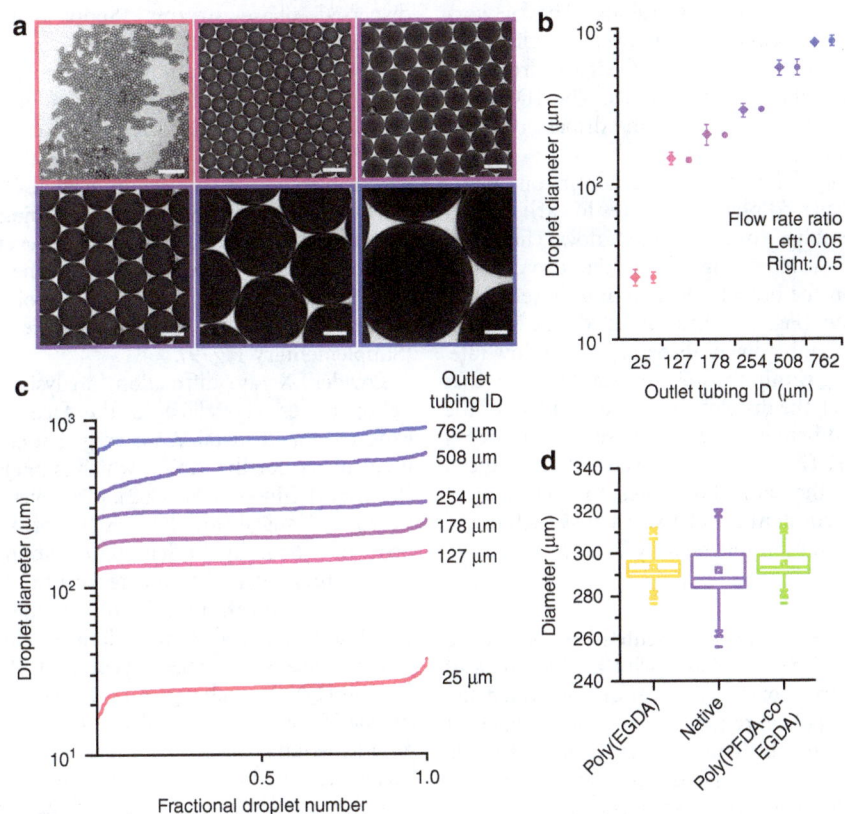

Figure 2 | Images of droplets and statistics on droplet sizes using various outlet tubing sizes. (**a**) Micrographs of the droplets formed using the six different sizes of outlet tubing listed. (**b**) Droplet diameter versus outlet tubing inner diameter for flow rate ratios of 0.05 (left, diamond) and 0.5 (right, circle). Error bars represent the s.d., some error bars are obscured by the symbols. (**c**) Plot of the droplet diameter versus fractional droplet number for various outlet tubing sizes. The solid lines represent the average droplet sizes for a single outlet size; each line includes values for flow rate ratios of 1:2 and 1:20 (dispersed to continuous phase). (**d**) Boxplot of the droplet size produced by droplet generators with the same outlet tubing (ID = 254 μm) and different surface chemistries (that is, hydrophilic, native and hydrophobic) on the channels, as modified by initiated chemical vapour deposition.

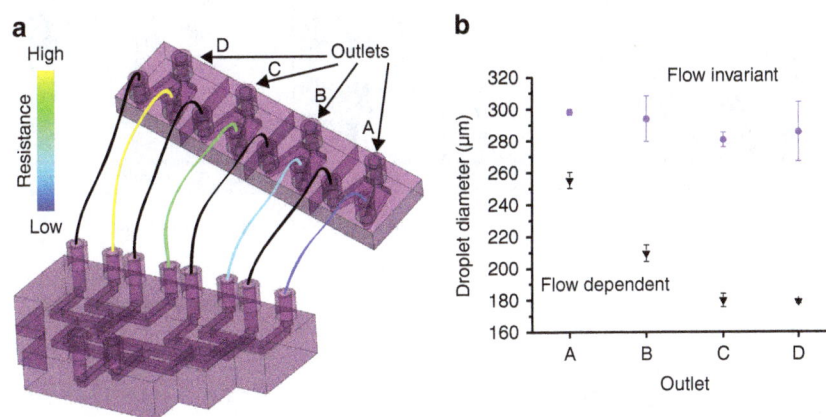

Figure 3 | Setup and performance of unbalanced parallel network. (**a**) Schematic of the parallel network assembled by connecting a distribution manifold to four droplet generators. The continuous phase was linked using low resistance jumper tubing (ID = 762 μm) and the dispersed phase was linked using various lengths of tubing (ID = 127 μm) to create a gradient of resistances across the four branches. (**b**) Droplet diameters (*n* > 1,000) produced by the four branches of the parallel network (left) by dispersed and continuous phase flow rates of 10 and 70 ml h[−1] (purple circles) and 30 and 210 ml h[−1] (black triangles) while operating in and beyond the flow invariant regime, respectively. Error bars represent the s.d.

because of their insensitivity to flow rate. For this reason, our droplet generator is uniquely suited for use in a parallelized network.

To demonstrate this, we construct a parallel network (*n* = 4) of droplet generators and deliver different flow rates to each device (Fig. 3a). We print a manifold to distribute the dispersed and

continuous phases to four droplet generators. The manifold connects to four independent droplet generators by jumper cables (that is, sections of poly(ether ether ketone) (PEEK) tubing). The jumper cables connecting the dispersed phases have lengths of 10, 12.5, 15 and 17.5 mm, resulting in relative resistances of 1x, 1.25x, 1.5x and 1.75x, respectively, because the resistance is linearly proportional to the length of cylindrical tubing. The network assembly creates the largest pressure drop over the dispersed phase jumper cables so there is minimal feedback from the continuous phase jumper cables or the outlets. The ID of the jumper cables are 127, 762 and 254 μm for the dispersed phase, continuous phase and outlets, respectively.

The network successfully delivers the same continuous phase flow rate and different dispersed phase flow rates to each droplet generator. When operating outside the flow invariant regime (that is, at high capillary number), the droplet size produced is dependent on the branch location and therefore the dispersed phase flow rate (Fig. 4b). As expected, the droplet size is smaller in branches with a lower dispersed phase flow rate. In contrast, when operating within the flow invariant regime (that is, at low capillary number), the droplet size is independent of the branch location despite a different dispersed phase flow rate being delivered to each channel (Fig. 4b). As expected, the balanced parallel network delivers the same flow rates to each droplet generator and produces a constant droplet size across each of the devices, in both flow regimes (Supplementary Fig. 7).

PtNP synthesis. The droplet generator presented here is suitable for chemical synthesis in dispersed phases with a wide range of solvent properties. As a proof-of-concept, we demonstrated the synthesis of PtNPs by a polyol reduction in droplet flows of 1-butyl-3-methylimidazolium bis(trifluoromethylsulfonyl)imide (BMIM-Tf$_2$N) ionic liquid solvent. There are few examples of droplet flows of ionic liquids because they represent an exceptional case of droplet flow behaviour as a result of their complex interfacial properties and high viscosity[1,24,25]. Here, BMIM-Tf$_2$N ionic liquid droplets are formed trivially in a modified droplet generator with three inlets to accommodate two reagent/

dispersed phase streams (Supplementary Fig. 8). The two reagent inlets supply (i) the potassium tetrachloroplatinate(II) (K$_2$PtCl$_4$) precursor and the reducing agent (that is, ethylene glycol) and (ii) the poly(vinylpyrrolidone) (PVP) in BMIM-Tf$_2$N. Droplets of the combined reagents are formed using PEEK tubing (ID = 762 μm) in the outlet. The reaction is initiated by flowing the droplets into a convection oven at 150 °C to quickly nucleate the PtNPs. The temperature in the tubing equilibrates in less than a second to trigger the nucleation event. Likewise, an abrupt cooling step quickly quenches the reaction and arrests nanoparticle growth. There is no observable clogging of or deposition on the channel surfaces after a 2-h reaction (Supplementary Fig. 9).

Powder X-ray diffraction analysis confirms the resulting nanoparticles crystallize in the face centered cubic structure expected for Pt metal. An average lattice parameter of $a = 3.87$ Å is calculated for the PtNPs, which is in close agreement with bulk Pt metal (PDF# 00-004-0802). Moreover, the diffraction peaks are broadened, suggesting the presence of small nanoparticles on the order of ∼6 nm by the Scherrer equation (Fig. 4b). We synthesize three rounds of PtNPs and recycle the BMIM-Tf$_2$N ionic liquid solvent between rounds. To recycle the ionic liquid solvent, the PtNPs are harvested from the ionic liquid and then it is washed to remove excess ethylene glycol and PVP. The purity of the BMIM-Tf$_2$N is unchanged, as evidenced by ^{1}H and ^{19}F nuclear magnetic resonance (NMR), upon recycling (Fig. 4c,d and Supplementary Note 1). Transmission electron microscopy images reveal that the PtNPs appear spherical and uniform in morphology across each reuse of the ionic liquid (Fig. 4a). For each condition, 500 particles are analysed and their average sizes are 5.65 ± 0.76, 5.96 ± 0.90 and 6.56 ± 0.92 nm for 1x, 2x and 3x recycled ionic liquid, respectively. The mean sizes are all within the standard deviation of each other.

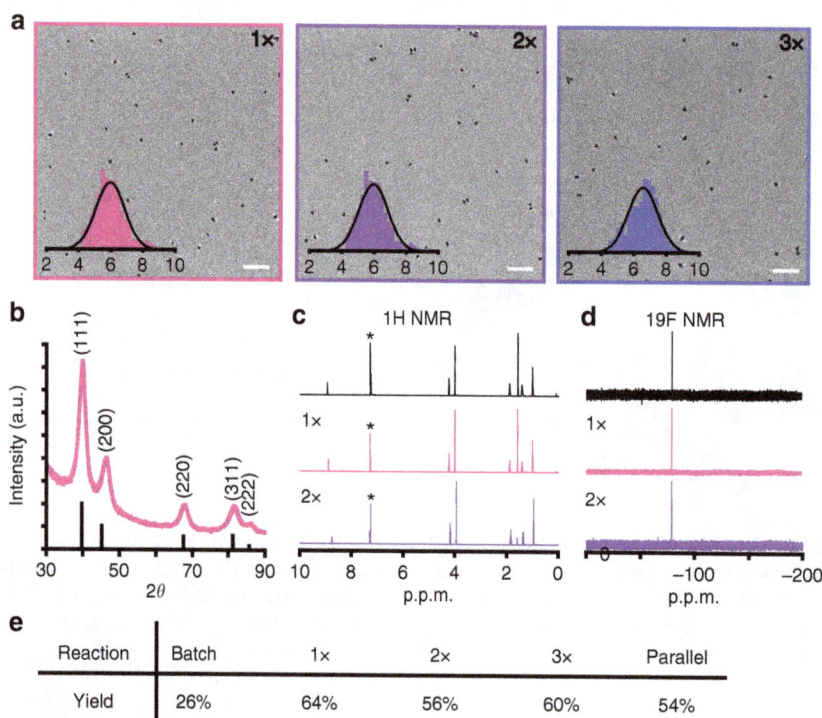

Reaction	Batch	1×	2×	3×	Parallel
Yield	26%	64%	56%	60%	54%

Figure 4 | Micrographs and characterization of PtNPs synthesized in ionic liquid solvent using 3D droplet generator. (a) Transmission electron microscopy (TEM) images of PtNPs produced using the microfluidic droplet generator using 1x recycled, 2x recycled and 3x recycled BMIM-Tf$_2$N ionic liquid. Scale bars represent 50 nm and histograms represent the NP diameters (*n* = 500) from multiple TEM images. **(b)** X-ray diffraction (XRD) of the PtNPs. **(c)** ^{1}H NMR spectra of the BMIM-NTf$_2$ ionic liquid (*residual solvent peak), and **(d)** ^{19}F NMR spectra of the BMIM-Tf$_2$N ionic liquid. Both spectra in black are of the ionic liquid as received. **(e)** Comparison of the overall yield from various reactions.

We provide a general device and platform for running reactions in parallelized droplet reactors. When PtNPs are synthesized in an $n = 4$ parallel network, all four branches produce high-quality particles, as evidenced by consistent size distributions that are within error of each other (Supplementary Fig. 10). In addition, translation of the PtNP synthesis from batch scale to a continuous flow droplet reactor results in significantly higher yields (Fig. 4e). For each run with the recycled ionic liquid and the overall parallel reaction, the continuous flow yield is about twice the batch yield. We attribute the greater yield to a more rapid and uniform heating profile, as well as improved mixing as compared with the batch reaction. The combined use of our droplet generator in a parallel network and a reusable solvent system provide an ideal platform for the efficient manufacturing of large quantities of nanomaterial product.

Discussion

We introduce a novel droplet generator that uses a 3D geometry to form droplets of controlled size. A key feature of this droplet formation geometry is that there is a broad regime of inlet flow rate ratios over which resulting droplet size is invariant to flow rate. Another advantage of this droplet formation format is that its inherent modularity makes it simple to select the size of droplets that will be formed. The size can easily be tuned by changing the ID of the outlet tubing. The ease of device fabrication and operation lowers the barrier-to-entry for first time users of microfluidic devices. An end user need not have an extensive fluid mechanics understanding to operate the device to achieve the desired and well-controlled droplet sizes.

The relationship between outlet tubing size and droplet size seems to underlie droplet size stability. We speculate that this relationship is due to the mode of droplet formation. Droplet breakup occurs as the two flows alternate to enter and fill the opening of the vertical cavity created by the outlet tubing. Droplets form at the interface between the outlet tubing and the horizontal flow when the dispersed phase is pinched off to form droplets (Fig. 1c,d and Supplementary Movie 1). When the dispersed phase fills the outlet it causes an upstream buildup of continuous phase pressure. Once the outlet is completely occluded by the dispersed phase, the droplets shear off, releasing the continuous phase pressure. As outlet tubing size increases, more dispersed phase accumulates before the continuous phase is completely occluded, explaining the correlation between outlet tubing size and terminal droplet size.

These droplet-forming devices are uniquely suited to high-throughput processing using microfluidics. When assembled in a parallel configuration, they are insensitive to small changes in flow that could arise because of feedback between channels. This is demonstrated by building a four-branched parallel network that produces droplets of similar size across the network despite having an intentional gradient of dispersed phase flow rates being delivered to each branch. The devices also resist clogging that could affect droplet formation.

We synthesize monodisperse PtNPs over multiple runs while using the same recycled ionic liquid solvent. The yield in the continuous flow reactions is $\sim 60\%$ and nearly twice the yield of an analogous batch reaction. Using this device infrastructure along with more sustainable chemistry provides an ideal platform for producing large quantities of precious metal nanoparticles. This can be easily extended to other applications requiring high-throughput synthesis in microfluidic droplets.

Methods

Device fabrication. Microfluidic chips were designed in ProEngineer, exported as sterolithography files and printed in Somos Watershed XC 11122 by FineLine Prototyping using high-resolution SLA printing technology. Devices were used as received. Inlet and outlets interfaced with OD = 1/16 inch tubing. The device shown in Fig. 1a has a channel height of 1 mm, inlet and outlet holes were 1.59 mm in diameter, the length of the main channel was 5 mm and the width of the main channel was 4 mm. The fully 3D printed droplet generator in Fig. 1b has the same dimensions as the device in Fig. 1a and the additional vertical cavity was printed with an internal diameter of 250 or 500 μm.

Droplet visualization. Water-in-oil droplets were formed using an aqueous phase of $Fe(SCN)_x^{(3-x)+}$ complex (for visualization) in deionized water, prepared by mixing 0.2 M KSCN (Sigma) with 0.067 M $Fe(NO_3)_3 \bullet 9H_2O$ (Sigma) in a 1:1 volumetric ratio. The oil phase was 1% (w/v) Span80 (Sigma) in hexanes (Sigma). Fluids were driven by syringe pumps (Harvard Apparatus). Droplets were collected in a glass bottom Petri dish (MatTex) containing 1 ml of the 1% (w/v) Span80 in hexanes solution and imaged on an inverted Zeiss microscope using a × 20 objective. At least 60 images of the collected droplets were captured for each flow condition. Droplet sizes were analysed using custom image processing code in Matlab, primarily using the imfindcircles function.

Droplet formation was monitored *in situ* using a Phantom V711 camera (Vision Research). Images were captured at 4,000 frames per second with a 240-μs exposure.

PtNP synthesis in microfluidic device. K_2PtCl_4 (99.9%; Strem), PVP (MW = 55,000; Aldrich), ethylene glycol (99.8%; Sigma-Aldrich) and BMIM-Tf$_2$N (99%; IoLiTec Lot # K00219.1.4.) were used as received. In a typical procedure, K_2PtCl_4 (156.0 mg) was added to ethylene glycol (10.0 ml) and bath sonicated until dissolved, affording a brown-red mixture. Separately, PVP (852.4 mg) was dissolved in BMIM-Tf$_2$N (30.0 ml) by heating at 130 °C for 10 min to give a clear solution.

PtNPs were synthesized using our 3D droplet-forming device by modifying the original design to incorporate three inlets (Supplementary Fig. 8). Two inlets supplied the K_2PtCl_4 in ethylene glycol at 10 ml h^{-1} and the PVP in BMIM-Tf$_2$N at 30 ml h^{-1}. The third inlet supplied a continuous phase of FC-40 (Sigma) at 90 ml h^{-1}. The outlet PEEK (ID = 0.03 inches) fed into 200 feet of perfluoroalkoxy tubing (McMaster-Carr) that was placed in a convection oven set to 150 °C. The effluent was collected into a receiving flask cooled in an ice bath to quench the reaction.

The PtNPs were transferred to two 50-ml centrifuge tubes and precipitated with acetone (30 ml) to afford a black suspension, which was briefly vortex stirred (~1 min), bath sonicated (3 min) and collected by centrifugation (6,000 r.p.m.; 5 min). The colourless supernatant was decanted and saved in order to wash the ionic liquid for further syntheses. The black nanoparticulate solid was redispersed in ethanol (10 ml), bath sonicated (3 min), precipitated with hexanes (30 ml) and centrifuged (6,000 r.p.m.; 5 min). This process was repeated 3 × in order to remove any residual organics (that is, PVP, BMIM-Tf$_2$N, and ethylene glycol). The purified PtNPs were re-dispersed in ethanol (5 ml) and remained colloidally stable for at least 5 months.

Recycling the ionic liquid. Equal volumes of hexanes, with respect to the ionic liquid, were added and vortex stirred before centrifugation (6,000 r.p.m.; 10 min). Upon phase separation, the hexanes layer was removed and additional hexanes were added; this extraction procedure was repeated 3 ×. The ionic liquid was dried in a rotary evaporator at 80 °C for ca 2 h. The purity of the washed ionic liquid was confirmed by ^{1}H and ^{19}F NMR spectroscopy.

Initiated chemical vapour deposition. Devices were coated as received with poly(1H,1H,2H,2H-perfluorodecyl acrylate-co-ethylene glycol diacrylate) (poly(PFDA-co-EGDA)) or poly(ethylene glycol diacrylate) (poly(EGDA))[1,26,27]. Briefly, the devices were coated using the iCVD process. Microfluidic reactors were placed on a temperature-controlled stage maintained at 30 °C within a pancake-shaped vacuum chamber (GVD Corp., 250 mm diameter, 48 mm height). The polymer coating was formed via a free-radical chain mechanism from the reaction of vapour phase precursors. Di-*tert*-butyl peroxide (Sigma) and PFDA (SynQuest) and/or EGDA (Monomer-Polymer) were introduced into the vacuum chamber at a pressure of 100 mTorr and the initiator molecules were thermally decomposed into free radicals at 250 °C by a resistively heated array of nichrome wire. The radicals and monomer molecules diffuse into the channels via the inlets and outlet, adsorb to the surface of the material, and polymerize to form a thin, conformal coating.

PtNP synthesis in batch. In a typical procedure, K_2PtCl_4 (39.0 mg) was added to ethylene glycol (2.5 ml) in a 23-ml vial and bath sonicated until dissolved, affording a brown-red mixture. Separately, PVP (213.1 mg) was dissolved in BMIM-Tf$_2$N (7.5 ml) by heating at 130 °C for 10 min to give a clear solution. The two solutions were then combined in a 25-ml round bottom flask and placed in a silicone oil bath preheated to 150 °C while stirring at 1,000 r.p.m. The reaction was quenched with an ice bath after 15 min. The washing of the particles and the recycling of the ionic liquid followed the same procedure as in the microfluidic device reaction.

Characterization. Transmission electron microscopy images were obtained using a JEOL JEM2100F (JEOL Ltd.) microscope operating at 200 kV. Samples were prepared on 400 mesh Cu grid coated with a lacey carbon film (Ted Pella, Inc.) by drop-casting a dilute suspension of PtNPs in ethanol. The size distribution of the PtNPs was determined by analysing 500 unique nanoparticles. Powder X-ray diffraction patterns were collected on a Rigaku Ultima IV diffractometer functioning at 40 mA and 40 kV with a Cu $K\alpha$ X-ray source ($\lambda = 1.5406$ Å). The step size and collection time were 0.015° and 10 s per step, respectively. ^1H and ^{19}F NMR spectra were obtained on Varian 600 spectrometer (600 and 564 MHz, respectively) with chemical shifts reported in p.p.m.

References

1. Lazarus, L. L. *et al.* Two-phase microfluidic droplet flows of ionic liquids for the synthesis of gold and silver nanoparticles. *ACS Appl. Mater. Interfaces* **4**, 3077–3083 (2012).
2. Lazarus, L. L., Yang, A. S., Chu, S., Brutchey, R. L. & Malmstadt, N. Flow-focused synthesis of monodisperse gold nanoparticles using ionic liquids on a microfluidic platform. *Lab Chip* **10**, 3377–3379 (2010).
3. Sebastian Cabeza, V., Kuhn, S., Kulkarni, A. A. & Jensen, K. F. Size controlled flow synthesis of gold nanoparticles using a segmented flow microfluidic platform. *Langmuir* **28**, 7007–7013 (2012).
4. Sebastian, V. *et al.* One-step continuous synthesis of biocompatible gold nanorods for optical coherence tomography. *Chem. Commun.* **48**, 6654–6656 (2012).
5. Mason, B. P., Price, K. E., Steinbacher, J. L., Bogdan, A. R. & McQuade, D. T. Greener approaches to organic synthesis using microreactor technology. *Chem. Rev.* **107**, 2300–2318 (2007).
6. Nightingale, A. M., Phillips, T. W., Bannock, J. H. & de Mello, J. C. Controlled multistep synthesis in a three-phase droplet reactor. *Nat. Commun* **5**, 3777 (2014).
7. Hartman, R. L., McMullen, J. P. & Jensen, K. F. Deciding whether to go with the flow: evaluating the merits of flow reactors for synthesis. *Angew. Chem. Int. Ed. Engl.* **50**, 7502–7519 (2011).
8. Phillips, T. W., Lignos, I. G., Maceiczyk, R. M., deMello, A. J. & deMello, J. C. Nanocrystal synthesis in microfluidic reactors: where next? *Lab Chip* **14**, 3172–3180 (2014).
9. Nightingale, A. M. *et al.* Large-scale synthesis of nanocrystals in a multichannel droplet reactor. *J. Mater. Chem. A* **1**, 4067 (2013).
10. Conchouso, D., Castro, D., Khan, S. A. & Foulds, I. G. Three-dimensional parallelization of microfluidic droplet generators for a litre per hour volume production of single emulsions. *Lab Chip* **14**, 3011–3020 (2014).
11. Hashimoto, M. *et al.* Formation of bubbles and droplets in parallel, coupled flow-focusing geometries. *Small* **4**, 1795–1805 (2008).
12. Song, H., Tice, J. D. & Ismagilov, R. F. A microfluidic system for controlling reaction networks in time. *Angew. Chem.* **115**, 792–796 (2003).
13. Casadevall i Solvas, X. & deMello, A. Droplet microfluidics: recent developments and future applications. *Chem. Commun.* **47**, 1936–1942 (2011).
14. Zhang, L. *et al.* Continuous and scalable production of well-controlled noble-metal nanocrystals in milliliter-sized droplet reactors. *Nano Lett.* **14**, 6626–6631 (2014).
15. Lohse, S. E., Eller, J. R., Sivapalan, S. T., Plews, M. R. & Murphy, C. J. A simple millifluidic benchtop reactor system for the high-throughput synthesis and functionalization of gold nanoparticles with different sizes and shapes. *ACS Nano* **7**, 4135–4150 (2013).
16. van Swaay, D. & Demello, A. Microfluidic methods for forming liposomes. *Lab Chip* **13**, 752–767 (2013).
17. Lorber, N. *et al.* Some recent advances in the design and the use of miniaturized droplet-based continuous process: applications in chemistry and high-pressure microflows. *Lab Chip* **11**, 779–787 (2011).
18. Teh, S. Y., Lin, R., Hung, L. H. & Lee, A. P. Droplet microfluidics. *Lab Chip* **8**, 198–220 (2008).
19. Dupont, J. & Scholten, J. D. On the structural and surface properties of transition-metal nanoparticles in ionic liquids. *Chem. Soc. Rev.* **39**, 1780–1804 (2010).
20. Welton, T. Room-temperature ionic liquids. Solvents for synthesis and catalysis. *Chem. Rev.* **99**, 2071–2084 (1999).
21. Christopher, G. F., Noharuddin, N. N., Taylor, J. A. & Anna, S. L. Experimental observations of the squeezing-to-dripping transition in T-shaped microfluidic junctions. *Phys. Rev. E Stat. Nonlin. Soft Matter Phys* **78**, 036317 (2008).
22. Au, A. K., Lee, W. & Folch, A. Mail-order microfluidics: evaluation of stereolithography for the production of microfluidic devices. *Lab Chip* **14**, 1294–1301 (2014).
23. Li, W. *et al.* Simultaneous generation of droplets with different dimensions in parallel integrated microfluidic droplet generators. *Soft Matter* **4**, 258 (2008).
24. Barikbin, Z. *et al.* Ionic liquid-based compound droplet microfluidics for 'on-drop' separations and sensing. *Lab Chip* **10**, 2458–2463 (2010).
25. Feng, X., Yi, Y., Yu, X., Pang, D. W. & Zhang, Z. L. Generation of water-ionic liquid droplet pairs in soybean oil on microfluidic chip. *Lab Chip* **10**, 313–319 (2010).
26. Riche, C. T., Zhang, C., Gupta, M. & Malmstadt, N. Fluoropolymer surface coatings to control droplets in microfluidic devices. *Lab Chip* **14**, 1834–1841 (2014).
27. Riche, C. T., Marin, B. C., Malmstadt, N. & Gupta, M. Vapor deposition of cross-linked fluoropolymer barrier coatings onto pre-assembled microfluidic devices. *Lab Chip* **11**, 3049–3052 (2011).

Acknowledgements

This article is based on work supported by the National Science Foundation under CMMI-1436872. C.T.R. was partially supported by a fellowship from the von Liebig center (CETCF). We thank Professor V. Eliasson and S. Koumlis for the use of the Phantom camera. We also thank the USC High Performance Computing Center for use of their cluster in performing image analysis.

Author contributions

C.T.R., E.J.R., N.M. and R.L.B. designed the experiments and analysed the data. C.T.R. and M.G. designed iCVD-based device coating methods. C.T.R. and E.J.R. performed experiments. C.T.R. and E.J.R. wrote the article text; all authors read and edited the text.

Additional information

11

Three-dimensional porous carbon composites containing high sulfur nanoparticle content for high-performance lithium–sulfur batteries

Guoxing Li[1], Jinhua Sun[1], Wenpeng Hou[1], Shidong Jiang[1], Yong Huang[1] & Jianxin Geng[1]

Sulfur is a promising cathode material for lithium–sulfur batteries because of its high theoretical capacity ($1,675\,mA\,h\,g^{-1}$); however, its low electrical conductivity and the instability of sulfur-based electrodes limit its practical application. Here we report a facile *in situ* method for preparing three-dimensional porous graphitic carbon composites containing sulfur nanoparticles (3D S@PGC). With this strategy, the sulfur content of the composites can be tuned to a high level (up to 90 wt%). Because of the high sulfur content, the nanoscale distribution of the sulfur particles, and the covalent bonding between the sulfur and the PGC, the developed 3D S@PGC cathodes exhibit excellent performance, with a high sulfur utilization, high specific capacity ($1,382$, $1,242$ and $1,115\,mA\,h\,g^{-1}$ at 0.5, 1 and 2 C, respectively), long cycling life (small capacity decay of 0.039% per cycle over 1,000 cycles at 2 C) and excellent rate capability at a high charge/discharge current.

[1] Technical Institute of Physics and Chemistry, Chinese Academy of Sciences, 29 Zhongguancun East Road, Haidian District, Beijing 100190, China. Correspondence and requests for materials should be addressed to J.G. (email: jianxingeng@mail.ipc.ac.cn).

Lithium–sulfur (Li–S) batteries have recently attracted great interest as promising electrochemical devices for energy conversion and storage applications because of the abundance, low cost, environmental friendliness and high theoretical capacity ($1,675 \, \text{mA h g}^{-1}$) of sulfur[1-4]. Despite these advantages, the practical application of Li–S batteries is still handicapped by the following problems: (1) the low electrical conductivities of sulfur ($5 \times 10^{-30} \, \text{S cm}^{-1}$ at $25 \, °\text{C}$), intermediate polysulphides and Li_2S; (2) the dissolution of lithium polysulphides, which results in a shuttling effect and in the deposition of insoluble lithium sulfide on the anode in each of the charge/discharge cycles and eventually the complete loss of capacity of the sulfur cathode; and (3) severe volume changes in the active electrode materials during the lithiation/delithiation processes[1,4-7], resulting in the pulverization of the electrode materials.

To overcome these problems, various carbon materials, including graphene[8-15], carbon nanotubes[16,17], porous carbon[18-26] and carbon nanofibres[27-29], have been tested in recent years as supporting materials for sulfur cathodes to improve the electrochemical performance of Li–S batteries. Carbon frameworks improve the electrical conductivity of sulfur cathodes and trap soluble polysulphides during cycling. In addition, yolk–shell structures such as a sulfur–TiO_2 yolk–shell[30] and a sulfur–polyaniline yolk–shell[31] have been developed to address the large volume changes of sulfur during the lithiation/delithiation processes. Recently, Choi et al. fabricated a polydopamine-coated S/C composite cathode with a high sulfur loading, which exhibited a high areal capacity ($9 \, \text{mA h cm}^{-2}$) (ref. 32). In a pioneering work, Pyun and co-workers prepared sulfur-containing polymers that exhibited high electrochemical activity and suitability as cathode materials for Li–S batteries[33,34]. However, despite these research efforts, no strategy has satisfactorily solved the aforementioned problems. Most seriously, the long-cycle stability under high charge/discharge rates remains a major challenge for sulfur-based cathodes, especially for composites with relatively high sulfur content. To date, conventional methods for the preparation of carbon–sulfur composites[8,16,19-21,35], in which carbon materials are impregnated with sulfur by diffusion after the carbon structures are prepared, still face challenges. These problems include the complexity of multistep operations, the low sulfur content of the composites and the out-diffusion of lithium polysulphide into the electrolyte during the charge/discharge cycles owing to the diffusion process used for the incorporation of sulfur into the carbon materials. Furthermore, certain unresolvable trade-offs have been found in previous studies. For example, a relatively high sulfur content in the sulfur/carbon hybrid structures is always accompanied by larger sulfur particles[11,30,36], which severely reduces the rate of sulfur utilization because of the long diffusion path for electrons and lithium ions[3]. Although a very high specific capacity ($>1,000 \, \text{mA h g}^{-1}$) can be obtained with an electrode that has a low sulfur content[37-40], the low sulfur content greatly reduces the overall volumetric capacity and energy density of the cathode. Therefore, it is crucial to design high-sulfur-content composites for use as cathode materials in Li–S batteries, in which the composite cathodes maintain a high sulfur utilization rate, a high specific capacitance, a long cycling life and good rate capability. This may be achievable by controlling the existing state and distribution of sulfur in the hybrid structures.

Herein, we report a facile and scalable strategy for the in situ synthesis of sulfur nanoparticles in three-dimensional (3D) porous graphitic carbon (PGC) (designated 3D S@PGC) with a tuneable sulfur content and demonstrate the utility of the 3D S@PGC as a cathode material for Li–S batteries. Compared with the conventional methods for the preparation of carbon–sulfur composites[8,16,19-21,35], our strategy facilitates access to composites that have the advantages of a high sulfur content (up to 90 wt%) and nanoscale distribution of the sulfur particles, as well as covalent bonding between the sulfur nanoparticles and the PGC network, which ensures the efficient utilization of the loaded sulfur. The sulfur content of the composite can be readily tuned by changing the Na_2S/glucose ratio. Because of the C–S bonds, unique interconnected hierarchical porous structures, high sulfur content and nanoscale sulfur particles, our 3D S@PGC (90% S) composite exhibits significantly improved electrochemical performance as a cathode material for Li–S batteries. In particular, this material has a high specific capacity ($1,382$, $1,242$ and $1,115 \, \text{mA h g}^{-1}$ at 0.5, 1 and 2 C, respectively), long cycling life (small capacity decay of 0.039% per cycle over 1,000 cycles at 2 C), and excellent rate capability at a high charge/discharge current.

Results

Synthesis of the 3D S@PGC composites. Figure 1 displays the scheme for the preparation of the 3D S@PGC composites. The self-stacking of water-soluble NaCl and Na_2S crystals was utilized to form a hard template for the 3D PGC network. Na_2S, which reacts with $Fe(NO_3)_3$ to form sulfur, was used as the sulfur precursor. The reaction is described by the equation $Na_2S + 2Fe(NO_3)_3 \rightarrow 2Fe(NO_3)_2 + 2NaNO_3 + S$. NaCl, Na_2S and glucose were first dissolved in deionized (DI) water to obtain a homogeneous solution, which was subsequently subjected to freeze-drying. During the freezing process, the sizes of the NaCl and Na_2S crystals were restricted, and the crystals were uniformly coated with an ultrathin glucose film. The NaCl crystals surrounded the Na_2S crystals because of the high NaCl/Na_2S molar ratio (10:1). The hybrid structure is hereafter referred to as 3D NaCl–Na_2S@glucose. Upon heating at a high temperature ($750 \, °\text{C}$) in an argon atmosphere, the glucose underwent carbonization to form graphitic carbon (GC) with micro- and

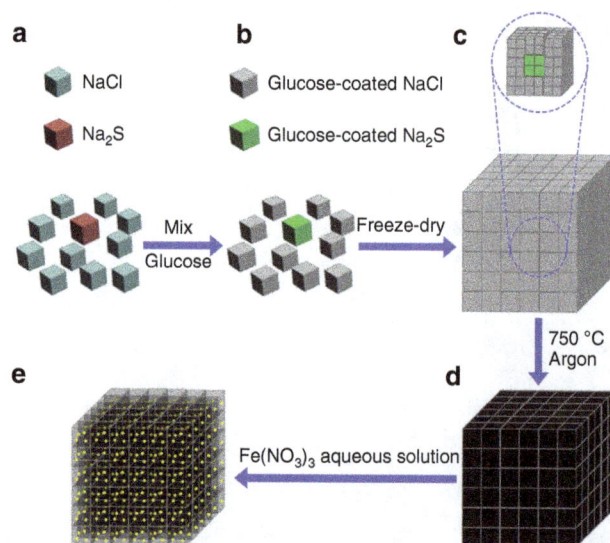

Figure 1 | Schematic illustration of an in situ strategy for the preparation of 3D S@PGC composites. (**a**) NaCl and Na_2S crystals. (**b**) Glucose-coated NaCl and Na_2S crystals. (**c**) Self-stacking of the glucose-coated NaCl and Na_2S crystals (3D NaCl–Na_2S@glucose), with Na_2S crystals surrounded by NaCl crystals. (**d**) Self-stacking of GC-coated NaCl and Na_2S crystals (3D NaCl–Na_2S@GC). (**e**) 3D S@PGC composite formed through the simultaneous dissolution of the NaCl crystals and oxidation of the Na_2S with $Fe(NO_3)_3$.

mesopores[41,42], thus leading to 3D NaCl–Na$_2$S@GC. Finally, the immersion of 3D NaCl–Na$_2$S@GC in an aqueous Fe(NO$_3$)$_3$ solution dissolved the NaCl template, leaving macropores in the composite. Concurrently, sulfur nanoparticles formed *in situ* and deposited on the walls of the macropores through the oxidation of Na$_2$S by Fe(NO$_3$)$_3$. During this process, the Na$_2$S that was dissolved in the aqueous Fe(NO$_3$)$_3$ solution in the 3D PGC macropores diffused into the micro- and mesopores because of capillarity, allowing the micro- and mesopores to fill with sulfur. Hence, the sulfur nanoparticles were uniformly distributed in the 3D PGC through an *in situ* chemical deposition process. According to previous reports[18,19], the PGC content, 3D porous architecture and sulfur content are crucial parameters that determine the properties of batteries such as the electrical conductivity of the electrode, electrolyte transport in the electrode and specific capacity. In our research, NaCl and Na$_2$S crystals were used as a template to fabricate the 3D porous architecture, and the sulfur content of the composites could be readily tuned by changing the Na$_2$S/glucose ratio. A 3D S@PGC composite with a sulfur content of up to 90 wt% [3D S@PGC (90% S)] was obtained by optimizing the Na$_2$S/glucose ratio (1:0.4). Therefore, our method has advantages over conventional methods[8,16,19–21,35] because the composition and structure of the composites can be easily controlled.

Structure and morphology of the 3D S@PGC composites. The crystallographic structure of the 3D S@PGC (90% S) composite was first analysed using X-ray diffraction (Fig. 2a). The X-ray diffraction pattern of elemental sulfur has three prominent Bragg reflections at 23.1, 25.9 and 27.8° (bottom trace), which can be indexed as the (222), (026) and (040) planes of the *fddd* orthorhombic structure[43]. As expected, the 3D S@PGC (90% S) composite produced a similar pattern, confirming the crystalline feature of the sulfur nanoparticles in the 3D PGC. The broad reflection peak, which can be clearly identified in the range of 15–35° (inset of Fig. 2a and Supplementary Fig. 1), is attributed to PGC. The peak for PGC in the 3D S@PGC (90% S) is weak because of the relatively low amount of PGC in the composite. A thermogravimetric analysis (TGA) indicated that the sulfur content of the 3D S@PGC (90% S) composite is ~90 wt% (Fig. 2b).

The morphology of the 3D S@PGC (90% S) composite was investigated via electron microscopy. Figure 3a displays an overview of the composite, with its honeycomb-like porous surface, obtained via scanning electron microscopy (SEM). At higher magnification (Fig. 3b), the unique network of the

composite, which is composed of interconnected submicron-sized macropores, can be seen. With a further increase in magnification (Fig. 3c), sulfur nanoparticles with sizes ranging from 18 to 54 nm, homogeneously and densely anchored onto the walls of the PGC network, can be observed (Supplementary Fig. 2). A comparison of these morphologies with those of the 3D NaCl–Na$_2$S@GC and the pure 3D PGC framework (Supplementary Figs 3 and 4) led us to conclude that the 3D self-stacking of the NaCl and Na$_2$S crystals had been well preserved after carbonization. Therefore, the SEM data support the formation mechanism of the 3D S@PGC composites (Fig. 1). An image obtained via transmission electron microscopy (TEM) reveals that the pores overlap each other and form a continuous 3D network with ultrathin GC walls (Fig. 3d). In good agreement with the SEM data, the ultrathin carbon walls are uniformly covered with sulfur nanoparticles (Fig. 3e). The composites that were subjected to an elongated period of sonication resulted in similar TEM images (Supplementary Fig. 5). The TEM results indicate that the attachment of the sulfur nanoparticles to the walls of the PGC network was robust because the materials used for TEM observation were subjected to long periods of sonication during the sample preparation. Moreover, the 3D porous structure remained intact, demonstrating the high mechanical flexibility of the 3D S@PGC composites. A high-resolution TEM image reveals the crystalline feature of the sulfur nanoparticles (Fig. 3f). Clear lattice fringes with an interlayer spacing of 0.38 nm corresponding to the (222) planes are readily observable, as the (222) Bragg reflection has the greatest intensity in the X-ray diffraction pattern (Fig. 2a). Energy-dispersive X-ray spectroscopy (EDS) elemental mapping confirmed the presence of carbon and sulfur in the 3D S@PGC (90% S) composite (Fig. 3g–i), as well as the homogenous distribution of sulfur nanoparticles in the PGC framework (Fig. 3i, sulfur mapping). A nanosized distribution is extremely important for the application of the sulfur particles as a cathode material for Li–S batteries, as the utilization rate of sulfur is higher for smaller sulfur particles because of the short diffusion path of the electrons and lithium ions[3]. Compared with previously reported hybrid structures that have a comparably high sulfur content (~70 wt%) (refs 10,11,30), the sulfur particles in the 3D S@PGC (90% S) were much smaller and had a much more uniform distribution. These features can be attributed to the simultaneous formation of the porous structures and the *in situ* deposition of sulfur nanoparticles through the oxidation of Na$_2$S.

To further investigate the interactions between sulfur and 3D PGC, X-ray photoelectron spectroscopy (XPS) was performed. For comparison, XPS data were also collected for pure 3D PGC,

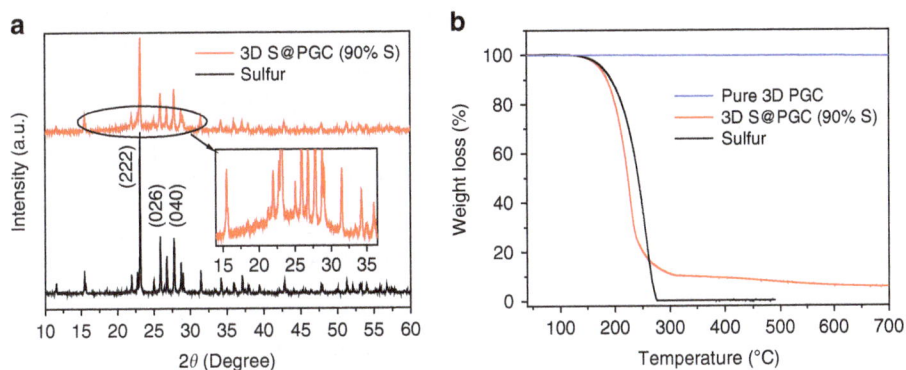

Figure 2 | Structural analysis of the 3D S@PGC (90% S) composite. (**a**) X-ray diffraction patterns of sulfur (bottom trace) and 3D S@PGC (90% S) composite (top trace), which confirm the crystalline feature of the sulfur nanoparticles in the 3D PGC. (**b**) TGA curves of sulfur, pure 3D PGC and the 3D S@PGC (90% S) composite. The 3D PGC apparently did not undergo weight loss up to 700 °C; the calculated sulfur content of the composite is thus ~90 wt%.

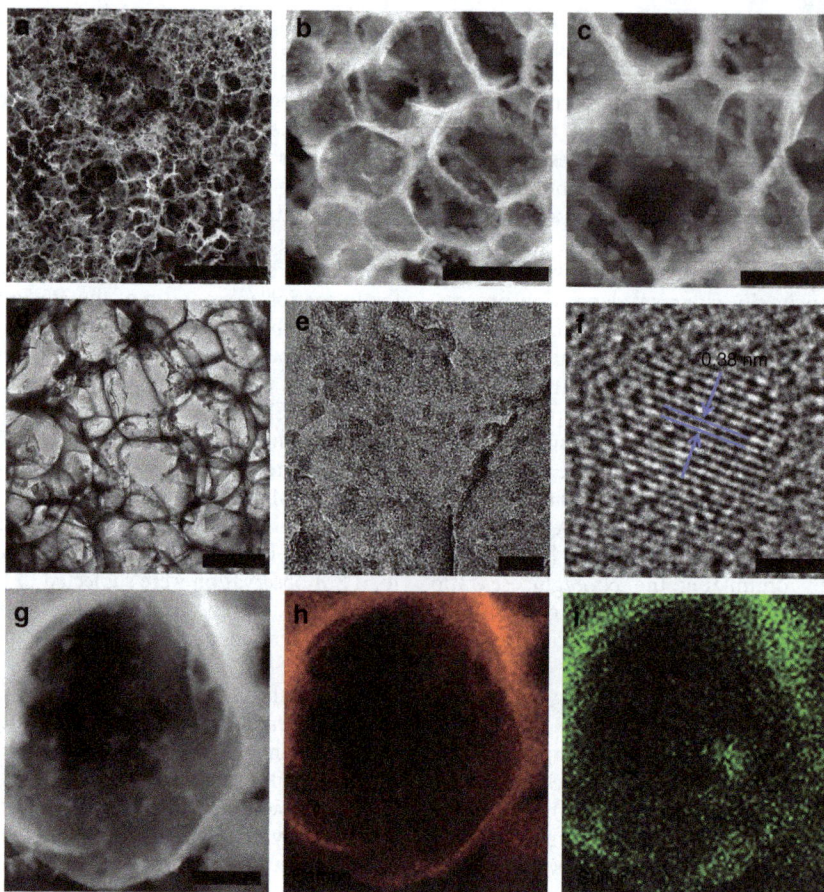

Figure 3 | Morphology of the 3D S@PGC (90% S) composite. (**a–c**) SEM images and (**d,e**) TEM images of the 3D S@PGC (90% S) composite at different magnifications. (**f**) HRTEM image of a sulfur nanoparticle in the composite. (**g**) TEM image of the 3D S@PGC (90% S) composite. (**h,i**) EDS elemental maps of (**h**) carbon and (**i**) sulfur, which were collected from the entire area shown in **g**. Scale bars in **a,b** and **c**: 20; 1; and 0.5 μm. Scale bars in **d,e,f** and **g**: 500; 50; 2; and 200 nm. The SEM and TEM images indicate that the composite possesses a 3D network consisting of interconnected submicron-sized macropores. From the SEM images, the sulfur nanoparticles anchored to the walls of the PGC network were calculated to have a size distribution of 18–54 nm. The EDS results indicate that the sulfur is uniformly distributed in the composite. HRTEM, high-resolution TEM.

which was prepared by immersing NaCl–Na$_2$S@GC in water to remove NaCl and Na$_2$S (Fig. 4a). The C 1s XPS spectrum of pure 3D PGC has a major peak at 284.7 eV, corresponding to sp^2 hybridized carbon, as well as three weak peaks at 286.4, 287.2 and 288.9 eV, which can be ascribed to C–O, C=O and O–C=O species, respectively[14]. The survey XPS spectrum of the 3D S@PGC (90% S) composite confirms the presence of sulfur in 3D PGC (Fig. 4b). In contrast to the C 1s XPS spectrum of pure 3D PGC, that of the 3D S@PGC (90% S) composite has an additional peak at 285.5 eV, which is ascribed to C–S bonds (Fig. 4c)[14]. This finding reveals the presence of covalent bonding between sulfur and PGC. The S 2p XPS peaks, that are characterized by an S 2$p_{3/2}$ and 2$p_{1/2}$ doublet with an energy separation of 1.2 eV, reconfirm the presence of C–S bonds (Fig. 4d), as the binding energy of the S 2$p_{3/2}$ peak (163.5 eV) is lower than that of elemental sulfur (164.0 eV)[14,44]. The weak peak at 168.6 eV is due to sulfate species formed by the oxidation of sulfur in air[14]. The presence of C–S bonds is also supported by Fourier transform infrared (FTIR) spectroscopy because the vibration characteristic of C–S bonds was detected at 671 cm^{-1} (Supplementary Fig. 6)[45]. The C–S bonds could be formed through the addition of various reactive intermediates, including free radicals (for example, HS$^\bullet$) and radical anions (for example, S$^{\bullet -}$ and S$_X^{\bullet -}$)[46], to the unsaturated carbon–carbon double bonds of the PGC as well as through the nucleophilic attack of transient negatively charged polysulphides (for example, S$_X^{2-}$) with

residual oxygen-containing functional groups present in the PGC (see Supplementary Note 1). Therefore, the C–S bonds were formed during the oxidation of Na$_2$S by Fe(NO$_3$)$_3$ because free radicals, radical anions and negatively charged polysulphides were the intermediate products of the oxidation reaction[46].

To reconfirm the presence of C–S bonds, the 3D S@PGC (90% S) composite was subjected to Soxhlet extraction using CS$_2$. The TGA curve of the extracted sample revealed a continual weight loss up to 700 °C (Supplementary Fig. 7); such a weight loss could be assigned to the removal of bonded sulfur[47], and the percentage of bonded sulfur was estimated to be ∼48 wt%. By comparing the sulfur content of the as-prepared 3D S@PGC (90% S) composite to that of the Soxhlet-extracted sample, the bonded and unbonded sulfur content were calculated to be ∼9 and 81 wt% in the 3D S@PGC (90% S) composite (Supplementary Table 1). The X-ray diffraction pattern of the extracted sample did not have any sulfur peaks (Supplementary Fig. 8), which is in line with the formation of C–S bonds[48]. Compared with previously reported physical strategies for confining sulfur[21,23,25,49,50], covalent bonding between sulfur nanoparticles and the PGC framework should effectively prevent the loss of the active materials and stabilize the cycling life of the corresponding Li–S batteries.

The sulfur content of our 3D S@PGC composites is readily tuneable. To tune the amount of sulfur, we used different Na$_2$S/glucose ratios (that is, 1:0.5, 1:0.45 and 1:0.4) during the

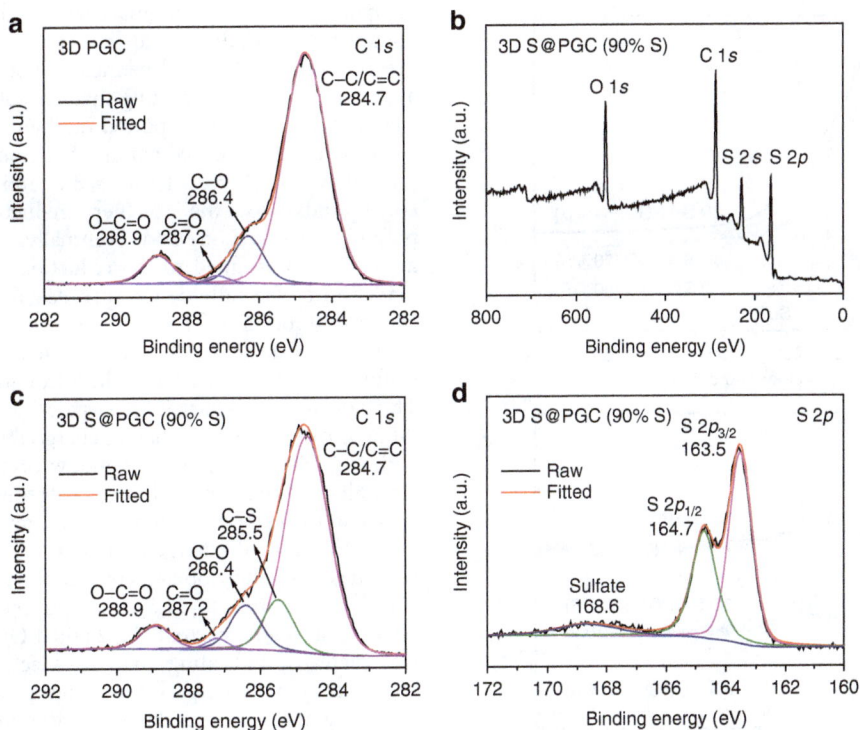

Figure 4 | XPS results for pure 3D PGC and the 3D S@PGC (90% S) composite. (**a**) C 1s XPS spectrum of pure PGC. (**b**) XPS survey spectrum of the 3D S@PGC (90% S) composite. (**c**) C 1s and (**d**) S 2p XPS spectra of the 3D S@PGC (90% S) composite. The data indicate the presence of C–S bonds in the 3D S@PGC (90% S) composite.

preparation while maintaining a constant Na_2S content (2.0 g). Figure 5 displays the TGA, X-ray diffraction and Raman spectroscopy data of 3D S@PGC composites prepared with different Na_2S/glucose ratios. The sulfur content calculated from the TGA curves for the 3D S@PGC composites prepared with the aforementioned ratios was $\sim$64, 70 and 90 wt%, respectively (Fig. 5a). The 3D S@PGC composites exhibited weight losses at different stages. Compared with pure sulfur, the 3D S@PGC composites exhibited slightly lower starting evaporation temperatures. The evaporation of sulfur in this temperature range was ascribed to the sulfur nanoparticles attached to the walls of the PGC without covalent bonding. This lowering of the evaporation temperature was mainly ascribed to the nanosized distribution of the sulfur particles, which possessed excess surface free energy and facilitated heat transfer because of the large contact area between the sulfur nanoparticles and the 3D PGC (ref. 51). The 3D S@PGC composites also exhibited evaporation temperature ranges that were higher than that of pure sulfur. This phenomenon was caused by the sulfur that filled the micro- and mesopores[52] and covalently bonded to the wall surfaces of the PGC[47]. The 3D S@PGC (64% S) composite exhibited the most obvious weight loss at a relatively higher temperature because the proportion of sulfur contained within the micro- and mesopores and bonded to the PGC relative to the total amount of sulfur was the highest among the three composites. The X-ray diffraction patterns of the composites contain reflection peaks for both sulfur (23.1°, 25.9° and 27.8°) and GC (broad peak from 15 to 35°) (Fig. 5b and Supplementary Fig. 1), indicating the presence of sulfur in the 3D PGC. With the decrease in the PGC content of the composites, the intensity of the PGC peak became weaker. Collectively, the sulfur content of the 3D S@PGC composites increased with the amount of Na_2S used in the synthetic procedure (Supplementary Table 1).

The structural features of sulfur and the carbon matrix of the 3D S@PGC composites were further investigated by Raman spectroscopy (Fig. 5c). Elemental sulfur produced characteristic peaks at 471, 216 and 151 cm^{-1}. In agreement with the XRD result, the Raman spectra of the 3D S@PGC composites contain characteristic peaks of sulfur. Furthermore, the Raman spectra of the composites present peaks at $\sim$1,350 and 1,588 cm^{-1}, which correspond to the D and G bands of carbon materials[53]. The presence of such bands in the Raman spectra suggests the conversion of glucose into GC. With the decrease in the PGC content of the composites, the D and G bands also became weaker.

The SEM and TEM images (Supplementary Figs 9 and 10) reveal that 3D S@PGC composites with different amounts of sulfur display similar morphologies, indicating the retention of the unique honeycomb-like network in the composites. Although the sizes of the sulfur nanoparticles calculated from the SEM images increased with the sulfur content (that is, 6–24 nm for 3D S@PGC (64% S), 12–35 nm for 3D S@PGC (70% S) and 18–54 nm for 3D S@PGC (90% S); Supplementary Fig. 2 and Supplementary Table 1), the sulfur particle size found in the 3D S@PGC (90% S) composite (18–54 nm) still resulted in high sulfur utilization, as discussed below. The sizes of the sulfur nanoparticles calculated from the Scherrer formula using the (222) Bragg reflection at 23.1° ($\sim$28, 33 and 44 nm for the 3D S@PGC (64% S), 3D S@PGC (70% S) and 3D S@PGC (90% S) composites, respectively) are in agreement with the values obtained from the SEM images. The FTIR spectra and XPS data confirm that C–S bonds also exist in the 3D S@PGC (70% S) and 3D S@PGC (64%) composites (Supplementary Figs 6 and 11). The pore structures of the 3D S@PGC composites were also characterized using N_2 physisorption measurements, providing insight into the micro- and mesopore structures of the composites (Supplementary Fig. 12 and Supplementary Note 2). All the 3D S@PGC composites had smaller surface areas and pore volumes than the corresponding PGC frameworks obtained by immersing NaCl–Na_2S@GC in water. The marked decreases in the specific

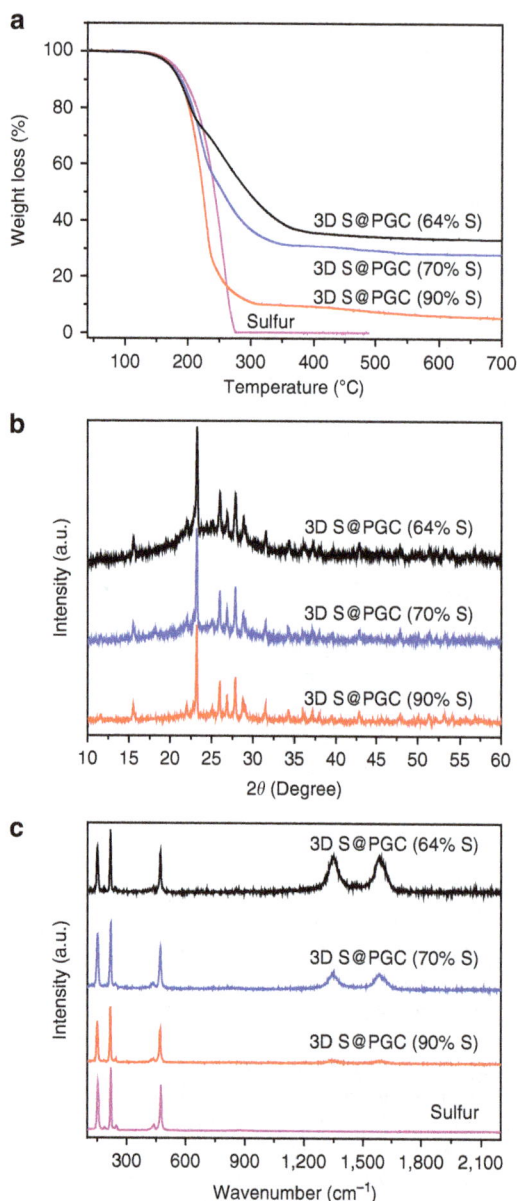

Figure 5 | Characterization of 3D S@PGC composites with various sulfur contents. (**a**) TGA curves, (**b**) X-ray diffraction patterns and (**c**) Raman spectra reveal that the sulfur content in the 3D S@PGC composites can be readily tuned by changing the Na_2S/glucose ratio and that the sulfur nanoparticles in the composites have the same crystalline structures.

surface area and total pore volume upon the deposition of sulfur nanoparticles indicated that the sulfur occupied the volumes of the pores in the 3D PGC frameworks. A comparison of the pore-size distributions of the 3D S@PGC composites and the corresponding 3D PGC frameworks indicated that the latter contained numerous micro- and mesopores, with sizes ranging from 1.5 to 25 nm, which were filled during formation of the composites. Combining the data from the SEM, TEM and N_2 physisorption analyses, we can conclude that the sulfur situates both in the micro-/mesopores and on the walls of the macropores of the 3D PGC.

The 3D S@PGC composites for Li–S battery cathode materials. Coin cells with Li foil as an anode were fabricated to evaluate the electrochemical performance of the 3D S@PGC (90% S)

composite as a cathode material. Profiles obtained by cyclic voltammetry conducted at a scan rate of $0.1\,mV\,s^{-1}$ are presented in Fig. 6a. Well-defined reduction peaks were observed at 2.37 and 2.04 V, indicating that the sulfur was reduced in two stages. According to a reported mechanism[15], the peak at 2.37 V corresponds to the reduction of elemental sulfur to lithium polysulphides (Li_2S_n, $4 \leq n \leq 8$), and the peak at 2.04 V corresponds to the further reduction of the lithium polysulphides to Li_2S_2 and, eventually, to Li_2S. The weak peak at $\sim 1.7\,V$ is assigned to the reduction of $LiNO_3$. Likewise, the oxidation of the cathode also proceeded through two stages: the conversion of Li_2S_2/Li_2S to Li_2S_n ($n > 2$), associated with the oxidation peak at 2.43 V, and the final formation of elemental sulfur, corresponding to the oxidation peak at 2.47 V. The galvanostatic charge/discharge behaviour of the 3D S@PGC (90% S) cathode was studied at charge/discharge rates of 0.5, 1, 2 and 4 C within a potential window of 1.5–3.0 V versus Li^+/Li (Fig. 6b–e). Consistent with the two reduction peaks at 2.37 and 2.04 V in the cathodic sweep, two plateaus were observed at ~ 2.3 and 2.0 V in the discharge process at 0.5 C, corresponding to the two-stage reduction of elemental sulfur to lithium polysulphides (Li_2S_{4-8}) and then to Li_2S_2/Li_2S, respectively. Discharge curves obtained at higher currents (1, 2 and 4 C) also clearly contain the two plateaus, indicating that the electrochemical reactions at higher charge/discharge rates follow processes similar to those occurring at lower rates. However, the non-conductive nature of Li_2S_2 and Li_2S subjects the conversion of Li_2S_{4-8} to Li_2S_2/Li_2S to higher polarization at higher charge/discharge rates, that is, decreased voltage at the second plateau[54]. As previously noted[16,47,55,56], the charge/discharge profiles can be sloppy at the ends. This phenomenon could be ascribed to the presence of C–S bonds[16,47] and the strong absorption of sulfur in the micro- and mesopores of the 3D PGC[55,56], as well as the irreversible reduction of $LiNO_3$ (refs 57,58).

The cycling performance of the 3D S@PGC (90% S) composite cathode was first studied at 0.5 C (Supplementary Fig. 13). The material achieved an initial discharge capacity as high as $1,382\,mA\,h\,g^{-1}$, which corresponds to 82.5% sulfur utilization based on the theoretical value of sulfur ($1,675\,mA\,h\,g^{-1}$). After 200 cycles, the calculated capacity retention was 62%. The high utilization rate of sulfur is because of the hierarchical porous structures and the nanosized sulfur particles distributed on the GC walls. The macropores highly favour the rapid access to the electrode interior by the electrolyte, while the nanosized sulfur particles and ultrathin GC walls with micro- and mesopores facilitate the efficient transport of ions into the deeper portions of the sulfur nanoparticles because of the short pathway[18,23,41]. Figure 6f displays the cycling performance of the 3D S@PGC (90% S) composite cathodes at charge/discharge rates of 1, 2 and 4 C. The initial discharge capacity was $1,242\,mA\,h\,g^{-1}$ at 1 C. After 200 cycles, a capacity of $917\,mA\,h\,g^{-1}$ remained, corresponding to a capacity retention of 74%. When the charge/discharge rate was raised to 2 C, the initial capacity obtained ($1,115\,mA\,h\,g^{-1}$) became lower than the corresponding value at 1 C, but the cycling stability improved. A capacity of $920\,mA\,h\,g^{-1}$ corresponding to a capacity retention of 83% was obtained after 200 cycles at 2 C. When the charge/discharge rate was further elevated to 4 C, the measured initial capacity was $638\,mA\,h\,g^{-1}$. After 200 cycles, the capacity remained at $548\,mA\,h\,g^{-1}$, which corresponds to a capacity retention of 86%. The capacities of the composite obtained at current densities of 0.5, 1 and 2 C (that is, 1,382, 1,242 and $1,115\,mA\,h\,g^{-1}$ at 0.5, 1 and 2 C, respectively) are much higher than those of previously reported sulfur–carbon composites ($430-1,100\,mA\,h\,g^{-1}$ at 0.5 C, $400-980\,mA\,h\,g^{-1}$ at 1 C and $450-900\,mA\,h\,g^{-1}$ at 2 C)[11,21,36,37,52,59–62]. Although some capacities reported at

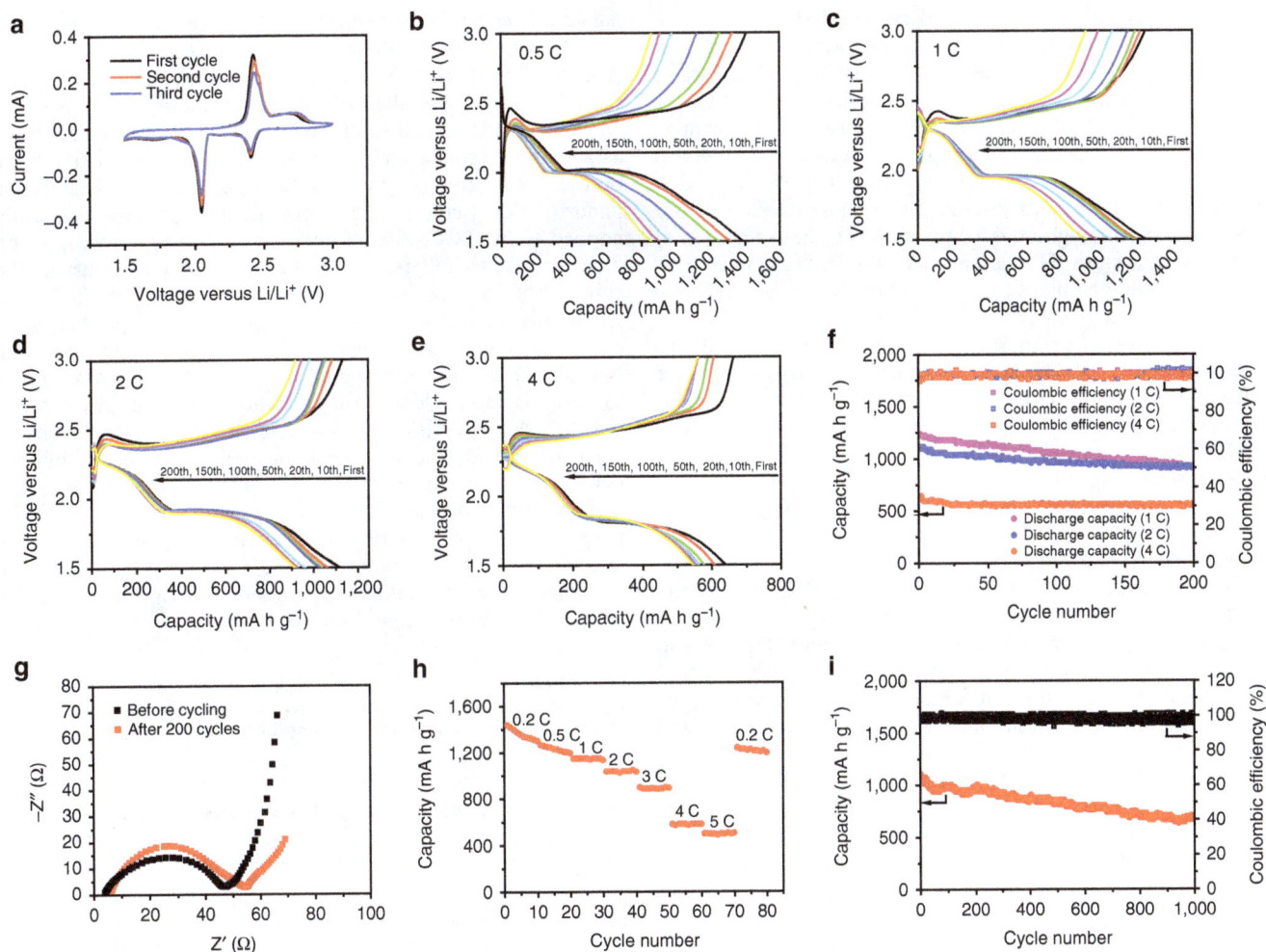

Figure 6 | Electrochemical performance of the 3D S@PGC (90% S) composite as a cathode material for Li–S batteries. (a) CV profiles of the 3D S@PGC (90% S) composite cathode. (b-e) Charge/discharge curves of the cathode at charge/discharge rates of (b) 0.5 C, (c) 1 C, (d) 2 C and (e) 4 C. (f) Cycling performance of the cathode at charge/discharge rates of 1, 2 and 4 C. (g) EIS curves of the cathode before and after 200 cycles at 2 C. (h) Rate performance of the cathode. (i) Cycling performance of the cathode over 1,000 cycles at a charge/discharge rate of 2 C. The data indicate that the 3D S@PGC (90% S) composite displays excellent electrochemical performance, in particular, a high sulfur utilization, high specific capacity (1,115 mA h g^{-1} at 2 C), long cycling life (small capacity decay of 0.039% per cycle over 1,000 cycles at 2 C), and excellent rate capability at a high charge/discharge current. CV, cyclic voltammetry. EIS, electrochemical impedance spectroscopy.

higher current densities (for example, 4 and 5 C) were relatively high (650–750 mA h g^{-1}), the sulfur content of their composites was relatively low (40–65 wt%)[21,60]. The S@PGC (90% S) composite cathode exhibited high coulombic efficiencies (>98%) at all the current rates tested, confirming that the shuttling effect of the polysulphides was efficiently suppressed because of the C–S bonding between the sulfur nanoparticles and 3D PGC as well as the unique porous structure of the 3D S@PGC composite. The use of LiNO$_3$ in the electrolyte also helped suppress polysulphide shuttling, improving the coulombic efficiency and enhancing the rate capability[63,64], as demonstrated by experiments without LiNO$_3$ in the electrolyte (Supplementary Fig. 14 and Supplementary Table 2). The relatively lower capacity retention at a lower charge/discharge rate was likely due to the incomplete oxidation of the insulating Li$_2$S$_2$ and Li$_2$S (refs 55,65) because the shuttling effect was efficiently suppressed and the coulombic efficiency was found to be higher than 98%.

The stability of the 3D S@PGC (90% S) composite cathode was evidenced by electrochemical impedance spectroscopy (EIS) before and after 200 cycles at 2 C (Fig. 6g). The corresponding Nyquist profiles were fitted to a widely used equivalent circuit

(Supplementary Fig. 15)[66]. The electrolyte resistance (R_e) and the charge-transfer resistance at the electrode/electrolyte interface (R_{ct}) were determined to be 3.6 and 43.2 Ω before cycling and 4.6 and 47.9 Ω after cycling, respectively. The slight changes in R_e and R_{ct} after the charge/discharge cycles, which could be ascribed to the formation of a passive layer on the carbon frameworks[67], suggest the high conductivity and good stability of the 3D S@PGC (90% S) cathode due to the stable 3D PGC framework and the C–S bonds between the sulfur nanoparticles and the 3D PGC framework. TEM observations also indicated the stability of the 3D S@PGC (90% S) composite electrode, as the porous structures of the 3D PGC remained intact and the sulfur nanoparticles were still firmly and homogeneously anchored to the PGC walls after 200 charge/discharge cycles at 2 C (Supplementary Fig. 16). In addition, the FTIR data of the electrode material indicated the retention of the C–S bonds after cycling (Supplementary Fig. 17), consistent with a previous study that reported the retention of C–S bonds after cycling[68].

The rate capability of the 3D S@PGC (90% S) composite is depicted in Fig. 6h. The discharge capacity gradually decreased as the current rate increased from 0.2 to 5 C. At the maximum charge/discharge rate tested (5 C), the specific capacity remained high (500 mA h g^{-1}) and the cycling remained stable; when the current rate was restored to 0.2 C, the composite recovered most of its capacity. Most importantly, the composite exhibited stable cycling performance over 1,000 charge/discharge cycles at 2 C (Fig. 6i). A high specific capacity (670 mA h g^{-1}) was retained after 1,000 cycles. The calculated capacity retention is 61%, which corresponds to a very small capacity decay (0.039% per cycle). This finding demonstrates the excellent cycling stability of our Li–S batteries. Although other composites have been reported to exhibit a long cycling life when used as cathodes in Li–S batteries, the sulfur content employed in such materials was relatively low (ranging from 30 to 80%)[30,37,63,69]. Therefore, 3D S@PGC is a high-sulfur-content (up to 90%) cathode material that exhibits excellent cycling stability at a high current density. The pure sulfur cathode (the control) used under the same conditions exhibited a much lower specific capacity and worse cycling stability than those of the 3D S@PGC (90% S) composite (Supplementary Fig. 18).

Discussion

The electrochemical performance of the 3D S@PGC (64% S) and 3D S@PGC (70% S) composites as cathodes in Li–S batteries were also evaluated. As summarized in Supplementary Fig. 19 and Supplementary Tables 3 and 4, 3D S@PGC composites with a lower sulfur content exhibited better performance: specifically, higher specific capacities, higher capacity retention and higher rate performance (Supplementary Note 3). These results are attributable to the smaller sizes of the sulfur nanoparticles in the 3D S@PGC composites with a relatively low sulfur content. Indeed, smaller sulfur nanoparticles have larger specific contact areas with the 3D PGC framework, which helps alleviate the shuttling effect and improves the cycle stability. Smaller particle sizes also facilitate electron and Li$^+$ diffusion and lead to better sulfur utilization and a higher specific capacity. The larger specific surface areas effectively reduce the discharging current densities and the Li$^+$ flux, thereby limiting the formation of a Li$_2$S blocking layer at high charge/discharge rates[51]. Although the 3D S@PGC composites with a lower sulfur content exhibited higher specific capacities calculated on the basis of sulfur, the relatively low sulfur content reduced the overall volumetric capacity and energy density of the corresponding cathodes. Therefore, 3D S@PGC composites with relatively high sulfur content may be promising candidates for use in practical applications.

The excellent overall electrochemical performance of the 3D S@PGC composites can be attributed to the following factors that stem from the design of the materials. First, the *in situ* chemical deposition method allows access to composites with high sulfur content (up to 90 wt%) and affords the nanoscale distribution of the sulfur particles in the resultant 3D PGC network. As described above, nanosized sulfur particles facilitate a high sulfur utilization rate (82.5% for 3D S@PGC (90% S), 84.5% for 3D S@PGC (70% S) and 86% for 3D S@PGC (64% S) at 0.5 C). Second, the C–S bonds formed between the sulfur nanoparticles and 3D PGC can effectively prevent agglomeration of the sulfur nanoparticles, minimize the loss of lithium polysulphides to the electrolyte and suppress the shuttling effect during the charge/discharge cycles. Third, the 3D PGC networks that display high electrical conductivities, large surface areas and high mechanical flexibility confer high electrical conductivity and structural integrity to the electrodes. The numerous walls between the interconnected macropores may function as multilayered barriers to further mitigate the dissolution of polysulphides into the electrolyte. Finally, the unique interconnected hierarchical pores in the 3D PGC network facilitate access to the sulfur nanoparticles by the electrolyte and preserve the rapid transport of Li$^+$ to the active material.

In conclusion, we report a new methodology that is facile and scalable and allows the *in situ* preparation of 3D S@PGC composites with a high sulfur content. The strategy utilizes Na$_2$S as a sulfur precursor and NaCl and Na$_2$S as a template for the porous structure of the resultant composite. The sulfur nanoparticles were homogenously distributed and covalently bonded to 3D PGC, as confirmed by various spectroscopic and microscopic techniques. Li–S batteries prepared using the composites as cathodes exhibited excellent performance; specifically, high sulfur utilization, high specific capacities, good cycling stabilities and high rate capabilities were observed. Notably, Li–S batteries prepared using 3D S@PGC (90% S) as a cathode displayed a long cycling stability, with a capacity decay of only 0.039% per cycle over 1,000 cycles at a high charge/discharge current (2 C). Overall, the methodology described herein offers a new avenue for the fabrication of cathode materials based on carbon–sulfur hybridized nanostructures for use in high-performance Li–S batteries. We believe that the strategy may also inspire the preparation of other 3D porous structures for use in other areas, including applications in catalysis, selective adsorption, separations and sensing.

Methods

Materials. All reagents were purchased from commercial sources and used without further purification. All solvents used were purified using standard procedures.

Representative synthesis of 3D S@PGC composites. In a typical synthesis, Na$_2$S · 9H$_2$O (2.0 g), NaCl (5.0 g) and glucose (0.8 g) were dissolved in DI water (15 ml). The resultant solution was frozen in liquid nitrogen, and the water in the mixture was removed via freeze-drying. The resultant gel was ground into a fine powder and then heated at 750 °C for 2 h under an atmosphere of argon. A black powder was obtained and subsequently stirred in an aqueous solution of Fe(NO$_3$)$_3$ (20 g Fe(NO$_3$)$_3$ · 9H$_2$O in 150 ml DI water) for 40 h to dissolve the residual NaCl crystals and to deposit the sulfur. Afterwards, the black powder product was washed several times with DI water and centrifuged to afford the desired composite. Composites with various sulfur contents were synthesized by using different amounts of glucose (0.9 and 1.0 g) in the aforementioned procedure.

Characterization. X-ray diffraction data were collected on a Bruker D8 Focus diffractometer using an incident wavelength of 0.154 nm (Cu Kα radiation) and a Lynx-Eye detector. Raman spectra were recorded on a Renishaw inVia-Reflex confocal Raman microscope at an excitation wavelength of 532 nm. TGA measurements were carried out using a TGA Q50 at a scanning rate of 10 °C min^{-1}. SEM observations were performed on a field-emission SEM (Hitachi S-4800) equipped with EDS. TEM images were obtained using a JEOL-2100F microscope operated under an accelerating voltage of 200 kV. EDS analysis was also performed on Tecnai F20 scanning transmission electron microscope operated at 200 keV using an Oxford detector with a beam current of ∼1 nA. N$_2$ adsorption–desorption isotherms and pore-size distribution were obtained at 77 K using a QuadraSorb SI MP apparatus. The total specific surface areas of the samples were calculated via the Brunauer–Emmett–Teller method. The pore-size distribution was calculated via the density functional theory model. XPS spectra were recorded on a PHI Quantera Scanning X-ray Microprobe using monochromated Al Kα radiation (1486.7 eV). FTIR spectra were recorded on an Excalibur 3100 spectrometer with a resolution of 0.2 cm^{-1} using KBr pellets.

Electrochemistry. The 3D S@PGC composites were combined with conductive carbon and poly(vinylidene fluoride) as a binder with a mass ratio of 80:10:10 and milled into a slurry with N-methylpyrrolidone. The slurry was then blade cast onto a carbon-coated Al foil and dried at 50 °C for 10 h in a vacuum oven. The loading density of sulfur was ca. 2.36 mg cm^{-2}. CR2032 coin cells were assembled in an argon-filled glove box employing the 3D S@PGC-coated Al foil as the cathode, a porous membrane (Celgard 3501) as the separator, and lithium foil as the reference/counter electrode. The electrolyte used was lithium bis(trifluoro methane)sulphonimide (0.38 M) and lithium nitrate (0.31 M) in a solvent mixture of 1,3-dioxolane and 1,2-dimethoxy ethane (1:1 v/v). Pristine sulfur electrodes were fabricated under similar conditions. Cyclic voltammetry curves were collected using a CHI 660E electrochemical workstation at a scan rate of 0.1 mV s^{-1} from 3.0 to 1.5 V. Cycling tests of the batteries were galvanostatically performed at various charge/discharge rates within a potential window of 1.5–3.0 V versus

Li^+/Li. The electrochemical impedance spectroscopy data were recorded using a Zennium 40088 electrochemical workstation by applying a sine wave with an amplitude of 10 mV over a frequency range of 100 kHz to 10 mHz.

References

1. Bruce, P. G., Freunberger, S. A., Hardwick, L. J. & Tarascon, J. M. Li-O$_2$ and Li-S batteries with high energy storage. *Nat. Mater.* **11**, 19–29 (2012).
2. Choi, N. S. *et al.* Challenges facing lithium batteries and electrical double-layer capacitors. *Angew. Chem. Int. Ed. Engl.* **51**, 9994–10024 (2012).
3. Evers, S. & Nazar, L. F. New approaches for high energy density lithium-sulfur battery cathodes. *Acc. Chem. Res.* **46**, 1135–1143 (2013).
4. Manthiram, A., Fu, Y. Z. & Su, Y. S. Challenges and prospects of lithium-sulfur batteries. *Acc. Chem. Res.* **46**, 1125–1134 (2013).
5. Yamin, H., Gorenshtein, A., Penciner, J., Sternberg, Y. & Peled, E. Lithium sulfur battery—oxidation reduction-mechanisms of polysulfides in THF solutions. *J. Electrochem. Soc.* **135**, 1045–1048 (1988).
6. Peled, E., Sternberg, Y., Gorenshtein, A. & Lavi, Y. Lithium-sulfur battery—evaluation of dioxolane-based electrolytes. *J. Electrochem. Soc.* **136**, 1621–1625 (1989).
7. Jin, B., Kim, J. U. & Gu, H. B. Electrochemical properties of lithium-sulfur batteries. *J. Power Sources* **117**, 148–152 (2003).
8. Chen, R. J. *et al.* Graphene-based three-dimensional hierarchical sandwich-type architecture for high-performance Li/S batteries. *Nano Lett.* **13**, 4642–4649 (2013).
9. Huang, J. Q. *et al.* Entrapment of sulfur in hierarchical porous graphene for lithium-sulfur batteries with high rate performance from −40 to 60 °C. *Nano Energy* **2**, 314–321 (2013).
10. Wang, H. L. *et al.* Graphene-wrapped sulfur particles as a rechargeable lithium-sulfur battery cathode material with high capacity and cycling stability. *Nano Lett.* **11**, 2644–2647 (2011).
11. Zhou, W. D. *et al.* Amylopectin wrapped graphene oxide/sulfur for improved cyclability of lithium-sulfur battery. *ACS Nano* **7**, 8801–8808 (2013).
12. Wang, C. *et al.* Macroporous free-standing nano-sulfur/reduced graphene oxide paper as stable cathode for lithium-sulfur battery. *Nano Energy* **11**, 678–686 (2015).
13. Wang, C. *et al.* Slurryless Li$_2$S/reduced graphene oxide cathode paper for high-performance lithium sulfur battery. *Nano Lett.* **15**, 1796–1802 (2015).
14. Wang, Z. Y. *et al.* Enhancing lithium-sulfur battery performance by strongly binding the discharge products on amino-functionalized reduced graphene oxide. *Nat. Commun.* **5**, 5002 (2014).
15. Zhao, M. Q. *et al.* Unstacked double-layer templated graphene for high-rate lithium-sulfur batteries. *Nat. Commun.* **5**, 3410 (2014).
16. Guo, J. C., Xu, Y. H. & Wang, C. S. Sulfur-impregnated disordered carbon nanotubes cathode for lithium-sulfur batteries. *Nano Lett.* **11**, 4288–4294 (2011).
17. Han, S. C. *et al.* Effect of multiwalled carbon nanotubes on electrochemical properties of lithium sulfur rechargeable batteries. *J. Electrochem. Soc.* **150**, A889–A893 (2003).
18. He, G., Ji, X. L. & Nazar, L. High 'C' rate Li-S cathodes: sulfur imbibed bimodal porous carbons. *Energy Environ. Sci.* **4**, 2878–2883 (2011).
19. Ji, X. L., Lee, K. T. & Nazar, L. F. A highly ordered nanostructured carbon-sulfur cathode for lithium-sulfur batteries. *Nat. Mater.* **8**, 500–506 (2009).
20. Zhang, B., Qin, X., Li, G. R. & Gao, X. P. Enhancement of long stability of sulfur cathode by encapsulating sulfur into micropores of carbon spheres. *Energy Environ. Sci.* **3**, 1531–1537 (2010).
21. Xin, S. *et al.* Smaller sulfur molecules promise better lithium-sulfur batteries. *J. Am. Chem. Soc.* **134**, 18510–18513 (2012).
22. Xin, S., Yin, Y. X., Guo, Y. G. & Wan, L. J. A high-energy room-temperature sodium-sulfur battery. *Adv. Mater.* **26**, 1261–1265 (2014).
23. Schuster, J. *et al.* Spherical ordered mesoporous carbon nanoparticles with high porosity for lithium-sulfur batteries. *Angew. Chem. Int. Ed. Engl.* **51**, 3591–3595 (2012).
24. Song, J. X. *et al.* Strong lithium polysulfide chemisorption on electroactive sites of nitrogen-doped carbon composites for high-performance lithium-sulfur battery cathodes. *Angew. Chem. Int. Ed. Engl.* **54**, 4325–4329 (2015).
25. Zhang, C. F., Wu, H. B., Yuan, C. Z., Guo, Z. P. & Lou, X. W. Confining sulfur in double-shelled hollow carbon spheres for lithium-sulfur batteries. *Angew. Chem. Int. Ed. Engl.* **51**, 9592–9595 (2012).
26. Su, Y. S. & Manthiram, A. Lithium-sulfur batteries with a microporous carbon paper as a bifunctional interlayer. *Nat. Commun.* **3**, 1166 (2012).
27. Choi, Y. J., Kim, K. W., Ahn, H. J. & Ahn, J. H. Improvement of cycle property of sulfur electrode for lithium/sulfur battery. *J. Alloy. Compd.* **449**, 313–316 (2008).
28. Ji, L. W. *et al.* Porous carbon nanofiber-sulfur composite electrodes for lithium/sulfur cells. *Energy Environ. Sci.* **4**, 5053–5059 (2011).
29. Zheng, G. Y., Yang, Y., Cha, J. J., Hong, S. S. & Cui, Y. Hollow carbon nanofiber-encapsulated sulfur cathodes for high specific capacity rechargeable lithium batteries. *Nano Lett.* **11**, 4462–4467 (2011).
30. Seh, Z. W. *et al.* Sulfur-TiO$_2$ yolk-shell nanoarchitecture with internal void space for long-cycle lithium-sulfur batteries. *Nat. Commun.* **4**, 1331 (2013).
31. Zhou, W. D., Yu, Y. C., Chen, H., DiSalvo, F. J. & Abruna, H. D. Yolk-shell structure of polyaniline-coated sulfur for lithium-sulfur batteries. *J. Am. Chem. Soc.* **135**, 16736–16743 (2013).
32. Kim, J. S., Hwang, T. H., Kim, B. G., Min, J. & Choi, J. W. A lithium-sulfur battery with a high areal energy density. *Adv. Funct. Mater.* **24**, 5359–5367 (2014).
33. Chung, W. J. *et al.* The use of elemental sulfur as an alternative feedstock for polymeric materials. *Nat. Chem* **5**, 518–524 (2013).
34. Griebel, J. J., Li, G. X., Glass, R. S., Char, K. & Pyun, J. Kilogram scale inverse vulcanization of elemental sulfur to prepare high capacity polymer electrodes for Li-S batteries. *J. Polym. Sci. A Polym. Chem* **53**, 173–177 (2015).
35. Chen, S. R. *et al.* Ordered mesoporous carbon/sulfur nanocomposite of high performances as cathode for lithium-sulfur battery. *Electrochim. Acta* **56**, 9549–9555 (2011).
36. Ji, L. W. *et al.* Graphene oxide as a sulfur immobilizer in high performance lithium/sulfur cells. *J. Am. Chem. Soc.* **133**, 18522–18525 (2011).
37. Lu, S. T., Cheng, Y. W., Wu, X. H. & Liu, J. Significantly improved long-cycle stability in high-rate Li-S batteries enabled by coaxial graphene wrapping over sulfur-coated carbon nanofibers. *Nano Lett.* **13**, 2485–2489 (2013).
38. Wang, D. W. *et al.* A microporous-mesoporous carbon with graphitic structure for a high-rate stable sulfur cathode in carbonate solvent-based Li-S batteries. *Phys. Chem. Chem. Phys.* **14**, 8703–8710 (2012).
39. Xin, S., Yin, Y. X., Wan, L. J. & Guo, Y. G. Encapsulation of sulfur in a hollow porous carbon substrate for superior Li-S batteries with long lifespan. *Part. Part. Syst. Char.* **30**, 321–325 (2013).
40. Zhang, Z. W. *et al.* 3D interconnected porous carbon aerogels as sulfur immobilizers for sulfur impregnation for lithium-sulfur batteries with high rate capability and cycling stability. *Adv. Funct. Mater.* **24**, 2500–2509 (2014).
41. Qin, J. *et al.* Graphene networks anchored with Sn@graphene as lithium ion battery anode. *ACS Nano* **8**, 1728–1738 (2014).
42. Zhu, T., Chen, J. S. & Lou, X. W. Glucose-assisted one-pot synthesis of FeOOH nanorods and their transformation to Fe$_3$O$_4$@carbon nanorods for application in lithium ion batteries. *J. Phys. Chem. C* **115**, 9814–9820 (2011).
43. Wang, Q. Q. *et al.* Improve rate capability of the sulfur cathode using a gelatin binder. *J. Electrochem. Soc.* **158**, A775–A779 (2011).
44. Zheng, S. *et al.* High performance C/S composite cathodes with conventional carbonate-based electrolytes in Li-S battery. *Sci. Rep.* **4**, 4842 (2014).
45. Yu, X. U. *et al.* Stable-cycle and high-capacity conductive sulfur-containing cathode materials for rechargeable lithium batteries. *J. Power Sources* **146**, 335–339 (2005).
46. Steudel, R. Mechanism for the formation of elemental sulfur from aqueous sulfide in chemical and microbiological desulfurization processes. *Ind. Eng. Chem. Res.* **35**, 1417–1423 (1996).
47. Zhang, Y. Z., Liu, S., Li, G. C., Li, G. R. & Gao, X. P. Sulfur/polyacrylonitrile/carbon multi-composites as cathode materials for lithium/sulfur battery in the concentrated electrolyte. *J. Mater. Chem. A* **2**, 4652–4659 (2014).
48. Zu, C. X. & Manthiram, A. Hydroxylated graphene-sulfur nanocomposites for high-rate lithium-sulfur batteries. *Adv. Energy Mater* **3**, 1008–1012 (2013).
49. Liang, C. D., Dudney, N. J. & Howe, J. Y. Hierarchically structured sulfur/carbon nanocomposite material for high-energy lithium battery. *Chem. Mater.* **21**, 4724–4730 (2009).
50. Ye, H., Yin, Y. X., Xin, S. & Guo, Y. G. Tuning the porous structure of carbon hosts for loading sulfur toward long lifespan cathode materials for Li-S batteries. *J. Mater. Chem. A* **1**, 6602–6608 (2013).
51. Chen, H. W. *et al.* Monodispersed sulfur nanoparticles for lithium sulfur batteries with theoretical performance. *Nano Lett.* **15**, 798–802 (2015).
52. Li, Z. *et al.* A highly ordered meso@microporous carbon-supported sulfur@smaller sulfur core-shell structured cathode for Li-S batteries. *ACS Nano* **8**, 9295–9303 (2014).
53. Sun, J. *et al.* Fluorine-doped SnO$_2$@graphene porous composite for high capacity lithium-ion batteries. *Chem. Mater.* **27**, 4594–4603 (2015).
54. Zhang, S. S. New insight into liquid electrolyte of rechargeable lithium/sulfur battery. *Electrochim. Acta* **97**, 226–230 (2013).
55. Li, G. C., Li, G. R., Ye, S. H. & Gao, X. P. A polyaniline-coated sulfur/carbon composite with an enhanced high-rate capability as a cathode material for lithium/sulfur batteries. *Adv. Energy Mater.* **2**, 1238–1245 (2012).
56. Li, G. C., Hu, J. J., Li, G. R., Ye, S. H. & Gao, X. P. Sulfur/activated-conductive carbon black composites as cathode materials for lithium/sulfur battery. *J. Power Sources* **240**, 598–605 (2013).
57. Zhang, S. S. Role of LiNO$_3$ in rechargeable lithium/sulfur battery. *Electrochim. Acta* **70**, 344–348 (2012).
58. Zhang, S. S. Effect of discharge cutoff voltage on reversibility of lithium/sulfur batteries with LiNO$_3$-contained electrolyte. *J. Electrochem. Soc.* **159**, A920–A923 (2012).

59. Li, B., Li, S. M., Liu, J. H., Wang, B. & Yang, S. B. Vertically aligned sulfur-graphene nanowalls on substrates for ultrafast lithium-sulfur batteries. *Nano Lett.* **15**, 3073–3079 (2015).

60. Zhou, G. M. *et al.* A graphene-pure-sulfur sandwich structure for ultrafast, long-life lithium-sulfur batteries. *Adv. Mater.* **26**, 625–631 (2014).

61. Jayaprakash, N., Shen, J., Moganty, S. S., Corona, A. & Archer, L. A. Porous hollow carbon@sulfur composites for high-power lithium-sulfur batteries. *Angew. Chem., Int. Ed. Engl.* **50**, 5904–5908 (2011).

62. Zhao, Y., Wu, W. L., Li, J. X., Xu, Z. C. & Guan, L. H. Encapsulating MWNTs into hollow porous carbon nanotubes: a tube-in-tube carbon nanostructure for high-performance lithium-sulfur batteries. *Adv. Mater.* **26**, 5113–5118 (2014).

63. Song, M. K., Zhang, Y. G. & Cairns, E. J. A long-life, high-rate lithium/sulfur cell: a multifaceted approach to enhancing cell performance. *Nano Lett.* **13**, 5891–5899 (2013).

64. Choudhury, S. *et al.* Nanoporous cathodes for high-energy Li-S batteries from gyroid block copolymer templates. *ACS Nano* **9**, 6147–6157 (2015).

65. Evers, S. & Nazar, L. F. Graphene-enveloped sulfur in a one pot reaction: a cathode with good coulombic efficiency and high practical sulfur content. *Chem. Commun.* **48**, 1233–1235 (2012).

66. Narayanan, S. R., Shen, D. H., Surampudi, S., Attia, A. I. & Halpert, G. Electrochemical impedance spectroscopy of lithium-titanium disulfide rechargeable cells. *J. Electrochem. Soc.* **140**, 1854–1861 (1993).

67. Zhao, M. Q. *et al.* Hierarchical vine-tree-like carbon nanotube architectures: *in situ* CVD self-assembly and their use as robust scaffolds for lithium-sulfur batteries. *Adv. Mater.* **26**, 7051–7058 (2014).

68. Fanous, J., Wegner, M., Grimminger, J., Andresen, A. & Buchmeiser, M. R. Structure-related electrochemistry of sulfur-poly(acrylonitrile) composite cathode materials for rechargeable lithium batteries. *Chem. Mater.* **23**, 5024–5028 (2011).

69. Liang, X. *et al.* A highly efficient polysulfide mediator for lithium-sulfur batteries. *Nat. Commun.* **6**, 5682 (2015).

Acknowledgements

This work was supported by the 'Hundred Talents Program' of Chinese Academy of Sciences and the National Natural Science Foundation of China (21404108, U1362106).

Author contributions

G.L. and J.G. conceived and designed the experiments. G.L. performed the laboratory experiments, characterization of materials and analysis of the results. S.J. conducted part of the TEM characterization. J.S. and W.H. participated in the discussion of electrochemical results. G.L. and J.G. co-wrote the paper and all authors discussed the results and commented on the manuscript. J.G. and Y.H. supervised the project.

Additional information

12

Spatial control of chemical processes on nanostructures through nano-localized water heating

Calum Jack[1], Affar S. Karimullah[1,2], Ryan Tullius[1], Larousse Khosravi Khorashad[3], Marion Rodier[1], Brian Fitzpatrick[1], Laurence D. Barron[1], Nikolaj Gadegaard[2], Adrian J. Lapthorn[1], Vincent M. Rotello[4], Graeme Cooke[1], Alexander O. Govorov[3] & Malcolm Kadodwala[1]

Optimal performance of nanophotonic devices, including sensors and solar cells, requires maximizing the interaction between light and matter. This efficiency is optimized when active moieties are localized in areas where electromagnetic (EM) fields are confined. Confinement of matter in these 'hotspots' has previously been accomplished through inefficient 'top-down' methods. Here we report a rapid 'bottom-up' approach to functionalize selective regions of plasmonic nanostructures that uses nano-localized heating of the surrounding water induced by pulsed laser irradiation. This localized heating is exploited in a chemical protection/deprotection strategy to allow selective regions of a nanostructure to be chemically modified. As an exemplar, we use the strategy to enhance the biosensing capabilities of a chiral plasmonic substrate. This novel spatially selective functionalization strategy provides new opportunities for efficient high-throughput control of chemistry on the nanoscale over macroscopic areas for device fabrication.

[1] School of Chemistry, University of Glasgow, Joseph Black Building, Glasgow G12 8QQ, UK. [2] School of Engineering, University of Glasgow, Rankine Building, Glasgow G12 8LT, UK. [3] Department of Physics and Astronomy, Ohio University, Athens, Ohio 45701, USA. [4] Department of Chemistry, University of Massachusetts, 710 North Pleasant Street, Amherst, Massachusetts 01003, USA. Correspondence and requests for materials should be addressed to A.S.K. (email: Affar.Karimullah@glasgow.ac.uk) or to M.K.(email: Malcolm.Kadodwala@glasgow.ac.uk).

Incorporating molecular functionality into engineered nanomaterials is a requirement for the development of new sensing[1-3], photovoltaic[4] and optical technologies[5-8]. Maximizing coupling between nanostructures and molecular species requires spatial control of surface chemistry on the nanoscale. Currently, there are a variety of techniques that can be used to spatially control chemical functionality on nanostructured substrates such as dip-pen lithography[9], inkjet printing[10] and direct laser patterning[11,12]. However, all these approaches are in essence 'top-down' methods for micrometre-scale substrate areas and each features limitations such as time-consuming processing and difficulties in achieving sub-50 nm resolutions. Furthermore, they require complex alignment procedures to be overlaid onto fabricated nanostructures.

In this study we show a high-throughput 'bottom-up' approach that uses the nano-localized heating of a liquid (water) surrounding a nanostructure, to spatially direct a protection/deprotection strategy[13]. This novel strategy enables molecular materials to be placed in selective regions of a nanostructure. The protection step involves the chemical passivation of the gold structure with a self-assembled monolayer (SAM) of a long-chain polyethylene glycol (PEG) thiol. The spatially selective deprotection step then is the thermally driven structural transformation of the PEG SAM.

Structurally complex plasmonic nanostructures can produce a range of electromagnetic (EM) fields that not only have different spatial extents but also differing intrinsic properties. Consequently, for sensing purposes, it would be advantageous to selectively place material in locations occupied by EM fields with the desired properties. To this end, we have employed the nano-localized heating-driven protection/deprotection strategy to functionalize chiral plasmonic nanostructures, to enable selective positioning of proteins in a region with EM fields of enhanced chiral asymmetry. The placement of proteins in regions with fields of optimal properties provides enhanced biosensing performance, enabling attomole ($\sim 6.0 \times 10^5$ molecules) rather than femtomole ($\sim 6.0 \times 10^8$ molecules) detection levels.

Results

Modelling of nano-localized thermoplasmonic water heating. To date, it is believed that spatially localized chemistry on individual nanostructures cannot be achieved through thermoplasmonic phenomena, although thermal gradients can be generated in ensembles of nanoparticles separated by a poorly conducting medium[14-16]. This assumption is based on the fact that although electric fields, created by plasmon excitation, may be highly localized, efficient thermal diffusion in metal nanostructures results in rapid and uniform temperature increase throughout a particular structure in less than a nanosecond. This rapid heat dissipation leads to a uniform temperature at the nanostructure surface and hence spatial uniformity of surface chemical/physical processes. In this study we demonstrate a new phenomenon, which leads to spatially localized thermally driven chemistry, allowing nanoscale precision placement of biomaterial on a nanostructure surface (Fig. 1). Localized chemistry in this study is the result of significant temperature gradients in the surrounding water and not thermal gradients across the nanostructure surface itself.

We demonstrate this new 'thermoplasmonic' effect using a chiral plasmonic nanomaterial formed using injection moulding, creating nanopatterned indentations in a polycarbonate slide (polycarbonate template) and coating the surface with a continuous Au film. The nanostructure in particular is a 'shuriken' structure (Fig. 2a), which is either left or right handed and is referred to as a templated plasmonic substrate (TPS)[17].

These structures are 500 nm end to end and have a pitch of 700 nm from centre to centre. In contrast to traditional electron beam or photolithography, the TPS offer low-cost high-throughput fabrication, effectively a disposable consumable ideally suited for technological exploitation. The TPS are incorporated into a liquid cell with a total volume of *ca.* 100 µl. Interfaces that will subsequently be referred to as the front face, back face, lower and upper surfaces are defined in Fig. 2 (further details in Supplementary Fig. 1). In our approach, the back face of the TPS was irradiated using a 1,064-nm Nd:YAG laser with an 8-ns pulse.

On irradiation, the plasmonic structure absorbs the energy based on its resonance condition and the spatial regions of the nanostructure with high current density will then act as heat sources[18,19]. To understand this pivotal step, we have modelled the spatially resolved time dynamics for thermal behaviour of the nanostructure and the surrounding water for an 8-ns laser pulse at a fluence of 15 mJ cm^{-2}. Unlike plasmonic heating with femtosecond pulses[20], the gold film, owing to its high thermal conductivity, does not generate large thermal gradients (Fig. 3a) with nanosecond laser pulse durations[16,19]. Although the plasmonic fields maybe highly localized, the heat generated diffuses in the structure over such time scales (Supplementary Figs 3–6). With thermal diffusivity almost three orders of magnitude smaller than gold, the surrounding water does generate thermal gradients, developing three regions of varying thermal behaviour by the time the 8-ns pulse ends. The water at the top surface has a low average temperature but the two regions of water within the indentation itself have high temperatures over significant distances. The water in the central

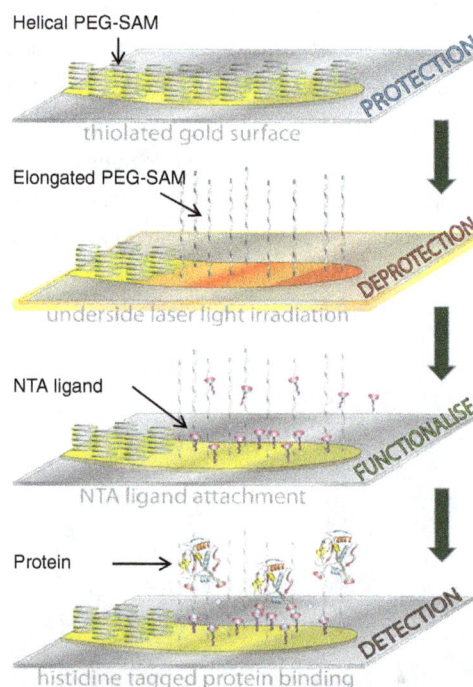

Figure 1 | Pictorial representation of the steps in the protection/deprotection strategy. The nanostructure is first protected using a thermally responsive PEG-SAM (helical form). The second step is the deprotection where the nanostructure is irradiated by a nanosecond pulse laser and the subsequent localized water temperatures will cause a transition in the PEG-SAM to its elongated form. The elongated PEG-SAM will not inhibit the NTA ligands from attaching to the surface, thereby functionalizing the selective regions. Proteins can then be positioned in the particular regions for detection.

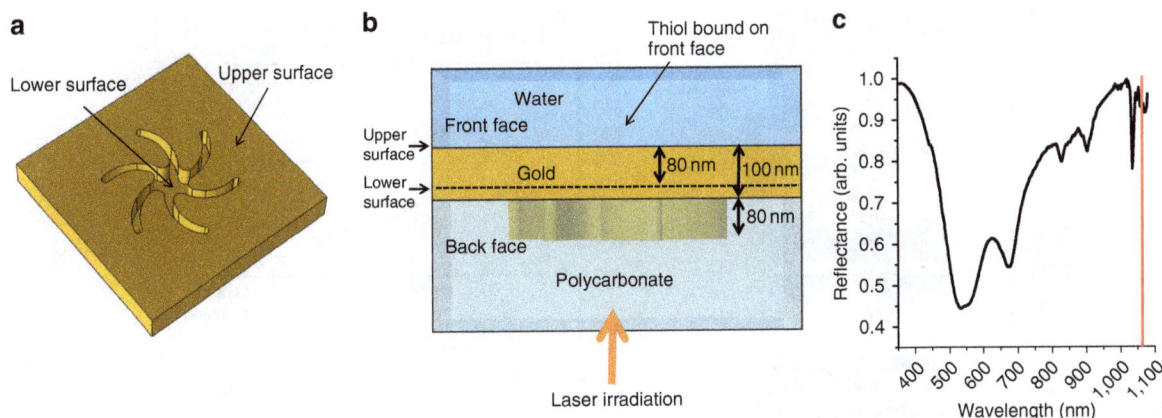

Figure 2 | TPS experimental geometry and reflectivity. (a) Perspective view of the front face of the gold layer on a TPS. **(b)** Side view of the TPS, which is irradiated using a nanosecond laser through the polycarbonate. **(c)** Reflectivity of the TPS measured through the polycarbonate. The red line indicates the wavelength of the nanosecond laser.

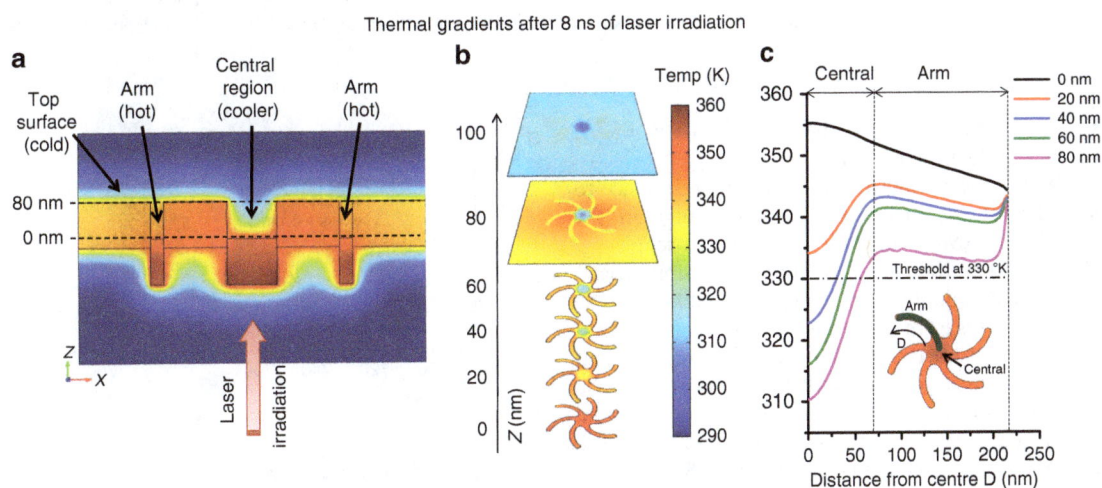

Figure 3 | Surface plot of the thermal gradients at 8 ns (pulse end). (a) Side view showing gold and water regions and **(b)** stacked surface plots of water regions for increasing height Z. **(c)** Graph showing the temperature changes along a parametric curve passing along an arm to the centre (as shown by green line in inset) of the nanostructure for varying height Z. The horizontal line at 330 K represents the threshold for the deprotection step discussed.

region of the indentation is cooler than that in the arms with increasing distance from the surface and the water temperature in the arms is relatively high and uniform with increasing height (Fig. 3a–c and Supplementary Fig. 3).

The time-dependant thermal behaviour shows that during the laser pulse, the water shows a steep increase in the average temperature for the two regions (Fig. 4) with the arm sections achieving higher temperatures than the central region, reaching a maximum at the end of the pulse (8 ns). This leads to thermal gradients in the water across the nanostructure that exist up to 12 ns after the pulse ends (that is, the arms are significantly hotter than the central region). After 12 ns, the water thermal gradients have completely disappeared and the water achieves thermal homogeneity with the metal as well. Equivalent thermal gradients do not exist in the metal (Fig. 4c). Thus, our simulations show thermal gradients in the water surrounding the nanostructure during the 8-ns laser pulse and a subsequent 4 ns afterwards. This phenomena derives from the disparity in the thermal diffusivities of water and Au. It is fundamentally different to the previously reported focusing of light into mesoscale volumes at water surfaces by Au nanoparticles, which creates a nanobubble of steam surrounding the resonantly heated nanoparticle surface[21].

Nano-localized thermoplasmonic chemical functionalization. The first step for our patterning process is the deposition of a protective SAM. To achieve spatially selective chemistry, we have used a thermally responsive SAM composed of a high-molecular-weight (6 kDa) PEG methyl ether thiol (PEG-thiol) polymer. The SAMs were formed by depositing PEG-thiol from solution onto the TPS; these substrates will subsequently be referred to as PEG-TPSs. The macromolecular structure of PEG displays a temperature dependency, which has been exploited previously to produce thermally responsive materials[22,23]. The macromolecular structure of PEG is governed by the conformation adopted by the $-(CH_2-CH_2-O)-$ subunits. At low temperatures or in polar environments a gauche conformation is favoured, whereas at higher temperatures or in non-polar environments the *trans* form is observed[24,25]. Hence, PEG thiols adopt a compact ordered helical structure within SAM at low temperature and in aqueous solutions, which inhibit the adsorption of biomaterials. Driven by the conformational transition of the $-(CH_2-CH_2-O)-$ subunits, PEG-SAM can undergo a transition from the helical state to an elongated form, which does not inhibit the adsorption of biomaterials[26]. To achieve this transition, Shima *et al.*[22] have shown that the PEG-SAM must be exposed to

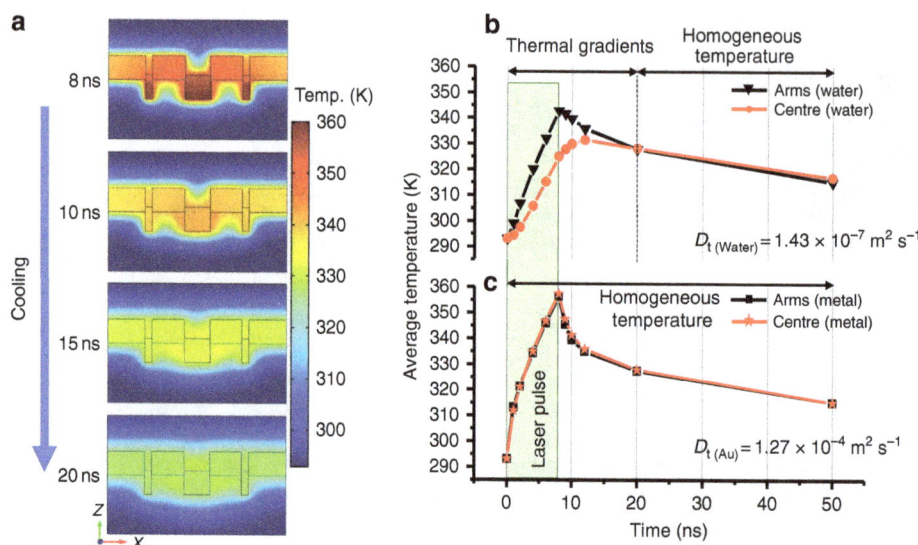

Figure 4 | Thermal behaviour of water on irradiation. (**a**) Simulated spatially resolved temperature dynamics after pulse (8 ns) laser irradiation. (**b**) Average temperature of water (10 nm < Z < 70 nm) and (**c**) metal (Z < 0 nm) in the arm and centre regions. The values D_t are the thermal diffusivity of water and Au.

temperatures over $\sim$330 K, along the entire length of its elongated structure, $\sim$35 nm (ref. 27).

The sensitivity of the wavelength of a plasmonic resonance to the refractive index of the near field enables change in the structure of the PEG-SAM to be monitored spectroscopically. We measured the shifts in the optical rotatory dispersion (ORD) of our left-handed plasmonic shurikens from the front face[17]. The helical and elongated *trans* form SAM have thickness of $\sim$13 and $\sim$35 nm, respectively[27]. Consequently, it would be expected that a transition from the helical to elongated *trans* form results in a significant increase in the thickness of the PEG-SAM layer and hence will cause a red shift in the plasmonic resonance. Indeed, when the PEG-TPSs are exposed either to water at 358 K (Fig. 5a) or a less polar solvent such as 1-butanol (see Supplementary Fig. 7), an irreversible red shift of $+5.0\pm0.2$ nm in the wavelength of the plasmonic resonance in ORD spectra is in fact observed. We attribute the irreversible nature of the transition to the kinetic effects of lateral interactions (such as intertwining) of the elongated *trans* PEG within the SAM that occur when temperatures are between 323 K (see Supplementary Fig. 7) and 358 K (Fig. 5a). When a PEG-TPS is heated to 358 K in air (for 48 h), no discernible shift within error (-0.2 ± 0.2 nm) is observed (Supplementary Fig. 7). Hence, heating a PEG-TPS in air only, causes no change, indicating that the process is water mediated (Supplementary Fig. 7D)[27].

The PEG-TPS were irradiated using a pulsed laser from the back and the energy dependence of the transition was evaluated. A plain TPS with no PEG-SAM shows no change (Fig. 5b) when irradiated using 8 ns pulse laser (15 mJ cm^{-2} fluence), indicating that the laser irradiation causes no deformation to the nanostructure. When a PEG-TPS is irradiated using 400 μs pulse laser, no observable changes occur (Fig. 5c) even at the highest fluence (20 mJ cm^{-2}) used. When the PEG-TPS is irradiated using 8 ns pulse laser (15 mJ cm^{-2} fluence), a small shift of $+1.0\pm0.2$ nm is observed, indicating that a small fraction of the PEG-SAM has been changed (Fig. 5d). The level of red shift saturates within 60 s of laser exposure with further radiation causing no measureable change (Supplementary Fig. 7). When a substrate that has had its PEG transformed to the elongated state using 1-butanol is irradiated with nanosecond pulses, no spectral changes are observed (Supplementary Fig. 7). These data clearly

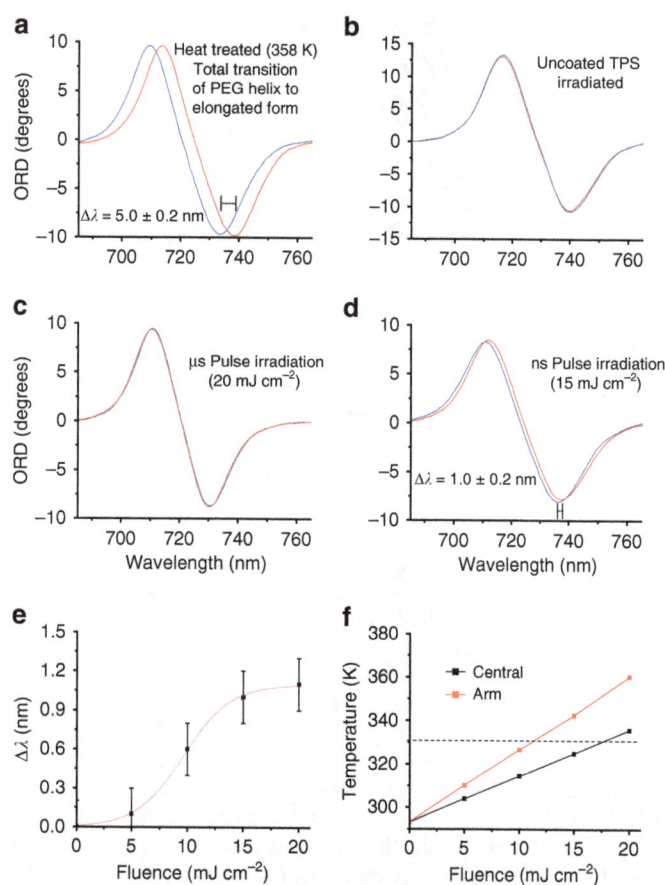

Figure 5 | TPS nanostructure ORD measurements. (**a**) PEG-TPS heat treated at 358 K. (**b**) A plain TPS irradiated with 8 ns pulse laser; PEG-TPS irradiated using (**c**) 400 μs pulse laser and (**d**) 8 ns pulse laser. (**e**) Fluence dependence of the resonance shift and (**f**) simulation results for average temperature within the nano indentation for varying fluence.

demonstrate that nanosecond laser heating causes only a partial yet irreversible change to the PEG-SAM. Furthermore, the effects of laser irradiation show dependency on the fluence, displaying stepwise behaviour (Fig. 5e). Nanosecond and microsecond pulses can be considered to be quasi continuous, as they have significantly longer durations than the electron–phonon relaxation time[28]. Consequently, they only generate linear hot electron-driven phenomenon rather than the nonlinear phenomenon associated with femtosecond pulses, which arise through high electron temperatures. Although microsecond pulses would generate smaller field intensities, they would occur for longer time scales. Thus, the time integrated total flux of hot electrons will be identical for both nano and microsecond pulses[29]. Hence, the experimental observations are consistent with a thermal process rather than either a direct optical or hot-electron-mediated excitation, as both would have no dependence on pulse duration and a linear dependency on fluence[30].

The small red shift (1.0 ± 0.2 nm) induced by nanosecond pulse laser irradiation compared with the shift (5.0 ± 0.2 nm) due to heating at 358 K indicates that only a small fraction (∼20%) of the total area in the near fields of the nanostructures of the PEG-SAM undergoes a transition to the elongated state. This limited transformation is consistent with our hypothesis that thermal gradients within the surrounding water can direct spatially selective chemistry. The irreversible transformation is a kinetically limited process that consequently requires the water temperature to be above the threshold for a sufficient time period, to enable the elongated PEG-thiol to intertwine with each other. Earlier studies of the dynamic relaxation for polymers in solution with similar molecular weight have shown this time period to be ∼5–10 ns (ref. 31). For shorter time periods, the transformation would be reversible. The results from our simulation show that at the top surface, the temperatures in the water are not high enough to achieve the conformational transition. Hence, the PEG-SAM here will remain in its helical conformation over the entire surface. At the central cooler region of the structure, the average water temperature barely crosses the threshold for <3 ns. The PEG-SAM here will not experience the required temperatures for a long-enough period and hence will also remain in a helical conformation. However, the arm region, is at a significantly higher average temperature (∼350 K) and above the threshold for ∼12 ns. Only in the arms is the average water temperature

higher than the threshold for long enough to achieve the irreversible transformation.

The dependence of the resonance shift on the laser fluence is also in agreement with our simulation results on the dependence of temperatures on laser fluence (Fig. 5f). The average temperature in the arm only increases above the threshold for the PEG transition for a fluence of $>12 \, \text{mJ cm}^{-2}$ and just about crosses that threshold in the centre at the highest fluence of $20 \, \text{mJ cm}^{-2}$, indicating that substantial transition of PEG-SAM will occur only in the arms.

To confirm the spatial localization of the PEG-SAM transformation and hence the validity of using thermal gradients in water to illicit spatially selective chemistry, we have collected atomic force microscopy (AFM) images. The significant height difference between the helical and the elongated *trans* form (∼22 nm) of the PEG can readily be detected by AFM. In Fig. 6 are AFM images collected from PEG-TPS before and after irradiation with nanosecond pulses. The AFM images show that after laser irradiation there is an average reduction in the depth of the arms by 20–30 nm in the region at the end of the arms. This decrease in depth is an expected change in SAM thickness on going from helical to elongated *trans*-PEG. The spatial localization of the change in SAM structure is also supported by scanning electron microscopy, which show change in contrast at the end of the arms post irradiation (see Supplementary Fig. 8).

Enhanced sensitivity of TPS spectroscopy to biomaterials. The elongated *trans*-PEG-SAM in the arms should allow adsorption of biomolecules, whereas the helical PEG-SAM will inhibit adsorption, essentially creating 'nanopockets' in the arms. This ability of the PEG-SAM to selectively vary spatial chemistry provides a powerful tool for enhancing the functionality of nanomaterials. As an exemplar of the potential applications, we used the nanopockets created on the TPSs, to achieve detection of proteins at the attomole sensitivity. Specifically, the chiral evanescent fields that occupy the nanopockets were used to perform a form of polarimetry, plasmonic polarimetry[32], which measures the asymmetry of the effective refractive indices of chiral media on the handedness of the applied chiral evanescent field[2]. The asymmetries in refractive indices of chiral media can be parameterized by $\Delta\Delta\lambda = \Delta\lambda_{\text{R}} - \Delta\lambda_{\text{L}}$, where $\Delta\lambda_{\text{R/L}}$ is the shift in

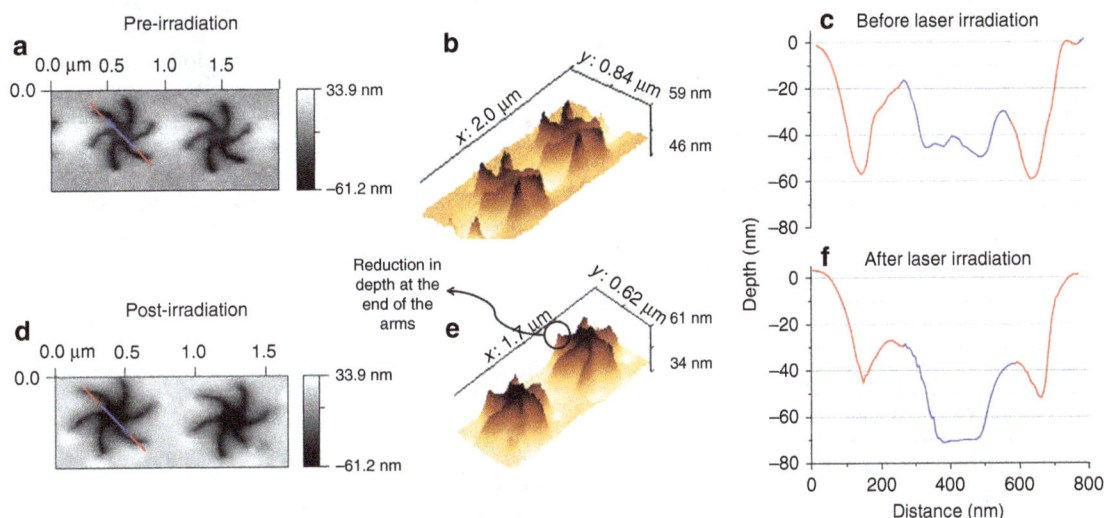

Figure 6 | AFM of pre and post irradiation of the PEG-TPS. (a,d) The height plots, **(b,e)** the inverted 3D plots and **(c,f)** the height data of along the lines shown in **a** and **d**. The red and blue colours correspond to the distances marked in the SEM(**a,d**).

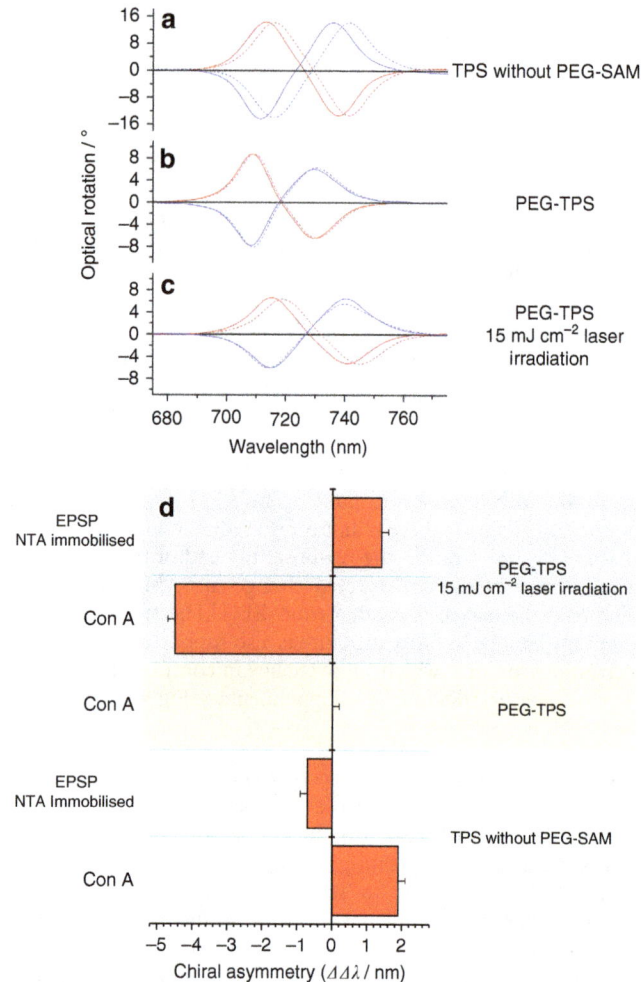

Figure 7 | ORD shifts from buffer (solid) to ConA (dotted). (a) TPS, **(b)** PEG-TPS but not irradiated and **(c)** irradiated PEG-TPS. Blue lines represent right-handed structures and red represents left-handed structures. **(d)** $\Delta\Delta\lambda$ results for ConA and EPSPS protein experiments.

the wavelengths of the ORD resonances induced by the adsorption of a chiral medium. The $\Delta\Delta\lambda$ parameter is analogous to the optical rotation measured with conventional optical polarimetry. We performed plasmonic polarimetry using bare PEG-TPSs and laser-irradiated PEG-TPSs. The ORD spectra collected for the left- and right-handed forms of the three different types of TPSs in buffer (reference value) and solution of a model protein, Concanavalin A (ConA), and the $\Delta\Delta\lambda$ values extracted from these spectra for the peak labelled are shown in Fig. 7d. ConA was chosen as an exemplar system, because previous work has shown that it displays a large plasmonic polarimetry response[2].

As expected, given that the helical PEG structure inhibits the adsorption of biomaterials, the ConA cannot be detected using non-irradiated PEG-TPSs. In contrast, $\Delta\Delta\lambda$ values of 1.9 and -4.5 nm are observed for the bare and laser-irradiated substrates, respectively. The change in sign of the $\Delta\Delta\lambda$ values on going from bare to laser-irradiated PEG-TPSs can be correlated with the sign of the overall integrated optical chirality of areas available for protein adsorption on both substrates (Supplementary Fig. 2). In the case of the bare substrate, protein will be exposed to evanescent fields with an overall negative integrated optical chirality. In contrast, for the laser-irradiated PEG-TPS, ConA will only be adsorbed in the arms of the structure and thus will experience fields with an overall positive

optical chirality. The $\Delta\Delta\lambda$ parameter decreases with decreasing laser fluence and hence smaller areas of thermally transformed PEG-SAM, as might be expected (see Supplementary Fig. 9).

In a second series of experiments, we functionalized the deprotected arms using a procedure first proposed by Sigal *et al.*[33], which enables recombinant histidine-tagged (His-tagged) proteins to be selectively immobilized to SAMs featuring a nitrilotriacetic acid (NTA) group co-ordinated to Ni(II). In particular, we have co-self-assembled a thiol-functionalized NTA derivative (NTA-thiol) and triethylene glycol mono-11-mercaptoundecyl ether (EG-thiol) spacer unit to create binding sites for the His-tagged protein 5-enolpyruvylshikimate-3-phosphate (EPSPS) synthase (see Supplementary Fig. 10) within the nanopockets. For comparison, we functionalized a bare TPS with the NTA-thiol/EG-thiol and immobilized the EPSPS across the whole surface. The $\Delta\Delta\lambda$ values obtained for EPSPS immobilized across the whole of the TPS and just in the nanopockets were -0.7 and 1.4 nm, respectively. The behaviour of EPSPS is qualitatively the same as for ConA, with laser-processed substrates having $\Delta\Delta\lambda$ values that are larger and of the opposite sign to those obtained when protein covers the entire surface.

In our experiments, we measure a single array of our nanostructures, which is roughly a square with 300 µm sides. Assuming that the irreversible transformation of PEG takes place in the arms of all nanostructures, we estimate that the total deprotected area in a single array would be 6.65×10^3 µm^2 (calculations in Supplementary Note 7). We only detect surface-bound molecules and hence can consider the total number of particles bound on our deprotected areas as those being detected. In the case of EPSPS, the molecules are His-tagged and chemically bound to the surface with no proteins present in the buffer during measurements. We estimate that for a plain TPS, the EPSPS contributing to the measured $\Delta\Delta\lambda$ values would be 1.46 femtomoles, whereas only 108 attomoles ($\sim 6.5 \times 10^7$ protein molecules) of EPSPS protein are being detected using our protection/deprotection strategy (calculations in Supplementary Note 7). In the case of ConA, we estimate that only 80 amol ($\sim 4.8 \times 10^7$ protein molecules) contribute to the measured $\Delta\Delta\lambda$ values (see Supplementary Note 7). Hence, a functionalized PEG-TPS allows us to detect attomole ($\sim 6.0 \times 10^5$ molecules) quantities and with enhanced performance from plasmonic polarimetry.

Discussion

In conclusion, we demonstrate a novel strategy to achieve spatially selective surface functionalization on a nanostructure. It is based on a protection/deprotection strategy, which is rapid and can chemically modify macroscopic areas of substrates. The spatially selective deprotection step is driven by the localized water heating-induced unravelling of an initially helical protective thiol in selective areas of a nanostructure. The unravelled regions are both less densely packed, thus can be subsequently functionalized with another thiol and do not inhibit biomolecule adsorption. The strategy has allowed us to enhance the sensitivity of a biosensor enabling analyte molecules to be selectively located in regions with EM fields of high net chiral asymmetry. Using thermal gradients in water (or any solvent) to control surface chemistry of nanoscale regions over macroscopic areas is a versatile route for the high-throughput production of advanced functional materials.

Methods

Fabrication of TPSs. The templated substrates were prepared by a combination of high-resolution electron beam lithography and injection moulding. In brief, clean silicon substrates were coated with 80 nm of PMMA (Elvacite 2041, Lucite International) and exposed in a Vistec VB6 UHR EWF lithography tool operating at 100 kV. After exposure, the substrates were developed and submitted for electroplating, where a 300-μm-thick nickel shim was formed[34]. The shim was trimmed and mounted in a custom-made tool capable of manufacturing ASA standard polymer slides. An Engel Victory Tech 28 tons injection moulding machine was used in fully automatic production mode in the manufacture of the polymer slides using polycarbonate (Makrolon DP2015) as feedstock. Polycarbonate is used as a substrate material, because it is known to have the best ability to replicate the nanofeatures and is commonly used in the industry for optical storage media[35]. This process enables us to make more than 200 substrates per hour. The injection-moulded substrates have the chiral nanostructures imparted in the plastic surface and are subsequently covered by a continuous 100-nm Au film to complete the TPS process.

PEG-SAM functionalization and ConA solutions. PEG-thiol (Sigma-Aldrich) was used to make a 833-μM solution in 95% ethanol and the TPS was allowed to self-assemble onto the nanostructure for 18 h. After being removed from solution, the TPSs were washed with water and dried under nitrogen. ConA (Sigma-Aldrich) was prepared in solutions at a concentration of 1 mg ml^{-1} (37.7 μM) using a 10-mM Tris/HCl buffer at pH 7.4. Surface plasmon resonance measurements to determine surface protein coverage were done using a Biacore 2000 Instrument (GE Lifesciences).

Preparation of the NTA-thiol and His-tagged EPSPS samples. NTA-thiol was purchased from Prochimia and EG-thiol was purchased from Sigma-Aldrich. The preparation for the NTA-thiol/EG-thiol monolayer was performed similar to that described by Sigal et al.[33].

An existing His-tagged *Escherica coli aroA* gene cloned into a pET 22b vector was transformed and overexpressed in BL21 star cells. One litre of liquid Luria broth media was used to grow the cells and protein expression was induced with isopropyl β-D-1-thiogalactopyranoside as an optical density of 0.6 at 600 nm. The cells were harvested 4 h after induction by centrifugation and resuspended in 50 mM Tris/HCl buffer pH 7.5. The cells were lysed using sonication and centrifuged at 25 K to remove all insoluble matter. The protein was purified from the cell lysate by NTA nickel affinity chromatography with the EPSPS protein eluted using 300 mM imidazole, 0.5 M NaCl in 20 mM Tris/HCl buffer pH 7.5, giving a protein solution with a concentration of 434.8 μM. This resulted in a yield of over 50 mg of protein. The protein was concentrated by centrifugal concentration and buffer exchanged by dialysis to make a final solution of 20 mg ml^{-1} EPSPS in 50 mM Tris/HCl buffer pH 7.5. The purity of the protein was assessed by SDS–PAGE and the enzyme activity was assessed using a phosphate release assay[36].

HEPES-buffered saline (HBS) used in the EPSPS experiments were 10 mM HEPES and 150 mM NaCl in water adjusted to pH 7.4.

The EPSPS solution was made using EPSP synthase in 50 mM Tris/HCl buffer of pH 7.5 with a concentration of 4 mg ml^{-1}.

After SAMs were fabricated from NTA-thiol/EG-thiol, measurements were taken using HBS for buffer values. The EPSP solution was left for 2 h and then rinsed with HBS before measurements were taken.

EM field simulations. Numerical simulations of EM fields and thermal heat transfer were performed using a commercial finite-element package (COMSOL v4.4, Wave optics module with Multiphysics and a heat transfer module). Permittivity values for gold were taken from Palik's optical constants[37]. Drude broadening was applied using the method described by Kuzyk et al.[38]. Earlier work by Bouillard et al.[39] shows that the variation in dielectric properties over the temperatures associated to our simulation are insignificant (<0.2%) and have hence been neglected. Periodic boundary conditions were used to emulate the array of nanostructures. Linear, polarized EM wave was applied at normal incidence through the polycarbonate substrate onto the structure. A subsequent heat transfer module was then used with the total heat dissipation from the EM model used as the heat source. A time-dependant function was applied to the total heat dissipation to create a heat source that would replicate a temporally square-shaped laser pulse as the source for EM heating. We used the appropriate thermal conductivities, heat capacity and density for the dielectrics and the metals. For the values used and further information on the simulations, see Supplementary Method and Supplementary Table 1.

ORD and laser irradiation. We have used a custom-made polarimeter that measures the reflected light from our samples. It uses a tungsten halogen light source (Thorlabs), Glan-Thompson polarizers (Thorlabs) and a × 10 objective (Olympus). The samples are positioned with the help of a camera (Thorlabs, DCC1645C) and the spectrum is measured using a compact spectrometer (Ocean optics USB4000). Using Stokes methods, we can measure the intensity of light at four angles of the analyser and calculate the optical rotation dispersion of our chiral plasmonic arrays.

A nanosecond (8 ns)-pulsed Nd:YAG laser (Spectra Physics Quanta Ray) operating at a 10-Hz repetition rate was used for the irradiation. The sample was irradiated at normal incidence S-polarized light, using an unfocussed beam with an area of 1 cm^2, for 1 min. Fluence was varied where stated. For all protein experiments, the fluence used was 15 mJ cm^{-2}.

References

1. Adato, R. & Altug, H. In-situ ultra-sensitive infrared absorption spectroscopy of biomolecule interactions in real time with plasmonic nanoantennas. *Nat. Commun.* **4,** 2154–2163 (2013).
2. Hendry, E. *et al.* Ultrasensitive detection and characterization of biomolecules using superchiral fields. *Nat. Nanotechnol.* **5,** 783–787 (2010).
3. Liu, N., Hentschel, M., Weiss, T., Alivisatos, A. P. & Giessen, H. Three-dimensional plasmon rulers. *Science* **332,** 1407–1410 (2011).
4. Atwater, H. A. & Polman, A. Plasmonics for improved photovoltaic devices. *Nat. Mater.* **9,** 205–213 (2010).
5. Pendry, J. B., Aubry, A., Smith, D. R. & Maier, S. A. Transformation optics and subwavelength control of light. *Science* **337,** 549–552 (2012).
6. Schurig, D. *et al.* Metamaterial electromagnetic cloak at microwave frequencies. *Science* **314,** 977–980 (2006).
7. Smith, D. R., Pendry, J. B. & Wiltshire, M. C. K. Metamaterials and negative refractive index. *Science* **305,** 788–792 (2004).
8. Gansel, J. K. *et al.* Gold helix photonic metamaterial as broadband circular polarizer. *Science* **325,** 1513–1515 (2009).
9. Salaita, K., Wang, Y. & Mirkin, C. A. Applications of dip-pen nanolithography. *Nat. Nanotechnol.* **2,** 145–155 (2007).
10. Park, J.-U. *et al.* High-resolution electrohydrodynamic jet printing. *Nat. Mater.* **6,** 782–789 (2007).
11. Shadnam, M. R., Kirkwood, S. E., Fedosejevs, R. & Amirfazli, A. Thermo-kinetics study of laser-induced desorption of self-assembled monolayers from gold: case of laser micropatterning. *J. Phys. Chem. B* **109,** 11996–12002 (2005).
12. Slater, J. H., Miller, J. S., Yu, S. S. & West, J. L. Fabrication of multifaceted micropatterned surfaces with laser scanning lithography. *Adv. Funct. Mater.* **21,** 2876–2888 (2011).
13. Jarowicki, K. & Kocienski, P. Protecting groups. *J. Chem. Soc. Perkin Trans.* **1,** 4005–4037 (1998).
14. Baldwin, C. L., Bigelow, N. W. & Masiello, D. J. Thermal signatures of plasmonic fano interferences: toward the achievement of nanolocalized temperature manipulation. *J. Phys. Chem. Lett.* **5,** 1347–1354 (2014).
15. Sanchot, A. *et al.* Plasmonic nanoparticle networks for light and heat concentration. *ACS Nano* **6,** 3434–3440 (2012).
16. Baffou, G., Quidant, R. & García de Abajo, F. J. Nanoscale control of optical heating in complex plasmonic systems. *ACS Nano* **4,** 709–716 (2010).
17. Karimullah, A. S. *et al.* Disposable plasmonics: plastic templated plasmonic metamaterials with tunable chirality. *Adv. Mater.* **27,** 5610–5616 (2015).
18. Govorov, A. O. & Richardson, H. H. Generating heat with metal nanoparticles. *Nano Today* **2,** 30–38 (2007).
19. Baffou, G. & Quidant, R. Thermo-plasmonics: using metallic nanostructures as nano-sources of heat. *Laser Photon. Rev.* **7,** 171–187 (2013).
20. Liu, L. *et al.* Highly localized heat generation by femtosecond laser induced plasmon excitation in Ag nanowires. *Appl. Phys. Lett.* **102,** 1–5 (2013).
21. Neumann, O. *et al.* Solar vapor generation enabled by nanoparticles. *ACS Nano* **7,** 42–49 (2013).
22. Shima, T. *et al.* Thermally driven polymorphic transition prompting a naked-eye-detectable bending and straightening motion of single crystals. *Angew. Chem. Int. Ed.* **53,** 7173–7178 (2014).
23. Love, J. C., Estroff, L. A., Kriebel, J. K., Nuzzo, R. G. & Whitesides, G. M. Self-assembled monolayers of thiolates on metals as a form of nanotechnology. *Chem. Rev.* **105,** 1103–1170 (2005).
24. Bjoerling, M., Karlstroem, G. & Linse, P. Conformational adaption of poly(ethylene oxide): a carbon-13 NMR study. *J. Phys. Chem.* **95,** 6706–6709 (1991).
25. Matsuura, H. & Fukuhara, K. Conformational analysis of poly(oxyethylene) chain in aqueous solution as a hydrophilic moiety of nonionic surfactants. *J. Mol. Struct.* **126,** 251–260 (1985).
26. Harder, P., Grunze, M. & Dahint, R. Molecular conformation in oligo (ethylene glycol)-terminated self-assembled monolayers on gold and silver surfaces determines their ability to resist protein adsorption. *J. Phys. Chem. B* **102,** 426–436 (1998).
27. Norman, A. I., Yiwei, F., Ho, D. L. & Greer, S. C. Folding and unfolding of polymer helices in solution. *Macromolecules* **40,** 2559–2567 (2007).
28. Gadzuk, J. W. The road to hot electron photochemistry at surfaces: a personal recollection. *J. Chem. Phys.* **137,** 091703 (2012).
29. Zhou, X.-L., Zhu, X.-Y. & White, J. M. Photochemistry at adsorbate/metal interfaces. *Surf. Sci. Rep.* **13,** 73–220 (1991).
30. Cao, L., Barsic, D. N., Guichard, A. R. & Brongersma, M. L. Plasmon-assisted local temperature control to pattern individual semiconductor nanowires and carbon nanotubes. *Nano Lett.* **7,** 3523–3527 (2007).
31. Bauer, D. R., Brauman, J. I. & Pecora, R. Depolarized Rayleigh spectroscopy studies of relaxation processes of polystyrenes in solution. *Macromolecules* **8,** 443–451 (1975).
32. Tullius, R. *et al.* 'Superchiral' spectroscopy: detection of protein higher order hierarchical structure with chiral plasmonic nanostructures. *J. Am. Chem. Soc.* **137,** 8380–8383 (2015).

33. Sigal, G. B., Bamdad, C., Barberis, A., Strominger, J. & Whitesides, G. M. A self-assembled monolayer for the binding and study of histidine-tagged proteins by surface plasmon resonance. *Anal. Chem.* **68,** 490–497 (1996).
34. Gadegaard, N., Mosler, S. & Larsen, N. B. Biomimetic polymer nanostructures by injection molding. *Macromol. Mater. Eng.* **288,** 76–83 (2003).
35. Monkkonen, K. *et al.* Replication of sub-micron features using amorphous thermoplastics. *Polym. Eng. Sci.* **42,** 1600–1608 (2002).
36. Oliveira, J. S., Mendes, M. A., Palma, M. S., Basso, L. A. & Santos, D. S. One-step purification of 5-enolpyruvylshikimate-3-phosphate synthase enzyme from *Mycobacterium tuberculosis. Protein Expr. Purif.* **28,** 287–292 (2003).
37. Palik, E. D. *Handbook of Optical Constants of Solids* (Elsevier Science, 1998).
38. Kuzyk, A. *et al.* Reconfigurable 3D plasmonic metamolecules. *Nat. Mater.* **13,** 1–5 (2014).
39. Bouillard, J. S. G., Dickson, W., O'Connor, D. P., Wurtz, G. A. & Zayats, A. V. Low-temperature plasmonics of metallic nanostructures. *Nano Lett.* **12,** 1561–1565 (2012).

Acknowledgements

We acknowledge financial support from the Engineering and Physical Sciences Research Council (EPSRC EP/K034936/1), National Science Foundation (NSF grant CHE-1307021) and JSPS Core to Core. A.K. thanks the Leverhulme Trust and technical support from the James Watt Nanofabrication Centre (JWNC). R.T. and C.J. thank the EPSRC for the award of scholarships. L.K.K. and A.O.G. were supported by the Volkswagen Foundation (Germany).

Author contributions

C.J., R.T., A.S.K. and M.R. performed the experimental work. A.S.K., L.K.K. and A.O.G. carried out COMSOL calculations for the EM fields and the local temperature maps. A.J.L. supervised protein purification and immobilization. B.F. facilitated the SAM formation. N.G. developed the fabrication process of the injection-moulded substrates. A.S.K., M.K., L.K.K. and A.O.G. discussed the mechanism of temperature increase and the origins of the observed chemical effects. M.K., A.S.K., L.D.B., A.J.L., V.R., G.C. and N.G. wrote the paper. M.K. conceived and designed the experiment.

Additional information

Competing financial interests: The authors declare no competing financial interests.

13

Near-field photocurrent nanoscopy on bare and encapsulated graphene

Achim Woessner[1], Pablo Alonso-González[2,3,*], Mark B. Lundeberg[1,*], Yuanda Gao[4], Jose E. Barrios-Vargas[5], Gabriele Navickaite[1], Qiong Ma[6], Davide Janner[1], Kenji Watanabe[7], Aron W. Cummings[5], Takashi Taniguchi[7], Valerio Pruneri[1,8], Stephan Roche[5,8], Pablo Jarillo-Herrero[6], James Hone[4], Rainer Hillenbrand[9,10] & Frank H.L. Koppens[1,8]

Optoelectronic devices utilizing graphene have demonstrated unique capabilities and performances beyond state-of-the-art technologies. However, requirements in terms of device quality and uniformity are demanding. A major roadblock towards high-performance devices are nanoscale variations of the graphene device properties, impacting their macroscopic behaviour. Here we present and apply non-invasive optoelectronic nanoscopy to measure the optical and electronic properties of graphene devices locally. This is achieved by combining scanning near-field infrared nanoscopy with electrical read-out, allowing infrared photocurrent mapping at length scales of tens of nanometres. Using this technique, we study the impact of edges and grain boundaries on the spatial carrier density profiles and local thermoelectric properties. Moreover, we show that the technique can readily be applied to encapsulated graphene devices. We observe charge build-up near the edges and demonstrate a solution to this issue.

[1] ICFO—Institut de Ciencies Fotoniques, The Barcelona Institute of Science and Technology, 08860 Barcelona, Spain. [2] CIC nanoGUNE, 20018 Donostia-San Sebastian, Spain. [3] Institute of Physics, Chinese Academy of Science, Beijing 100190, China. [4] Department of Mechanical Engineering, Columbia University, New York, New York 10027, USA. [5] Catalan Institute of Nanoscience and Nanotechnology (ICN2), CSIC and The Barcelona Institute of Science and Technology, Campus UAB, 08193 Barcelona, Spain. [6] Department of Physics, Massachusetts Institute of Technology, Cambridge, Massachusetts 02139, USA. [7] National Institute for Materials Science, 1-1 Namiki, Tsukuba 305-0044, Japan. [8] ICREA-Institució Catalana de Recerca i Estudis Avançats, 08010 Barcelona, Spain. [9] CIC nanoGUNE and UPV/EHU, 20018 Donostia-San Sebastian, Spain. [10] IKERBASQUE, Basque Foundation for Science, 48011 Bilbao, Spain. * These authors contributed equally to this work. Correspondence and requests for materials should be addressed to F.H.L.K. (email: frank.koppens@icfo.eu).

As large scale integration and wafer scale device processing capabilities of graphene have become available[1-8], technological implementations of electronic and optoelectronic graphene devices are within reach[9-11]. At the same time, to achieve high device performance, any imperfections at the nanometer or even atomic scale need to be minimized or even eliminated. For example, in large area graphene, grown by chemical vapour deposition (CVD), grain boundaries are the stitching regions between different monocrystalline parts of graphene and act as carrier scatterers, limiting the graphene mobility and uniformity[12,13]. In addition, even perfectly monocrystalline graphene is still highly sensitive to its environment, and on typical substrates charge–density inhomogeneities (charge puddles)[14-19] and additional doping near contacts, defects and edges arise, which reduce the device performance as well. Therefore, it is important to efficiently probe the nanoscale optoelectronic properties of graphene devices and to understand their microscopic physical behaviour.

A major challenge is that many of the available characterization techniques are invasive[20], need specifically designed device structures[13,21,22], image only very small areas[14,15,21,23-25], rely on high doping of the graphene,[26] require unhindered electrical access of the probe to the graphene[14,15,21,23,24] or lack the desired nanometer resolution[27] and are expensive and difficult to implement. For the direct quality control of graphene devices, a method that can image electrical and optical properties of graphene devices, at nanoscale resolution, without any special preparation and without modifying the devices is required.

Here we demonstrate fully non-invasive room-temperature scanning near-field photocurrent nanoscopy[28-38] for the first time applied on graphene with infrared frequencies and use it to study the nanoscale optoelectronic properties of graphene devices that can later be used for real applications. This technique is based on electrical probing of the photoresponse due to strongly localized heating. We apply this technique to study the microscopic physics of grain boundaries and charge–density inhomogeneities. In the case of grain boundaries, we were able to identify the magnitude of their Seebeck coefficient, while for charge–density inhomogeneities, we show how they influence the global charge neutrality point of graphene devices. In addition, we study encapsulated graphene devices[39,40], where the encapsulation would prevent many other scanning probe techniques from accessing local properties of graphene. In these devices, we find a charge build-up near the edges and show that using local metal gates instead of a global backgate effectively suppresses this type of edge doping. In general, this technique operates most effectively with mid-infrared light because it does not lead to photodoping[41] and it is more stable in operation, compared with visible light.

Results

Measurement principle. The measurement principle is sketched in Fig. 1a. The setup is based on a scattering-type scanning near-field optical microscope (s-SNOM)[26,42] augmented with electrical contact to the sample to measure currents *in situ*[28-38]. In contrast to conventional s-SNOM, we do not need to measure the outscattered light but rather directly measure current induced by the near-field as explained in the following. A 10.6-μm mid-infrared laser illuminates a metallized atomic force microscope probe, tapping at its mechanical resonance frequency. Part of the incoming light, polarized parallel to the shaft of the probe, excites a strong electric field at the tip apex due to an antenna effect[43]. The spatial extent of this near-field is on the order of 25 nm, limited only by the tip radius and much smaller than the free space wavelength of the impinging light[43].

The near and far fields impinging on the device induce charge flows in the device (by mechanisms discussed below), and drive currents into an external current amplifier via contacts on the device. We isolate the part of the current that is induced by near fields by demodulating the current at the second harmonic of the tip tapping frequency[43]. This demodulated current is denoted I_{PC} and referred to as near-field photocurrent and is obtained together with near-field optical and topography information. A typical map of I_{PC}, obtained by scanning the tip over a CVD graphene device, is shown in Fig. 1b.

We can assess the spatial resolution of the photocurrent maps by comparing a region near the edge (Fig. 1d) with a topographic image from the same region (Fig. 1c). As can be seen, I_{PC} falls to zero for tip locations away from the graphene on a similar length scale as the topography, demonstrating the successful isolation of near-field contributions. In Fig. 1e, we quantify the resolution by observing the change in I_{PC} as the tip is moved over the edge of graphene. The full-width at half maximum of the photocurrent peak at this location is ∼100 nm, matching the rise distance in the topographic signal. This resolution is far below any limits relating to the 10.6 μm free space light wavelength.

Photothermoelectric photocurrent generation mechanism. As to the physical mechanism of the photocurrent, we consider the photothermoelectric effect that has been shown to dominate the photoresponse of graphene[11,44-48]: the light (in this case, the tip-enhanced near-field) locally heats the graphene, and this heat acts via non-uniformities in Seebeck coefficient S to drive charge currents within the device and into the contacts (see Methods section). Therefore, we interpret the variations of I_{PC} in terms of microscopic variations in S. The Seebeck coefficient, which depends on material properties such as carrier density and mobility, is a measure of the electromotive force driven by a temperature difference in a material. A complete description of I_{PC} needs to take into account the carrier cooling length[45,46] and overall sample geometry[49]. The carrier cooling length $l_{cool} = \sqrt{\kappa/g}$, where κ the sheet thermal conductivity in plane and g the interfacial thermal conductivity out of plane to the heat sinking substrate, describes how far heat propagates through the charge carriers, before dissipating to the environment (see Supplementary Fig. 2)[46]. A quantitative model of the thermoelectric photocurrent mechanism can be found in the Methods section.

Grain boundary characterization. We first discuss the application of this infrared near-field photocurrent technique to grain boundaries. They are not visible in the simultaneously acquired topography, and are responsible for some of the line-shaped features in the photocurrent map in Fig. 1b. Some of the other features stem from large scale inhomogeneities of the sample. The region within the green frame is shown with higher resolution in Fig. 1d, exhibiting a strong photocurrent signal that changes sign along a sharp boundary, yet the graphene is topographically flat in the vicinity of this boundary (Fig. 1c). We show now that this type of feature indicates a grain boundary.

Figure 2a shows a line profile of I_{PC} across the boundary feature identified in Fig. 1d. This antisymmetric I_{PC} can be explained by a localized deviation in S at the boundary, that is, a line defect within an otherwise uniform thermoelectric medium (Supplementary Figs 3,4 and Supplementary Note 1). Indeed, grain boundaries behave as localized lines of strongly modified electronic properties, within otherwise uniform graphene[20,21,26,50,51]. We remark that the decay of the photocurrent away from the boundary extends over more than

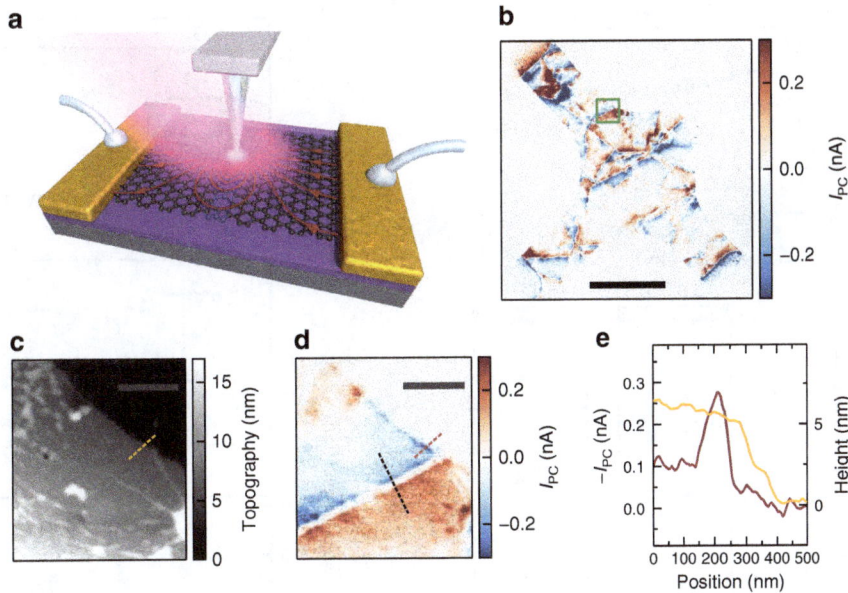

Figure 1 | Near-field photocurrent working principle and photocurrent from grain boundaries. (**a**) Sketch of the scattering-type scanning near-field optical microscope setup. A mid-infrared laser illuminates the atomic force microscope tip, which generates a locally concentrated optical field, which is absorbed by the graphene generating a position dependent photocurrent. The blue region in the graphene lattice represents a grain boundary with a modified Seebeck coefficient. The arrows sketch the photocurrent flow pattern. For each position only the magnitude and direction of the current are measured. The sketch is not to scale. (**b**) I_{PC} map at at backgate voltage $V_{BG} = 0$ V of a single layer CVD graphene device (Supplementary Fig. 1) with three contacts: top left (drain), right (source) and bottom left (ground). Both grain boundaries and wrinkles show characteristic photocurrent patterns. (scale bar, 5 μm) The green box indicates the measurement region in **c,d**. (**c**) Topography of etched CVD graphene does not show grain boundary but only wrinkles and other inhomogeneities due to the transfer process. (scale bar, 500 nm) (**d**) I_{PC} at $V_{BG} = 0$ V clearly shows a grain boundary and the expected sign change around it. The black dashed line indicates the measurement positions in Fig. 2a,d. (scale bar, 500 nm) (**e**) Topography (orange) and I_{PC} (red) measured at the orange line in **c** and the red line in **d**, respectively.

100 nm, which is due to a larger hot carrier cooling. We find in this case $l_{cool} = 140$ nm.

To gain more insight in the Seebeck coefficient at the grain boundary, we tune the carrier density by a global gate (Fig. 2d). We observe that the antisymmetric spatial photocurrent profile changes sign as the backgate voltage V_{BG} passes the peak in resistance, that is, the global charge neutrality point V_D. The Seebeck coefficient S_G of graphene itself changes sign at the charge neutrality point[44,45,52,53] (Fig. 2c). Thus, after calibrating the sign to the known sign of the contact photocurrent, our data implies that the Seebeck coefficient of the grain boundary S_{GB} is always smaller in magnitude than S_G, since $I_{PC}(V_{BG}) \propto S_G(V_{BG}) - S_{GB}(V_{BG})$.

Using a polycrystalline graphene model, we compute the resistance due to grain boundaries using a Kubo transport formalism and real space simulations[54]. S_{GB} is the ratio of the first- and zero-order Onsager coefficients (Supplementary Note 2). Indeed, we find that S_{GB} is always smaller in magnitude and has a similar lineshape as S_G in the carrier density range measured (Fig. 2c). Figure 2e shows a simulation of the photocurrent for the calculated Seebeck coefficients, which is in agreement with the measurements (Supplementary Figs 5,6).

Charge puddle characterization. We next examine near-field photocurrent in a typical two-probe exfoliated graphene device (Fig. 3). A strong photocurrent is obtained with the tip near the metal contacts, similar to previous near- and far-field measurements[34,47,55]. In addition, an apparently random pattern of photocurrent is present throughout the device, as in high-resolution far-field measurements[55] but at a much finer scale.

The random photocurrent pattern between the contacts in Fig. 3a indicates random variations in Seebeck coefficient over

short length scales (Supplementary Fig. 7). Random variations of the Seebeck coefficient are indeed expected since it depends on carrier density[52], which in turn has fine-scaled inhomogeneities (charge puddles)[14–18]. The photocurrent variations can thus be used to gain insight in the charge puddle distribution. A more detailed view of the photocurrent due to charge puddles in Fig. 3b shows that the length scale that can be resolved is on the order of hundreds of nanometres.

Quantitatively, from the autocorrelation of the photocurrent in comparison with a photothermoelectric model taking into account the size of the charge puddles in Fig. 3c we extract $l_{cool} \sim 200$ nm. The charge puddles are modelled to have a size of ~ 20 nm, in accordance with measurements of graphene on silicon oxide (SiO$_2$; refs 15–18).

By changing the gate voltage we study the carrier density profile with high spatial resolution (Fig. 4) and highlight the possibility of spatially resolving the charge neutrality point for a large device. I_{PC} from charge puddles is largest around the charge neutrality point and varies with position. This is consistent with the very high sensitivity of the Seebeck coefficient to changes in carrier density, near-zero density (Fig. 4b). The magnitude of photocurrent from charge puddles depends on the difference of Seebeck coefficient between two adjacent charge puddles or in other words the strongest photocurrent from charge puddle appears at the position of highest Seebeck gradient. This allows us to map the local carrier density offset (charge inhomogeneity) throughout the device, as indicated by the extremum of photocurrent in a scan of photocurrent versus gate voltage (Fig. 4c). The photocurrent from adjacent charge puddles with a given charge carrier density offset does not change sign when sweeping through the charge neutrality point. This is because the difference in Seebeck coefficient between these puddles does not change sign.

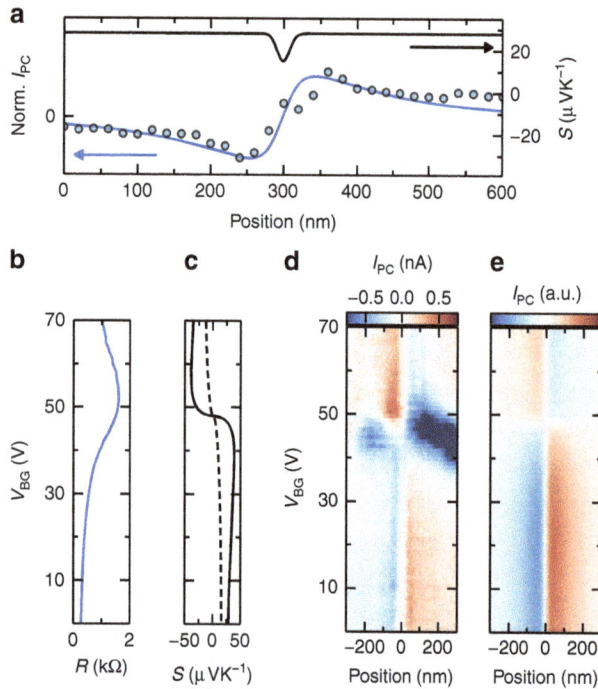

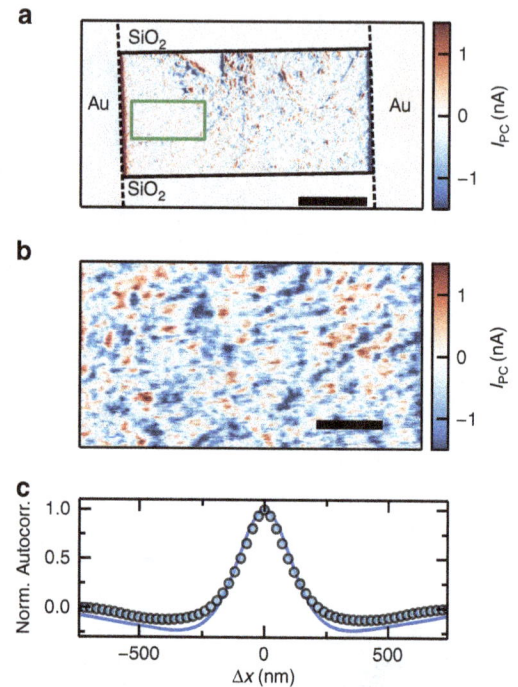

Figure 2 | Photocurrent profile at a grain boundary and its gate voltage dependence. (**a**) Photocurrent profile, measured at the black dashed line in Fig. 1d, perpendicular to the grain boundary at $V_{BG} = 0$ V shows good agreement with the photothermoelectric model with $l_{cool} = 140$ nm. The modelled spatial Seebeck profile (with FWHM 20 nm) is shown in black. (**b**) Two-probe device resistance as a function of V_{BG}. (**c**) Simulated Seebeck coefficient S_G for pristine graphene (solid line) and S_{GB} for polycrystalline graphene with an average grain size of 25 nm (dashed line; Supplementary Note 2). (**d**) Backgate dependent photocurrent profile perpendicular to the grain boundary shows that the grain boundary changes its sign at the charge neutrality point. (**e**) Simulated backgate dependent photocurrent profile based on the Seebeck profiles in **c** normalized to the simulated photocurrent maximum.

Figure 3 | Photocurrent from charge puddles. (**a**) Near-field photocurrent map of an exfoliated graphene device on 300 nm SiO_2 at $V_{BG} = 20$ V. The dashed lines indicate the position of the contacts and solid lines the graphene edges. The green box indicates the measurement region in **b** (scale bar, 5 μm). (**b**) Detailed photocurrent map at the charge neutrality point of the device ($V_{BG} = 7$ V) reveals the charge puddles and the high spatial resolution of the technique. (scale bar, 1 μm) (**c**) Autocorrelation of the photocurrent from charge puddles at V_D (data points) compared to photocurrent expected from a random charge puddle distribution and $l_{cool} = 200$ nm (blue curve). Autocorrelation is taken along the source drain current path.

We can thus resolve the local charge neutrality point at a given position of the device (green curve, Fig. 4a), which can be different from the global charge neutrality point V_D, the backgate voltage V_{BG} at which the resistance is maximum (blue curve, Fig. 4a). We show that the global charge neutrality point (blue curve, Fig. 4a) is determined by an average of the gate voltages at which the local charge neutrality points appear (red curve, Fig. 4a). Spatially resolved puddle photocurrent can be much narrower (green curve, Fig. 4a) than the average of all possible current paths (red curve, Fig. 4a) This indicates that the graphene locally has less inhomogeneity. Thus the technique gives insight not only in the global but also in the local behaviour of the device.

Characterization of encapsulated devices. Finally, we apply this technique to a graphene device encapsulated between two layers of hexagonal boron nitride (h-BN), using the polymer-free van der Waals assembly technique[39,40] as sketched in Fig. 5e. This device lies on top of an oxidized silicon wafer, used as a backgate. The stack is etched into a triangle and electrically side-contacted by two metal contacts[39]. Recent studies in vacuum and low temperatures[56,57] have shown that the edges affect where current flows in the device, in particular near charge neutrality. In the following we study the build-up of edge doping in ambient conditions and provide a solution to this.

While monitoring the photocurrent of such encapsulated devices, we observe indications of strong carrier density variations near the edges over micrometre scales. These variations are influenced by lighting conditions (Supplementary Fig. 8), gate voltages, and temperature, and evolve over timescales ranging from minutes to weeks. As an example, Fig. 5a–d shows a progression of photocurrent maps, taken after annealing the device at 200 °C for 30 min in ambient conditions to temporarily remove charge density variations near the edges. Initially in Fig. 5a we see very small photocurrents indicating a flat carrier density landscape. After some time (hours), in the dark with only gate voltages smaller than 3 V applied, a small doping gradient between the contacts builds up. This gradient leads to the stronger photocurrent shown in Fig. 5b. The local charge neutrality point, indicated by the maximum of photocurrent, is at the same position close to the edge of the device as further inside the bulk. After keeping the device for 3 h in ambient conditions we can see a change of the local charge neutrality point at the edge of the graphene compared to the bulk in Fig. 5c. The edge is slightly more p-type compared to the bulk. Finally, we apply high gate voltages, of in this case 50 V for ~20 h, to increase the edge doping. A strong p-doping at the edge and an n-doping in the bulk of the graphene is induced in Fig. 5d. This indicates that electric field accelerates the speed and increases this type of edge doping.

We exploit the observed edge doping to create a natural p–n junction along the edge of the device. For this we

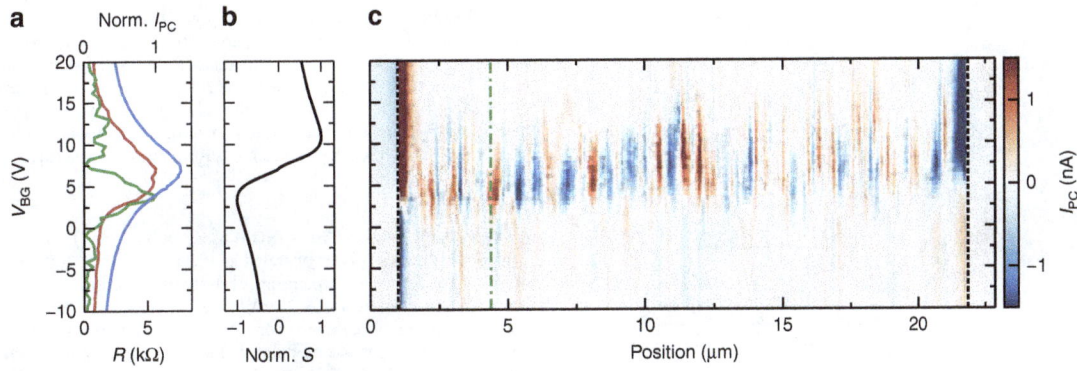

Figure 4 | Dependence of photocurrent profiles on backgate voltage reveals doping inhomogeneities. (a) Backgate dependence of the resistance of the device measured simultaneously to the photocurrent in blue. The red curve shows the normalized root mean square of the photocurrent across the device. The green curve shows a single normalized photocurrent backgate trace, corresponding to the green dashed dotted line in **c**. **(b)** Backgate dependent Seebeck coefficient of graphene, calculated from the gate dependent resistance in **a** using the Mott formula[52]. **(c)** Backgate dependence of the photocurrent across the device. Graphene is between the black dashed lines, which indicate the edges of the metal contacts.

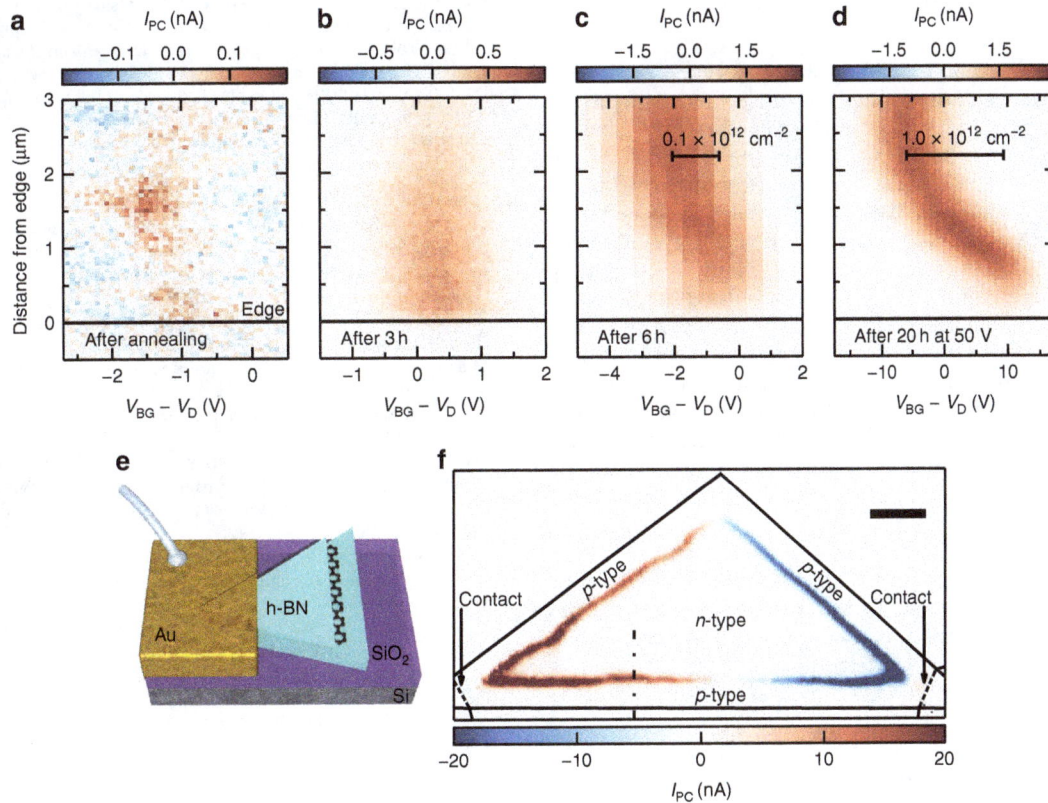

Figure 5 | Near-field photocurrent maps revealing edge doping in encapsulated graphene. (a) Spatial photocurrent profile versus backgate voltage V_{BG} (minus voltage of the resistance maximum V_D) near the edge of encapsulated graphene. These data are taken directly after annealing the device. ($V_D = -0.3$ V). **(b)** The same scan on the same device after three hours in air, ($V_D = -0.5$ V) and **c**, after annealing and applying V_{BG} up to 3 V ($V_D = 3$ V). **(d)** The same scan after approximately 20 h in air and after applying V_{BG} up to 50 V ($V_D = 30$ V). **(e)** Sketch of the device, a stack of h-BN(46 nm)/ Graphene/h-BN(7 nm) on a Si/SiO$_2$(300 nm) wafer used as global backgate. In this sketch we only show one of the two contacts used for electrical measurements. **(f)** Photocurrent close to the resistance maximum at $V_{BG} = -28$ V shows a triangular photocurrent pattern, due to edge p-doping. The dashed dotted line indicates where the measurements in **a–d** were taken. (scale bar, 2 μm).

apply a backgate voltage at which the edge of the graphene is p-type and the bulk n-type. This is similar to the situation in Fig. 5d at $V_{BG} = V_D$. We observe photocurrent at the junction in Fig. 5f around the whole device, indicating that the edge doping is uniform around the graphene. The photocurrent decays gradually towards the midline between the electrodes as a result of how the triangular geometry modifies the

ability of the contacts to capture photocurrents[49]. The sign change in the middle of the device is because in the current direction the junction changes from a p–n junction to an n–p junction (Supplementary Fig. 9 and Supplementary Note 2). We are able to temporarily reset the edge doping by annealing the device on a hotplate at 200 °C for 30 min as we show in Fig. 5a.

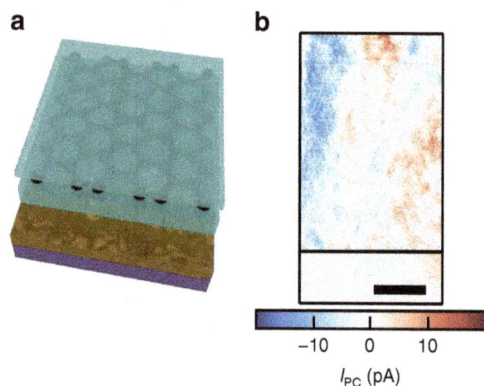

Figure 6 | Edge doping is efficiently suppressed using devices with local metal gates. (**a**) Sketch of the device with local metal gate with the two h-BN layers in light blue and the local metal gate in gold. The layers are from bottom to top SiO_2(300 nm)/AuPd(15 nm)/h-BN(42 nm)/Graphene/h-BN(13 nm). (**b**) Photocurrent from charge puddles in encapsulated graphene on a metal gate close to the charge neutrality point. The electrical contacts are on the left and right outside of this figure. (scale bar, 2 μm).

Local gates prevent charge build-up at the device edges. While we have not been able to precisely identify the origin of the edge doping, we suspect that water molecules are able to penetrate between the boron nitride and the SiO_2 due to the surface of the h-BN being not completely conformal with the underlying SiO_2 substrate. This water molecule penetration then leads to trapped charges is responsible for the observed edge doping. We also present here a technique to completely eliminate the edge doping. We place encapsulated graphene on top of a local conductive gate, such as a 15 nm AuPd alloy sketched in Fig. 6a. As the photocurrent measurement in Fig. 6b shows, we find that edge doping is efficiently suppressed even after extended periods of time at ambient conditions and high gate voltages. Furthermore, such devices efficiently suppress the photodoping effect observed for devices where the h-BN is in contact with SiO_2 (ref. 41; Supplementary Fig. 10 and Supplementary Note 1).

In the device with a local metal gate used to suppress both edge- and photodoping, we find small features due to charge puddles on top of a slowly varying background photocurrent, due to large-scale carrier density inhomogeneities. The size of the features due to charge puddles determined by autocorrelation is ~800 nm. The long length scale of those features is either due to the longer cooling length of the encapsulated graphene compared to the graphene on SiO_2 or due to larger charge puddle size in the encapsulated devices. Further work is required to clearly distinguish these effects.

Discussion

To conclude, we have demonstrated that scanning near-field photocurrent nanoscopy is a versatile technique to characterize the electronic and optoelectronic and even previously inaccessible properties of relevant graphene devices. This technique is highly promising for spatially resolved quality control of regular graphene devices without the need for special device structures and can therefore be readily applied.

Methods

Photocurrent model. Photocurrent I_{PC} in graphene as generated by the photothermoelectric effect and is described as:[45–47]

$$I_{PC} = \frac{1}{RW} \int \int \frac{\partial T}{\partial x} S \, dxdy \qquad (1)$$

where R is the total resistance including graphene, contacts and circuitry, W the device width and x the current flow direction. This is valid for rectangular graphene devices and special care needs to be taken for arbitrary shapes, such as in Fig. 5 (ref. 49). For the temperature profile $T(x)$ we consider that the heat spreads in two dimensions with heat sinking to lattice and substrate, producing a $T(x)$ profile described by a modified Bessel function of the second kind, with a finite tip size correction (Supplementary Note 2). A 25-nm finite tip-size correction was used for all simulations.

Measurement details. The s-SNOM used was a NeaSNOM from Neaspec GmbH, equipped with a CO_2 laser operated at 10.6 μm, away from the phonon resonance of SiO_2, which can lead to strong substrate contributions to the photocurrent[48]. The laser power used was ~20 mW. The probes were commercially available metallized atomic force microscopy probes with an apex radius of approximately 25 nm. The tip height was modulated at a frequency of approximately 250 kHz with a 60–80 nm amplitude. A Femto DLPCA-200 current pre-amplifier was used. The probe tip was electrically grounded. Because of the different device geometries and the fact that all the measurements were taken at different times and slightly different device conditions the absolute values of the photocurrents are not comparable.

Device fabrication. The CVD graphene, grown on copper, was transferred onto a self-assembled monolayer[58] on 285 nm of SiO_2 to stabilize the charge neutrality point. The contacts were defined using optical lithography with Ti (5 nm)/Pd (35 nm). The graphene was transferred onto deposited contacts.

The exfoliated graphene device was fabricated on a Si/SiO_2(300 nm) wafer, used as backgate. The Cr(0.8 nm)/Au(80 nm) contacts were defined using electron beam lithography.

The Si/SiO_2(300 nm)/h-BN(46 nm)/Graphene/h-BN(7 nm) and the Si/SiO_2(300 nm)/AuPd(15 nm)/h-BN(42 nm)/Graphene/h-BN(13 nm) stacks, were fabricated using the polymer-free van der Waals assembling technique[39].

References

1. Li, X. *et al.* Large-area synthesis of high-quality and uniform graphene films on copper foils. *Science* **324**, 1312–1314 (2009).
2. Reina, A. *et al.* Large area, few-layer graphene films on arbitrary substrates by chemical vapor deposition. *Nano Lett.* **9**, 30–35 (2009).
3. Bae, S. *et al.* Roll-to-roll production of 30-inch graphene films for transparent electrodes. *Nat. Nanotech.* **5**, 574–578 (2010).
4. Bonaccorso, F. *et al.* Production and processing of graphene and 2d crystals. *Mater. Today* **15**, 564–589 (2012).
5. Ren, W. & Cheng, H.-M. The global growth of graphene. *Nat. Nanotech.* **9**, 726–730 (2014).
6. Lee, J.-H. *et al.* Wafer-scale growth of single-crystal monolayer graphene on reusable hydrogen-terminated germanium. *Science* **344**, 286–289 (2014).
7. Gao, L. *et al.* Face-to-face transfer of wafer-scale graphene films. *Nature* **505**, 190–194 (2014).
8. de Heer, W. A. *et al.* Large area and structured epitaxial graphene produced by confinement controlled sublimation of silicon carbide. *Proc. Natl Acad. Sci. USA* **108**, 16900–16905 (2011).
9. Ferrari, A. C. *et al.* Science and technology roadmap for graphene, related two-dimensional crystals, and hybrid systems. *Nanoscale* **7**, 4598–4810 (2014).
10. Akinwande, D., Petrone, N. & Hone, J. Two-dimensional flexible nanoelectronics. *Nat. Commun.* **5**, 5678 (2014).
11. Koppens, F. H. L. *et al.* Photodetectors based on graphene, other two-dimensional materials and hybrid systems. *Nat. Nanotech.* **9**, 780–793 (2014).
12. Yazyev, O. V. & Louie, S. G. Electronic transport in polycrystalline graphene. *Nat. Mater.* **9**, 806–809 (2010).
13. Yu, Q. *et al.* Control and characterization of individual grains and grain boundaries in graphene grown by chemical vapour deposition. *Nat. Mater.* **10**, 443–449 (2011).
14. Martin, J. *et al.* Observation of electron-hole puddles in graphene using a scanning single-electron transistor. *Nat. Phys.* **4**, 144–148 (2008).
15. Chen, J.-H., Jang, C., Xiao, S., Ishigami, M. & Fuhrer, M. S. Intrinsic and extrinsic performance limits of graphene devices on SiO_2. *Nat. Nanotech.* **3**, 206–209 (2008).
16. Zhang, Y., Brar, V. W., Girit, C., Zettl, A. & Crommie, M. F. Origin of spatial charge inhomogeneity in graphene. *Nat. Phys.* **5**, 722–726 (2009).
17. Decker, R. *et al.* Local electronic properties of graphene on a BN substrate via scanning tunneling microscopy. *Nano Lett.* **11**, 2291–2295 (2011).
18. Xue, J. *et al.* Scanning tunnelling microscopy and spectroscopy of ultra-flat graphene on hexagonal boron nitride. *Nat. Mater.* **10**, 282–285 (2011).
19. Burson, K. M. *et al.* Direct imaging of charged impurity density in common graphene substrates. *Nano Lett.* **13**, 3576–3580 (2013).
20. Duong, D. L. *et al.* Probing graphene grain boundaries with optical microscopy. *Nature* **490**, 235–239 (2012).
21. Huang, P. Y. *et al.* Grains and grain boundaries in single-layer graphene atomic patchwork quilts. *Nature* **469**, 389–392 (2011).

22. Yasaei, P. *et al.* Bimodal phonon scattering in graphene grain boundaries. *Nano Lett.* **15**, 4532–4540 (2015).

23. Deshpande, A., Bao, W., Miao, F., Lau, C. & LeRoy, B. Spatially resolved spectroscopy of monolayer graphene on SiO$_2$. *Phys. Rev. B* **79**, 205411 (2009).

24. Gibertini, M., Tomadin, A., Guinea, F., Katsnelson, M. I. & Polini, M. Electron-hole puddles in the absence of charged impurities. *Phys. Rev. B* **85**, 201405 (2012).

25. Cho, S. *et al.* Thermoelectric imaging of structural disorder in epitaxial graphene. *Nat. Mater.* **12**, 913–918 (2013).

26. Fei, Z. *et al.* Electronic and plasmonic phenomena at graphene grain boundaries. *Nat. Nanotech.* **8**, 821–825 (2013).

27. Ferrari, A. C. & Basko, D. M. Raman spectroscopy as a versatile tool for studying the properties of graphene. *Nat. Nanotech.* **8**, 235–246 (2013).

28. Hsu, J. W. P., Fitzgerald, E. A., Xie, Y. H. & Silverman, P. J. Near-field scanning optical microscopy imaging of individual threading dislocations on relaxed GexSi1-x films. *Appl. Phys. Lett.* **65**, 344–346 (1994).

29. Buratto, S. K. *et al.* Imaging InGaAsP quantum-well lasers using near-field scanning optical microscopy. *J. Appl. Phys.* **76**, 7720 (1994).

30. Goldberg, B. B., Ünlü, M. S., Herzog, W. D., Ghaemi, H. F. & Towe, E. Near-field optical studies of semiconductor heterostructures and laser diodes. *IEEE J. Sel. Top. Quantum Electron.* **1**, 1073–1081 (1995).

31. Richter, A., Tomm, J. W., Lienau, C. & Luft, J. Optical near-field photocurrent spectroscopy: A new technique for analyzing microscopic aging processes in optoelectronic devices. *Appl. Phys. Lett.* **69**, 3981 (1996).

32. McNeill, C. R. *et al.* Direct photocurrent mapping of organic solar cells using a near-field scanning optical microscope. *Nano Lett.* **4**, 219–223 (2004).

33. Gu, Y. *et al.* Near-field scanning photocurrent microscopy of a nanowire photodetector. *Appl. Phys. Lett.* **87**, 043111 (2005).

34. Mueller, T., Xia, F., Freitag, M., Tsang, J. & Avouris, P. Role of contacts in graphene transistors: a scanning photocurrent study. *Phys. Rev. B* **79**, 245430 (2009).

35. Rauhut, N. *et al.* Antenna-enhanced photocurrent microscopy on single-walled carbon nanotubes at 30 nm resolution. *ACS Nano* **6**, 6416–6421 (2012).

36. Mauser, N. & Hartschuh, A. Tip-enhanced near-field optical microscopy. *Chem. Soc. Rev.* **43**, 1248–1262 (2014).

37. Mauser, N. *et al.* Antenna-enhanced optoelectronic probing of carbon nanotubes. *Nano Lett.* **14**, 3773–3778 (2014).

38. Grover, S., Dubey, S., Mathew, J. P. & Deshmukh, M. M. Limits on the bolometric response of graphene due to flicker noise. *Appl. Phys. Lett.* **106**, 051113 (2015).

39. Wang, L. *et al.* One-dimensional electrical contact to a two-dimensional material. *Science* **342**, 614–617 (2013).

40. Kretinin, A. V. *et al.* Electronic properties of graphene encapsulated with different two-dimensional atomic crystals. *Nano Lett.* **14**, 3270–3276 (2014).

41. Ju, L. *et al.* Photoinduced doping in heterostructures of graphene and boron nitride. *Nat. Nanotech.* **9**, 348–352 (2014).

42. Fei, Z. *et al.* Infrared nanoscopy of dirac plasmons at the graphene-SiO$_2$ interface. *Nano Lett.* **11**, 4701–4705 (2011).

43. Keilmann, F. & Hillenbrand, R. Near-field microscopy by elastic light scattering from a tip. *Phil. Trans. R Soc. A* **362**, 787–805 (2004).

44. Lemme, M. C. *et al.* Gate-activated photoresponse in a graphene p-n junction. *Nano Lett.* **11**, 4134–4137 (2011).

45. Gabor, N. M. *et al.* Hot carrier-assisted intrinsic photoresponse in graphene. *Science* **334**, 648–652 (2011).

46. Song, J. C. W., Rudner, M. S., Marcus, C. M. & Levitov, L. S. Hot carrier transport and photocurrent response in graphene. *Nano Lett.* **11**, 4688–4692 (2011).

47. Tielrooij, K. J. *et al.* Hot-carrier photocurrent effects at graphene' metal interfaces. *J. Phys. Condens. Matter* **27**, 164207 (2015).

48. Badioli, M. *et al.* Phonon-mediated mid-infrared photoresponse of graphene. *Nano Lett.* **14**, 6374–6381 (2014).

49. Song, J. C. W. & Levitov, L. S. Shockley-Ramo theorem and long-range photocurrent response in gapless materials. *Phys. Rev. B* **90**, 075415 (2014).

50. Tsen, A. W. *et al.* Tailoring electrical transport across grain boundaries in polycrystalline graphene. *Science* **336**, 1143–1146 (2012).

51. Van Tuan, D. *et al.* Scaling properties of charge transport in polycrystalline graphene. *Nano Lett.* **13**, 1730–1735 (2013).

52. Zuev, Y., Chang, W. & Kim, P. Thermoelectric and magnetothermoelectric transport measurements of graphene. *Phys. Rev. Lett.* **102**, 096807 (2009).

53. Wei, P., Bao, W., Pu, Y., Lau, C. N. & Shi, J. Anomalous thermoelectric transport of dirac particles in graphene. *Phys. Rev. Lett.* **102**, 166808 (2009).

54. Cummings, A. W. *et al.* Charge transport in polycrystalline graphene: challenges and opportunities. *Adv. Mater.* **26**, 5079–5094 (2014).

55. Lee, E. J. H., Balasubramanian, K., Weitz, R. T., Burghard, M. & Kern, K. Contact and edge effects in graphene devices. *Nat. Nanotech.* **3**, 486–490 (2008).

56. Ben Shalom, M. *et al.* Quantum oscillations of the critical current and high-field superconducting proximity in ballistic graphene. *Nat. Phys.* doi:10.1038/nphys3592 (2015).

57. Allen, M. T. *et al.* Spatially resolved edge currents and guided-wave electronic states in graphene. *Nat. Phys.* **12**, 128–133 (2016).

58. Chen, S. Y., Ho, P. H., Shiue, R. J., Chen, C. W. & Wang, W. H. Transport/magnetotransport of high-performance graphene transistors on organic molecule-functionalized substrates. *Nano Lett.* **12**, 964–969 (2012).

Acknowledgements

We thank Stijn Goossens, Klaas-Jan Tielrooij and Misha Fogler for many fruitful discussions and Fabien Vialla for the assistance in rendering the graphics in Fig. 1a. Open source software was used (www.matplotlib.org, www.python.org, www.povray.org). F.H.L.K. acknowledges financial support from the Spanish Ministry of Economy and Competitiveness, through the 'Severo Ochoa' Programme for Centres of Excellence in R&D (SEV-2015-0522), support by Fundacio Cellex Barcelona, the ERC Career integration grant (294056, GRANOP), the ERC starting grant (307806, CarbonLight), the Government of Catalonia trough the SGR grant (2014-SGR-1535), the Mineco grants Ramón y Cajal (RYC-2012-12281) and Plan Nacional (FIS2013-47161-P), and project GRASP (FP7-ICT-2013-613024-GRASP). F.H.L.K. and R.H. acknowledge support by the E.C. (European Commission) under Graphene Flagship (contract no. CNECT-ICT-604391). D.J. acknowledges support from the Ramón y Cajal Fellowship Program. Y.G. and J.H. acknowledge support from the US Office of Naval Research N00014-13-1-0662. We acknowledge financial support from the Spanish Ministry of Economy and Competitiveness and 'Fondo Europeo de Desarrollo Regional' through Grant TEC2013-46168-R. Q.M. and P.J.-H. have been supported by AFOSR Grant No. FA9550-11-1-0225 and the Packard Fellowship program. This work made use of the Materials Research Science and Engineering Center Shared Experimental Facilities supported by the National Science Foundation (NSF) (Grant No. DMR-0819762) and of Harvard's Center for Nanoscale Systems, supported by the NSF (Grant No. ECS-0335765). S.R. acknowledges the Spanish Ministry of Economy and Competitiveness for funding (MAT2012-33911), the Secretaria de Universidades e Investigacion del Departamento de Economia y Conocimiento de la Generalidad de Catalunya and the Severo Ochoa Program (MINECO SEV-2013-0295). J.E.B.-V. acknowledges support from SECITI (Mexico, D.F.).

Author contributions

A.W. performed the experiments, analysed the data and wrote the manuscript. P.A.-G. helped with measurements, interpretation and discussion of the results. M.B.L. helped with data analysis, measurements, interpretation, discussion of the results and manuscript writing. Y.G. fabricated the h-BN/graphene/h-BN devices. G.N. and D.J. fabricated the CVD graphene devices. Q.M. fabricated the exfoliated graphene devices. K.W. and T.T. synthesized the h-BN. J.E.B.-V., A.W.C. and S.R. performed the simulations of the Seebeck coefficient at grain boundaries. R.H. and F.H.L.K. supervised the work, discussed the results and co-wrote the manuscript. All authors contributed to the scientific discussion and manuscript revisions.

Additional information

14

High-performance thermoelectric nanocomposites from nanocrystal building blocks

Maria Ibáñez[1,2,3], Zhishan Luo[3], Aziz Genç[4], Laura Piveteau[1,2], Silvia Ortega[3], Doris Cadavid[3], Oleksandr Dobrozhan[3], Yu Liu[3], Maarten Nachtegaal[5], Mona Zebarjadi[6], Jordi Arbiol[4,7], Maksym V. Kovalenko[1,2] & Andreu Cabot[3,7]

The efficient conversion between thermal and electrical energy by means of durable, silent and scalable solid-state thermoelectric devices has been a long standing goal. While nanocrystalline materials have already led to substantially higher thermoelectric efficiencies, further improvements are expected to arise from precise chemical engineering of nanoscale building blocks and interfaces. Here we present a simple and versatile bottom–up strategy based on the assembly of colloidal nanocrystals to produce consolidated yet nanostructured thermoelectric materials. In the case study on the PbS–Ag system, Ag nanodomains not only contribute to block phonon propagation, but also provide electrons to the PbS host semiconductor and reduce the PbS intergrain energy barriers for charge transport. Thus, PbS–Ag nanocomposites exhibit reduced thermal conductivities and higher charge carrier concentrations and mobilities than PbS nanomaterial. Such improvements of the material transport properties provide thermoelectric figures of merit up to 1.7 at 850 K.

[1] Department of Chemistry and Applied Biosciences, Institute of Inorganic Chemistry, ETH Zürich, Vladimir Prelog Weg 1, CH-8093 Zurich, Switzerland. [2] Laboratory for Thin Films and Photovoltaics, Empa-Swiss Federal Laboratories for Materials Science and Technology, Dübendorf, Überlandstrasse 129, CH-8600 Dübendorf, Switzerland. [3] Advanced Materials Department, Catalonia Energy Research Institute - IREC, Sant Adria de Besos, Jardins de les Dones de Negre n.1, Pl. 2, 08930 Barcelona, Spain. [4] Department of Advanced Electron Nanoscopy, Catalan Institute of Nanoscience and Nanotechnology (ICN2), CSIC and The Barcelona Institute of Science and Technology, Campus UAB, Bellaterra, 08193 Barcelona, Spain. [5] Paul Scherrer Institute, 5232 Villigen PSI, Switzerland. [6] Department of Mechanical and Aerospace Engineering, Rutgers University, 98 Brett Rd, Piscataway, New Jersey 08854-8058, USA. [7] Institució Catalana de Recerca i Estudis Avançats, ICREA, Passeig de Lluís Companys, 23 08010 Barcelona, Spain. Correspondence and requests for materials should be addressed to A.C. (email: acabot@irec.cat).

Thermoelectric devices allow direct conversion of heat into electricity and *vice versa*, holding great potential for heat management, precise temperature control and energy harvesting from ubiquitous temperature gradients. The efficiency of thermoelectric devices is primarily governed by three interrelated material parameters: the electrical conductivity, σ, the Seebeck coefficient or thermopower, S, and the thermal conductivity, κ. These parameters are grouped into a dimensionless figure of merit, ZT, defined as $ZT = \sigma S^2 T \kappa^{-1}$ where T is the absolute temperature. While there is no known limitation to the maximum thermoelectric energy conversion efficiency other than the Carnot limit, current thermoelectric materials struggle to simultaneously display high σ and S, and low κ, which prevents their widespread implementation[1].

Control of the chemical composition and crystallinity of thermoelectric materials at the nanoscale via engineering of multicomponent nanomaterials (nanocomposites) has proven to be effective for the reduction of thermal conductivity by promoting phonon scattering at grain boundaries[2-7]. The remaining major challenge facing the next generation of high-efficiency thermoelectric materials is the enhancement of the thermoelectric power factor ($PF = S^2\sigma$) while keeping a low thermal conductivity. Strategies to accomplish this goal have focused on increasing the average energy per carrier through energy filtering[8], carrier localization in narrow bands in quantum confined structures[9], or the introduction of resonant levels[10]. At the same time, the electrical conductivity must be optimized by properly adjusting the concentration of charge carriers and maximizing their mobility.

These improvements of thermoelectric properties have been mainly demonstrated at the thin-film level, oftentimes using precise but expensive vacuum-based materials growth techniques[11]. However, practical applications of thermoelectric materials demand inexpensive and, for withstanding relatively high temperature gradients, macroscopic devices. In this regard, current approaches to produce bulk nanocomposites, such as ball-milling or the precipitation of secondary phases from metastable solid solutions, lack precision control over the distribution of phases and/or are limited in compositional versatility. Thus, novel cost-effective and general strategies to produce bulk nanocomposites with high accuracy and versatility need to be developed.

Here we demonstrate that a simple route to engineering nanocomposites by the assembly of precisely designed nanocrystal building blocks is able to reach high thermoelectric efficiencies. We propose to blend semiconductor nanocrystals with metallic nanocrystals forming Ohmic contact with the host semiconductor. In such nanocomposites, metallic nanocrystals control charge carrier concentration through charge spill over to the host semiconductor. The goal of this configuration is to reach large charge carrier concentrations without deteriorating the mobility of charge carriers[12-16]. In this three-dimensional (3D) modulation doping strategy[17-20], composition, size and distribution of semiconductor and metal nanodomains control the nanocomposite transport properties. Consider a slab of a semiconductor sandwiched between two metallic plates. Transfer of charges at the interfaces will cause band bending, extending over charge-screening length. If the width of a semiconductor slab is on the order of this screening length, there will be an overlap of the bended bands and therefore carriers will not be confined to the interfaces, but they will be able to travel through the bulk of a semiconducting region. Similarly, in the 3D case, the size of the semiconducting nanocrystals should be on the order of the screening length, enabling charge transport with minimum scattering. The position of the quasi Fermi level relative to the conduction band of the semiconductor at the

metal–semiconductor interface should be adjusted to align the bands and minimize scattering rates. This can be done by selecting metallic nanocrystals of the appropriate material and size and adjusting their volume fraction. The bottom-up approach presented in this work (Fig. 1) offers sufficient materials versatility to harness the key benefits of such 3D modulation doping by selecting materials with appropriate Fermi levels and allowing a facile control of nanocrystals size and volume fraction. For this study, PbS was selected as inexpensive host semiconductor that comprises Earth-abundant elements (that is, contrary to tellurides), and holds great potential for reaching high thermoelectric efficiencies; with reported ZT values of up to 1.3 at 923 K in PbS–CdS[21]. Silver is chosen as a nanoscopic metallic dopant owing to its low work function (4.26–4.9 eV) (refs 22–24) needed for efficient injection of electrons into the PbS conduction band. The PbS–Ag nanocomposites derived from colloidal PbS and Ag nanocrystals exhibit high electrical conductivities due to (i) injection of electrons from Ag nanoinclusions to the host PbS and (ii) improved charge carrier mobility. The simultaneous combination of high electrical conductivity, relatively large Seebeck coefficients, and reduced thermal conductivities provides thermoelectric figures of merit up to 1.7 at 850 K.

Results

PbS–Ag nanocomposites. PbS–Ag nanocomposites were produced by combining cubic PbS nanocrystals (*ca.* 11 nm, Fig. 2a) and spherical Ag nanocrystals (*ca.* 3 nm Fig. 2b), followed by the evaporation of a solvent. Thereby obtained powdered nanocrystal blend was annealed to remove residual organic compounds and then hot-pressed into pellets. This simple procedure yielded nanocomposites with a highly homogeneous distribution of metallic Ag nanodomains at the interfaces of PbS grains, as evidenced by high-resolution transmission electron microscopy (HRTEM), high-angle annular dark field scanning transmission electron microscopy and energy-dispersive X-ray spectroscopy (Fig. 2c,d). Further atomistic insights into bonding and chemical identities of constituents were obtained by X-ray absorption

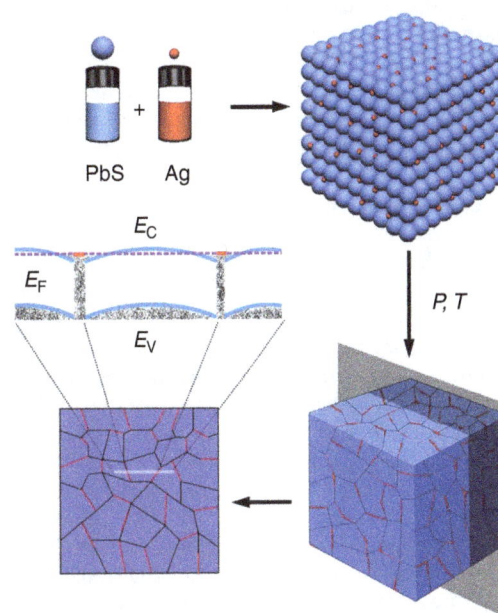

Figure 1 | Bottom-up Design. Bottom-up assembly process to produce PbS–Ag TE nanocomposites from the assembly of PbS (blue) and Ag (red) NCs, and the corresponding band alignment of the resulting nanocomposite.

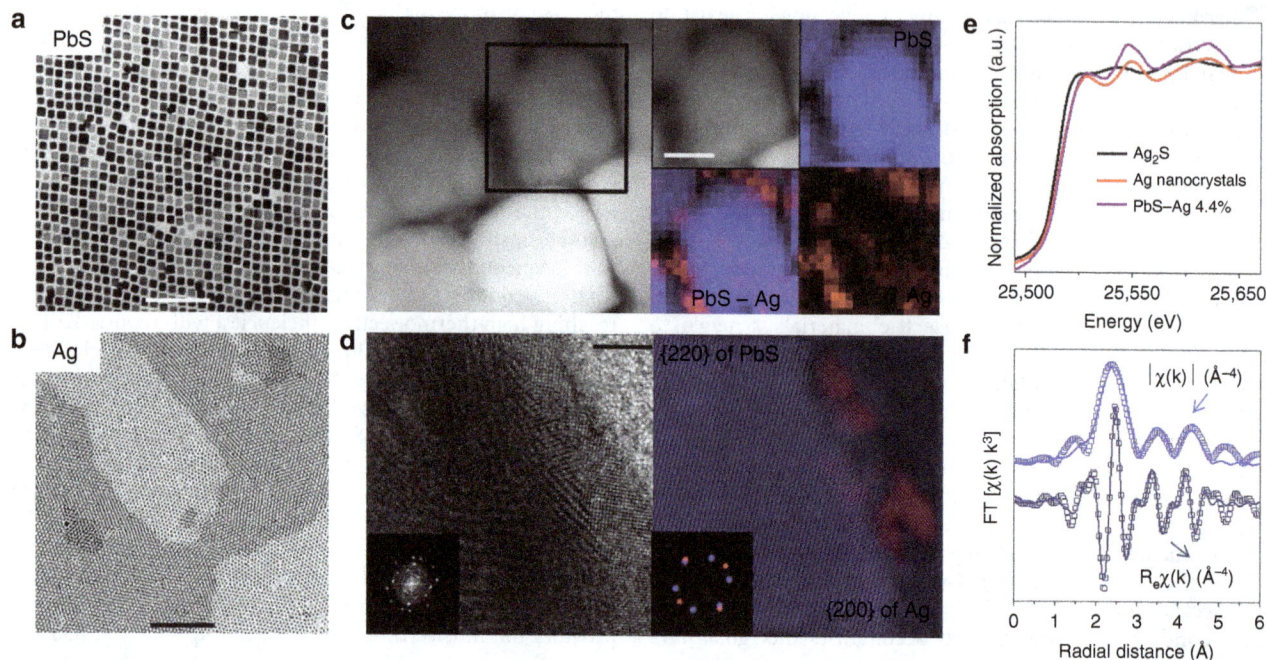

Figure 2 | Structural and compositional characterization of initial NCs and resulting PbS-Ag 4.4 mol% nanocomposite. TEM micrographs of (**a**) PbS and (**b**) Ag NCs; (**c**) HAADF-STEM micrograph and elemental EDX maps; and (**d**) HRTEM micrograph of the PbS–Ag interface with the corresponding power spectrum (inset) and filtered colourful composite image for the {220} family of planes of PbS (blue) and {200} family of planes of Ag (red). In the filtered images, PbS and Ag are visualized along their [111] and [001] zone axes, respectively (Supplementary Note 1); (**e**) Ag K-edge XANES spectra of Ag_2S reference (black), Ag NCs (red) and PbS-Ag 4.4 mol% nanocomposite (purple); (**f**) Fourier transform magnitude, $|\chi_{(k)}|$, and real part, $_{Re}\chi_{(k)}$, of the Pb L_3-edge EXAFS spectrum, the experimental data are given by the dotted line, the best fit by the solid line. Scale bars, 100 nm (**a**), 100 nm (**b**), 50 nm (**c**), 5 nm (**d**).

spectroscopy (XAS) at the Ag K edge (25515 eV) and Pb L_3 edge (13035 eV), respectively. Linear combination fitting of the X-ray absorption near edge structure (XANES) around the Ag absorption K edge, using references of Ag_2S and Ag metal, confirmed that at least 97% of Ag retained its metallic state (Fig. 2e, Supplementary Fig. 1 and Table 1). Only a small percentage of Ag may had diffused as Ag^+ within the PbS matrix. In addition, fitting of the Pb L_3 edge extended X-ray absorption fine structure spectra indicated interatomic distances and coordination numbers characteristics of PbS[25] (Fig. 2f). Metallic lead, lead oxide or lead sulfate species did not noticeably contribute to the spectra, suggesting that the presence of these phases can be disregarded (Supplementary Table 2).

Thermoelectric properties. To determine the effect of the Ag content on thermoelectric performance, a series of nanocomposites with Ag concentrations up to 5 mol% were prepared and analysed. Ag-free PbS nanomaterials exhibited relatively low electrical conductivities, which greatly increased from $0.07\,S\,cm^{-1}$ at room temperature (RT) up to $46\,S\,cm^{-1}$ at 850 K due to band to band charge carrier thermal excitation (Fig. 3a). This increase was accompanied by a sign inversion of the Seebeck coefficient from positive to negative at around 470 K (Fig. 3b). On the contrary, PbS-Ag nanocomposites possess significantly higher, Ag concentration-dependent, electrical conductivities. At RT, electrical conductivities of up to $660\,S\,cm^{-1}$ were measured for Ag concentrations above 4 mol% (Fig. 3a). Furthermore, over the whole studied temperature range of 300–850 K, PbS-Ag nanocomposites exhibit negative Seebeck coefficients. Unlike to electrical conductivity, Seebeck coefficient decreases with increasing Ag content. Hence an optimal concentration of ca. 4.4–4.6 mol% Ag nanocrystals was established for maximizing the PF (Fig. 3c). Overall, around 20 PbS-Ag pellets were produced, all showing PFs above $1\,mW\,m^{-1}\,K^{-2}$ at 850 K, with a champion

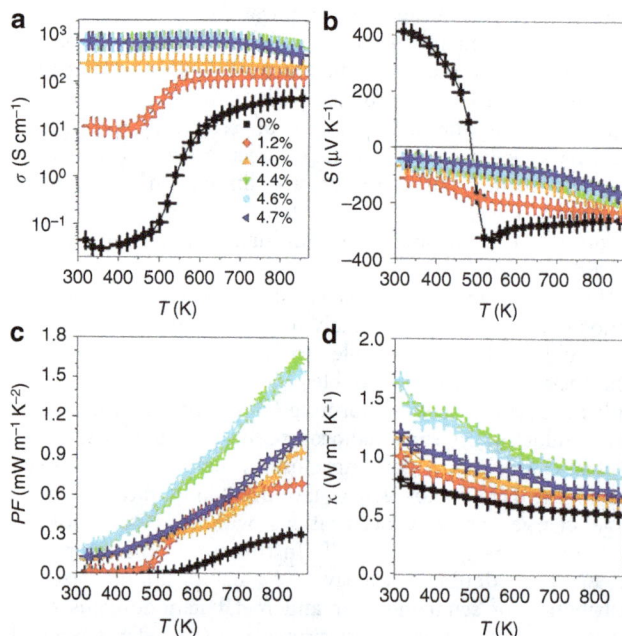

Figure 3 | Thermoelectric characterization of PbS-Ag nanocomposites. Temperature dependence of the (**a**) electrical conductivity, σ; (**b**) Seebeck coefficient, S; (**c**) thermal conductivity, κ; and (**d**) power factor, PF. Error bars were estimated from the repeatability of the experimental result; 3-5 measurements were carried out for each material.

value of $1.68\,mW\,m^{-1}\,K^{-2}$ at 850 K for a PbS-Ag 4.4 mol%. This is a sixfold increase over identically prepared Ag-free samples, and a significant improvement by 23–47% over previously reported PbS-based nanocomposites[21,26].

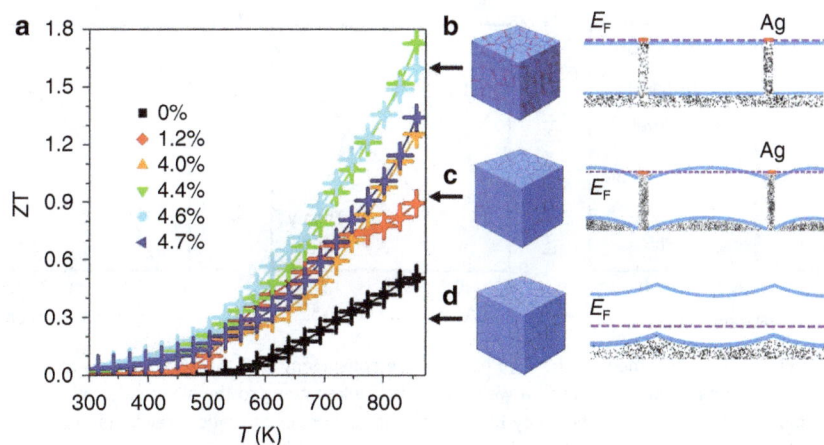

Figure 4 | Figure of merit and schematic representation of the electron energy band alignment. (**a**) Figure of merit, ZT, of PbS and PbS–Ag pellets; (**b**) band alignment in PbS–Ag nanocomposite with a larger amount of Ag and flat bands across the whole PbS domains; (**c**) band aligment in PbS–Ag nanocomposite with a low Ag volume fraction, showing electron energy wells in between PbS domains; (**d**) band alignment in bare PbS with an upward band-bending at the PbS intergrains. Error bars were estimated from the repeatability of the experimental result; 3-5 measurements were carried out for each material.

The thermal conductivities of the PbS–Ag nanocomposites monotonically increased with the Ag content (Fig. 3d) due to the increase of the electronic contribution (Supplementary Fig. 2 and Methods), yet not exceeding in the high-temperature range the values reported for PbS-based nanocomposites produced by co-precipitation of secondary phases[21,26,27]. Thus, beyond injecting charge carriers and facilitating charge transport between PbS nanocrystals, Ag nanodomains also assisted in blocking phonon propagation. The relatively low thermal conductivities may also in part result from phonon scattering at nanodomains within the large PbS grains as observed by extensive HRTEM analysis (Supplementary Figs 3 and 4 and Note 2).

Overall, the outstanding electrical properties along with low thermal conductivities of PbS–Ag samples resulted in thermoelectric figures of merit of up to $ZT = 1.7$ at 850 K (Fig. 4a). This value corresponds to a 30% increase over the highest figure of merit obtained for PbS to date ($ZT = 1.3$, $Pb_{0.975}Na_{0.025}S + 3\%$ CdS)[21] and is comparable to the highest thermoelectric figures of merit reported for other lead chalcogenide materials (Supplementary Table 3)[3,21,26–30].

Discussion

The high PFs displayed by PbS–Ag nanocomposites are at the origin of their outstanding thermoelectric figures of merit. Such high PFs could not be explained by a simple weighed sum of the properties of two randomly distributed compounds (Supplementary Table 4 and Supplementary Discussion). Neither can we explain the high electrical conductivity by percolation transport through Ag domains, as much lower Seebeck coefficients would be expected for a metallic conductor. It must be also pointed out that doping of PbS with Ag^+ ions cannot explain these transport properties either, since previous studies have demonstrated Ag^+ to be a p-type dopant for PbS[31–34]. Simultaneous combination of high electrical conductivity and relatively large Seebeck coefficients can be explained by an efficient injection of electrons from the metal to the conduction band of the semiconductor (Fig. 4b–d). In this regard, RT Hall charge carrier concentration measurements evidenced an increase in concentration of majority charge carriers from $p = 1 \times 10^{16}$ in the bare PbS nanomaterial to $n = 3 \times 10^{19}$ in PbS–Ag 4.4 mol% samples (Fig. 5 and Supplementary Table 5). This is consistent with the initial Ag Fermi level above that of the intrinsic PbS.

What is surprising is that the obtained charge carrier mobilities also increased with the Ag introduction, from $20\,cm^2\,V^{-1}\,s^{-1}$ for bare PbS to $90\,cm^2\,V^{-1}\,s^{-1}$ for PbS–Ag 4.4 mol% (Supplementary Table 5). In a simple modulation doping scenario, one would expect the injection of charge carriers from Ag nanodomains into PbS to have little negative impact on the charge carrier mobility. However, in PbS–Ag nanocomposites, the actual effect of Ag was found to facilitate charge transport through the material. We attribute this to a reduction of the energy barriers between PbS crystal domains (Fig. 4b–d). To determine the band alignment, we used the Anderson model to align the vacuum levels of Ag and PbS and then solve the Poisson equation self consistently assuming a parabolic two-band model for PbS (Fig. 5a). We used the Ag work function, which depends on its crystallographic surface and domain size, as a fitting parameter to fit the experimentally measured Hall data (Fig. 5b). The obtained fitted parameter was 4.4 eV, which is in the correct range (4.26–4.9 eV) (refs 22,23).

At a bulk metal–semiconductor junction, the Fermi level is pinned by the metallic layer due to the large carrier density in the metallic layer (Fig. 5a). However, for small nanocrystals, the number of electrons is limited and at low metal concentrations it may be not enough to completely pin the Fermi level. To simulate charge transfer from Ag nanocrystals to PbS, we assumed Ag nanocrystals were spheres of radius 1.5 nm (to replicate the 3-nm size of Ag nanodomains observed) embedded within another sphere made out of PbS. The radius of PbS sphere was determined by the volume fraction of each sample (Fig. 5b, inset). Figure 5c shows the Fermi level with respect to the bottom of the conduction band. This is an indicator for the effective well depth for the electrons as marked in Fig. 5a. As can be seen in Fig. 5c, when the Ag fraction increases, the Fermi level also increases, lowering the effective well depth in between PbS grains, which reaches zero at around a 0.5 % Ag volume fraction, that is, 1.5 mol%. At this point and beyond, there is no effective well in the path of the electrons (Fig. 4b). Therefore, as the Ag fraction increases, the electron-nanocrystal scattering decreases, which is consistent with the observed enhancement in electron mobility.

Note that the high PFs found in PbS–Ag nanocomposites cannot be obtained by a conventional doping strategy, where the introduction of ionic impurities would lead to increased scattering of charge carriers. This is shown in Fig. 6a,b, where the thermoelectric properties of PbS nanomaterials with different

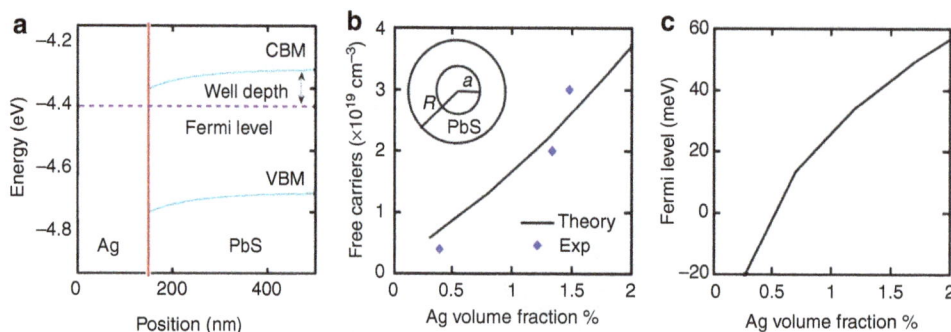

Figure 5 | Theoretical calculations. (**a**) Band alignment at the Ag-PbS interface considering bulk Ag and bulk PbS at 300 K. The Fermi level is pinned at -4.4 eV (silver work function). Electrons in PbS experience a well, as marked, when going to the silver side. (**b**) Carrier concentration versus silver volume fraction. Experimentally measured Hall data are shown by blue dots. Free carrier concentration, calculated using Anderson model and Poisson solver for the geometry shown in the inset of the figure, is shown by solid line. The inner sphere represents a silver NC ($a = 1.5$ nm) and the outer sphere represent PbS host matrix with radius R (silver volume fraction $= (a/R)^3$. (**c**) Fermi level plotted with respect to the conduction band minimum of PbS (far away from the interface). When negative, this value corresponds to the well depth experienced by electrons when moving between PbS grains.

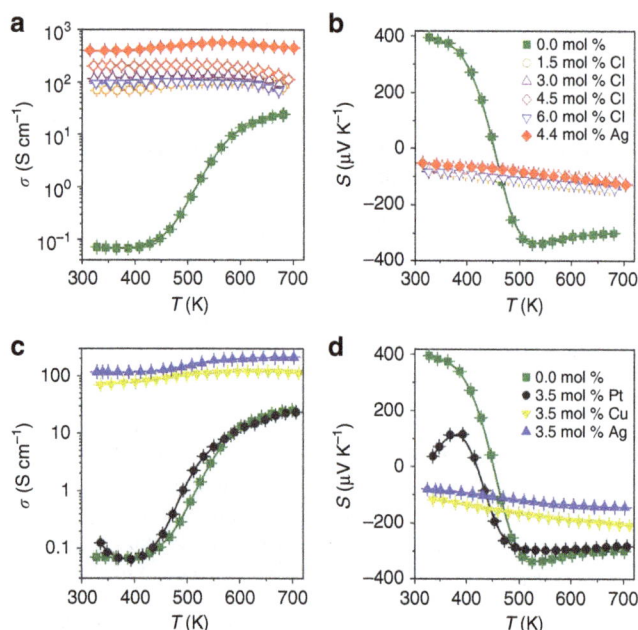

Figure 6 | Electrical properties of PbS nanomaterials with different doping strategies. (**a**) σ and (**b**) S of x mol% PbS ($x = 1.5$, 3.0, 4.5 and 6.0) doped with Cl (open symbols) compared with pure PbS (0.0 mol %) and PbS-Ag (4.4 mol%). (**c**) σ and (**d**) S of PbS (0.0 mol%) and PbS-X (3.5 mol%) with X = Pt, Cu and Ag.

chlorine concentrations are presented. By introducing Cl$^-$ ions, a common dopant used in n-type PbS, the electron density can be increased, which translates into an enhancement of the electrical conductivity[35]. However, the highest electrical conductivities reached by halide doping remained below the values obtained for PbS–Ag nanocomposites. Thus, despite that PbS:Cl (4.5 mol%) nanomaterials had charge carrier concentrations in the same order of magnitude and similar Seebeck coefficients as the PbS–Ag (4.4 mol%) nanocomposites, the later showed significantly higher electrical conductivities and, therefore, higher PFs.

Besides PbS–Ag nanocomposites, a variety of other compositions were accomplished from colloidal nanocrystal building blocks, often showing similar synergistic effects on charge transport. For instance, we have carefully examined composites of PbS with nanoscopic Cu (Supplementary Fig. 5) and Pt (Supplementary Fig. 6). Like Ag, Cu has a low work function and hence is able to inject charges into the conduction band of PbS.

On the contrary, Pt has a much higher work function and does not exhibit efficient charge transfer (Fig. 6c,d).

Colloidal nanocrystals can be prepared with unmatched control over size, composition, shape, crystal phase and surface chemistry and benefit from facile handling and mixing in stable dispersions[36]. The current availability of a rich palette of such building blocks lends the opportunity to create a plethora of different nanocomposites by simply blending nanocrystals of various materials in the appropriate proportions and, afterward, consolidating them into arbitrarily shaped composites. Thus the facile bottom–up approach used here allows engineering a nearly endless variety of nanocomposites, which will allow a high-throughput screening of materials in the effort to maximize the thermoelectric energy conversion efficiency of durable, silent and scalable thermoelectric devices.

Methods

Nanocomposite preparation. *Chemicals.* Lead(II) oxide (PbO, 99.9%), copper(I) acetate (CuOAc, 97%), silver nitrate (AgNO$_3$, $\geq$99.8%), iron(III) nitrate non-ahydrate (Fe(NO$_3$)$_3 \cdot$9H$_2$O, 99.99%), platinum acetylacetonate (Pt(acac)$_2$, 97%), manganese(0) carbonyl (Mn$_2$(CO)$_{10}$, 98%), elemental sulfur (99.998%), oleic acid (OA, tech. 90%), 1-octadecene (ODE, 90%), oleylamine (OLA, tech. 70%) and benzyl ether (BE, 98%) were purchased from Aldrich. Tri-n-octylamine (TOA, 97%) was purchased from Across. Tetradecylphosphonic acid (TDPA, 97%) was purchased from PlasmaChem. Hexane, toluene, ethanol, anhydrous chloroform and anhydrous methanol were obtained from various sources. All chemicals, except OLA, were used as received without further purification. OLA was distilled to remove impurities.

Synthesis of nanocrystals. All syntheses were carried out using standard air-free techniques; a vacuum/dry argon Schlenk line was used for synthesis and an argon-filled glove box for storing and handling air and moisture-sensitive chemicals.

PbS nanocrystals with a mean edge size of 11 nm were prepared similarly to previously reported procedures[35]. In a typical synthesis, PbO (4.46 g, 20 mmol) and OA (50 ml, 0.158 mol) were mixed with 100 ml of ODE. This mixture was degassed at RT and 100 °C for 0.5 h each to form the lead oleate complex. Then the solution was flushed with Ar, and the temperature was raised to 210 °C. At this temperature, a sulfur precursor, prepared by dissolving elemental sulfur (0.64 g, 20 mmol) in distilled OLA (20 ml, 0.061 mol), was rapidly injected. The reaction mixture was maintained between 195 °C and 210 °C for 5 min and then quickly cooled down to RT using a water bath. Ag nanocrystals with an average diameter of 2–3 nm were produced using a modified approach of that reported by Wang et al.[37] In a typical reaction, AgNO$_3$ (0.17 g, 1 mmol), Fe(NO$_3$)$_3 \cdot$9H$_2$O (0.04 g, 0.01 mmol), OA (10 ml, 0.0316 mol) and OLA (10 ml, 0.0305 mol) were mixed and placed under Ar at RT for 30 min. Afterwards the reaction mixture was heated to 120 °C at the rate of 5 °C min^{-1} and kept at this temperature for an additional 60 min. Cu nanocrystals with an average diameter of 5–6 nm were prepared following the approach developed by Yang et al.[38] In a typical synthesis, TOA (50 ml, 0.114 mol) was heated in a 100 ml three-neck flask to 130 °C for 30 min under Ar atmosphere. After cooling to RT, CuOAc (613 mg, 5 mmol) and TDPA (696 mg, 2.5 mmol) were added to the flask. The mixture was heated to 180 °C and maintained at this temperature for 30 min. Then, the reaction temperature was further increased to 270 °C and held for another 30 min. Cu nanocrystals are highly air sensitive and easily oxidized. To avoid any possible oxidation, the nanocrystals were purified in an Ar filled glove box. Pt nanocrystals with an average diameter of 6 nm were prepared using the method developed by Murray et al.[39] In a typical synthesis, Pt(acac)$_2$ (80 mg, 0.20 mmol) was dissolved in BE (10 ml, 52.6 mmol), OLA (7.36 ml, 22.37 mmol) and OA (1.25 ml, 3.94 mmol) under Ar atmosphere for 30 min at 60 °C. The precursor mixture was heated to 160 °C and a solution of 80 mg Mn$_2$(CO)$_{10}$ in 1 ml of chloroform was rapidly injected. Afterwards, the temperature was heated to 200 °C and held for an additional 30 min at this reaction temperature. Finally, the crude solution was cooled to RT.

Blending of nanocrystals. In this work we prepared PbS-metal semiconductor nanocomposites with different metal concentrations. The blending of nanocrystals was performed by wetting 1 g of dried PbS nanocrystals (a powder) with different amounts of a 0.093 M solution of metallic nanocrystals in anhydrous chloroform. Subsequently, the solvent was allowed to evaporate under Ar atmosphere. The concentration of metallic nanocrystals in chloroform was initially estimated by mass (considering a 30% of organic ligand) and later verified by inductively coupled plasma (ICP) spectroscopy. The values were found to differ slightly between the estimation from the weight and from ICP. The final quantities reported in this work correspond to the values obtained from ICP measurements.

Pellet fabrication. Before powder consolidation in a hot press, the nanocrystal powders were treated thermally to decompose the remaining organic ligands present at the nanocrystal surface. All nanocrystal powders were heated to 450 °C at 10 °C min^{-1} and held at this temperature for 1 h under an Ar flow. After cooling to RT, the nanocrystal powders were pressed using a custom-made hot press. In this system, heat is provided by an induction coil operated in the RF range and applied directly to a graphite die acting as a susceptor. Before hot pressing, coarse powders were ground into fine powder using a mortar inside the glove box and then loaded into a 10-mm diameter graphite die lined with 0.13-mm thick graphite paper. The filled die was placed in the hot press system. The densification profile applied an axial pressure of 45 MPa before heating the die to between 420 and 440 °C. The temperature was held between 420 and 440 °C for 4 min. The pressure was then removed and the die cooled to RT. The resulting pellets were >85% dense compared with theoretical maximum into air-stable monoliths measuring ~1-mm thick by 10 mm in diameter. The density of the pressed pellets was measured by the Archimedes method.

Structural characterization. The size and shape of the initial nanocrystals were examined by TEM using a ZEISS LIBRA 120 instrument, operating at 120 kV. Structural and compositional characterizations of the nanocomposites were examined after thermoelectric characterization. TEM Samples were mechanically thinned to 20–30 μm and further thinned to electron transparency by Ar$^+$ polishing using a Gatan Precision Ion Polishing System. HRTEM and scanning transmission electron microscopy studies were conducted by using a FEI Tecnai F20 field emission gun microscope operated at 200 kV, which is equipped high-angle annular dark field and energy-dispersive X-ray spectroscopy detectors. ICP atomic emission spectrometry was used for elemental analysis of the nanocomposites, especially to determine the ratio between Pb and Ag, Pt or Cu. ICP atomic emission spectrometry measurements were carried out using Perkin–Elmer Optima instrument, model 3200RL, under standard operating conditions. Samples were prepared by microwave-assisted digestion of the dried materials in a mixture of HNO$_3$ and H$_2$O$_2$ in a closed container. X-ray powder diffraction analyses were collected directly on the as-synthesized nanocrystals and final pellets using a Bruker AXS D8 Advance X-ray diffractometer with Ni-filtered (2 μm thickness) Cu K$_\alpha$ radiation (λ = 1.5406 Å) operating at 40 kV and 40 mA (Supplementary Fig. 7). A LynxEye linear position-sensitive detector was used in reflection geometry. XAS measurements were carried out at the X10DA (SuperXAS) beamline at the Swiss Light Source, Villigen, Switzerland, which operated with a ring current of ~400 mA in top-up mode. The polychromatic radiation from the superbend magnet, with a magnetic field of 2.9 T and critical energy of 11.9 keV, was monochromatized using a channel cut Si(311) crystal monochromator. Spectra were collected on pressed pellets optimized to 1 absorption length at the Ag K edge (25515 eV) and Pb L$_3$ edge (13035 eV) in transmission mode. XAS data were treated with the Demeter software suite[40]. For all samples, three spectra were acquired and merged. These averaged XAS data were background-subtracted and normalized. Linear combination fitting of the Ag K edge X-ray absorption near edge structure spectra was performed over the energy range from 25,498 to 25,598 eV. The goal was to determine composition of Ag-species present in the PbS-Ag sample. For this, reference spectra of metallic Ag NPs and Ag$_2$S were combined linearly. Fourier transformation of the Pb L$_3$ edge extended X-ray absorption fine structure spectra was performed over the k range of 3 – 9 Å^{-1} yielding a pseudo radial structure function. The R-range from 1 to 4.8 Å was fitted using theoretical single scattering paths of bulk PbS (S6)[25] based on crystallographic data.

Thermoelectric characterization. *Electric properties.* The pressed samples were polished, maintaining the disk-shape morphology. Final pellets had a 10-mm diameter and were ~1 mm thick. The Seebeck coefficient was measured using a static DC method. Electrical resistivity data were obtained by a standard four-probe method. Both the Seebeck coefficient and the electrical resistivity were simultaneously measured with accuracies better than 1% in a LSR-3 LINSEIS system from RT to 850 K, under helium atmosphere. Samples were held between two alumel

electrodes and two probe thermocouples with spring-loaded pressure contacts. A resistive heater on the lower electrode created temperature differentials in the sample to determine the Seebeck coefficient. Samples measured up to 850 K were spray coated with boron nitride to minimize out-degassing except where needed for electrical contact with the thermocouples, heater and voltage probes. In addition, before high-temperature measurements, samples were heated within the LINSEIS system in a He atmosphere up to 850 K at 3 K min^{-1} and hold at this temperature for 10 min with the boron nitride coating. Such preliminary treatment warrants sample stability for all the cycles tested (Supplementary Figs 8 and 9 and Discussion). Carrier concentration and mobility were estimated from Hall Effect measurements, which were performed at RT using an Ecopia HMS-3000 set-up with golden spring-loaded contacts positioned at the edges of plates in the Van der Pauw configuration.

Thermal properties. An XFA 600 Xenon Flash Apparatus was used to determine the thermal diffusivities of all samples with an accuracy of *ca.* 6%. Total thermal conductivity (κ) was calculated using the relation $\kappa = DC_p\rho$, where D is the thermal diffusivity, C_p is the heat capacity and ρ is the mass density of the pellet. The ρ values were calculated using the Archimedes method. The specific heat (C_p) of the samples was measured using a Differential Scanning Calorimeter DSC 204 F1 Phoenix from NETZSCH (Supplementary Fig. 10). The electronic contribution of the thermal conductivity was calculated using the Wiedemann–Franz law (Supplementary Fig. 11 and Methods).

References

1. Bell, L. E. Cooling, heating, generating power, and recovering waste heat with thermoelectric systems. *Science* **321**, 1457–1461 (2008).
2. Vineis, C. J., Shakouri, A., Majumdar, A. & Kanatzidis, M. G. Nanostructured thermoelectrics: big efficiency gains from small features. *Adv. Mater.* **22**, 3970–3980 (2010).
3. Biswas, K. et al. High-performance bulk thermoelectrics with all-scale hierarchical architectures. *Nature* **489**, 414–418 (2012).
4. Ibáñez, M. et al. Core-shell nanoparticles as building blocks for the bottom-up production of functional nanocomposites: PbTe-PbS thermoelectric properties. *ACS Nano* **7**, 2573–2586 (2013).
5. Hsu, K. F. et al. Cubic AgPb$_m$SbTe$_{2+m}$: bulk thermoelectric materials with high figure of merit. *Science* **303**, 818–821 (2004).
6. Kim, S. I. et al. Dense dislocation arrays embedded in grain boundaries for high-performance bulk thermoelectrics. *Science* **348**, 109–114 (2015).
7. Ibáñez, M. et al. Crystallographic control at the nanoscale to enhance functionality: polytypic Cu$_2$GeSe$_3$ nanoparticles as thermoelectric materials. *Chem. Mater.* **24**, 4615–4622 (2012).
8. Bahk, J.-H., Bian, Z. & Shakouri, A. Electron energy filtering by a nonplanar potential to enhance the thermoelectric power factor in bulk materials. *Phys. Rev. B* **87**, 075204 (2013).
9. Ohta, H. et al. Giant thermoelectric Seebeck coefficient of a two-dimensional electron gas in SrTiO3. *Nat. Mater.* **6**, 129–134 (2007).
10. Heremans, J. P. et al. Enhancement of thermoelectric efficiency in PbTe by distortion of the electronic density of states. *Science* **321**, 554–557 (2008).
11. Harman, T. C., Taylor, P. J., Walsh, M. P. & LaForge, B. E. Quantum dot superlattice thermoelectric materials and devices. *Science* **297**, 2229–2232 (2002).
12. García de Arquer, F. P., Lasanta, T., Bernechea, M. & Konstantatos, G. Tailoring the electronic properties of colloidal quantum dots in metal–semiconductor nanocomposites for high performance photodetectors. *Small* **11**, 2636–2641 (2015).
13. Zebarjadi, M., Esfarjani, K., Dresselhaus, M. S., Ren, Z. F. & Chen, G. Perspectives on thermoelectrics: from fundamentals to device applications. *Energ. Environ. Sci.* **5**, 5147–5162 (2012).
14. Dingle, R., Störmer, H. L., Gossard, A. C. & Wiegmann, W. Electron mobilities in modulation-doped semiconductor heterojunction superlattices. *Appl. Phys. Lett.* **33**, 665–667 (1978).
15. Hicks, L. D. & Dresselhaus, M. S. Thermoelectric figure of merit of a one-dimensional conductor. *Phys. Rev. B* **47**, 16631–16634 (1993).
16. Moon, J., Kim, J.-H., Chen, Z. C. Y., Xiang, J. & Chen, R. Gate-modulated thermoelectric power factor of hole gas in Ge–Si core–shell nanowires. *Nano Lett.* **13**, 1196–1202 (2013).
17. Zebarjadi, M. et al. Power factor enhancement by modulation doping in bulk nanocomposites. *Nano Lett.* **11**, 2225–2230 (2011).
18. Yu, B. et al. Enhancement of thermoelectric properties by modulation-doping in silicon germanium alloy nanocomposites. *Nano Lett.* **12**, 2077–2082 (2012).
19. Pei, Y.-L., Wu, H., Wu, D., Zheng, F. & He, J. High thermoelectric performance realized in a BiCuSeO system by improving carrier mobility through 3D modulation doping. *J. Am. Chem. Soc.* **136**, 13902–13908 (2014).
20. Wu, D. et al. Significantly enhanced thermoelectric performance in n-type heterogeneous BiAgSeS composites. *Adv. Funct. Mater.* **24**, 7763–7771 (2014).
21. Zhao, L.-D. et al. Raising the thermoelectric performance of p-type PbS with endotaxial nanostructuring and valence-band offset engineering using CdS and ZnS. *J. Am. Chem. Soc.* **134**, 16327–16336 (2012).
22. Chelvayohan, M. & Mee, C. H. B. Work function measurements on (110), (100) and (111) surfaces of silver. *J. Phys. C* **15**, 2305 (1982).

23. Dweydari, A. W. & Mee, C. H. B. Work function measurements on (100) and (110) surfaces of silver. *Phys. Status Solidi A* **27**, 223–230 (1975).

24. Knapp, R. A. Photoelectric Properties of Lead Sulfide in the Near and Vacuum Ultraviolet. *Phys. Rev.* **132**, 1891–1897 (1963).

25. Noda, Y., Ohba, S., Sato, S. & Saito, Y. Charge distribution and atomic thermal vibration in lead chalcogenide crystals. *Acta Crystallogr. Sect. B* **39**, 312–317 (1983).

26. Zhao, L.-D. *et al.* High performance thermoelectrics from earth-abundant materials: enhanced figure of merit in PbS by second phase nanostructures. *J. Am. Chem. Soc.* **133**, 20476–20487 (2011).

27. Zhao, L.-D. *et al.* Thermoelectrics with earth abundant elements: high performance p-type PbS nanostructured with SrS and CaS. *J. Am. Chem. Soc.* **134**, 7902–7912 (2012).

28. Wu, H. J. *et al.* Broad temperature plateau for thermoelectric figure of merit $ZT > 2$ in phase-separated $PbTe_{0.7}S_{0.3}$. *Nat. Commun.* **5**, 4515 (2014).

29. Korkosz, R. J. *et al.* High ZT in p-Type $(PbTe)_{1-2x}(PbSe)_x(PbS)_x$ thermoelectric materials. *J. Am. Chem. Soc.* **136**, 3225–3237 (2014).

30. Girard, S. N. *et al.* High performance Na-doped PbTe–PbS thermoelectric materials: electronic density of states modification and shape-controlled nanostructures. *J. Am. Chem. Soc.* **133**, 16588–16597 (2011).

31. Yun, Z. *et al.* Thermoelectric transport properties of p-type silver-doped PbS with in situ Ag_2S nanoprecipitates. *J. Phys. D Appl. Phys* **47**, 115303 (2014).

32. Shanyu, W. *et al.* Exploring the doping effects of Ag in p-type PbSe compounds with enhanced thermoelectric performance. *J. of Phys. D Appl. Phys* **44**, 475304 (2011).

33. Dow, H. S. *et al.* Effect of Ag or Sb addition on the thermoelectric properties of PbTe. *J. Appl. Phys.* **108**, 113709 (2010).

34. Ryu, B. *et al.* Defects responsible for abnormal n-type conductivity in Ag-excess doped PbTe thermoelectrics. *J. Appl. Phys.* **118**, 015705 (2015).

35. Ibáñez, M. *et al.* Electron doping in bottom-up engineered thermoelectric nanomaterials through HCl-mediated ligand displacement. *J. Am. Chem. Soc.* **137**, 4046–4049 (2015).

36. Kovalenko, M. V. *et al.* Prospects of nanoscience with nanocrystals. *ACS Nano* **9**, 1012–1057 (2015).

37. Li, L., Hu, F., Xu, D., Shen, S. & Wang, Q. Metal ion redox potential plays an important role in high-yield synthesis of monodisperse silver nanoparticles. *Chem. Commun.* **48**, 4728–4730 (2012).

38. Hung, L.-I., Tsung, C.-K., Huang, W. & Yang, P. Room-temperature formation of hollow Cu2O nanoparticles. *Adv. Mater.* **22**, 1910–1914 (2010).

39. Kang, Y. *et al.* Shape-controlled synthesis of Pt nanocrystals: the role of metal carbonyls. *ACS Nano* **7**, 645–653 (2012).

40. Ravel, B. & Newville, M. ATHENA, ARTEMIS, HEPHAESTUS: data analysis for X-ray absorption spectroscopy using IFEFFIT. *J. Synchrotron Radiat.* **12**, 537–541 (2005).

Acknowledgements

At IREC, work was supported by European Regional Development Funds and the Framework 7 program under project UNION (FP7-NMP 310250). M.I. and S.O. thank AGAUR for their Beatriu i Pinós post-doctoral grant (2013 BP-A00344) and the PhD grant, respectively. A.G. thanks to the Turkish Ministry of National Education for the PhD grant. A.G. and J.A. acknowledge the Spanish MINECO MAT2014-51480-ERC (e-ATOM) and the ICN_2 Severo Ochoa Excellence Program. Z.L. and Y.L. thanks the China Scholarship Council for their PhD grant. IREC and ICN_2 groups acknowledge the funding from Generalitat de Catalunya 2014SGR1638. The work performed at Rutgers was supported by NSF grant number 1400246. M.V.K. acknowledges partial financial support by the European Union (EU) via FP7 ERC Starting Grant 2012 (Project NANOSOLID, GA No. 306733). We thank Prof. Yaroslav Romanyuk for the use of his Ecopia HMS-3000 set-up for Hall Effect measurements and Dr Nicholas Stadie for reading the manuscript. The Swiss Light Source is thanked for the provision of beamtime at the SuperXAS beamline.

Author contributions

The manuscript was prepared through the contribution of all authors. M.I., Z.L., S.O., O.D. and D.C. produced the nanocomposite and performed the thermoelectric characterization. A.G., J.A., L.P. and M.N. performed structural nanocomposite characterization. M.Z. performed band alignment calculations. M.I., M.K. and A.C. planned and supervised the work and had major input in the writing of the manuscript.

Additional information

15

Layer-dependent quantum cooperation of electron and hole states in the anomalous semimetal WTe$_2$

Pranab Kumar Das[1,2], D. Di Sante[3,4], I. Vobornik[1], J. Fujii[1], T. Okuda[5], E. Bruyer[3], A. Gyenis[6], B.E. Feldman[6], J. Tao[7], R. Ciancio[1], G. Rossi[1,8], M.N. Ali[9], S. Picozzi[3], A. Yadzani[6], G. Panaccione[1,*] & R.J. Cava[9,*]

The behaviour of electrons and holes in a crystal lattice is a fundamental quantum phenomenon, accounting for a rich variety of material properties. Boosted by the remarkable electronic and physical properties of two-dimensional materials such as graphene and topological insulators, transition metal dichalcogenides have recently received renewed attention. In this context, the anomalous bulk properties of semimetallic WTe$_2$ have attracted considerable interest. Here we report angle- and spin-resolved photoemission spectroscopy of WTe$_2$ single crystals, through which we disentangle the role of W and Te atoms in the formation of the band structure and identify the interplay of charge, spin and orbital degrees of freedom. Supported by first-principles calculations and high-resolution surface topography, we reveal the existence of a layer-dependent behaviour. The balance of electron and hole states is found only when considering at least three Te–W–Te layers, showing that the behaviour of WTe$_2$ is not strictly two dimensional.

[1] Istituto Officina dei Materiali (IOM)-CNR, Laboratorio TASC, in Area Science Park, S.S.14, Km 163.5, I-34149 Trieste, Italy. [2] International Centre for Theoretical Physics (ICTP), Strada Costiera 11, I-34100 Trieste, Italy. [3] Consiglio Nazionale delle Ricerche—CNR-SPIN, I-67100 L'Aquila, Italy. [4] Department of Physical and Chemical Sciences, University of L'Aquila, Via Vetoio, I-67100 L'Aquila, Italy. [5] Hiroshima Synchrotron Radiation Center (HSRC), Hiroshima University, 2-313 Kagamiyama, Higashi-Hiroshima 739-0046, Japan. [6] Joseph Henry Laboratories and Department of Physics, Princeton University, Princeton, New Jersey 08544, USA. [7] Department of Condensed Matter Physics and Materials Science, Brookhaven National Laboratory, Upton, New York 11973, USA. [8] Dipartimento di Fisica, Università di Milano, Via Celoria 16, I-20133 Milano, Italy. [9] Department of Chemistry, Princeton University, Princeton, New Jersey 08544, USA. * These authors contributed equally to this work. Correspondence and requests for materials should be addressed to R.J.C. (email: rcava@Princeton.EDU).

Transition metal dichalcogenides (TMDs) are a group of layered materials with chemical formula MX_2, where M is a transition metal and X can be S, Se or Te. Their properties span from pure insulators to good metals, and they also exhibit various low-temperature phenomena such as metal–insulator transitions, superconductivity and charge density waves[1]. An important aspect of TMDs is the presence of anisotropic bonding with different strengths: the X–M–X building blocks are stacked along the crystallographic c-direction, and while the inter-layer interaction is mainly of weak van der Waals type, the intra-layer bonding between the atoms is strong and covalent. Dimensionality is thus expected to play a significant role in TMDs, because the transition from a single or a few layers to bulk implies significant change in the symmetry of the orbitals and in quantum confinement, producing important differences in the electronic structure.

Among TMDs, WTe_2 is special because it displays an additional structural distortion: the tungsten atoms form zigzag chains along the crystallographic a axis, producing a quasi one-dimensional arrangement. Moreover, WTe_2 is a semimetal with a reduced density of states at the Fermi level coming from a small overlap between valence and conduction bands without a band gap. It exhibits an extremely large uniaxial positive magnetoresistance with no saturation up to a magnetic field as high as 60 T (refs 2,3), which has been attributed to perfect electron and hole compensation. In addition, the presence of two heavy elements points to the importance of spin–orbit coupling (SOC) in determining the details of the Fermi surface and/or the relevant low-energy excitations of the system.

Here, by combining results of scanning tunnelling microscopy (STM), spin- and angle-resolved photoemission spectroscopy (ARPES), and layer-resolved ab initio calculations, we explored the details and the evolution of electron and hole states in WTe_2. Our data reveal significant differences between surface and bulk electronic properties, with a clear evolution as a function of depth from the surface. The Fermi surface measured by ARPES is significantly reduced with respect to the one calculated for the bulk, indicating the presence of a reduced, yet still balanced, number of electrons and holes at the surface and near-surface region. The importance of SOC is directly shown by spin-resolved ARPES measurements, and the data are consistent with our theoretical calculations.

Results

STM measurements to characterize the surface properties. To visualize the surface quality of WTe_2, we performed high-resolution scanning tunnelling microscopy (Fig. 1). The STM images reveal that the top layer is composed of two inequivalent Te atoms, labelled Te1 and Te2 in Fig. 1b (see also Supplementary Note 1, and Supplementary Fig. 1). The surface is atomically ordered with extremely high quality and low impurity concentration, with approximately one underlying defect per 3,000 atoms observed (Fig. 1a), corresponding to $\approx 9 \times 10^{11}$ defects cm^{-2}. In addition, the topographic image and its Fourier transform (Fig. 1c) clearly show that the electronic structure of the surface has an angular distortion, similar to that reported previously[4,5]. We observe the distortion at multiple tip-sample biases, in multiple samples, and with different STM tips, indicating that this finding is not related to thermal drift, tip artifacts or a particular energy (see Supplementary Note 2, and Supplementary Figs 2–4). Comparison with TEM data suggest that the distortion is due to surface reconstruction and model calculations for a single WTe_2 monolayer show that it does not significantly affect the band structure (see Supplementary Note 2, and Supplementary Fig. 5).

ARPES results and DFT calculations. We next explore the detailed electronic structure of WTe_2 and the related layer dependence of the electron and hole states. In Fig. 2b we show the experimental E versus k ARPES spectra overlaid with bulk theoretical band structures calculated using density functional

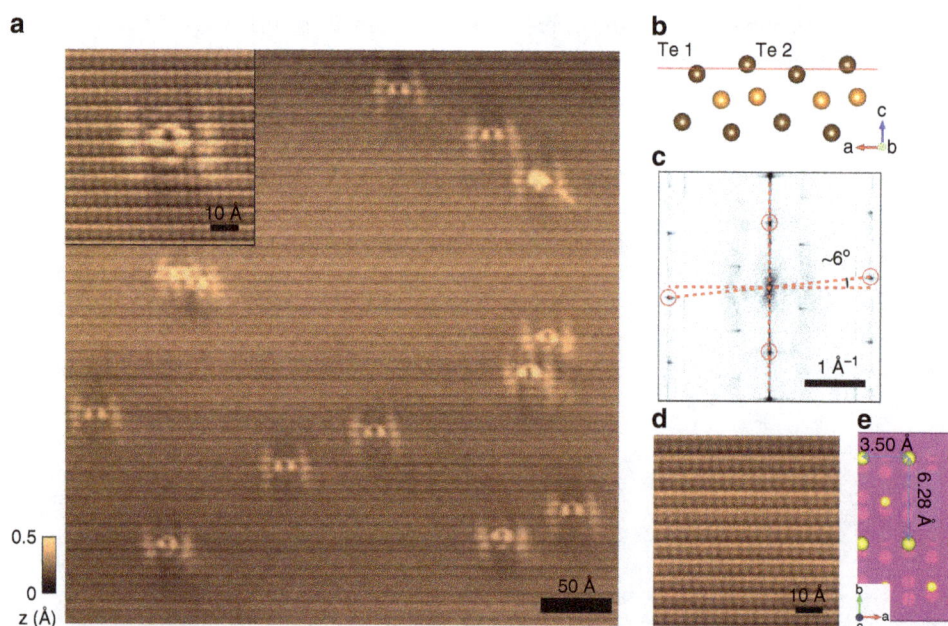

Figure 1 | Surface topography and structure of WTe₂. (**a**) Topographic image of a 400 Å × 400 Å area on the (001) surface of WTe_2 with several 'tie-fighter'-like impurities ($V_{bias} = -80$ mV, $I_{tunneling} = 100$ pA and T = 30 K). Inset: Zoom-in on an individual defect. (**b**) Side schematic view of the WTe_2 lattice highlighting the two inequivalent Te atoms. (**c**) Fourier transform of the topograph in **a** reveals Bragg peaks corresponding to the atomic corrugation (red circles indicate the first-order Bragg peaks) and the anomalous angle between the crystallographic axes on the surface (6° in this sample). The degree of the angular distortion varied a few degrees between different samples. (**d**) Zoom-in on a clean area displays the surface Te atoms. (**e**) Top schematic view of the WTe_2 (001) surface highlighting the surface Te atoms and the unit cell dimensions.

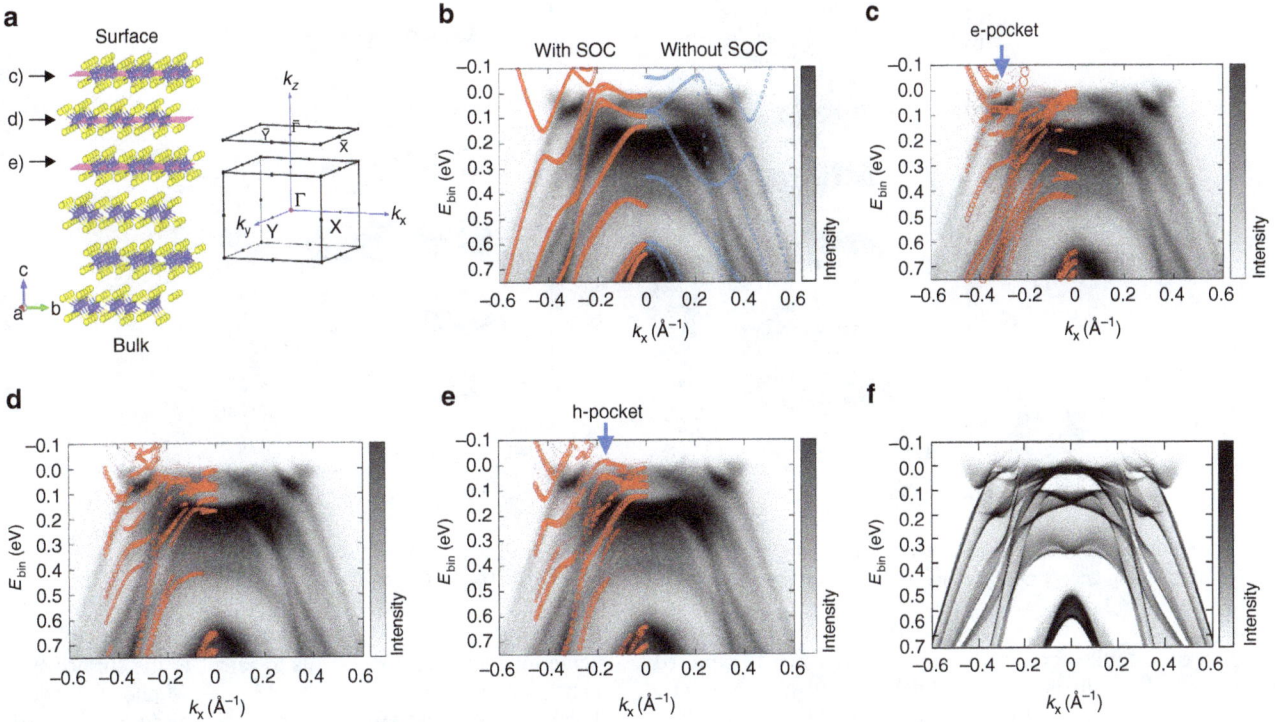

Figure 2 | Evolution of band structure with number of layers. (a) Crystal structure of WTe$_2$ with the bulk and surface Brillouin zones on the right. On the left, the slab model used in the theoretical calculations, consisting of six Te–W–Te layers; each layer is formed by hexagonally packed W sandwiched between two Te layers. **(b–e)** ARPES measurements ($hv = 68$ eV, $T = 77$ K) of the electronic structure along the ΓX high symmetry direction (along the W chains); **(b)** bulk electronic structures as calculated with SOC (red bands at negative momenta) and without SOC (blue bands at positive momenta); **(c)** theoretical bands projected on the topmost WTe$_2$ planes; **(d,e)** theoretical bands projected on second and third plane, respectively. In **c,e**, blue arrows mark the positions of the theoretical electron and hole pockets, respectively. In **b,e**, the size of the circles is proportional to the weight of the layer-resolved orbital character, calculated as the sum of the orbital characters of all the atoms belonging to the respective layer. **(f)** The theoretical surface spectral function $A(\mathbf{k},E)$.

theory (DFT) with (red bands on the left) and without (blue bands on the right) SOC, along the reciprocal space line XΓX. Following the standard procedure and assuming WTe$_2$ as a pure 2D-layered system, calculations are performed with out-of-plane **k**-vector component $k_z = 0$. We observe well-defined electron and hole pockets at the Fermi energy along the tungsten chain direction (ΓX), confirming previous results[6]. However, a detailed comparison of the observations to bulk theoretical band structure as described above presents some inconsistencies. The theoretical position of the electron pocket is significantly further away from the zone centre and its maximum binding energy is larger than the measured one by more than a factor two. The presence of heavy atoms like W and Te suggests that SOC plays an important role in the interpretation of WTe$_2$ spectral features. The changes in the theoretical band structure associated with SOC (Fig. 2b, red curves) illustrate that relativistic effects cannot be neglected in WTe$_2$, and although we obtain somewhat better results, a quantitative agreement is not found in either case. We also note a non-negligible k_z band dispersion[2,6] when comparing experimental spectra with theoretical bands for reciprocal lines parallel to ΓX at different k_z momenta, ranging from the Brillouin zone centre ($k_z = 0$) to the Brillouin zone edge ($k_z = \pi/c$) (see Supplementary Figs 6–9). However, including k_z dispersion does not result in an improved agreement between calculations and experiment, since none of the calculated band structures for any of the given k_z values can fully reproduce the measured dispersion (see Supplementary Fig. 6).

We find a significant improvement, however, when using a more realistic model that takes into account the contribution of individual layers (that is, all the bulk k_z momenta at the same

time projected on the surface Brillouin zone). This model is based on a supercell made by van der Waals-bonded WTe$_2$ planes stacked along the [001] direction (Fig. 2a). From the large number of bands in the supercell calculations, we are able to isolate the individual contribution of each WTe$_2$ plane in the stack by projecting the electronic band structure onto the atoms belonging to that given plane only. We note that the experimental ARPES spectra correspond to a weighted spectral intensity from the different layers probed, and the contributions of deeper layers are exponentially attenuated by the inelastic mean free path of photoelectrons. Superimposing the theoretical bands on the experimental spectra, and projecting onto the first, second and third WTe$_2$ planes (Fig. 2c–e, respectively), we filter the layer-dependent information about the electronic structure because we are able to correlate individual features of the layer-resolved calculations in the experimental spectra. In this analysis, the size of the plotted circles is proportional to the contribution of the given WTe$_2$ plane: the bigger the circle, the larger the contribution of that plane to the spectral features. We thus observe in Fig. 2c that the electron pocket is located at $0.35 \, \text{Å}^{-1}$ from the zone centre (highlighted by the blue arrow) and is present even when only the first Te–W–Te layer is considered. The hole pocket (blue arrow in panel e) starts appearing only when taking into account the presence of the third layer, and thus has a more bulk-like character. Furthermore, we cannot exclude the existence of a small hole pocket at Γ, as recently proposed by theoretical findings[7] for an isolated WTe$_2$ monolayer (note that in the same work the electron pockets are closer to the zone centre, similar to our case); a similar hole pocket has also recently been observed in other ARPES experiments[8]. In our case (Fig. 2), the

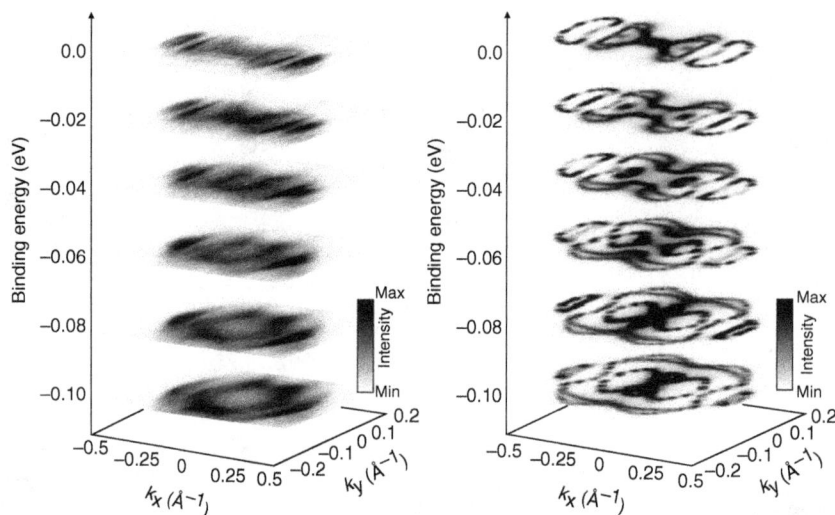

Figure 3 | Fermi surface topography. Constant energy cuts from E_F down to 100 meV binding energy (in steps of 20 meV). Left: experimental spectra measured at a photon energy of 68 eV and at 77 K, right: theoretical calculations for bulk WTe$_2$ at $k_z = 0$.

hole pocket is fully occupied and the range of explored temperatures is well below the Lifshitz transition recently reported at above 160 K (ref. 9). Given that the hole pocket is very close to E_F, the difference could be due to temperature induced shift of the chemical potential as proposed in ref. 9. Moreover, hole pockets close to each other on either side of the Γ point would justify the unusual magnetic breakdown observed in quantum oscillations[10].

We further stress that our slab calculations do not reveal the existence of distinct surface states. This is supported by the comparison of bulk theoretical band structures for different k_z momenta with the electronic states of the slab (see Supplementary Fig. 7). In fact, all bulk bands lie within the continuum of the slab's bulk band structure projected onto the surface Brillouin zone. Indeed, the bands shown in Fig. 2c–e overlaid to ARPES spectra are bulk states showing substantial spectral weight on the topmost surface layers; this seems in analogy to what is observed in MoS$_2$ and many other systems, like iron pnictides, with cleaved neutral surfaces[11,12].

The present analysis not only provides a quantitative agreement between experiment and theory in terms of dispersion and momentum space locations of electronic states, but clearly shows the presence of a layer-dependent evolution of electron and hole states: a bulk-like electronic structure characterized by the electron-hole charge balance is obtained only when more than two Te–W–Te layers are taken into account. Our results clearly display some analogies with other 'less than 3D' materials: in the case of topological insulators, for example, it has been theoretically predicted and experimentally confirmed not only that the critical thickness of six quintuple layers is needed to set the topological properties of the surface but also that the spin–orbital texture of a topological insulator evolves in a layer-dependent manner, extending over several nanometers from the surface[13,14].

Given that a nearly equal concentration of electrons and holes is a necessary condition for non-saturating magnetoresistance to be observed in two-component systems[15,16], the present observation has important implications towards the realization of devices based on few-layer TMDs[17].

To further check the robustness and the reliability of our theoretical interpretation, we used two different approaches: (i) an *ab initio* tight-binding model for a 40 layers slab (thickness about 42 nm, see Supplementary Fig. 10); (ii) a renormalization

scheme for semi-infinite systems to calculate the surface spectral function, shown in Fig. 2f (ref. 18; see 'Methods' section for details). These checks aimed at excluding spurious effects arising from the interaction between the two extreme surfaces (always present in slab calculations if the number of layers is not large enough). The theoretical bands of Fig. 2f provide a direct link with the measured E versus k intensity maps (except for dipole matrix elements, neglected in our calculations). By using both methods, most of the experimental features are well reproduced, in particular the hole and electron pockets near E_F as well as the bands at higher binding energy. Moreover, no remarkable differences are observed in comparison with the DFT results for the six-layer-thick slab.

Figure 3 shows the experimental (left panel) and bulk theoretical (right panel) constant energy cuts from E_F down to 100 meV binding energy, in steps of 20 meV. The theoretical results qualitatively reproduce the main features of the Fermi surface, including the shape and the distribution of the pockets and a finite intensity at Γ (that is, confirming the presence of a possible hole pocket at Γ). However, the calculated bulk features have a significantly larger area than the measured Fermi surface, providing further evidence that simply recasting bulk calculations is insufficient to explain the experimentally observed ARPES features. The area of the Fermi surface calculated by the bulk theoretical model is larger than the area of the observed Fermi surface, which implies that the total number of carriers is larger in the bulk than on the surface. In spite of this difference, the comparison shows that the balance between electrons and holes is maintained in both cases, to within the sensitivity of our technique (see also Supplementary Fig. 9). This, in turn, indicates that the balancing between electrons and holes, that is, their 'quantum-cooperation' over different layers, is the dominant factor that determines the macroscopic properties of the system such as also the non-saturating magnetoresistance.

Spin-resolved ARPES results. To determine the role of SOC and obtain insight into the spin texture, we performed spin-resolved measurements at a number of **k** points in the Brillouin zone. Spin-resolved ARPES spectra are presented in Fig. 4, as measured at the hole pocket (panels a, c; blue and yellow circles), close to Γ on the ΓX line (panel b, red circle) and close to Γ, but off the ΓX line (panel d, green circle). Panel f shows the calculated spin-resolved band structure along ΓX.

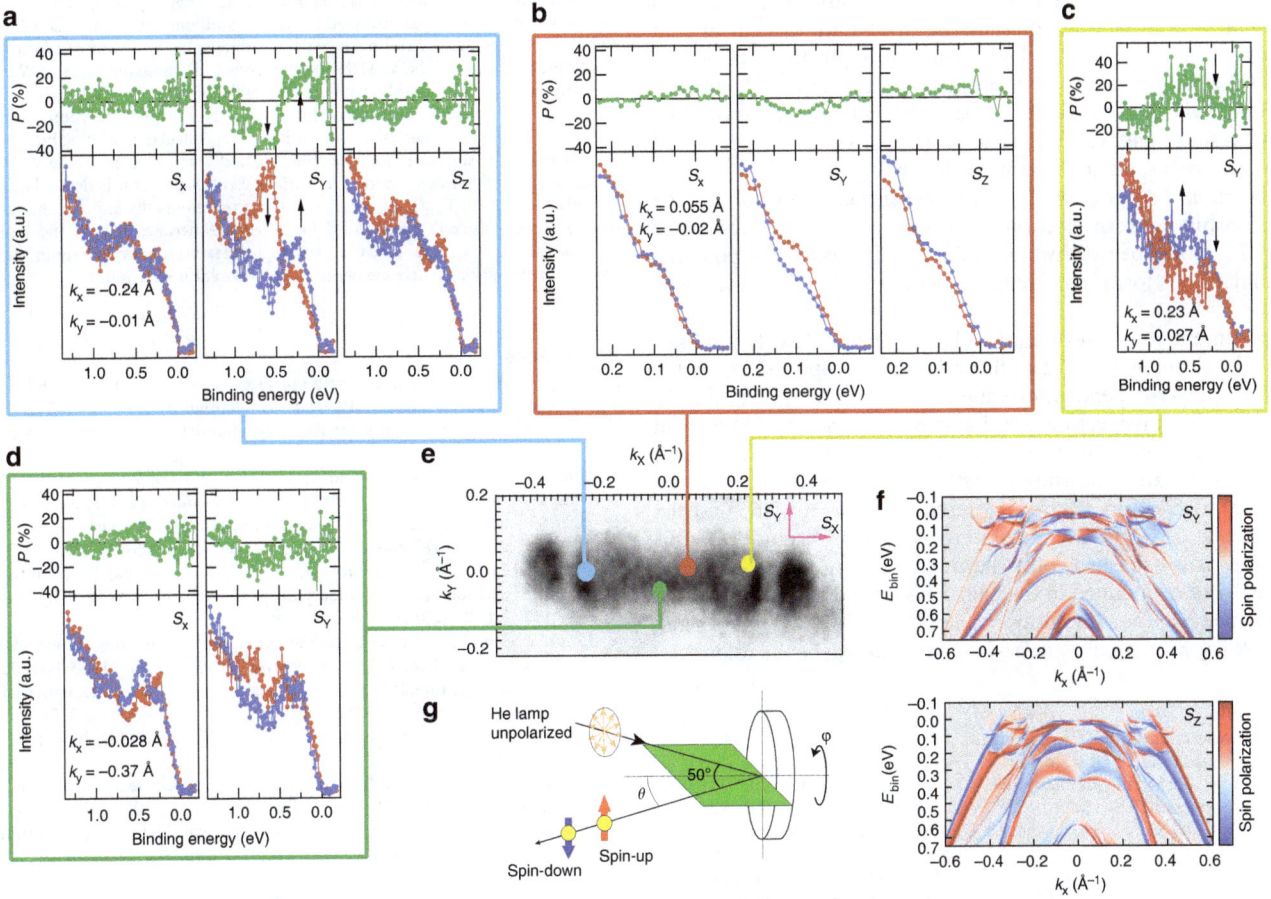

Figure 4 | Experimental and theoretical spin polarized band structure of large SOC and non-centro-symmetric WTe$_2$. (a–d) Measured spin-resolved ARPES spectra (red and blue curves) and spin polarization (green curves) at four distinct **k** points as indicated in **e**; (**e**) experimental Fermi surface with the corresponding spin measurement positions; (**f**) calculated band structure along k_x direction, upper/lower panel shows S_y/S_z spin orientation. The colour scales represent the degree of spin polarization; red, blue and white refer to up, down and no-spin polarization, respectively; (**g**) experimental geometry of spin-resolved ARPES setup. In **a**, S_y, the down-spin spectrum (red curve) has a larger peak at 0.6 eV (down arrow) and the up-spin spectrum (blue curve) has a larger peak at 0.2 eV (up arrow), whereas at **c**, on the contrary, the up-spin spectrum has a dominating peak at 0.6 eV and the down-spin spectrum has a dominating peak at 0.2 eV, supporting time-reversal effect.

We measure a sizeable spin polarization (P) for an extended binding energy range. It reaches more than 35% for P_y at a binding energy of 0.55 eV (panel a, black arrows). This indicates that there must not only be a broken space inversion symmetry but also a significant influence of SOC in WTe$_2$. The spin texture is quite complex, with large oscillations in value and sign of P in a narrow energy window, clearly visible in both P_y and P_z (Fig. 4a). This observation is confirmed by the calculation in Fig. 4f. Experimental error cannot exclude the presence of a finite spin polarization at E_F, but it is negligible with respect to those observed at higher binding energies. The fine spectral structures in the calculation are not visible in the experimental spin polarization measurements, due to the experimental energy and angular resolution. However, we obtain notable qualitative agreement, as detailed below. The P_x component is zero along the ΓX line (Fig. 4a,b), while we observe non-zero P_x component for points away from ΓX (Fig. 4d). This means that the electron spin along ΓX is perpendicular to the W chains, in agreement with calculations that predict only large P_y and P_z spin polarizations (Fig. 4f). The results shown in Fig. 4a,b (at the hole pocket) and Fig. 4f reveal the presence of both in-plane and out-of-plane components of the spin polarization, in contrast with ordinary Rashba systems where P is only in-plane[19]. The spin polarization changes sign upon crossing the Brillouin zone centre, that is, the sign of P is reversed at $\mathbf{k}' = -\mathbf{k}$,

as experimentally confirmed in Fig. 4a,c, which show data taken at positive and negative k_x values. This indicates that time reversal symmetry is preserved, that is, $[E(\mathbf{k}, \uparrow) = E(-\mathbf{k}, \downarrow)]$ and that the observed spin polarization of bands has nonmagnetic origin. A large spin polarization of electronic bands has been recently reported in the semiconducting TMD WSe$_2$, where P occurs due to the local asymmetry of layers[20] (consecutive Se–W–Se layers have opposite net dipole moment, which modulate the spin texture strongly even though the global inversion symmetry is preserved in the crystal). Unlike WSe$_2$, the crystal structure of WTe$_2$ is non-centrosymmetric. Therefore, by symmetry consideration one naturally expects a lifting of the spin degeneracy via spin–orbit coupling. The amplitude of spin polarization depends on many different factors (that is, orbital character, band gap, electric fields and so on), but primarily on the strength of the SOC; therefore, our results indicate that SOC is clearly reflected in the spectral function of WTe$_2$, as also recently reported by CD-ARPES results[8].

Discussion

Considering our present results from a more general perspective, it is important to underline that exact carrier compensation is a necessary, yet not sufficient, condition for a non-saturating quadratic behaviour of magnetoresistance[15]. Bismuth provides an example where compensation is observed, but quadratic dependence of resistance in a magnetic field and non-saturation are not[15,21]. Other prerequisites for behaviour similar to that observed in WTe$_2$ are: (a) low density of impurities and defects and (b) a carrier density far from the quantum limit[22,23]. Here we have shown that condition (a) is nearly fulfilled as shown by STM results, and condition (b) is supported by the reported low resistivity and high mobility in WTe$_2$ (refs 2,7,24). These observations suggest that future experiments exploring the relationship between electronic structure and magnetoresistance in WTe$_2$ would be worthwhile and that WTe$_2$ could be a potential candidate to form an excitonic dielectric in the Abrikosov sense[22,25].

In summary, our theoretical and experimental ARPES findings provide clear evidence that the electronic properties of WTe$_2$ display a layer-dependent evolution from surface to bulk, that is, it cannot be considered a priori as a non-interacting 2D-layered system, in agreement also with recent temperature-dependent magnetoresistance measurements[26]. The balance between the hole and electron states, representing one of the crucial conditions for the non-saturating magnetoresistance in this system, is established only beyond finite number of layers (three) and maintained in the bulk. This consideration provides a fundamental input for future exploitation of TMDs in general, and WTe$_2$ in particular, in devices and heterogeneous interfaces.

Methods

ARPES experiments. ARPES experiments were performed at APE-IOM beamline[27] at the ELETTRA Sincrotrone Trieste. High quality surfaces were obtained by cleaving the samples in UHV at a base pressure of 1×10^{-10} mbar. The crystallographic orientation was examined by low-energy electron diffraction (LEED) patterns. Core levels, valence-band and Fermi surface measurements were performed using a high-resolution VG-SCIENTA DA30 electron analyzer, in a photon energy range of 20–100 eV, with an angular and energy resolution better than 0.2 deg and 20 meV, respectively. The low temperature data are collected at 16 K using a liquid helium cooled cryostat.

STM. Before the measurements, the samples were cleaved in ultra-high vacuum at room temperature and immediately transferred to our home-built variable temperature STM. Measurements were performed at 28 K.

Spin-resolved ARPES. Spin-resolved ARPES experiments were performed at HiSOR[28]. The spin-resolved spectra were measured by means of VLEED spin detector using Fe-O target; two VLEED detectors positioned orthogonally are able to measure the x, y (in-plane) and z (out of plane) components of the spin polarization. Standard He laboratory light source ($hv = 21.22$ eV) was used as incident beam at various temperatures down to 10 K. The asymmetry of the spin polarization was quantified by reversing the current through a coil. The analyser resolution of SR-ARPES was 60 meV and the angular resolution was 1.5 degrees. The spin asymmetry A is given by $A = (I_+ - I_-)/(I_+ + I_-)$. The actual polarization P depends on the instrument and detector setup (target), and is given by $P = A/S_{eff}$ where S_{eff} is the effective Sherman function value corresponding to the detector and instrumental setup used. Here we have used $S_{eff} = 0.2$. Next the up and down spin component $S_\uparrow, S_\downarrow$ can be calculated following the expression $S_\uparrow (S_\downarrow) = (1 \pm P)(I_+ + I_-)/2$.

Density functional theory calculations. Supercell calculations were performed using the generalized gradient approximation (GGA) as implemented in the DFT code Vienna Ab-Initio Simulation Package (VASP)[29,30]. Atomic positions were fully relaxed starting from data in ref. 31. We used the projector augmented wave method by explicitly treating six valence electrons both for W and Te, while d electrons of Te were kept within the core of the PBE pseudopotentials. Integration over the first Brillouin zone was made with a $12 \times 8 \times 1$ Monkhorst-Pack k-mesh centred at Γ ($24 \times 12 \times 2$ for bulk calculations). For all the simulations, a 400 eV plane-wave energy cut-off was used. Spin–orbit coupling has been self-consistently taken into account. Dipole corrections, as implemented in VASP, were applied along the z direction to counteract any spurious electric field that might arise from periodic boundary conditions in the presence of a dipole moment normal to the surface in a slab geometry (with 20 Å-thick layer of vacuum). To model surface as well as more bulk features, we considered a slab containing six planes of WTe$_2$ stacked onto each other along the [001] direction. The effects of van der Waals interactions between WTe$_2$ planes has been properly taken into account by the DFT-D2 method of Grimme[32], but no significant differences in the electronic properties have been detected with respect to GGA calculations.

To calculate bulk Fermi surfaces reported in Fig. 3 (right panel), we adopt a two step procedure: (i) first, we projected the bulk Hamiltonian onto a basis made of s and d W-centered, and s and p Te-centered orbitals, for a total of 112 Wannier functions, by means of the WANNIER90 package; (ii) subsequently, the Wannier Hamiltonian is used to build up the bulk Green's function as $G(\mathbf{k},w) = 1/(w - H(\mathbf{k}) + i\delta)$, which, in turn, gives the Fermi surface maps at the chosen binding energy, w. Theoretical spectral functions have been calculated within the framework of the surface renormalization method[18] based on the same Wannier Hamiltonian as described before. The surface spectral function is defined as $A(\mathbf{k},w) = -(1/\pi)ImG_{surf}(\mathbf{k},w)$, where $G_{surf}(\mathbf{k},w)$ represents the angular and energy resolved surface Green's function. Major differences between $A(\mathbf{k},w)$ and ARPES data, as, for example, line intensities and high-$\mathbf{k}$ features, possibly rely in the absence of transition matrix elements in the theoretical description.

References

1. Wang, Q. H. et al. Electronics and optoelectronics of two-dimensional transition metal dichalcogenides. *Nat. Nanotechnol.* **7**, 699–712 (2012).
2. Ali, M. N. et al. Large, non-saturating magnetoresistance in WTe$_2$. *Nature* **514**, 205–208 (2014).
3. Crossley, A., Myhra, S. & Sofield, C. J. STM analysis of WTe$_2$ surfaces—correlation with crystal and electronic structures. *Surf. Sci.* **318**, 39–45 (1994).
4. Hlaa, S. W., Marinković, V., Prodana, A. & Muševiča, I. STM/AFM investigations of β-MoTe2, α-MoTe2 and WTe2. *Surf. Sci.* **352–354**, 105–111 (1996).
5. Pletikosić, I., Ali, M. N., Fedorov, A. V., Cava, R. J. & Valla, T. Electronic structure basis for the extraordinary magnetoresistance in WTe$_2$. *Phys. Rev. Lett.* **113**, 216601 (2014).
6. Lv, H. Y. et al. Perfect charge compensation in WTe2 for the extraordinary magnetoresistance: from bulk to monolayer. *Europhys. Lett.* **110**, 37004 (2015).
7. Jiang, J. et al. Signature of strong spin-orbital coupling in the large nonsaturating magnetoresistance material WTe$_2$. *Phys. Rev. Lett.* **115**, 166601 (2015).
8. Wu, Y. et al. Temperature induced Lifshitz transition in WTe$_2$. *Phys. Rev. Lett.* **115**, 166602 (2015).
9. Zhu, Z. et al. Quantum oscillations, termoelectric coefficients, and the Fermi surface of semimetallic WTe$_2$. *Phys. Rev. Lett.* **114**, 176601 (2015).
10. Gehlmann, M. et al. Quasi 2D electronic states with high spin-polarization in centrosymmetric MoS2 bulk crystals. Preprint at http://arxiv.org/abs/1510.04101 (2015).
11. Lankau, A. et al. Absence of surface states for LiFeAs investigated using density functional calculations. *Phys. Rev. B* **82**, 184518 (2010).
12. Zhang, Y., He, K. & Chang, C. Crossover of the three-dimensional topological insulator Bi$_2$Se$_3$ to the two-dimensional limit. *Nat. Phys.* **6**, 584–588 (2010).
13. Zhu, Z. H. & Levy, G. Rashba spin-splitting control at the surface of the topological insulator Bi$_2$Se$_3$. *Phys. Rev. Lett.* **107**, 186405 (2011).
14. Alekseev, P. S. et al. Magnetoresistance in two-component systems. *Phys. Rev. Lett.* **114**, 156601 (2015).
15. Parish, M. M. & Littlewood, P. B. Non-saturating magnetoresistance in heavily disordered semiconductors. *Nature* **426**, 162–165 (2003).
16. Wang, L. et al. Tuning Magnetotransportin a Compensated Semimetal at the Atomic Scale. Preprint at http://arxiv.org/abs/1510.04827 (2015).
17. Henk, J. & Schattke, W. A subroutine package for computing Green's functions of relaxed surfaces by the renormalization method. *Comput. Phys. Commun.* **77**, 69–83 (1993).
18. Hasan, M. Z. & Kane, C. L. Colloquium: topological insulators. *Rev. Mod. Phys.* **82**, 3045 (2010).
19. Riley, J. M. et al. Direct observation of spin-polarized bulk bands in an inversion-symmetric semiconductor. *Nat. Phys.* **10**, 835–839 (2014).
20. Liang, T. et al. Evidence for massive bulk Dirac fermions in Pb$_{1-x}$Sn$_x$Se from Nernst and thermopower experiments. *Nat. Commun.* **4**, 2696 (2013).
21. Abrikosov, A. A. The transformation of a semimetal into an exciton dielectric in a strong magnetic field. *Sov. Phys. Usp.* **15**, 662–663 (1973).
22. Fenton, E. W. Electrical resistivity of semimetals in the extreme quantum limit. *J. Low Temp. Phys.* **7**, 415–432 (1972).
23. Ali, M. N. et al. Correlation of crystal quality and extreme magnetoresistance of WTe$_2$. *Europhys. Lett.* **110**, 67002 (2015).
24. Abrikosov, A. A. Quantum linear magnetoresistance. *Europhys. Lett.* **49**, 789–793 (2000).
25. Thoutam, L. R. et al. Temperature-dependent three-dimensional anisotropy of the magnetoresistance in WTe$_2$. *Phys. Rev. Lett.* **115**, 046602 (2015).
26. Panaccione, G. et al. Advanced photoelectric effect experiment beamline at Elettra: a surface science laboratory coupled with synchrotron radiation. *Rev. Sci. Instrum.* **80**, 043105 (2009).
27. Okuda, T. et al. Efficient spin resolved spectroscopy observation machine at Hiroshima Synchrotron Radiation Center. *Rev. Sci. Instrum.* **82**, 103302 (2011).
28. Kresse, G. & Furthmüller, J. Efficient iterative schemes for *ab initio* total-energy calculations using a plane-wave basis set. *Phys. Rev. B* **54**, 11169 (1996).
29. Kresse, G. & Joubert, D. From ultrasoft pseudopotentials to the projector augmented-wave method. *Phys. Rev. B* **59**, 1758 (1999).

30. Mar, A., Jobic, S. & Ibers, A. Metal-metal vs tellurium-tellurium bonding in WTe$_2$ and its ternary variants TaIrTe$_4$ and NbIrTe$_4$. *J. Am. Chem. Soc.* **114**, 8963 (1992).

31. Grimme, S. Semiempirical GGA-type density functional constructed with a long-range dispersion correction. *Comp. Chem.* **27**, 1787 (2006).

32. Mostofi, A. A. *et al.* wannier90: a tool for obtaining maximally-localised Wannier functions. *Comput. Phys. Commun.* **178**, 685 (2008).

Acknowledgements

This work has been partly performed in the framework of the nanoscience foundry and fine analysis (NFFA-MIUR Italy) project. The electron diffraction study at Brookhaven National Laboratory was supported by the DOE BES, by the Materials Sciences and Engineering Division under contract DE-AC02–98CH10886, and through the use of the Center for Functional Nanomaterials. The work at Princeton was supported by the National Science Foundation MRSEC program grant DMR-1420541, with STM support from NSF-DMR-1104612, ARO-W911NF-1–0262, ARO-MURI program W911NF-12-1-0461 and DARPA-SPWAR Meso program N6601-11-1-4110. D.D.S. and S.P. acknowledge the CARIPLO Foundation through the MAGISTER project Rif. 2013-0726. This work was partly supported by the Italian Ministry of Research through the project PRIN Interfacce di ossidi: nuove proprietà emergenti, multifunzionalità e dispositivi per elettronica e energia (OXIDE).

Author contributions

R.J.C. and G.P. conceived the experiment and wrote the paper with contributions from D.D.S. and I.V. All the authors discussed the results, commented on the manuscript and prepared written contributions. R.J.C. and M.N.A. grew the crystals and characterized the samples. J.T. performed the electron diffraction experiments. A.G., B.E.F. and A.Y. performed STM experiments and analysis. D.D.S., E.B. and S.P. performed the calculations. P.K.D., I.V., G.P., G.R., J.F., R.C. and T.O. performed the synchrotron radiation experiments and analysed the data.

Additional information

Competing financial interests: The authors declare no competing financial interests.

16

1s-intraexcitonic dynamics in monolayer MoS$_2$ probed by ultrafast mid-infrared spectroscopy

Soonyoung Cha[1,*], Ji Ho Sung[2,3,*], Sangwan Sim[1], Jun Park[1], Hoseok Heo[2,3], Moon-Ho Jo[2,3] & Hyunyong Choi[1]

The 1s exciton—the ground state of a bound electron-hole pair—is central to understanding the photoresponse of monolayer transition metal dichalcogenides. Above the 1s exciton, recent visible and near-infrared investigations have revealed that the excited excitons are much richer, exhibiting a series of Rydberg-like states. A natural question is then how the internal excitonic transitions are interrelated on photoexcitation. Accessing these intraexcitonic transitions, however, demands a fundamentally different experimental tool capable of probing optical transitions from 1s 'bright' to np 'dark' states. Here we employ ultrafast mid-infrared spectroscopy to explore the 1s intraexcitonic transitions in monolayer MoS$_2$. We observed twofold 1s→3p intraexcitonic transitions within the A and B excitons and 1s→2p transition between the A and B excitons. Our results revealed that it takes about 0.7 ps for the 1s A exciton to reach quasi-equilibrium; a characteristic time that is associated with a rapid population transfer from the 1s B exciton, providing rich characteristics of many-body exciton dynamics in two-dimensional materials.

[1] School of Electrical and Electronic Engineering, Yonsei University, Seoul 120-749, Korea. [2] Center for Artificial Low Dimensional Electronic Systems, Institute for Basic Science (IBS), Pohang University of Science and Technology (POSTECH), Pohang 790-784, Korea. [3] Division of Advanced Materials Science, Pohang University of Science and Technology (POSTECH), Pohang 790-784, Korea. * These authors contributed equally to this work. Correspondence and requests for materials should be addressed to M-H.J. (email: mhjo@postech.ac.kr) or to H.C. (email: hychoi@yonsei.ac.kr).

Photogenerated electron-hole (e–h) pairs in solids create bound states, whose elementary quasiparticle state is called 1s exciton in a Wannier–Mott exciton model. Since the optoelectronic response is governed by the light-induced dynamic behaviour of this elementary ground state, knowledge of the 1s exciton response to the optical stimuli has been a crucial issue in many optoelectronic applications, such as phototransistors[1,2], photovoltaics[3], light-emitting diodes[4,5], van der Waals heterostructure-based optoelectronics[6–8] and valleytronic device applications[5,9–12]. In transition metal dichalcogenides (TMDCs), this is particularly the case as the two-dimensional (2D) materials approach a monolayer limit, where the reduced dielectric screening results in a strong Coulomb interaction[13–21], leading to an unusually large 1s exciton binding energy E_{bind}, typically a few hundreds of meV below the electronic bandgap of a few eV (refs 16,22–27).

Above the fundamental 1s exciton, theories predicted the presence of densely spaced exciton states in monolayer MoS₂ with 1s exciton E_{bind} of 0.4–0.54 eV (refs 18–21,28–31), whose (s-like) bright and (p-like) dark exciton characters were later confirmed by a series of seminal experiments via linear one-photon absorption[21,22], two-photon photoluminescence excitation (PLE)[22–24,32] and nonlinear wave-mixing spectroscopy[25,32], whereby E_{bind} was experimentally measured to be between 0.22 (ref. 26) and 0.44 eV (refs 23,26); the reported E_{bind}, however, shows somewhat discrepancy depending on the measurement methods and is varied from samples to samples[23,26,27]. These experimental techniques, although they are appropriate to clarify the optical state of the excitons, may address indirectly the dynamic transient information between the 1s 'bright' and the excited np 'dark' exciton (n is the principle quantum number); we denoted the exciton states in analogy to the hydrogen series[21]. By contrast, if one measures the $1s \to np$ transitions, then the data should describe the internal excitonic transients, directly providing the transient optical nature of the fundamental 1s exciton dynamics. This, so called intraexcitonic spectroscopy[33], fundamentally differs from band-to-band and other time-resolved spectroscopies[8,34–36], and the technique can not only explain the transient response of the 1s exciton, but more importantly, may provide experimental manoeuvre in exploiting the photoinduced excitonic responses to the TMDC-based optoelectronic devices. For example, knowledge of the 1s and np exciton energies and their associated dynamics afford the first-order quantitative information on the exciton dissociation energy, where in an ideal case at least E_{bind}/e (e is the electron charge) of an external

or internal potential is required to dissociate the bound e–h pairs. In addition, because intraexcitonic spectroscopy can access the p-like dark excitons, one may design a scheme of coupling an infrared (IR) light to the 2D TMDC materials, via below-gap two-photon excitation, for the light-harnessing applications.

Here we explore the 1s intraexcitonic transient dynamics in monolayer MoS₂ by using time-resolved mid-IR spectroscopy. Inspired by a theoretical GW–Bethe–Salpeter result[19], where the fundamental $1s \to 2p$ transitions are predicted to be 0.32 and 0.3 eV for the A and B exciton in isolated, suspended monolayer MoS₂, we employed an ultrafast mid-IR spectroscopy (0.23–0.37 eV probe) in conjunction with an ultrafast white-light continuum spectroscopy (Fig. 1a). The mid-IR measurements show that there are two $1s \to 3p$ transitions for A and B exciton and $1s \to 2p$ between 1s A and 2p B exciton. The time-dependent IR absorption rapidly subdues over broad probe–photon energies, representing the transient absorption from the 1s to the quasi-continuum states after pump excitation.

Results

Time-resolved intraexcitonic and band-to-band dynamics.
The samples used in our experiment were monolayer MoS₂, grown by chemical vapour deposition method, and were transferred to a CaF₂ substrate (see Supplementary Note 1 for the sample characterization). As schematically shown in Fig. 1a, the sample was non-resonantly excited by a 70 fs, 3.1 eV pump pulse, and the corresponding differential-transmission changes $\Delta T/T_0$ were measured in a vacuum cryostat (Methods). The 3.1 eV pump excites carriers into the quasi-continuum of the A and B excitons[37,38] or into the band-nesting C-band near the Γ point[39,40]. The former generates the unbound e–h plasma above the A and B excitons and the latter case experiences a rapid inter-valley scattering into K and K′ valley. Nevertheless, both cases generate carriers in much higher energy compared with the A or B exciton resonance. Figure 1b shows a direct comparison of two representative data measured by mid-IR probe (0.35 eV) and interband A-exciton probe (1.86 eV) with the same pump fluence $F = 24.4 \mu J cm^{-2}$ (equivalent to e–h pair density of $7.4 \times 10^{12} cm^{-2}$ given 15% absorption)[8,41] measured at 77 K. The polarization of pump and probe beam are linear and orthogonal with respect to each other, such that we do not account for the recently discovered valley-exciton-locked selection rule[32]. The fact that two $\Delta T/T_0$ transients exhibit an opposite sign implies the kinetic origin of the photoresponses is indeed different. For the 1.86 eV dynamics, the increased

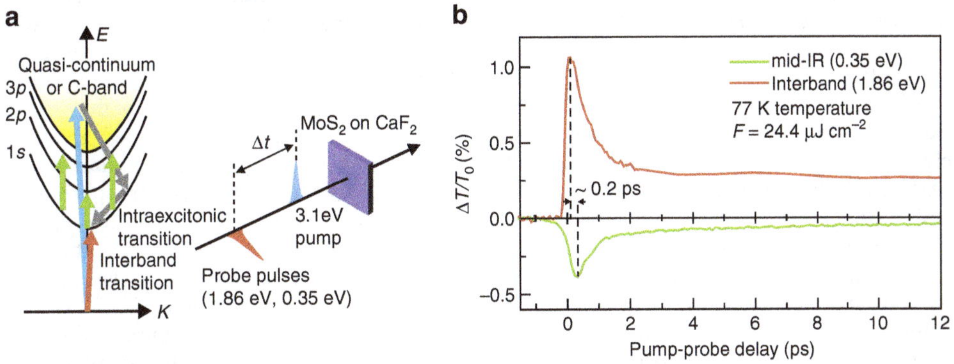

Figure 1 | Ultrafast intraexcitonic and band-to-band spectroscopy in monolayer MoS₂. (a) Schematic illustration of the ultrafast intraexcitonic (green) and conventional band-to-band (red) interband spectroscopy. The 3.1 eV optical pump (blue arrow) creates e-h pairs from ground to quasi-continuum states or C-band. The mid-IR pulse (green) measures the $1s \to np$ transitions, while the white-light continuum pulse measures the ground-to-1s transition. (b) Transient dynamics of $\Delta T/T_0$ measured at two representative energies of 1.86 eV (red) and 0.35 eV (green). The positive sign of $\Delta T/T_0$ is observed for 1.86 eV probe, while 0.35 eV probe exhibits a negative sign. A clear temporal delay (~0.2 ps) between the two rising transients was observed. The experimentally determined temporal error bar for each time-zero (20 fs for 0.35 eV and 8 fs for 1.86 eV) is discussed in Supplementary Fig. 8.

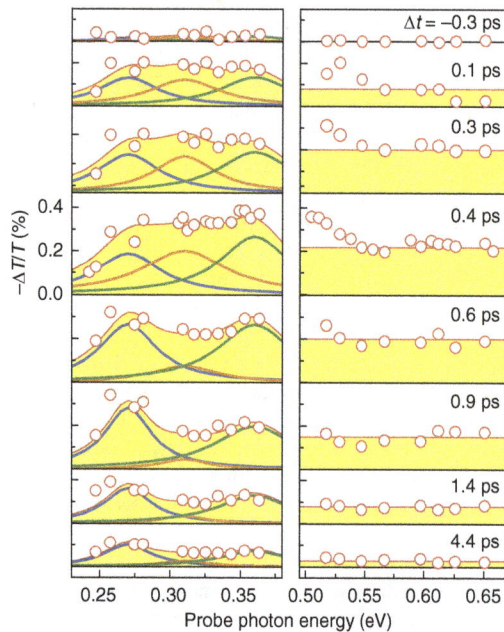

Figure 2 | Temporally and spectrally resolved transitions from 1s state.
Left panel is the mid-IR transient and right panel is the IR transient dynamics. The transient spectra (red dots) are fitted by three Lorentzian oscillators (blue, orange and green) for the mid-IR range and a step function (red) for the IR range. The red solid in left panel is the sum of three intraexcitonic oscillators. We note that it was not possible to fit all the mid-IR spectra if only two oscillators were used.

probe transmission is typically attributed to the ground-state bleaching[35,42,43], where the increased occupation probability of electrons in conduction band and holes in the valence band leads to the reduced probe absorption, that is, increased $\Delta T/T_0 > 0$. On the other hand, given that the 0.35 eV probe is far below the band-to-band A-exciton energy, one may attribute the decreased $\Delta T/T_0 < 0$ (increased probe absorption) to the transition within the bands, that is, intraband free-carrier absorption. However, considering that the mid-IR peak signal is only 36.2% of the 1.86 eV one, we can exclude this possibility because the intraband oscillator strength is much smaller than the interband one, usually by an order of magnitude, as revealed by prior works on the quasi-2D quantum wells[44] or recent 2D MoS_2 (ref. 45); as discussed later in Figs 2–4, we provide compelling experimental evidences to support our rationale (see also Supplementary Note 2 (refs 46–48)). The increased probe absorption of the mid-IR suggests that there exists an occupied state below the electronic gap.

We find that there exists a clear time departure between the two rising dynamics, where the onset of the 0.35 eV probe peak appears ~0.2 ps later than the 1.86 eV probe (dashed line in Fig. 1b). Immediately after the pump, the 1.86 eV probe rapidly increases, while the 0.35 eV dynamics emerge rather slowly. We observed that this rapid upsurge of the 1.86 eV is not a local spectral behaviour, but being presented in a broad range of high-energy probes (Supplementary Notes 3 and 4), evidencing the quasi-instantaneous ground-state bleaching[35]. Understanding this high-energy dynamics has been a subject to debate; different investigations have proposed different kinetic origins of the 1s exciton, such as exciton linewidth broadening[35], stimulated emission[42], dynamic bandgap renormalization[49] and biexciton formation[38]. As discussed, more details in Supplementary Note 4, both earlier[50–52] and recent studies[35,42,53], have shown that the 3.1 eV photoexcitation into the quasi-continuum of unbound states generates a significant amount of free-carriers. Because the

exciton formation occurs after exciton–free carrier scattering, the initial decaying kinetics of mid-IR is slightly delayed compared with the rising transient of the interband one, explaining the observed time-delay between the two transients of Fig. 1b. Since the mid-IR probe can resonantly measure the internal exciton dynamics, the measured intraexcitonic transients are expected to provide pure population dynamics of the ground 1s exciton[33,54].

Temperally and spectrally resolved intraexcitonic dynamics. Figure 2 is the temporally and spectrally resolved mid-IR dynamics. Here we probed not only the broad mid-IR transients (0.23–0.37 eV), but also measured the IR dynamics (0.47–0.67 eV). This scheme affords a simultaneous access to the dynamic transitions from the 1s ground exciton to the higher lying np excitons or quasi-continuum states. The measured $-\Delta T/T_0$ spectra show peculiar energy-dependent behaviours. At $\Delta t \leq 0.4$ ps, the $-\Delta T/T_0$ spectra are strongly reshaped, exhibiting a relatively small increase of differential absorption (not absolute absorption) near 0.27 eV compared with the increased absorption around 0.3–0.5 eV. The increased absorption is more prominent at $\Delta t > 0.4$ ps, where one can see that $-\Delta T/T_0$ is gradually larger near 0.27 eV with increasing Δt, and the differential absorption at 0.3–0.5 eV is concurrently smaller with increasing Δt. Between $0.4 < \Delta t \leq 0.9$ ps, $-\Delta T/T_0$ above 0.3 eV is rapidly vanished, while the absorption resonance below 0.3 eV is accordingly increased. After $\Delta t > 0.9$ ps, the absorption resonance below 0.3 eV keeps reserved and it slowly subdues with featureless IR spectra above 0.3 eV.

For a quantitative analysis of the observed transient spectra, we use the following model consisted of multi-oscillator components[55]:

$$\alpha(E) = \sum_n \frac{\hbar e^2 S_{1s-np}}{2mc\varepsilon_0 \sqrt{\varepsilon}} \frac{\hbar\Gamma}{(E_n - E)^2 + (\hbar\Gamma)^2} + \Theta(E - E_{\text{bind}}). \quad (1)$$

The term of summation represents the intraexcitonic absorption from 1s to either A or B excitonic np state and the second term $\Theta(E - E_{\text{bind}})$ is a step-like transition[21] from 1s to the continuum with E_{bind} of 0.44 eV. In the equation, ε (= 4.2) (refs 14,27) and ε_0 are the dielectric constant of monolayer MoS_2 and the vacuum dielectric constant, respectively. There, the absorption amplitude $S_{1s \rightarrow np}$, or the spectral weight of the intraexcitonic $1s \rightarrow np$ transition, is proportional to the product of the oscillator strength $f_{1s \rightarrow np}$ and the ground exciton density n_{1s} (refs 39,40,55,56). Because the 3.1 eV pump excitation creates e–h plasma in the band nesting resonance, an accurate estimation of n_{1s} requires both theoretical study of intervalley scatterings and the corresponding ultrafast measurements, which is beyond the scope of our ultrafast mid-IR intraexcitonic spectroscopy. In fact, the spectral weight from $1s \rightarrow np$ is not only proportional to the population, but also depends on the probability of finding an empty final np state. As discussed about the transient spectra dynamics above, the photoexcited unbound e–h pairs experience rapid relaxation and start to form a ground-state exciton within ~0.4 ps. It is strictly true that the np exciton population is negligible only at $\Delta t \geq 0.4$ ps. Similar studies on 1D and quasi-2D quantum-well structures have shown that the contribution from $np \rightarrow$ continuum is negligible[33,54,56–58]. We found that the spectral fit matches well the measured data when we used up to three mid-IR oscillators, with the following transition energy E_n of $E_1 = 0.27$ eV, $E_2 = 0.31$ eV, and $E_3 = 0.36$ eV. On the basis that the observed E_ns do not vary Δt, we fix E_n to fit the time-resolved mid-IR spectra, but vary $S_{1s \rightarrow np}$ and the phenomenological exciton broadening parameter Γ. For the IR transients, the spectra are featureless representing the step-function-like 1s to the continuum transition[21]; this featureless IR spectrum is

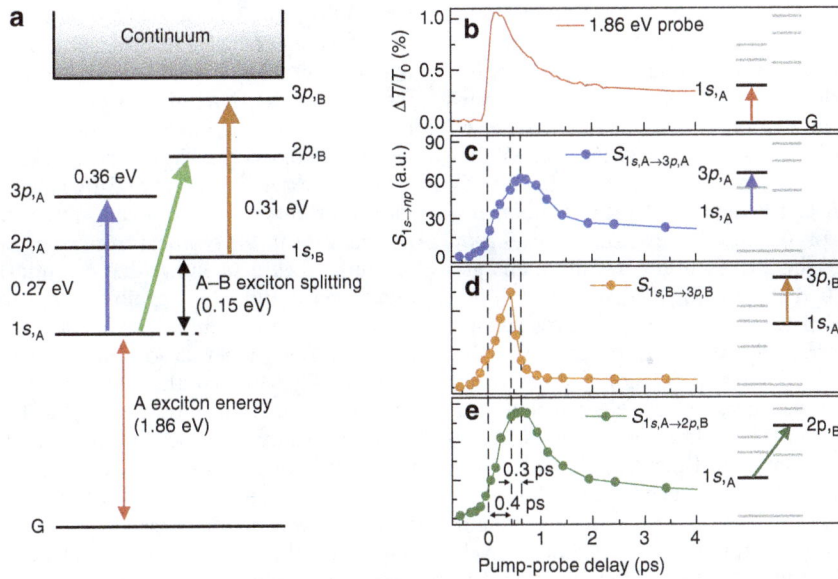

Figure 3 | Schematic for 1s intraexcitonic transition and relevant spectral weight. (a) Energy diagram of the ground state (G), and the fundamental excitons ($1s_A$ and $1s_B$) and the higher excited np dark excitons in shown. Transition energies of three oscillators are indicated by blue (0.27 eV), orange (0.31 eV) and green (0.36 eV) arrows. The transient band-to-band dynamics (b) is directly compared with the intraexcitonic absorption dynamics (c–e). Transient dynamics of the intraexcitonic spectral weight parameter $S_{1s \to np}$ for each three oscillator are shown at each row: (c) $1s_A \to 3p_A$, (d) $1s_B \to 3p_B$ and (e) $1s_A \to 2p_B$, respectively. Dashed lines show the maximum $S_{1s \to np}$ peak for each intraexcitonic transition.

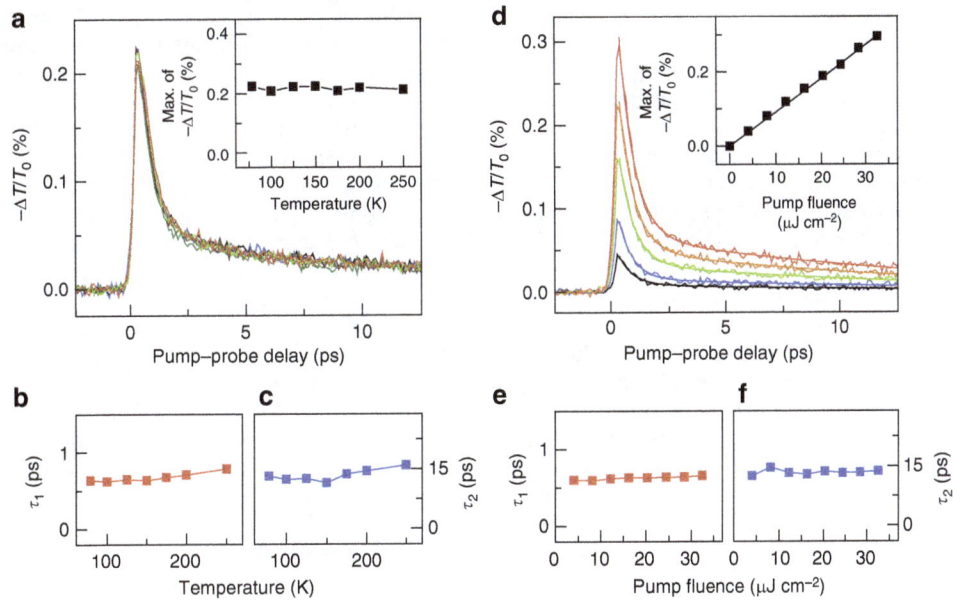

Figure 4 | Temperature- and fluence-dependent mid-IR dynamics. (a) Temperature-dependent $-\Delta T/T_0$ dynamics measured at 0.6 eV. Inset: the peak value of $-\Delta T/T_0$ is plotted as a function of the temperature. No temperature-dependent dynamics were observed, thereby the free-carrier absorption can be excluded in the analysis of Figs 1–3. We performed fitting using a biexponential function. The summarized results are shown in **b** for the fast τ_1 and in **c** for the slow decay component τ_2, where both components are temperature-independent. (d) Fluence-dependent $-\Delta T/T_0$ dynamics measured at 0.6 eV probe. Inset: the peak $-\Delta T/T_0$ shows a linear fluence dependence, such that no higher-order nonlinear exciton dynamics were observed. Fluence-dependent fast τ_1 (**e**) and slow decay component τ_2 (**f**). Solid line for each trace is the corresponding biexponential fit. Both τ_1 and τ_2 are independent of the pump fluence, implying that no absorption occurs from the defect states.

deviated from the step-like absorption at $\Delta t \leq 0.4$ ps, which might be due to the time-dependent thermalization process after the pump excitation.

Given that the magnitude of $S_{1s \to np}$ differs only by a factor of 2 for each E_n, these transitions cannot simply be assigned to the phenomena taking place within single exciton Rydberg series of the A (or B) exciton branch. This is because even when the nonhydrogenic excitonic nature of a monolayer TMDC is

considered, where a strongly (weakly) screened Coulomb potential is dominant when $n \leq 2$ ($n \geq 3$) (ref. 21), the spectral weight should be substantially decreased by nearly an order of magnitude with increasing n, which is too large to account for our measurement results. Of course, care should be taken to estimate the precise strength of intraexcitonic transition in a monolayer TMDC because the nonhydrogenic exciton is dominant when $n \leq 2$, thus the $1s \to np$ transition can deviate from the hydrogenic excitonic nature for $n \geq 3$. This is because any intraexcitonic $1s \rightarrow np$ transition depends both on the wavefunction of the nth exciton as well as the $1s$ exciton state, and the latter certainly deviates from the 2D-hydrogen model. Recent PLE[23] revealed that the energy levels of the exciton Rydberg series are 1.88 eV ($1s$), 2.05 eV ($2s$) and 2.15 eV ($3s$) for A exciton and 2.03 eV ($1s$), 2.24 eV ($2s$) and 2.34 eV ($3s$) for B exciton. By considering 0.15 eV energy splitting between A and B excitons and the difference of reduced exciton masses of $0.25m_0$ (A exciton) and $0.28m_0$ (B exciton)[18], we estimated the intraexcitonic transition energies of 0.27 eV for $E_{1A \rightarrow 3A}$, 0.31 eV for $E_{1B \rightarrow 3B}$ and 0.36 eV for $E_{1A \rightarrow 2B}$, which are exactly matched our measured intraexcitonic transition energy E_n. Interestingly, these values are somewhat deviated from the GW–Bethe–Salpeter prediction[19], possibly due to the substrate dielectric screening effect. Nevertheless, our measurements agree well with the experimental PLE investigation due to similar dielectric constant of CaF_2 and fused silica[23] as a substrate. Although there is a small difference (~ 20 meV) for the A exciton energy between PLE (1.88 eV) and our photocurrent spectra and ultrafast absorption measurement (1.86 eV, Supplementary Notes 1 and 3), the difference is very marginal[19,28,30,31] and the intraexcitonic spectroscopy can measure the energy difference between $1s$ and np, regardless of the A-exciton resonance. In accordance with PLE and our mid-IR measurements, we expect the fundamental $1s \rightarrow 2p$ would be 0.17 eV for the A exciton and 0.21 eV for the B exciton, and this is beyond our capability of tuning the mid-IR spectrum. Therefore, as schematically shown in Fig. 3a, we understand our intraexcitonic transition energy of E_1 as $1s_{,A} \rightarrow 3p_{,A}$ within A exciton, E_2 as $1s_{,B} \rightarrow 3p_{,B}$ within B exciton and E_3 as $1s_{,A} \rightarrow 2p_{,B}$ between A and B exciton. Indeed, our energy assignment well-corroborates a recent many-body Bethe–Salpeter prediction on the nonhydrogenic characters of excited excitons[19,20,28,30,31], underscoring a distinct capability of our intraexcitonic spectroscopy in measuring the relative energy difference between $1s$ and np. For the exciton broadening parameter Γ, since the effective mass of A and B exciton is different, Γ ($=28.2$ meV) for $1s_{,A} \rightarrow 3p_{,A}$, Γ ($=37.4$ meV) for $1s_{,B} \rightarrow 3p_{,B}$ and Γ ($=30$ meV) for $1s_{,A} \rightarrow 2p_{,B}$ are slightly different due to the different exciton dispersion.

Dynamics of $1s \rightarrow np$ intraexcitonic spectral weights. For further analysis, we show the temporal dynamics of $S_{1s,A \rightarrow 3p,A}$ (Fig. 3c, blue), $S_{1s,B \rightarrow 3p,B}$ (Fig. 3d, orange) and $S_{1s,A \rightarrow 3p,A}$ (Fig. 3e, green). We identify three different kinetic regimes: immediately after the pump, the rising transients of all three spectral weights show similar behaviours, representing the hot-carrier relaxation from the quasi-continuum to the A and B exciton branch. This kinetics clearly differs from the dynamics of 1.86 eV probe (Fig. 3b), where the latter arises from the quasi-instantaneous bleaching dynamics. At $0.4 \leq \Delta t \leq 0.7$ ps, the dynamics of $S_{1s,B \rightarrow 3p,B}$ rapidly decrease, while the peak $S_{1s,A \rightarrow 3p,A}$ and $S_{1s,A \rightarrow 3p,B}$ emerge ~ 0.3 ps later. Because the $1s$ B exciton is 0.15 eV higher than that of A exciton (Fig. 3a), the $1s$ B exciton serves as a population supplier to the energetically lower $1s$ A exciton, thereby the two transients show a complementary dynamics. At longer $\Delta t > 0.7$ ps, because the $1s$ A excitons are thermalized and reaches a quasi-equilibrium condition, the dynamics of $S_{1s,A \rightarrow 3p,A}$ nearly follows that of $S_{1s,A \rightarrow 2p,B}$. This highlights that although $S_{1s,A \rightarrow 3p,A}$ and $S_{1s,A \rightarrow 3p,B}$ are spectrally separated apart, that is 0.27 and 0.36 eV, respectively, both transients are closely interrelated because these absorptions originate from the same $1s_{,A}$ ground state exciton.

Discussion

At an elevated temperature, the free-carrier absorption from $1s$, $2s$, $2p$, $3s$, $3p$... may contribute to the increased probe absorption

with $\Delta T/T_0 < 0$. This scenario typically shows a strong temperature dependence of the relaxation rate, in which the higher temperature the larger the electron–phonon scattering rate, resulting in the dynamics to be highly temperature dependent. Here given that the formation time scale of the $1s$ exciton is very fast within 0.4 ps (see Figs 2 and 3) and the Drude scattering rate cannot be extended to the mid-IR range (Supplementary Note 2), the contribution of $np \rightarrow$ continuum transition may be very insignificant to the temperature-dependent mid-IR intraexcitonic response. Figure 4a shows that our mid-IR transients, fitted by a biexponential function, exhibit nearly temperature independent of the relaxation components (Fig. 4b,c). This implies that the effect of free-carrier absorption is negligible. We additionally show in Fig. 4b that the recombination of excitons arises on sub-ps and tens of ps time scale. At $T = 77$ K, the mid-IR peak $|\Delta T/T_0|$ linearly increases with F up to 32.5 μJ cm^{-2} (equivalent to e–h pair density of 9.86×10^{12} cm^{-2}) (refs 8,41). The linear F-dependence reflects that there exists no high-order nonlinear excitonic interaction, ensuring that our mid-IR transients represent the first-order population dynamics. A recent below-gap-probe study[45] reported very similar relaxation times to our results. These time components were explained using defect-assisted exciton recombination. Given that we observed negligible F-dependent relaxation dynamics (Fig. 4e,f), we can infer that our mid-IR decay transients do not arise from the photoinduced absorption of filled e–h pair in the localized states, but arise from the exciton capture into the defects.

In summary, we report the experimental observation of the $1s$ intraexcitonic transition. Recently, Poellmann et al.[47] investigated a similar investigation of intraexcitonic transition in monolayer WSe$_2$, reporting the presence of strong absorption in a 2D TMDC, whose fundamental optical absorption originates from the $1s$ ground exciton. Our ultrafast mid-IR measurements reveal twofold $1s \rightarrow 3p$ transition energies to be 0.27 eV and 0.31 eV for A and B exciton, respectively. We additionally uncover an intraexcitonic relaxation channel of $1s \rightarrow 2p$ to be 0.36 eV between $1s$ A and $2p$ B exciton. The large exciton-binding energy due to the non-local dielectric screening ensures not only $1s \rightarrow 2p$ transition[47] to be observable, but also a higher-order transition of $1s \rightarrow 3p$ in a monolayer 2D TMDC at an elevated temperature, which cannot be accessible using conventional interband spectroscopy, or any in quasi-2D quantum-well structures. In addition, looking to the future, the availability of electric-gate tuning may enable to investigate the coherent many-body inter-excitonic correlations among exciton, biexciton[38,59] and trion[12,13,18,48] in a time-resolved controlled manner, which is non-trivial to study in other low-dimensional inorganic semiconductor structures.

Methods

Ultrafast optical pump-probe spectroscopy. Using 250 kHz, 50 fs Ti/sapphire laser system (Coherent RegA 9050), optical parametric amplifier (Coherent OPA 9850) yields signal (0.77–1.12 eV) and idler (0.47 eV–0.67 eV) pulses that are used to generate mid-IR pulse (0.23–0.37 eV) via difference frequency generator (Coherent DFG). The idler and DFG output serve as the probe pulse in the IR and mid-IR range, respectively. The chirp of mid-IR pulse is discussed in Supplementary Note 5. High-energy interband response was measured by using a white-light continuum (1.76–2.03 eV) generated by focusing 1.55 eV pulses into a 1 mm sapphire disk. For the group-delay dispersion (GDD) of the white-light continuum pulse, we compensated using a pair of prism, and further checked the GDD-induced delay via cross-correlation of the white-light pulse and 1.55 eV pulse, whose details are explained in the Supplementary Note 3. The 3.1 eV pump pulse was created by second harmonic generation of 1.55 eV pulse in a 1-mm-thick beta barium borate (BBO) crystal. Due to the combination of OPA and DFG, where both signal and idler from OPA were used to generate the mid-IR DFG output, the only available seed pulse for 3.1 eV pump pulse was 1.55 eV in our system, so that the mid-IR measurement with resonant pump excitation at either A or B $1s$ exciton was not possible in our current system. For the each mid-IR or IR measurement, pump pulse and probe pulse are simultaneously focused on the sample in the cryostat equipped with two CaF_2 windows, and pump–probe delay is controlled by a mechanical delay stage (Newport M-IMS300LM). The spot size of our pump and probe beams were 100 μm, and 50 μm, respectively, which were simultaneously focused using $f = 50$ mm lens before the temperature-controlled vacuum cryostat. In our optical geometry, the 3.1 eV pump passes through a mechanical delay stage, so called 'pump delay'. Because the pump delay is recorded in a computer as an absolute length, we performed cross-correlation measurement to estimate the probe delay using BBO (visible upconversion) and KTA (mid-IR upconversion) crystals. More detailed information for determining the pump–probe 'time-zero' is explained in Supplementary Note 6. Differential transmission signal ($\Delta T/T_0$) was recorded in a lock-in amplifier (Stanford Research Systems SR850) with 10 kHz chopping frequency (Scitec 300CD). The schematics of mid-IR and IR setup are illustrated in the Supplementary Fig. 9.

References

1. Lopez-Sanchez, O., Lembke, D., Kayci, M., Radenovic, A. & Kis, A. Ultrasensitive photodetectors based on monolayer MoS₂. *Nat. Nano* **8**, 497–501 (2013).
2. Wang, Q. H., Kalantar-Zadeh, K., Kis, A., Coleman, J. N. & Strano, M. S. Electronics and optoelectronics of two-dimensional transition metal dichalcogenides. *Nat. Nano* **7**, 699–712 (2012).
3. Furchi, M. M., Pospischil, A., Libisch, F., Burgdorfer, J. & Mueller, T. Photovoltaic effect in an electrically tunable van der Waals heterojunction. *Nano Lett.* **14**, 4785–4791 (2014).
4. Cheng, R. *et al.* Electroluminescence and Photocurrent generation from atomically sharp WSe₂/MoS₂ heterojunction p–n diodes. *Nano Lett.* **14**, 5590–5597 (2014).
5. Zhang, Y., Oka, T., Suzuki, R., Ye, J. & Iwasa, Y. Electrically switchable chiral light-emitting transistor. *Science* **344**, 725–728 (2014).
6. Britnell, L. *et al.* Strong light-matter interactions in heterostructures of atomically thin films. *Science* **340**, 1311–1314 (2013).
7. Lee, C.-H. *et al.* Atomically thin pn junctions with van der Waals heterointerfaces. *Nat. Nano* **9**, 676–681 (2014).
8. He, J. *et al.* Electron transfer and coupling in graphene–tungsten disulfide van der Waals heterostructures. *Nat. Commun.* **5**, 5622 (2014).
9. Gong, Z. *et al.* Magnetoelectric effects and valley-controlled spin quantum gates in transition metal dichalcogenide bilayers. *Nat. Commun.* **4**, 2053 (2013).
10. Xu, X., Yao, W., Xiao, D. & Heinz, T. F. Spin and pseudospins in layered transition metal dichalcogenides. *Nat. Phys.* **10**, 343–350 (2014).
11. Mak, K. F., McGill, K. L., Park, J. & McEuen, P. L. The valley Hall effect in MoS₂ transistors. *Science* **344**, 1489–1492 (2014).
12. Jones, A. M. *et al.* Optical generation of excitonic valley coherence in monolayer WSe₂. *Nat. Nano* **8**, 634–638 (2013).
13. Mak, K. F., Lee, C., Hone, J., Shan, J. & Heinz, T. F. Atomically thin MoS₂: a new direct-gap semiconductor. *Phys. Rev. Lett.* **105**, 136805 (2010).
14. Cheiwchanchamnangij, T. & Lambrecht, W. R. Quasiparticle band structure calculation of monolayer, bilayer, and bulk MoS₂. *Phys. Rev. B* **85**, 205302 (2012).
15. Mak, K. F. *et al.* Tightly bound trions in monolayer MoS₂. *Nat. Mater.* **12**, 207–211 (2013).
16. Ugeda, M. M. *et al.* Giant bandgap renormalization and excitonic effects in a monolayer transition metal dichalcogenide semiconductor. *Nat. Mater.* **13**, 1091–1095 (2014).
17. Zhang, Y. *et al.* Direct observation of the transition from indirect to direct bandgap in atomically thin epitaxial MoSe₂. *Nat. Nano* **9**, 111–115 (2014).
18. Berkelbach, T. C., Hybertsen, M. S. & Reichman, D. R. Theory of neutral and charged excitons in monolayer transition metal dichalcogenides. *Phys. Rev. B* **88**, 045318 (2013).
19. Qiu, D. Y., Felipe, H. & Louie, S. G. Optical spectrum of MoS₂: many-body effects and diversity of exciton states. *Phys. Rev. Lett.* **111**, 216805 (2013).
20. Berghäuser, G. & Malic, E. Analytical approach to excitonic properties of MoS₂. *Phys. Rev. B* **89**, 125309 (2014).
21. Chernikov, A. *et al.* Exciton binding energy and nonhydrogenic rydberg series in monolayer WS₂. *Phys. Rev. Lett.* **113**, 076802 (2014).
22. He, K. *et al.* Tightly bound excitons in monolayer WSe₂. *Phys. Rev. Lett.* **113**, 026803 (2014).
23. Hill, H. M. *et al.* Observation of excitonic Rydberg states in monolayer MoS₂ and WS₂ by photoluminescence excitation spectroscopy. *Nano Lett.* **15**, 2992–2997 (2015).
24. Ye, Z. *et al.* Probing excitonic dark states in single-layer tungsten disulphide. *Nature* **513**, 214–218 (2014).
25. Wang, G. *et al.* Giant enhancement of the optical second-harmonic emission of WSe₂ monolayers by laser excitation at exciton resonances. *Phys. Rev. Lett.* **114**, 097403 (2015).
26. Zhang, C., Johnson, A., Hsu, C.-L., Li, L.-J. & Shih, C.-K. Direct imaging of band profile in single layer MoS₂ on graphite: quasiparticle energy gap, metallic edge states, and edge band bending. *Nano Lett.* **14**, 2443–2447 (2014).
27. Klots, A. R. *et al.* Probing excitonic states in suspended two-dimensional semiconductors by photocurrent spectroscopy. *Sci. Rep.* **4**, 6608 (2014).
28. Berkelbach, T. C., Hybertsen, M. S. & Reichman, D. R. Bright and dark singlet excitons via linear and two-photon spectroscopy in monolayer transition metal dichalcogenides, Preprint at http://arxiv.org/abs/1505.07127 (2015).
29. Stroucken, T. & Koch, S. W. Evidence for optically bright p-excitons in transition-metal dichalchogenides, Preprint at http://arxiv.org/abs/1404.4238v2 (2014).
30. Wu, F., Qu, F. & MacDonald, A. H. Exciton band structure of monolayer MoS₂. *Phys. Rev. B* **91**, 075310 (2015).
31. Chaves, A., Low, T., Avouris, P., Çakır, D. & Peeters, F. Anisotropic exciton Stark shift in black phosphorus. *Phys. Rev. B* **91**, 155311 (2015).
32. Xiao, J. *et al.* Optical selection rule based on valley-exciton locking for 2D valleytronics, Preprint at http://arxiv.org/abs/1504.04947 (2014).
33. Kaindl, R. A., Carnahan, M. A., Hägele, D., Lövenich, R. & Chemla, D. S. Ultrafast terahertz probes of transient conducting and insulating phases in an electron–hole gas. *Nature* **423**, 734–738 (2003).
34. Shi, H. *et al.* Exciton dynamics in suspended monolayer and few-layer MoS₂ 2D crystals. *ACS Nano* **7**, 1072–1080 (2013).
35. Sim, S. *et al.* Exciton dynamics in atomically thin MoS₂: interexcitonic interaction and broadening kinetics. *Phys. Rev. B* **88**, 075434 (2013).
36. Hong, X. *et al.* Ultrafast charge transfer in atomically thin MoS₂/WS₂ heterostructures. *Nat. Nano* **9**, 682–686 (2014).
37. Cabo, A. G. *et al.* Observation of ultrafast free carrier dynamics in single layer MoS₂. *Nano Lett.* **15**, 5883–5887 (2015).
38. Sie, E. J., Frenzel, A. J., Lee, Y.-H., Kong, J. & Gedik, N. Intervalley biexcitons and many-body effects in monolayer MoS₂. *Phys. Rev. B* **92**, 125417 (2015).
39. Carvalho, A., Ribeiro, R. & Neto, A. C. Band nesting and the optical response of two-dimensional semiconducting transition metal dichalcogenides. *Phys. Rev. B* **88**, 115205 (2013).
40. Kozawa, D. *et al.* Photocarrier relaxation pathway in two-dimensional semiconducting transition metal dichalcogenides. *Nat. Commun.* **5**, 4543 (2014).
41. Li, Y. *et al.* Measurement of the optical dielectric function of monolayer transition-metal dichalcogenides: MoS₂, MoSe₂, WS₂, and WSe₂. *Phys. Rev. B* **90**, 205422 (2014).
42. Mai, C. *et al.* Many-body effects in valleytronics: direct measurement of valley lifetimes in single-layer MoS₂. *Nano Lett.* **14**, 202–206 (2013).
43. Mai, C. *et al.* Exciton valley relaxation in a single layer of WS₂ measured by ultrafast spectroscopy. *Phys. Rev. B* **90**, 041414 (2014).
44. Klimov, V., Schwarz, C. J., McBranch, D., Leatherdale, C. & Bawendi, M. Ultrafast dynamics of inter-and intraband transitions in semiconductor nanocrystals: Implications for quantum-dot lasers. *Phys. Rev. B* **60**, R2177 (1999).
45. Wang, H., Zhang, C. & Rana, F. Ultrafast dynamics of defect-assisted electron–hole recombination in monolayer MoS₂. *Nano Lett.* **15**, 339–345 (2015).
46. Shen, C.-C. *et al.* Charge dynamics and electronic structures of monolayer MoS₂ films grown by chemical vapor deposition. *Appl. Phys. Exp.* **6**, 125801 (2013).
47. Poellmann, C. *et al.* Resonant internal quantum transitions and femtosecond radiative decay of excitons in monolayer WSe₂. *Nat. Mater.* **14**, 889–893 (2015).
48. Lui, C. H. *et al.* Trion-induced negative photoconductivity in monolayer MoS₂. *Phys. Rev. Lett.* **113**, 166801 (2014).
49. Wang, Q. *et al.* Valley carrier dynamics in monolayer molybdenum disulfide from helicity-resolved ultrafast pump–probe spectroscopy. *ACS Nano* **7**, 11087–11093 (2013).
50. Wake, D. R., Yoon, H. W., Wolfe, J. P. & Morkoç, H. Response of excitonic absorption spectra to photoexcited carriers in GaAs quantum wells. *Phys. Rev. B* **46**, 13452–13460 (1992).
51. Knox, W. H. *et al.* Femtosecond excitation of nonthermal carrier populations in GaAs quantum wells. *Phys. Rev. Lett.* **56**, 1191–1193 (1986).
52. Knox, W. H., Chemla, D. S., Livescu, G., Cunningham, J. E. & Henry, J. E. Femtosecond carrier thermalization in dense Fermi seas. *Phys. Rev. Lett.* **61**, 1290–1293 (1988).
53. Chernikov, A., Ruppert, C., Hill, H. M., Rigosi, A. F. & Heinz, T. F. Population inversion and giant bandgap renormalization in atomically thin WS₂ layers. *Nat. Photon.* **9**, 466–470 (2015).
54. Wang, J., Graham, M. W., Ma, Y., Fleming, G. R. & Kaindl, R. A. Ultrafast spectroscopy of midinfrared internal exciton transitions in separated single-walled carbon nanotubes. *Phys. Rev. Lett.* **104**, 177401 (2010).
55. Jörger, M., Fleck, T., Klingshirn, C. & Von Baltz, R. Midinfrared properties of cuprous oxide: High-order lattice vibrations and intraexcitonic transitions of the 1s paraexciton. *Phys. Rev. B* **71**, 235210 (2005).
56. Huber, R., Kaindl, R. A., Schmid, B. A. & Chemla, D. S. Broadband terahertz study of excitonic resonances in the high-density regime in GaAs. Al$_x$Ga$_{1-x}$As quantum wells. *Phys. Rev. B* **72**, 161314 (2005).

57. Suzuki, T. & Shimano, R. Time-resolved formation of excitons and electron-hole droplets in Si studied using terahertz spectroscopy. *Phys. Rev. Lett.* **103**, 057401 (2009).

58. Kaindl, R. A., Hägele, D., Carnahan, M. A. & Chemla, D. S. Transient terahertz spectroscopy of excitons and unbound carriers in quasi-two-dimensional electron-hole gases. *Phys. Rev. B* **79**, 045320 (2009).

59. You, Y. *et al.* Observation of biexcitons in monolayer WSe$_2$. *Nat. Phys.* **11**, 477–481 (2015).

Acknowledgements

S. Cha, S. Sim, J. Park and H. Choi were supported by the National Research Foundation of Korea (NRF) through the government of Korea (MSIP) (Grants No. NRF-2011-0013255, NRF-2009-0083512, NRF-2015R1A2A1A10052520), Global Frontier Program (2014M3A6B3063709), the Yonsei University Yonsei-SNU Collaborative Research Fund of 2014 and the Yonsei University Future-leading Research Initiative of 2014. J.H.S., H.H. and M.-H.J. were supported by Institute for Basic Science (IBS), Korea under the contract number of IBS-R014-G1-2016-a00.

Author contributions

S.C. and H.C. conceived the idea and designed the experiments. H.H., J.H.S. and M.-H.J. fabricated and characterized vapour-phase-grown MoS$_2$ monolayer crystals; and S.C., S.S. and J.P. conducted the ultrafast optical pump–probe spectroscopy. S.C., J.H.S. and S.S. analysed the results. All authors discussed the results and prepared the manuscript.

Additional information

Competing financial interests: The authors declare no competing financial interests.

17

A metallic mosaic phase and the origin of Mott-insulating state in 1T-TaS$_2$

Liguo Ma[1,2,*], Cun Ye[1,2,*], Yijun Yu[1,2], Xiu Fang Lu[2,3,4], Xiaohai Niu[1,2], Sejoong Kim[5], Donglai Feng[1,2], David Tománek[6], Young-Woo Son[5], Xian Hui Chen[2,3,4] & Yuanbo Zhang[1,2]

Electron–electron and electron–phonon interactions are two major driving forces that stabilize various charge-ordered phases of matter. In layered compound 1T-TaS$_2$, the intricate interplay between the two generates a Mott-insulating ground state with a peculiar charge-density-wave (CDW) order. The delicate balance also makes it possible to use external perturbations to create and manipulate novel phases in this material. Here, we study a mosaic CDW phase induced by voltage pulses, and find that the new phase exhibits electronic structures entirely different from that of the original Mott ground state. The mosaic phase consists of nanometre-sized domains characterized by well-defined phase shifts of the CDW order parameter in the topmost layer, and by altered stacking relative to the layers underneath. We discover that the nature of the new phase is dictated by the stacking order, and our results shed fresh light on the origin of the Mott phase in 1T-TaS$_2$.

[1] State Key Laboratory of Surface Physics and Department of Physics, Fudan University, Shanghai 200433, China. [2] Collaborative Innovation Center of Advanced Microstructures, Nanjing 210093, China. [3] Hefei National Laboratory for Physical Science at Microscale and Department of Physics, University of Science and Technology of China, Hefei, Anhui 230026, China. [4] Key Laboratory of Strongly Coupled Quantum Matter Physics, Chinese Academy of Sciences, School of Physical Sciences, University of Science and Technology of China, Hefei 230026, China. [5] Korea Institute for Advanced Study, Hoegiro 85, Seoul 02455, Korea. [6] Physics and Astronomy Department, Michigan State University, East Lansing, Michigan 48824, USA. * These authors contributed equally to this work. Correspondence and requests for materials should be addressed to Y.Z. (email: zhyb@fudan.edu.cn).

When the correlation between electrons become predominant, the interaction may lead to the localization of electrons in materials with half-filled energy bands, and turn the otherwise metallic systems into insulators[1]. Such insulator (the so called Mott insulator, MI), therefore, serves as an ideal starting point for the study of strongly correlated electron systems. Indeed, MI and its transition to metallic state (commonly referred to as metal–insulator transition) form the basis of our understanding of various magnetic phenomena[2,3] and high-temperature superconductivity[2,4,5].

The correlation effects play an important role in layered transition metal dichalcogenide 1T-TaS$_2$, which is believed to turn into a MI after a series of charge-density-wave (CDW) phase transitions as the temperature is lowered[6-10]. The insulating ground state of 1T-TaS$_2$, however, differs from typical MIs in that it resides inside a commensurate CDW (CCDW) state. As a result the localization centres in 1T-TaS$_2$ are CDW superlattices, instead of atomic sites found in conventional MIs; there is also no apparent magnetic ordering accompanying the insulating ground state in 1T-TaS$_2$ (refs 8,11–13). Meanwhile, because of the close proximity of the various competing charge-ordered phases in energy, external perturbations can effectively modulate the CCDW (and thus the Mott phase) in 1T-TaS$_2$ and induce a myriad of phase transitions[14-30]. 1T-TaS$_2$ is therefore well suited to be a test bed for MI and other related strongly correlated phases.

In this study, we use voltage pulses from the tip of a scanning tunnelling microscope (STM) to create a mosaic CDW state out of the insulating ground state of 1T-TaS$_2$ following a procedure described in refs 18,31. We found that the mosaic state exhibits a metallic behaviour that is fundamentally different from the parent insulating state. Atomically resolved mapping of the mosaic metallic (MM) phase uncovers the root of such difference: each domain in the top layer of the mosaic phase is characterized by well-defined phase shift of the CDW order parameter with respect to neighbouring domains, and to the layer underneath; the altered stacking of CDW superlattice dictates whether the new phase is insulating or MM phase. Our results therefore provide fresh insight to the origin of the insulating ground state in 1T-TaS$_2$ which has so far been shrouded in controversies[32-34]. Moreover, we find that the MM phase created at low temperature is metastable in nature: it switches back into the insulating phase after a thermal cycle. Such observation links the MM phase to the metastable phases of 1T-TaS$_2$ induced by ultra-fast laser pulses[24] and current excitation[25,28-30]. Our result may therefore provide a microscopic understanding for those novel phases that are of importance in practical applications.

Results

Voltage-pulse-induced MM phase. 1T-TaS$_2$ bulk crystal has a layered structure, with each unit layer composed of a triangular lattice of Ta atoms, sandwiched by S atoms in an octahedral coordination. The Ta lattice is susceptible to in-plane David-star deformation where 12 Ta atoms contract towards a central Ta site[6] as illustrated in Fig. 1a. Below 180 K the crystal enters a CCDW phase, where the David-stars become fully interlocked, forming a $\sqrt{13} \times \sqrt{13}$ superlattice[8]. Such commensurate lattice modulation is accompanied by electronic reconstructions, which split the Ta 5d-band into several submanifolds, leaving exactly one conduction electron per David-star. Strong electron–electron interaction further localize these electrons, and leads to an insulating ground state[8].

Displayed in Fig. 1b is the surface of 1T-TaS$_2$ in the MI-CCDW phase imaged by STM at 6.5 K. A prominent CCDW superlattice is clearly resolved with each of the bright spots corresponding to a

CDW cluster. Close examination of individual cluster reveals the position of the S atoms in the topmost layer (Fig. 1b, inset), which bulges vertically to accommodate the distortion of the Ta lattice, and forms truncated triangles located directly above the David-stars (marked red, Fig. 1b, inset). The in-plane CDW charge modulation $\delta\rho$ can be described by a set of complex order parameters, $\phi_i(\mathbf{r}) = \Delta_i(\mathbf{r}) \cdot e^{\theta_i(\mathbf{r})}$, such that $\delta\rho = \mathrm{Re}\left[\sum_{i=1,2,3} \exp(i\mathbf{Q}_i \cdot \mathbf{r}) \cdot \phi_i(\mathbf{r}) \right]$. Here $\mathbf{Q}_i$ ($i = 1, 2, 3$) are the three in-plane reciprocal lattice vectors associated with the $\sqrt{13} \times \sqrt{13}$ supermodulation, and $\theta_i(\mathbf{r})$ and $\Delta_i(\mathbf{r})$ represent the phase and amplitude of the CDW charge order, respectively. The regular triangular superlattice seen in Fig. 1b indicates a ground state with uniform ϕ_i in 1T-TaS$_2$.

The insulating ground state makes drastic transition to a mosaic state when subjected to a voltage pulse applied across the tip-sample junction at low temperatures. A patch of such pulse-generated mosaic state is presented in Fig. 1c. Inside the patch the originally homogeneous CCDW superlattice disintegrates into nanometre-sized domains separated by well-defined domain wall textures. Within the domains the commensurate David-star configuration is strictly preserved (Fig. 1c, inset), whereas the phase of the CDW order $\theta_i(\mathbf{r})$ undergoes abrupt change across the domain walls, which we shall discuss later.

A detailed study of the pulse parameter is summarized in Supplementary Fig. 1. Empirically, both positive and negative pulses are capable of triggering the transition (Supplementary Fig. 2), and higher pulse voltages tend to create MM states on a larger area, ranging from tens of nanometres to sub-micrometres in diameter (Supplementary Fig. 3). Such mosaic phase is distinctly different from the nearly commensurate CDW phase existing at higher temperatures (refs 9,35 and Supplementary Fig. 4), but shows strong similarity to the supercooled nearly commensurate CDW state at low temperatures (Supplementary Fig. 5).

The electronic structure of the new mosaic state is fundamentally different from that of the original insulating ground state. Whereas the pristine CCDW ground state is an insulator featuring a 430-meV energy gap (Fig. 2c, black), the mosaic state is of a metallic nature with finite local density of states (DOS) around the Fermi level both inside the domains (Fig. 2c, red) and on the domain walls (Fig. 2c, blue). Such a distinction is clearly captured by the differential conductance (dI/dV) spectral waterfall acquired across a MM–insulator interface shown in Fig. 2b, where a sharp metal–insulator transition occurs within a superlattice unit cell. The transition from MM to insulator is accompanied by prominent deformation of the energy bands on the insulator side, which we attribute to the combined space charging and tip-induced band-bending effect similar to that at a semiconductor–metal interface (Supplementary Figs 6 and 7).

Metastable nature of the MM phase. Even though the MM state appears stable at low temperatures, we find that the state is in fact metastable in nature. Figure 3a displays a typical pulse-induced MM patch surrounded by the pristine insulating state. No change of the MM patch was observed after weeks of intensive imaging and spectroscopic measurements at $T = 6.5$ K. On increasing the temperature, however, the MM patch becomes unstable and the domain structure melts away. Figure 3b display the same area of the sample surface as shown in Fig. 3a, but at an elevated temperature of $T = 46$ K. The dense aggregation of domain walls dissolves, leaving an ordered CCDW superlattice decorated with sparse boundary lines, part of which traces the low-temperature domain walls (Fig. 3b, dashed lines). Meanwhile, the insulating

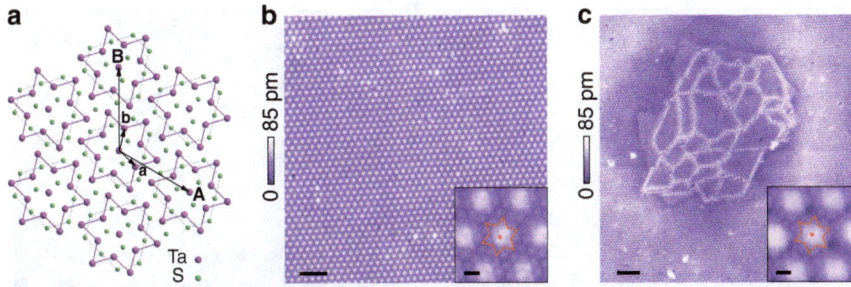

Figure 1 | MM state induced from CCDW ground state in 1T-TaS2 by voltage pulses. (a) Schematic of a monolayer 1T-TaS$_2$ crystal viewed from top. Only the topmost layer of S atoms are shown here for clarity. The interlocked clusters of Ta atoms (the David-star) in the CCDW state are sketched in purple, and the grey lines outline the S atoms accompanying each David-star. A and B are in-plane basis vectors of the CCDW superlattice, whereas a and b are the basis of the underlying atomic lattice. (b) STM topography of the cleaved surface of pristine 1T-TaS$_2$. The $\sqrt{13} \times \sqrt{13}$ triangular CCDW superlattice is resolved at 6.5 K. Scale bar, 5 nm. Inset: a zoomed-in view of the CCDW order. Individual S atoms are resolved as small blobs on each CDW cluster, which enable us to locate the Ta atoms underneath in David-star formation (red). Scale bar, 0.5 nm. (c) Topographical image of an MM patch generated by a 2.8-V voltage pulse in a MI background. Scale bar, 10 nm. Inset: CCDW charge order of the MM state with the same David-star formation. Scale bar, 0.5 nm. STM images were taken under the tunnelling condition $V_t = 0.5$ V and $I_t = 0.1$ nA (main panels), or $V_t = 0.04$ V and $I_t = 2$ nA (insets).

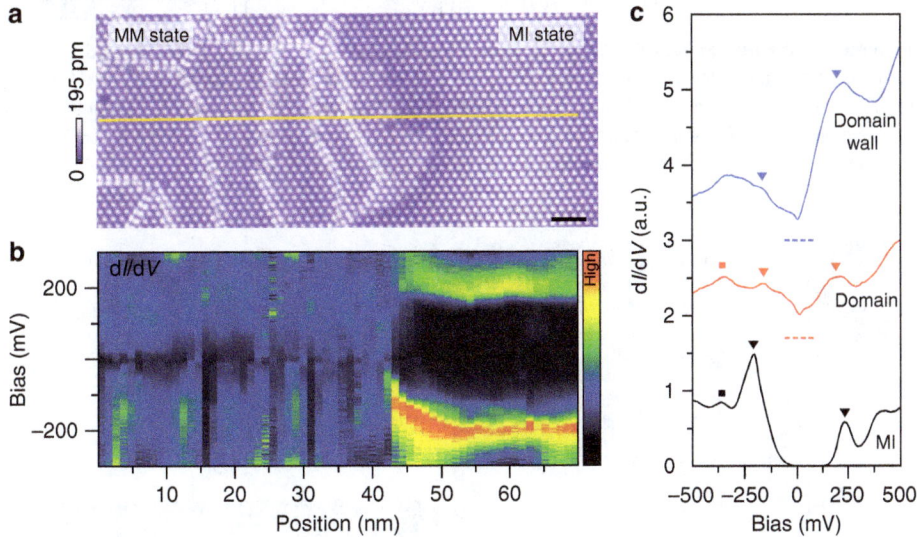

Figure 2 | Electronic structure of the MM state. (a) STM topography of the MM state interfaced with MI state. Scale bar, 5 nm. Image was recorded under tunnelling condition $V_t = 0.3$ V and $I_t = 0.1$ nA. (b) Differential conductance (dI/dV) as a function of sample bias (vertical axis) and distance (horizontal axis) measured along the yellow line in a. (c) Spatially averaged dI/dV spectra acquired in MI state (black), MM domains (red) and domain walls (blue). Curves are vertically shifted for clarity. The squares signify the spectral peaks corresponding to the submanifolds associated with the CCDW formation[42]. The triangles label the position of the Hubbard bands in MI state (black), or the edges of the V-shaped DOS suppression in MM state, which are reminiscent of the Hubbard bands (red and blue).

behaviour fully recovers over the entire surface except on those boundary lines. Cooling down the sample again does not bring back the MM state, and the insulating state (with boundary lines) persists to low temperatures. Such hysteretic behaviour unambiguously demonstrates that the MM state is metastable. The metastable nature of the MM state is further corroborated by the state's fragility at elevated temperatures: the border of the metallic state gradually recedes when perturbed by repeated scanning of an STM tip at $T = 30$ K (tunnelling voltage and current is $V_t = 150$ mV and $I_t = 10$ pA, respectively) (Fig. 3c–f). Finally, we note that the MM phase observed in our experiment may be intimately linked to the metastable metallic states induced by various macroscopic techniques, including both optical excitation[24] and carrier injection[25,28–30], from the ground state of 1T-TaS$_2$. Our STM study may therefore provide crucial microscopic understanding of those phases for the first time.

Phase configuration of the MM domains. A key feature of the MM state is the constant phase of the order parameter θ_i inside each domain, and abrupt change in θ_i across the domain walls. Shown in Fig. 4a–c,e–g are two most common types of domain wall observed in the MM state. The atomically resolved STM images recorded at low bias ($V_t = 15$ mV) enable us to precisely determine the positions of the David-stars on the domain wall. It turns out that the superlattices of David-stars in the neighbouring domains are shifted against each other by a lattice vector of the underlying two-dimensional atomic crystal, **T**, so that the phase difference between the two domains can be written as

$$\Delta\theta_i = \mathbf{Q}_i \cdot \mathbf{T} \qquad (1)$$

T takes the value of $-\mathbf{a} + \mathbf{b}$ and $-\mathbf{b}$ (a and b are basis vectors of 1T-TaS$_2$ two-dimensional crystal) for domain walls shown in Fig. 4a–c,e–g, respectively. As a result two columns of David-stars form the domain wall in an edge-sharing (Fig. 4d) or

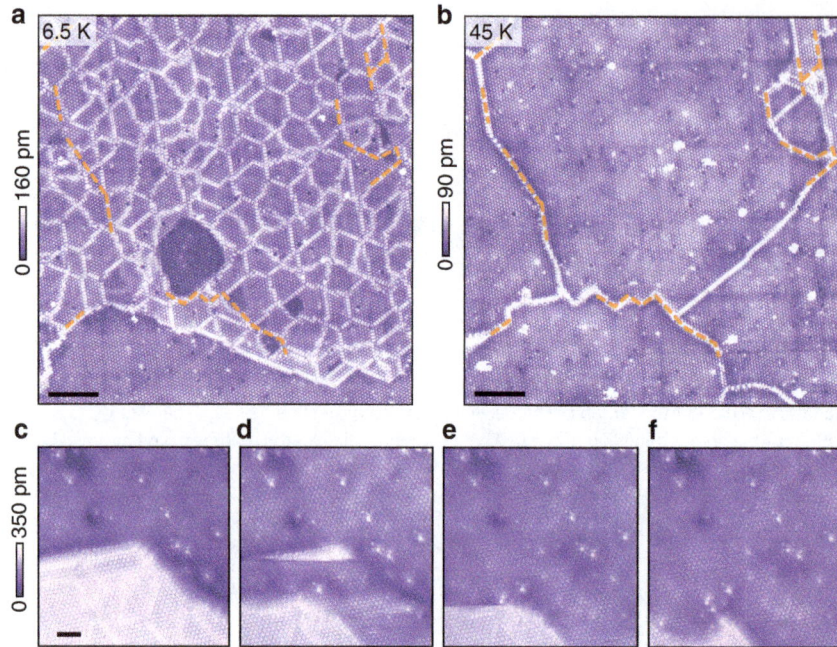

Figure 3 | The metastable nature of the MM state. (a) Low-temperature STM topography of the MM state recorded at $T = 6.5$ K. Scale bar, 20 nm. **(b)** Topographical image taken over the same area as in **a**, but at an elevated temperature of $T = 46$ K. Scale bar, 20 nm. The ramping rate of temperature was kept under 0.2 K min^{-1}. Part of the sparse residual boundaries (marked by red dash lines) is inherited from original domain walls in the low-temperature MM state as shown in **a**. **(c-f)** An MM patch gradually converted to MI state by repeated scanning at $T = 30$ K. Scale bar, 5 nm. **c-f** are in chronological order, and the data were taken over a 6-min period. STM scanning condition: $V_t = 0.15$ V and $I_t = 0.01$ nA.

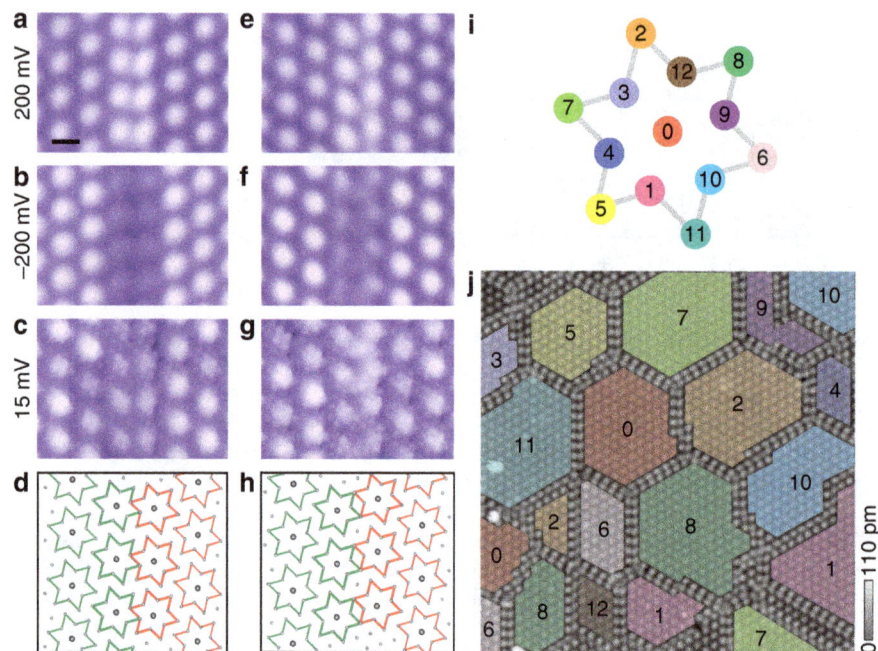

Figure 4 | In-plane domain/phase configuration of the CCDW order in the MM phase. (a-c,e-g) STM topography two common types of CDW domain walls recorded at varying sample biases. Scale bar, 1 nm. The atomically resolved images at low bias enabled us to establish the structure model of both types of domain walls **(d,h)**. **(i)** David-star of Ta atoms in the unit cell of the CCDW superlattice. The atoms are numbered following the convention described in ref. 36. Each number represents 1 of the 12 possible relative translations (and corresponding phase differences) of the CCDW superlattices in neighbouring domains. Two consecutive translations can be represented by the difference of the two numbers. **(j)** Topography of an MM state showing multiple domains. The phase of each mosaic domains (relative to the domain **0**) are determined from analysis of domain walls using procedures similar to **a**. The phases are coded by the numbers defined in **i** (see text).

corner-sharing (Fig. 4h) configuration. No rotation of the David-star triangular lattice was observed on the domain wall, nor was any defect in the underlying atomic crystal.

In fact, surveys of the MM state reveal that equation (1) describes all the domain walls observed in our experiment (Supplementary Figs 8 and 9), which enables us to completely determine the phase configuration of the MM domains. The relative translation $\mathbf{T}$ in general takes the form $\mathbf{T} = m\mathbf{a} + n\mathbf{b}$, where m and n take integer values such that a $\mathbf{T}$ sitting at the central Ta atom runs through 12 other Ta on the same David-star (Fig. 4i). There are therefore 12 possible types of domain walls with associated phase difference $(\Delta\theta_1, \Delta\theta_2, \Delta\theta_3) = 2\pi(3m + n, -4m + 3n, m - 4n)/13$ (Supplementary Fig. 8). Equivalently, the 12 possible $\mathbf{T}$ can be labelled by the Ta atom that characterizes the translation (Fig. 4i). Here the Ta atoms (and therefore the $\mathbf{T}$) are numbered **0** ... **12** following the convention adopted in ref. 36, which has the added advantage that two consecutive translations are represented by the difference of the two numbers. Armed with above analysis, we are able to completely delineate the phase configuration of an MM state, such as the one shown in Fig. 4j. Here the phase of all the domains is referenced to the domain **0**, and the relative phase shift between two domains is readily unscrambled by taking the difference of their domain numbers. Finally, we note that out of the 12 possibilities only 4 types of domain walls are observed in our experiment, with varying frequency of occurrence (Supplementary Fig. 9), indicating subtle differences in energy associated with each type of domain walls.

CCDW stacking order and the electronic structure. The domain and associated phase structure distinguishes the MM state from its parent MI state, even though the two states share the same CCDW superlattice. There are two main points to notice. First, the MI state is parasitic to the CCDW[27], rather than a requisite as suggested by some of the previous works[37,38]. Second, phase shift of the CCDW order in one atomic layer implies a shifted CCDW superlattice relative to other layers. In fact, not only do we see phase shifts in the topmost layer, clear signature of random domain wall networks are also observed in the second layer (Fig. 5a). One example of such domain wall is shown in Fig. 5c, where a flat monolayer H-TaS$_2$ patch (also induced by the same voltage pulse) enables us to see through the top layer and resolve the domain wall's atomic structure. We find that the domain wall in the second layer also corresponds to a phase shift described by equation (1) (Fig. 5d). The presence of randomly distributed domains and phases in two adjacent layers therefore incurs a randomized stacking of the CCDW superlattices.

We are now poised to address the central question: how does the MI state emerge from CCDW, against competing metallic state in 1T-TaS$_2$? An important clue comes from rare occurrences such as the one shown in Fig. 5b, where a small insulating patch appears inside metallic random MM domains. Close examination reveals that the insulating patch is surrounded partly by the topmost-layer domain walls, and partly by the domain wall in the second layer. Such a domain wall configuration leads to a distinct stacking order of CCDW superlattice that restores the insulating state inside the patch, in contrast to the surrounding domains still in the MM state. The existence of uniform insulating patch with sharp boundaries also rules out carrier doping and disorder potential[39,40] as the cause of the insulator–metal transition. Our observations therefore point to interlayer stacking as a decisive factor in determining the electronic structure of 1T-TaS$_2$ ground state.

Unambiguous evidence directly linking the electronic structure and the CCDW stacking order comes from a reversible, tip-induced metal-to-insulator transition on a monolayer terrace on 1T-TaS$_2$ surface as shown in Fig. 6a. We were able to use gentle voltage pulses to reversibly switch the upper terrace from metallic (Fig. 6b) to insulating (Fig. 6c) and back to metallic (Fig. 6d) state at 6.5 K, while keeping the lower terrace intact. The CCDW lattice shift was carefully determined for each of the three states. It turns out that switching from metallic to insulating state corresponds to a CCDW stacking shift of $8 \rightarrow 5$ (Fig. 6e,f),

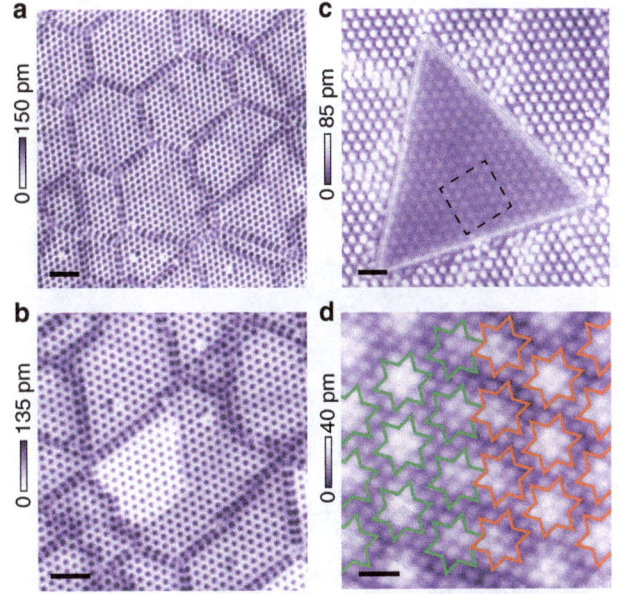

Figure 5 | Altered stacking order of the CCDW superlattice in the MM state. (a) STM topography of a 50×50-nm MM area. STM imaging condition: $V_t = 0.2$ V and $I_t = 0.1$ nA. Apart from the well-defined domain walls in the top layer, domain walls in the layer underneath are clearly resolved as networks of random filamentary features. Here an inverted colour scale is adopted for better contrast. Scale bar, 5 nm. (b) An MI patch (bright region) surrounded by MM domains in the MM state. The patch is encircled by domain walls in either the top layer or the second layer. STM imaging condition: $V_t = 0.3$ V and $I_t = 0.02$ nA. Scale bar, 5 nm. (c) A triangular patch of H-phase TaS$_2$ accidently created in 1T-TaS$_2$ by a voltage pulse. The H-phase is induced only on the topmost atomic layer, and the flat surface enabled us to see through the top layer and to observe the $\sqrt{13} \times \sqrt{13}$ CCDW superlattice as well as domain walls in the layer underneath[15,18]. Scale bar, 3 nm. (d) Zoomed-in STM image of the area marked by black dashed square in c showing atomically resolved domain wall structure in the second layer. The domain wall has a type **2** configuration. Scale bar, 8 Å.

following the notation defined in Fig. 4i, with the lower terrace taken to be the reference (**0**). Such correspondence is reversible and reproducible such that the metallic nature of the upper terrace is recovered when the stacking sequence is switched back to **8** (Fig. 6g). Our experiment therefore establish the direct link between the CCDW stacking order and the electronic structure of the domains. Finally we note that the stacking orders of $0 \rightarrow 8$ and $0 \rightarrow 5$ produce metallic and insulating states, respectively, in contradiction with the seeming degeneracy of the two stacking orders. Such degeneracy is however lifted if the CCDW unit cell deviates from a perfect David-star (slight distortions indeed appear in our *ab initio* calculations), or if there is a domain boundary in the lower layer under the step edge, which gives an additional lattice shift. Next-nearest-neighbour coupling with the third layer may also be able to lift the degeneracy.

Discussion

The importance of the interlayer stacking can be understood from a three-dimensional Hubbard model with intra- and interlayer hopping taken into account. The one-band Hamiltonian of 1T-TaS$_2$ CCDW ground state can be written as:

$$H = -\sum_{\langle ij \rangle, \sigma} t_{ij} \left(c_{i\sigma}^{\dagger} c_{j\sigma} + h.c. \right) + U \sum_i n_{i\uparrow} n_{i\downarrow} \qquad (2)$$

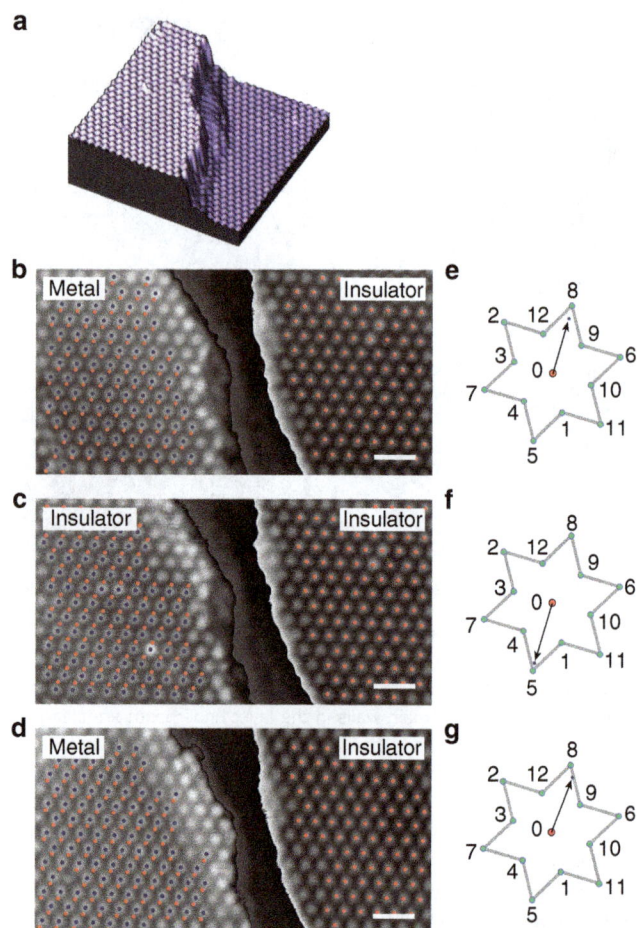

Figure 6 | Reversible switching of metal–insulator domain and the corresponding stacking order. (**a**) Three-dimensional rendering of a 30 × 30-nm STM image of a monolayer terrace on 1T-TaS$_2$ surface. The height of the terrace is 5.93 Å. Image was recorded at 6.5 K. (**b-d**) STM topography of the same area shown in **a**. The upper terrace was switched from metallic state (**b**) to insulating state (**c**) and back to metallic state (**d**), by gentle voltage pulses (~1.5-2 V, 50-ms duration). All other domains were kept intact during the pulses. The red (blue) dot array marks the CCDW lattice on the lower (upper) terrace, respectively. The relative CCDW lattice shift can be extracted by extrapolating the position of the array on the lower terrace to the upper terrace, and projecting the relative displacement onto a CCDW unit cell. Scale bars, 3 nm. (**e-g**) The CCDW lattice shift between the upper and lower terraces extracted from the analysis shown in **b-d**, respectively. The shifts (arrows and black dots) are projected relative to **0** onto the CCDW unit cell (David-star). Reversible switching of the metal–insulator domain is directly linked to the reversible switching of the CCDW stacking order.

where t_{ij} is the effective hopping between the David-stars and U the on-site Coulomb repulsion on one David-star. Because of the flat pancake shape of the David-star, the interlayer distance of the CCDW superlattice (5.9 Å) is significantly shorter than the in-plane separation between the centres of neighbouring David-stars (12.1 Å; ref. 7). Various experimental and theoretical studies[33,34,41,42] have suggested the importance of interlayer coupling. Indeed, our angle-resolved photoemission spectroscopy measurements on the pristine crystal reveals a bandwidth of $W \sim 50$ meV for the lower Hubbard band in both $\mathbf{k}_z$ and $\mathbf{k}_{||}$ direction (Supplementary Fig. 10). This observation indicates that the effective out-of-plane hopping factor $t_\perp$ is comparable to its in-plane counterpart $t_{||}$. We note that $U/W \sim 8$ for pristine 1T-TaS$_2$, a typical value for a MI ground state.

A MI to metal transition occurs on increasing the bandwidth W with respect to the Hubbard U, where W is determined by the effective hopping factors $t_\perp$ and $t_{||}$ (with coordination number taken into account). As the stacking order of the CCDW is varied, the variation in the separation between the David-stars in neighbouring layer (as well as coordination number) could bring drastic change in $t_\perp$. We speculate that in certain stacking configurations, the bandwidth W is driven beyond certain critical value, and the MI insulator to metal transition becomes a possibility. Such a speculation is supported by recent density functional theory calculations suggesting that with an altered interlayer stacking sequence, $t_\perp$ may experience an order of magnitude increase, and brings W to the same order of magnitude as U (ref. 34).

The rare occurrence of the insulating domain among the randomly stacked MM domains (Fig. 5b) implies that only a small number of stacking order yields insulating states with the rest metallic in nature. The exact interlayer stacking order of the various states, however, is not directly determined due to an intrinsic limitation of STM; tunnelling microscopy could only resolve the atomic structure of the topmost layer. Here we point out that even the stacking order in the pristine CCDW ground state remains elusive[36,43–46], and we call for further experimental and theoretical work to clarify the exact stacking order of the various charge-ordered states. Finally we note that an alternative scenario, where orbital order, instead of electron–electron interaction, dominates the electronic structure of the CCDW phase, has been proposed recently[34]. In this scenario, the stacking order (which determines the orbital order) would also dictates the electronic state of 1T-TaS$_2$. However, calculations within this framework invariably predict a dispersive band (therefore a finite DOS) crossing Fermi level in the out-of-plane direction in the pristine insulating state. The lack of such dispersive band in angle-resolved photoemission spectroscopy measurements (Supplementary Fig. 10) and the observation of a fully gapped DOS near Fermi level (Fig. 2c, black curve) are evidences against the orbital-order scenario. It remains to be seen whether theories based on orbital order could reconcile these conflicts and provide a complete, quantitative description of the various CDW states observed in 1T-TaS$_2$.

In summary, we studied a mosaic, metallic state induced from the Mott-insulating CCDW phase in 1T-TaS$_2$ by voltage pulses. The mosaic phase features a fragmented in-plane phase distribution of the CDW order parameter, and exhibits metallic behaviour. We discovered that the relative phase shift between adjacent layers leads to an altered local stacking order, which dictates whether the resulting structure is a MI or a metal. Our results therefore shed fresh light to the origin of the insulating ground state of 1T-TaS$_2$, and uncover the interlayer coupling as the root. In addition, the MM phase bears strong similarities to the metastable metallic states induced by various external perturbations such as ultra-fast laser pulses[24] and current excitations[25,28–30]. Our study may provide a microscopic understanding for those novel phases that are of importance in practical applications.

Methods

Sample preparation and STM measurements. High-quality 1T-TaS$_2$ single crystals were grown using a standard chemical vapour transport method. The samples were cleaved in high vacuum at room temperature, and subsequently cooled down for STM measurements. STM experiments were performed in a commercial low-temperature STM (Createc Fischer & Co. GmbH) operated in ultrahigh vacuum. Electrochemically etched polycrystalline tungsten tips calibrated on clean Au(111) surfaces were used for all our STM measurements. Typical tip-calibration data are shown in Supplementary Fig. 11. The STM topography was taken in the constant-current mode, and the dI/dV spectra were collected using a standard lock-in technique, with a modulation frequency of 789.1 Hz. Before we opened the feedback loop to apply voltage pulses, the tip was parked above the sample surface under typical tunnelling condition of $V_t = 0.3$–0.5 V and $I_t = 0.3$–1 nA. The pulse duration was fixed at 50 ms.

Calculation of the energy associated with various stacking order. Fully relaxed David-star geometry for single and bilayer 1T-TaS$_2$ was obtained from density functional theory calculations using Quantum Espresso Code[47]. We adopted the PBE (Perdew-Burke-Ernzerhof) generalized gradient approximation[48] for the exchange-correlation functional, the norm-conserving pseudo-potential[49], and an energy cutoff of 55 Ry and 6×6 k-points for the supercell. In the relaxed bilayer 1T-TaS$_2$, the centre of David-star in the top layer is found to shift laterally by **a** relative to the centre in the bottom layer. For trilayer 1T-TaS$_2$, the binding energy is calculated for two cases: (1) the centre of the David-star in the third (bottom) layer shifting laterally by **2a** with respect to the centre in the first (top) layer, that is, ABC stacking; and (2) the centre in the third layer aligning with that in the topmost layer, that is, ABA stacking. We computed the energy associated with various stacking orders as the David-star in one layer is displaced relative to that in other layers, and found the energy difference is on the order of 6.4 meV.

References

1. Mott, N. F. Metal-insulator transition. *Rev. Mod. Phys.* **40**, 677–683 (1968).
2. Imada, M., Fujimori, A. & Tokura, Y. Metal-insulator transitions. *Rev. Mod. Phys.* **70**, 1039 (1998).
3. Dagotto, E. Complexity in strongly correlated electronic systems. *Science* **309**, 257–262 (2005).
4. Dagotto, E. Correlated electrons in high-temperature superconductors. *Rev. Mod. Phys.* **66**, 763 (1994).
5. Lee, P. A. & Wen, X.-G. Doping a Mott insulator: physics of high-temperature superconductivity. *Rev. Mod. Phys.* **78**, 17–85 (2006).
6. Wilson, J. A., Di Salvo, F. J. & Mahajan, S. Charge-density waves and superlattices in the metallic layered transition metal dichalcogenides. *Adv. Phys.* **24**, 117–201 (1975).
7. Fazekas, P. in *Modern Trends in the Theory of Condensed Matter: Lecture Notes in Physics* vol. 115, 328–338 (Springer, 1980).
8. Fazekas, P. & Tosatti, E. Charge carrier localization in pure and doped 1T-TaS$_2$. *Physica B + C* **99**, 183–187 (1980).
9. Thomson, R. E., Burk, B., Zettl, A. & Clarke, J. Scanning tunneling microscopy of the charge-density-wave structure in 1T-TaS$_2$. *Phys. Rev. B* **49**, 16899–16916 (1994).
10. Kim, J.-J., Yamaguchi, W., Hasegawa, T. & Kitazawa, K. Observation of Mott localization gap using low temperature scanning tunneling spectroscopy in commensurate 1T-TaS$_2$. *Phys. Rev. Lett.* **73**, 2103–2106 (1994).
11. DiSalvo, F. J. & Waszczak, J. V. Paramagnetic moments and localization in 1T-TaS$_2$. *Phys. Rev. B* **22**, 4241–4246 (1980).
12. Furukawa, A., Kimishima, Y., Oda, Y., Nagano, H. & Ōnuki, Y. Magnetic behavior of 1T-TaS$_2$ in the Anderson localized state. *J. Phys. Soc. Jpn* **54**, 4300–4305 (1985).
13. Perfetti, L., Gloor, T. A., Mila, F., Berger, H. & Grioni, M. Unexpected periodicity in the quasi-two-dimensional Mott insulator 1T-TaS$_2$ revealed by angle-resolved photoemission. *Phys. Rev. B* **71**, 153101 (2005).
14. Pettenkofer, C. & Jaegermann, W. Charge-density-wave transformation induced by Na intercalation into 1T-TaS$_2$. *Phys. Rev. B* **50**, 8816–8823 (1994).
15. Kim, J.-J. *et al.* Observation of a phase transition from the T phase to the H phase induced by a STM tip in 1T-TaS$_2$. *Phys. Rev. B* **56**, R15573–R15576 (1997).
16. Endo, T., Yamaguchi, W., Shiino, O., Hasegawa, T. & Kitazawa, K. Anomalous domain structure in 1T-TaS$_{2-x}$Se$_x$ observed using scanning tunneling microscopy. *Surf. Sci.* **453**, 1–8 (2000).
17. Yamaguchi, W., Shiino, O., Endo, T., Kitazawa, K. & Hasegawa, T. Microtip-assisted metal–insulator transition in a layered chalcogenide. *Appl. Phys. Lett.* **76**, 517–519 (2000).
18. Kim, P., Zhang, J. & Lieber, C. M. Charge density wave formation in nanocrystals. *Solid State Phys.* **55**, 119–157 (2001).
19. Sipos, B. *et al.* From Mott state to superconductivity in 1T-TaS$_2$. *Nat. Mater.* **7**, 960–965 (2008).
20. Rossnagel, K. Suppression and emergence of charge-density waves at the surfaces of layered 1T-TiSe$_2$ and 1T-TaS$_2$ by in situ Rb deposition. *New J. Phys.* **12**, 125018 (2010).
21. Ang, R. *et al.* Real-space coexistence of the melted mott state and superconductivity in fe-substituted 1T-TaS$_2$. *Phys. Rev. Lett.* **109**, 176403 (2012).
22. Ritschel, T. *et al.* Pressure dependence of the charge density wave in 1T-TaS$_2$ and its relation to superconductivity. *Phys. Rev. B* **87**, 125135 (2013).
23. Ang, R. *et al.* Superconductivity and bandwidth-controlled Mott metal-insulator transition in 1T-TaS$_{2-x}$Se$_x$. *Phys. Rev. B* **88**, 115145 (2013).
24. Stojchevska, L. *et al.* Ultrafast switching to a stable hidden quantum state in an electronic crystal. *Science* **344**, 177–180 (2014).
25. Vaskivskyi, I. *et al.* Fast non-thermal switching between macroscopic charge-ordered quantum states induced by charge injection. Preprint at http://arxiv.org/abs/1409.3794 (2014).
26. Yoshida, M. *et al.* Controlling charge-density-wave states in nano-thick crystals of 1T-TaS$_2$. *Sci. Rep.* **4**, 7302 (2014).
27. Yu, Y. *et al.* Gate-tunable phase transitions in thin flakes of 1T-TaS$_2$. *Nat. Nanotechnol.* **10**, 270–276 (2015).
28. Hollander, M. J. *et al.* Electrically driven reversible insulator–metal phase transition in 1T-TaS$_2$. *Nano Lett.* **15**, 1861–1866 (2015).
29. Tsen, A. W. *et al.* Structure and control of charge density waves in two-dimensional 1T-TaS$_2$. *Proc. Natl Acad. Sci. USA* **112**, 15054–15059 (2015).
30. Yoshida, M., Suzuki, R., Zhang, Y., Nakano, M. & Iwasa, Y. Memristive phase switching in two-dimensional 1T-TaS$_2$ crystals. *Science Advances* **1**, e1500606 (2015).
31. Cho, D. *et al.* Nanoscale manipulation of the Mott insulating state coupled to charge order in 1T-TaS$_2$. *Nat. Commun.* **7**, 10453 (2016).
32. Perfetti, L. *et al.* Time evolution of the electronic structure of 1T-TaS$_2$ through the insulator-metal transition. *Phys. Rev. Lett.* **97**, 067402 (2006).
33. Darancet, P., Millis, A. J. & Marianetti, C. A. Three-dimensional metallic and two-dimensional insulating behavior in octahedral tantalum dichalcogenides. *Phys. Rev. B* **90**, 045134 (2014).
34. Ritschel, T. *et al.* Orbital textures and charge density waves in transition metal dichalcogenides. *Nat. Phys.* **11**, 328–331 (2015).
35. Wu, X. L. & Lieber, C. M. Hexagonal domain-like charge density wave phase of TaS$_2$ determined by scanning tunneling microscopy. *Science* **243**, 1703–1705 (1989).
36. Nakanishi, K. & Shiba, H. Theory of three-dimensional orderings of charge-density waves in 1T-TaX$_2$ (X: S, Se). *J. Phys. Soc. Jpn* **53**, 1103–1113 (1984).
37. Eichberger, M. *et al.* Snapshots of cooperative atomic motions in the optical suppression of charge density waves. *Nature* **468**, 799–802 (2010).
38. Petersen, J. C. *et al.* Clocking the melting transition of charge and lattice order in 1T-TaS$_2$ with ultrafast extreme-ultraviolet angle-resolved photoemission spectroscopy. *Phys. Rev. Lett.* **107**, 177402 (2011).
39. Heidarian, D. & Trivedi, N. Inhomogeneous metallic phase in a disordered mott insulator in two dimensions. *Phys. Rev. Lett.* **93**, 126401 (2004).
40. Lahoud, E., Meetei, O. N., Chaska, K. B., Kanigel, A. & Trivedi, N. Emergence of a novel pseudogap metallic state in a disordered 2D Mott insulator. *Phys. Rev. Lett.* **112**, 206402 (2014).
41. Spijkerman, A., de Boer, J. L., Meetsma, A., Wiegers, G. A. & van Smaalen, S. X-ray crystal-structure refinement of the nearly commensurate phase of 1T-TaS$_2$ in $(3 + 2)$-dimensional superspace. *Phys. Rev. B* **56**, 13757–13767 (1997).
42. Bovet, M. *et al.* Interplane coupling in the quasi-two-dimensional 1T-TaS$_2$. *Phys. Rev. B* **67**, 125105 (2003).
43. Scruby, C. B., Williams, P. M. & Parry, G. S. The role of charge density waves in structural transformations of 1T-TaS$_2$. *Philos. Mag.* **31**, 255–274 (1975).
44. Tanda, S., Sambongi, T., Tani, T. & Tanaka, S. X-ray study of charge density wave structure in 1T-TaS$_2$. *J. Phys. Soc. Jpn* **53**, 476–479 (1984).
45. Naito, M., Nishihara, H. & Tanaka, S. Nuclear quadrupole resonance in the charge density wave state of 1T-TaS$_2$. *J. Phys. Soc. Jpn* **53**, 1610–1613 (1984).
46. Ishiguro, T. & Sato, H. Electron microscopy of phase transformations in 1T-TaS$_2$. *Phys. Rev. B* **44**, 2046–2060 (1991).
47. Giannozzi, P. *et al.* QUANTUM ESPRESSO: a modular and open-source software project for quantum simulations of materials. *J. Phys. Condens. Matter* **21**, 395502 (2009).
48. Perdew, J. P., Burke, K. & Ernzerhof, M. Generalized gradient approximation made simple. *Phys. Rev. Lett.* **77**, 3865–3868 (1996).
49. Troullier, N. & Martins, J. L. Efficient pseudopotentials for plane-wave calculations. *Phys. Rev. B* **43**, 1993–2006 (1991).

Acknowledgements

We thank S.W. Cheong and Y.-H. Cho for providing some of the 1T-TaS$_2$ crystals used in this study, and P. Kim, J. Guan for discussions. L.M., C.Y., Y.Y. and Y.Z. acknowledge financial support from the National Basic Research Program of China (973 Program; grant no. 2013CB921902) and from the NSF of China (grant nos. 11425415 and 11421404). X.F.L. and X.H.C. are supported by the 'Strategic Priority Research Program (B)' of the Chinese Academy of Sciences (grant no. XDB04040100) and the NSF of China (grant no. 11190021). Y.-W.S. is supported by the NRF of Korea grant funded by MEST (no. 2015001948). D.T. received partial financial support from Fudan University, the NSF/AFOSR EFRI 2-DARE grant number #EFMA-1433459, and acknowledges the hospitality of Fudan University while performing this research. X.N. and D.F. are supported by National Basic Research Program of China (973 Program; grant no. 2012CB921402).

Author contributions

Y.Z. conceived the project; L.M., C.Y. and Y.Y. performed the STM experiments and analysed the data; X.F.L. and X.-H.C. grew 1T-TaS$_2$ single crystals; X.N. and D.F. performed angle-resolved photoemission spectroscopy measurement on bulk 1T-TaS$_2$ crystal; S.K. and Y.-W.S. carried out *ab initio* calculations; D.T. contributed to the understanding of the metastable nature of the MM phase; L.M., C.Y. and Y.Z. wrote the paper and all authors commented on it.

Additional information

Competing financial interests: The authors declare no competing financial interests.

18

Super-resolution microscopy reveals structural diversity in molecular exchange among peptide amphiphile nanofibres

Ricardo M.P. da Silva[1,2,3], Daan van der Zwaag[2], Lorenzo Albertazzi[2,4], Sungsoo S. Lee[5], E.W. Meijer[2] & Samuel I. Stupp[1,5,6,7,8]

The dynamic behaviour of supramolecular systems is an important dimension of their potential functions. Here, we report on the use of stochastic optical reconstruction microscopy to study the molecular exchange of peptide amphiphile nanofibres, supramolecular systems known to have important biomedical functions. Solutions of nanofibres labelled with different dyes (Cy3 and Cy5) were mixed, and the distribution of dyes inserting into initially single-colour nanofibres was quantified using correlative image analysis. Our observations are consistent with an exchange mechanism involving monomers or small clusters of molecules inserting randomly into a fibre. Different exchange rates are observed within the same fibre, suggesting that local cohesive structures exist on the basis of β-sheet discontinuous domains. The results reported here show that peptide amphiphile supramolecular systems can be dynamic and that their intermolecular interactions affect exchange patterns. This information can be used to generate useful aggregate morphologies for improved biomedical function.

[1] Simpson Querrey Institute for BioNanotechnology (SQI), Northwestern University, Chicago, Illinois 60611, USA. [2] Laboratory of Macromolecular and Organic Chemistry and Institute for Complex Molecular Systems, Eindhoven University of Technology, Eindhoven MB 5600, The Netherlands. [3] Craniofacial Development & Stem Cell Biology, King's College London, London, SE1 9RT, UK. [4] Nanoscopy for Nanomedicine Group, Institute for Bioengineering of Catalonia (IBEC), Barcelona 08028, Spain. [5] Department of Materials Science and Engineering, Northwestern University, Evanston, Illinois 60208, USA. [6] Department of Chemistry, Northwestern University, Evanston, Illinois 60208, USA. [7] Department of Medicine, Northwestern University, Chicago, Illinois 60611, USA. [8] Department of Biomedical Engineering, Northwestern University, Evanston, Illinois 60208, USA. Correspondence and requests for materials should be addressed to E.W.M. (email: e.w.meijer@tue.nl) or to S.I.S. (email: s-stupp@northwestern.edu).

Reversible supramolecular interactions are ubiquitous in nature, controlling the self-assembly of ordered functional structures that need to be dynamic to perform their biological functions. One-dimensional cytoskeletal filaments such as actin and tubulin are typical examples of structures that use dynamics to mediate the adaptive behaviour of cells, resulting in cell motility, shape change, cell division, signalling and muscular contraction at larger length scales[1-6]. Artificial supramolecular materials could offer this bio-inspired dynamic behaviour, thus allowing enhanced interaction with natural systems and increased biomedical functionality. Since many natural processes are carefully regulated, optimization of an artificial supramolecular material requires a detailed understanding of its dynamic properties, for example its exchange kinetics.

Molecular mixing experiments typically assess exchange kinetics by utilizing Förster resonance energy transfer (FRET) between a pair of donor and acceptor fluorophores[7,8], and alternatively, radio-labelled molecules[9] or time-resolved small-angle neutron scattering[10,11]. Although these ensemble experiments can provide the timescale of the processes, they cannot distinguish different mechanisms. Moreover, they fail to detect the structural diversity among fibres, or within an individual fibre, for example, the occurrence of segregated domains. Local variations of molecular composition can have important biological implications, as they can greatly influence the signalling potency through multivalency effects[12-14], and thus understanding the exchange heterogeneity is important to design materials in which function is connected with dynamics for adaptive or responsive behaviour[15].

Super-resolution techniques are powerful tools to reveal the spatial distribution of molecules at the nanoscale, but these techniques have thus far been mainly applied to imaging fine details of cellular structures[16,17]. For instance, a resolution of ~20 nm can be achieved using stochastic optical reconstruction microscopy (STORM)[18], which is an order of magnitude below the diffraction limit and near the molecular scale. The enhanced resolution is achieved through the accurate localization of single fluorescent molecules; to identify individual fluorophores only a sparse subset of labels should be active at any given time. This sparse population of fluorophores is obtained using probes that can be photo-switched to a temporary non-fluorescent 'off' state by light. By repeatedly activating different subsets and overlaying the resulting localizations, an image can be reconstructed. We have previously reported how to apply STORM to probe the dynamics of supramolecular fibres[19]. Using two-colour STORM and quantitative image analysis, we were able to resolve the monomer distribution along the fibre backbone. By following the monomer distribution during the molecular exchange process, we were able to infer the exchange mechanism.

Peptide amphiphiles (PAs) that self-assemble into high aspect ratio objects offer exciting opportunities for regenerative medicine and other therapeutic applications[20-24]. As illustrated in Fig. 1a, this class of molecules is composed of an unbranched alkyl chain linked to a peptide segment, which can be further subdivided in several domains. A segment with propensity to form β-sheets is conjugated to the alkyl tail, followed by a charged segment for solubility. Additional domains can be conjugated to the canonical structure to introduce biofunctionality in the nanofibres. Whereas the hydrophobic collapse of the aliphatic tails induces self-assembly, experimental[25,26] and theoretical[27] evidence suggests that the formation of directional hydrogen bonds within the β-sheet domain is an additional important component of the driving force for assembly of the molecules into one-dimensional filamentous shapes. The facile incorporation of multiple bioactive signals at controlled concentrations[14,28], together with their structural resemblance with extracellular matrix fibres makes PA assemblies useful as bioactive artificial extracellular matrix components for cell signalling. Furthermore, they are also intrinsically biocompatible and biodegradable and can therefore disappear easily after fulfilling their biological functions. PAs have been extensively studied as a platform for applications that include bone, cartilage, enamel, neuronal regeneration, angiogenesis for ischaemic disease, targeted drug delivery and cancer therapeutics[20-24,29].

Ensemble measurements of PA dynamics revealed partial self-healing using rheological techniques[30] or the existence of a critical aggregation concentration[31], while kinetic spectroscopy showed pathway selection of PAs into different morphologies[32,33]. However, these approaches do not consider structural heterogeneity and its effect on dynamic molecular exchange at the level of individual filaments. In this paper, we use STORM to image individual PA nanofibres, addressing the distribution of molecules along the fibre during exchange and therefore analysing diversity among supramolecular nanofibres.

Results

Nanofibre design and preparation. The PA molecule studied in this work is shown in Fig. 1a. It consists of a palmitic acid tail, six alanines in the β-sheet-forming region, followed by three glutamic acids as charged solubilizing moieties. This PA molecule self-assembles in water, forming nanofibres around 7 nm in diameter and lengths in the range of micrometres as observed by cryogenic transmission electron microscopy (cryoTEM; Fig. 1e). Amino acids with different propensities to form β-sheets affect considerably the internal order of PA assemblies, as well as nanofibre stiffness[26]. Since alanine has a weaker tendency to form β-sheets than valine[34], placing this amino acid in the β-sheet-forming segment was expected to yield relatively dynamic PA fibres. Circular dichroism spectroscopy of this PA is consistent with the typical β-sheet conformation found for other PAs[26], at physiological pH and ionic strength (Fig. 1c). To perform imaging by fluorescence microscopy, PA molecules were labelled with cyanine dyes, namely Cy3 and Cy5 (Fig. 1b). Water soluble sulfonated dyes with a net charge of −1 were used to preserve the amphiphilic asymmetry of the PA molecule and, since sulfonate groups are fully ionized in the vicinity of physiological pH, their behaviour is insensitive to pH in the range of interest. The Cy3 and Cy5 dyes have been chosen for their suitable photophysical properties for STORM imaging; moreover, they constitute a good FRET pair with a Förster radius of 50 Å (ref. 35). Single-colour labelled nanofibres were created by mixing a stock solution of either Cy3-PA or Cy5-PA with a stock solution of unlabelled PA, lyophilization, brief co-dissolution in trifluoroacetic acid (TFA) and immediate TFA removal, lyophilization and final reconstitution in the aqueous working solution to form labelled supramolecular aggregates. The degree of labelling was accurately controlled by premixing the different PAs at the desired ratios. Fluorescently labelled nanofibres revealed a morphology that was indistinguishable from their non-labelled counterparts, as shown by cryoTEM (Fig. 1f).

Ensemble molecular exchange kinetics. The timescale of molecular exchange in PA-based nanofibres was measured using FRET kinetic experiments. Two sets of PA nanofibres were separately pre-assembled from a mixture of non-labelled PAs with either Cy3-PA (0.5%) or Cy5-PA (0.5%), as illustrated in Fig. 2a. Next, the two solutions were mixed and the FRET ratio, defined as the ratio between the fluorescence intensities of Cy5 acceptor and Cy3 donor, was monitored over time. As shown in Fig. 2b, the FRET ratio increases with time, reaching a plateau after several hours. This means that PA molecules are able to migrate between

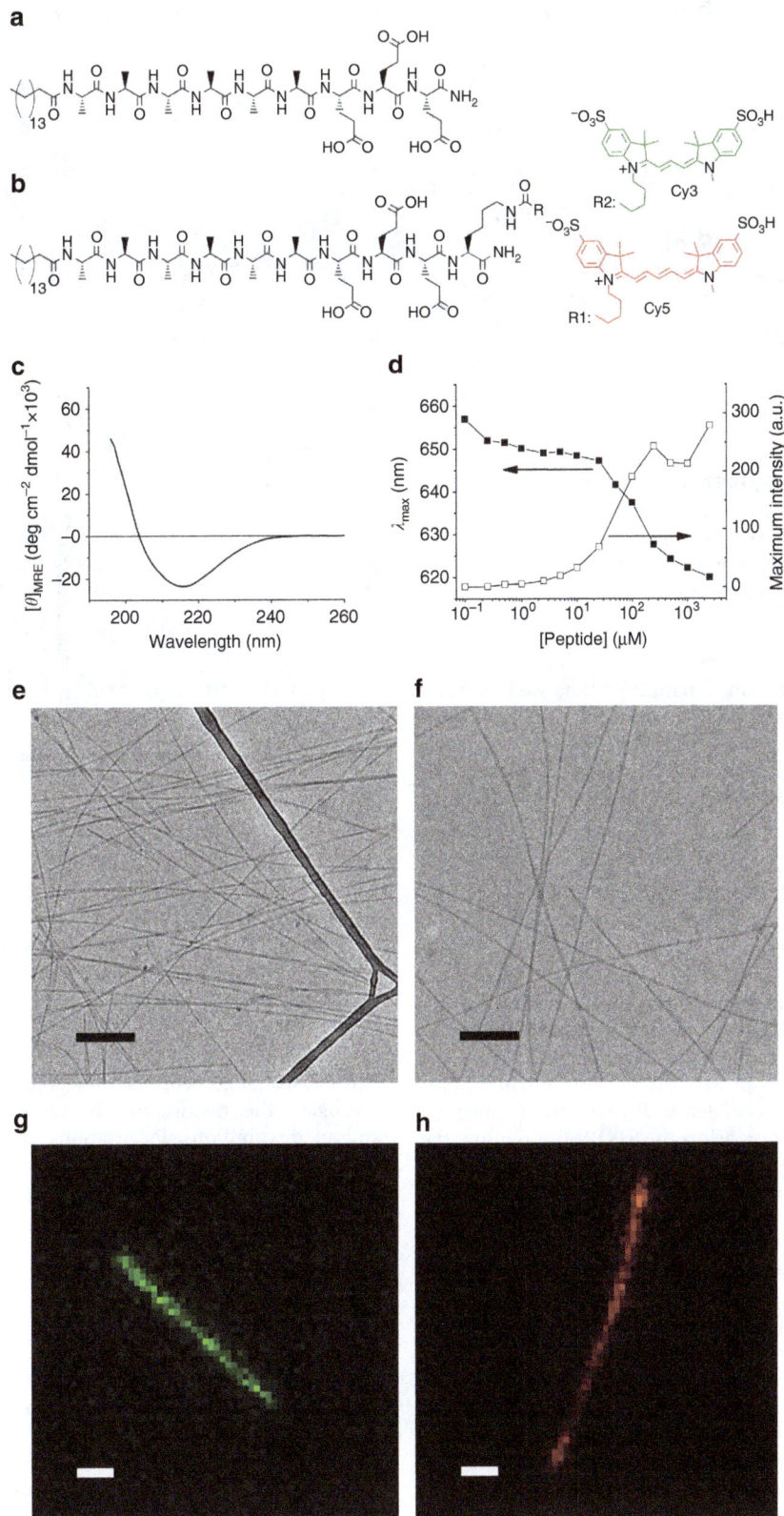

Figure 1 | PA self-assembly. Molecular structure of (**a**) non-labelled PA and (**b**) PA molecules labelled with photo-switchable sulfonated cyanine dyes, namely Cy3 (green) and Cy5 (red). (**c**) Circular dichroism spectrum and (**d**) Nile Red assay of non-labelled PA. CryoTEM images of nanofibres self-assembled at pH 7.5 and NaCl 150 mM (**e**) from non-labelled PA alone and (**f**) from a molecular mixture of non-labelled and Cy5-labelled PAs (scale bar, 200 nm). Diffraction-limited fluorescence microscopy images of Cy3- (**g**) and Cy5-labelled (**h**) PA nanofibres (scale bar, 1 μm).

nanofibres, resulting in mixed fluorophore fibres. Figure 2b depicts kinetic experiments at different temperatures, showing that the exchange rate is remarkably faster at 37 °C than at 20 °C. On the other hand, we observed a limited effect of concentration on the exchange rate (Fig. 2c), proving that the exchange process is not diffusion limited in this concentration range. These results resemble observations on self-assembled polymeric micelles, in which unimer expulsion and insertion is thought to be the

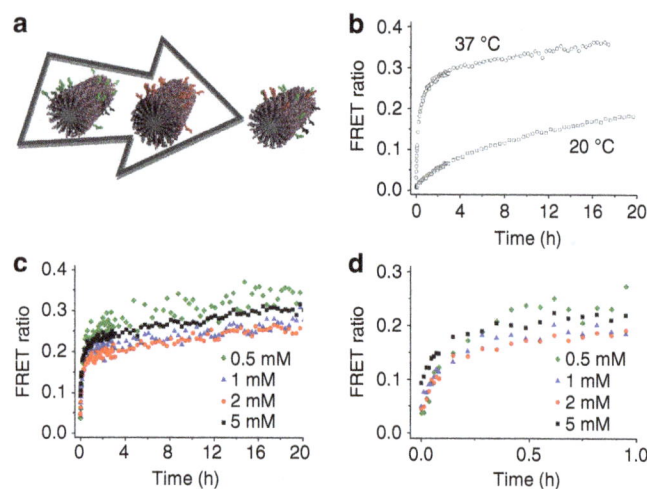

Figure 2 | Molecular exchange by FRET. (a) Schematic representation of a molecular exchange kinetic measurement. The molecular exchange progress over time is estimated by means of FRET ratio (dividing Cy5 by Cy3 fluorescence intensities), either at **(b)** a constant PA concentration (0.5 mM) or at **(c,d)** a constant temperature (37 °C), where **d** shows the FRET ratio for short timescale.

rate-determining step of system dynamics[10,11], as well as other synthetic supramolecular polymers[7]. So, while FRET measurements provide useful information about the timescale of the exchange, FRET is an ensemble technique, and it does not provide the spatially resolved information required to elucidate the mechanism of exchange.

Imaging PA nanofibres. Due to its high spatial resolution and multicolour imaging ability, STORM microscopy can be used to investigate the spatial details of the exchange process in PA nanofibres. Cy3- or Cy5-labelled PA nanofibres have been immobilized on a glass slide by physisorption and subsequently have been imaged using STORM (see the 'Methods' Section for details). The resulting images show well-reconstructed aggregates (Fig. 3a,b), indicating that both dyes display appropriate photo-switching behaviour when associated to PA structures. Since the actual width of the nanofibre is below the STORM resolution, the apparent thickness of the fibre can be used to provide an estimate for the experimental resolution of these measurements, approximately equal to 50 nm (Supplementary Fig. 1). The fibre characteristics observed by STORM, for example, rigidity and fibre length, match those reported by the cryoTEM images in Fig. 1e,f.

To analyse the monomer distribution inside PA nanofibres, which is crucial to understand the exchange mechanism, we utilized an image analysis routine previously developed in the Meijer laboratory[19]. This method removes background localizations, identifies the contour of the fibre backbone to study its mechanical properties (Supplementary Fig. 2) and computes the localization density along the polymer. Figure 3c,d display the localization density plots for Cy5- and Cy3-labelled PA, respectively. These profiles contain information about the distribution of the dye-functionalized PA molecules in a nanofibre. As can be clearly observed, the number of localizations shows fluctuations along the fibre. These fluctuations can be attributed to two causes: (i) a heterogeneous distribution of monomers or (ii) the stochastic processes taking place during STORM image acquisition, for example, fluorophore blinking[36] and fluorophore bleaching. Therefore, variations in the density of localizations cannot be unequivocally attributed to changes in local monomer concentration. To address this issue,

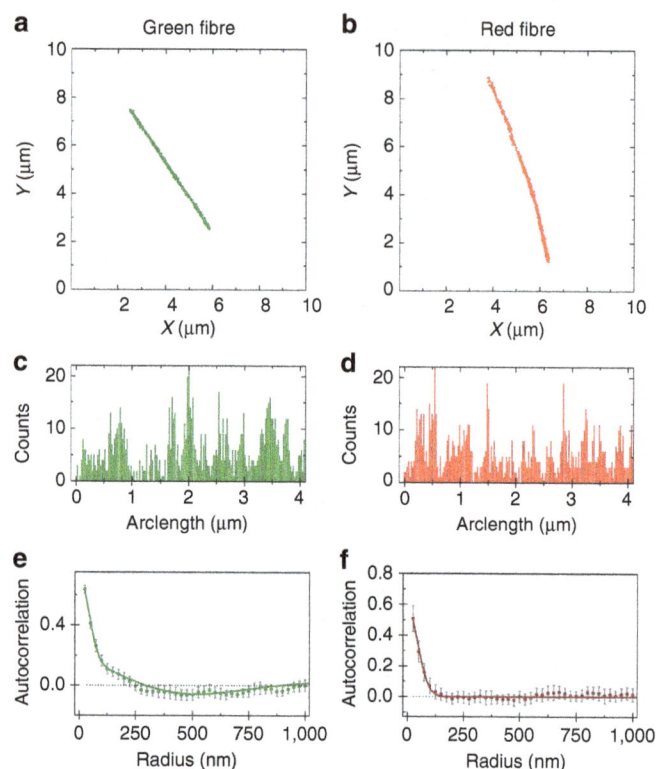

Figure 3 | STORM analysis of single-colour nanofibres. PA nanofibres were labelled with 5 mol% of either Cy3-PA (**a,c,e**) or Cy5-PA (**b,d,f**) before the molecular exchange experiment. Localization maps after applying a clustering algorithm, background cleaning and backbone finding (**a,b**). Each fibre localization distribution profile was determined along the fibre backbone (arc length) using a bin size of 25 nm (**c,d**). Averaged autocorrelation (**e,f**) was computed from distribution profiles of a large set of fibre images ($n \geq 19$). Solid lines correspond to model fittings (equation (1) for red channel and equation (2) for green channel) and error bars represent 95% confidence level.

we use a spatial autocorrelation algorithm, a powerful method to investigate the distribution of these localizations. In purely random distributions, localizations at distance r apart are not correlated with each other, by definition giving an autocorrelation $g(r)$ with a value of zero. Positive values of $g(r)$ represent an increased probability that a second localization is found at a certain distance (r) from a first given localization, while negative values indicate a decreased probability. The autocorrelation curves of fibres in single-colour samples are shown in Fig. 3e,f. The autocorrelation of Cy5-labelled nanofibres (Fig. 3f) is strongly positive in the short range (<100 nm) and zero for longer distances. This short-ranged contribution is the typical signature for multiple localizations of the same dye (overcounting) and is present in all STORM images. As previously described, the overcounting contribution in the spatial autocorrelation traces can be modelled using equation (1) (ref. 37),

$$g(r) = A_1 exp\left(-r^2/4\sigma^2\right) \qquad (1)$$

where the standard deviation (σ) provides an estimate of the resolution achieved. Equation (1) provides a good fit to the autocorrelations of Cy5-labelled single-colour nanofibres (Fig. 3f). Therefore, the localization density fluctuations visible in the STORM images for these PA nanofibres (Fig. 3d) are due to stochastic processes inherent to the technique, while the monomer distribution is homogeneous. On the other hand, the autocorrelation function for the Cy3-channel (Fig. 3e) is not well-described by Equation (1), since the Cy3-functionalized fibres display a consistent anti-correlation ($g(r) < 0$) at intermediate distances (~ 500 nm). The autocorrelation behaviour of these fibres was correctly described using a micro-emulsion model, that accounts for the existence of 'microdomains' richer in the fluorescent monomer, according to equation (2) (ref. 37):

$$g(r) = A_1 exp\left(-r^2/4\sigma^2\right) + A_2 cos(\pi r/2r_0)exp\left(-r/\alpha\right) \quad (2)$$

The first term of this equation is equal to equation (1), since the overcounting phenomenon is still present. In the second term, the parameter r_0 is the average size of the 'microdomains' and α is the coherence length of the domains. As defined by equation (2), 'microdomain' corresponds to an extended fibre region that is enriched in fluorescently labelled molecules compared with the fibre average, thus increasing the likelihood of finding those molecules in that particular region. In this context, 'microdomain' does not mean that molecules are completely clustered and confined to those regions.

An improvement in the goodness of the fit was not observed for the Cy5-labelled nanofibres, therefore it is reasonable to adopt the simpler model with less parameters, which excludes the existence of 'microdomains'. On the other hand, in the case of Cy3-labelled nanofibres a considerable improvement on the χ^2-test (> 3.5 times lower) is obtained for the model with the 'microdomain' component, compared with the model that only considers the contribution of the overcounting. A 'microdomain' size of ~ 300 nm and coherence length of ~ 500 nm were obtained from the fit (Fig. 3e), showing a regular fluctuation of localization density in the Cy3-fibres at the submicron length scale. The difference observed between the distributions of Cy3 and Cy5-PAs is surprising, because of the great chemical and structural resemblance of these dyes. These results show that a small change in the PA structure may have noticeable effects on the self-assembly process, and that the STORM-based autocorrelation analysis can be used to investigate structure at the single-aggregate level. In addition, it shows that Cy5-functionalized PA is suitable for use in quantitative exchange experiments, since the Cy5 dye does not affect the molecular distribution of the PA monomers.

Fibre bundling and multicolour imaging. Since the final morphology in supramolecular polymers is highly sensitive to not only the molecular structure, but also the polymerization conditions, it is very important to design a correct sample preparation protocol to obtain the desired structure. A typical problem that may occur is the aggregation of fibres into bundles, a phenomenon that is hard to detect using traditional techniques. Avoiding fibre bundling is crucial when investigating molecular mixing, because bundles of nanofibres with different colours confound the observation of exchanging monomers. In this work we show that STORM-based analysis can be applied to detect PA fibre bundling. When nanofibre preparation was performed at higher PA concentration ($[PA]_{final} > 1$ μM), extremely long and curved fibrillar structures were observed in diffraction-limited fluorescence microscopy (Supplementary Fig. 3). However, the increased resolution attained using STORM imaging revealed finer details of the adsorbed structures, showing extensive fibre bundling. What appeared to be curvature at low resolution was actually the intersection of smaller stiff fibres with different orientations. Two-colour STORM has been performed on PA samples of different concentrations and ionic strengths to unambiguously prove the presence or absence of bundled fibres. Two aqueous solutions of PA nanofibres labelled with either Cy3 or Cy5 were mixed together and immediately adsorbed onto a glass coverslip at room temperature, freezing all kinetics. For this short mixing timescale (< 1 min), molecular exchange is negligible according to the FRET measurements shown in Fig. 2b. Therefore, single fibres should be either fully green or fully red. If fibres are adsorbed at high ionic strength and low concentration (< 0.1 μM), this behaviour was indeed observed (Fig. 4a). On the contrary, as shown in Fig. 4b, fibres adsorbed at concentrations higher than 1 μM can be observed in both channels simultaneously, indicating bundles of PA nanofibres. It is also possible to observe that some of the fibres visible in the green channel are not perfectly aligned with the overlapping fibres visible in the red channel (Fig. 4b). This provides further evidence that at this concentration bundles are observed instead of fully mixed fibres. Therefore, STORM allows us to verify the absence of fibre bundling and therefore select the optimal sample preparation to perform the molecular exchange experiments.

Molecular exchange kinetics and mechanisms. The ability of STORM to image individual nanofibres with high resolution and to study the distribution of different molecular species inside aggregates (*vide supra*), makes it the perfect tool to study exchange in PA samples. We prepared samples for these experiments in similar manner to the FRET measurements. First, two sets of single-colour PA nanofibres with either Cy3-PA (5%) or Cy5-PA (5%) were separately preassembled in aqueous buffer solution, followed by a 16-h aging period. Subsequently the two aged solutions were brought to 37 °C and mixed, allowing molecular exchange. Aliquots of the solution were withdrawn over the course of 48 h and nanofibres were adsorbed onto a glass

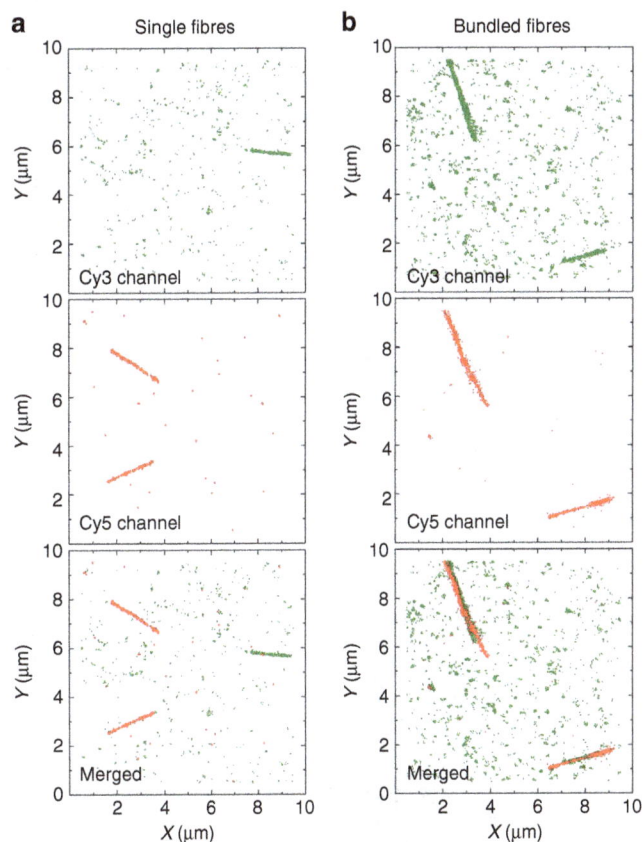

Figure 4 | Single fibre attachment. STORM images (with background) of PA nanofibres attached at a final PA concentration of either (**a**) 0.1 μM or (**b**) 1 μM, showing single fibres or bundled fibres, respectively. Attachment was performed immediately on a glass coverslip after dilution (NaCl 1 M).

coverslip at room temperature to freeze the exchange. Two-colour STORM images of assemblies were then acquired at the different time points. Fig. 5 shows representative PA nanofibres at $t = 1$ min and after 48 h of mixing; as clearly shown in the images and in the corresponding localization density plot (Fig. 5b) the fibres initially containing a single label fully mix during the course of the experiment.

We verified that the exchange process is bidirectional by following green fibres incorporating red monomers (Supplementary Fig. 4) and vice versa (Supplementary Fig. 5). The intermediate time points provide insight into the exchange mechanism. Visual inspection of STORM images (Supplementary Fig. 6) showed insertion of labelled monomers over the entire length of the nanofibres throughout the exchange experiment. This finding suggests an exchange mechanism on the basis of expulsion and reinclusion of PA molecules. Polymerization–depolymerization at the nanofibre ends would have resulted in preferential exchange in those regions, while fragmentation–recombination would have

resulted in a block-like structure; neither of these mechanisms is consistent with the acquired STORM images.

To confirm this observation and further quantify the exchange, we studied the change in distribution of both dyes with time in initially green-labelled nanofibres, thus tracking the insertion of Cy5-PA. Figure 6 shows the auto- and cross-correlation plots for the green and red channels at the different time points. The green channel displays the previously observed clustered distribution described by equation (2) rendering it less amenable for quantitative analysis. Therefore, we monitored the autocorrelation of the red channel in time to analyse monomer exchange. The red channel autocorrelation, shown in Fig. 6b, can be fit with equation (1) for all time points, suggesting a random exchange of monomers. However, for later time points (that is, after incubation longer than 6 h) the quality of the fit deteriorates, indicating some non-random variations of the Cy5-PA concentration in the nanofibres other than the single decay originated from overcounting. However, the formation of regular 'microdomains' does not seem plausible, because equation (2) did not improve the quality of the fit or yield reasonable fitting parameter values. Since clustering is not observed in single-colour Cy5-PA nanofibres (Fig. 3f), this observation points to a heterogeneous exchange and indicates the presence of structural variations along the fibre or between fibres. This heterogeneity can also be perceived visually by observing single fibre images (Supplementary Fig. 6). The cross-correlation between both channels was computed to assess how the distribution of one labelled PA would affect the distribution of the other, as shown in Fig. 6c. A cross-correlation of zero was found for the entire range, indicating that the distribution of Cy5-PA is not influenced by the presence of Cy3-PA, and vice versa. In other words, the heterogeneous exchange of Cy5-PA is not due to the pre-existing clusters of Cy3-PA, but rather a consequence of structural variation in the PA nanofibre.

The observation of this heterogeneity in exchange, not detected by the FRET experiment, was possible due to the ability of STORM to evaluate the progress of molecular exchange for each individual supramolecular nanofibre. The variability of the exchange progress between different aggregates could be measured, thus allowing us to further probe the heterogeneity in the system during and after molecular exchange. The linear density of localizations, defined as the number of localizations per nanometre arc length, has been computed and could be used as a rough estimate for the ensemble concentration of labelled molecules[19]. The density of Cy5-PA in originally single-colour green fibres was measured as a function of time, showing an increase in the average value as molecular exchange proceeded (Fig. 7a). However, the standard deviation also increased markedly, another indication for the presence of intrinsic structural diversity. In Fig. 7b, it is possible to observe how the distribution of the Cy5-PA linear density gets progressively wider as molecular exchange evolves in time. This can be confirmed visually by inspecting STORM images acquired after 48 h mixing time (Fig. 7c-f), which displayed a range of different behaviours. It was possible to observe a large subset of fibres that had a Cy5-PA density consistent with full mixing (Fig. 7c), as well as fibres comprised of domains of Cy3-PA only (Fig. 7d,e), and both subsets co-existed with a smaller subpopulation of nanofibres that had undergone very little exchange all over their length (Fig. 7f). The existence of regions displaying minimal or no exchange after 48 h implies that full equilibration of these regions will take weeks to months, making them persistent for most practical purposes. This analysis suggests that the population of fibres is considerably more heterogeneous after 48 h than would be expected on the basis of the FRET ensemble measurements.

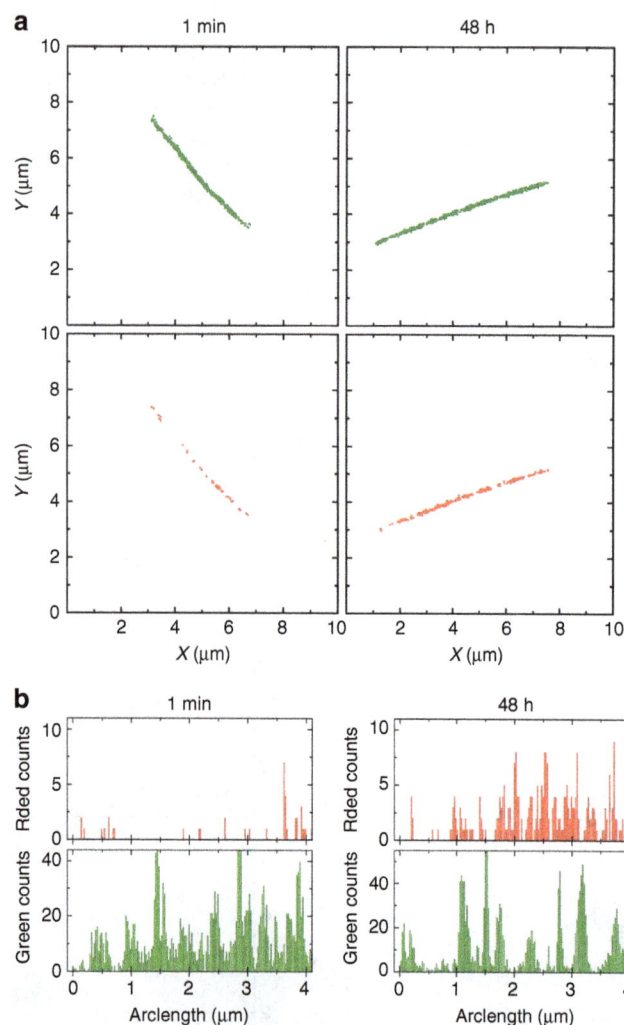

Figure 5 | Quantitative analysis of STORM imaging during a molecular exchange experiment. (a) Localization maps of PA nanofibres immobilized on a glass coverslip at different time points, after applying a clustering algorithm, background cleaning and backbone finding. The better reconstruction of the fibres in the green channel reflects the selection of initially Cy3-labelled fibres for this experiment. **(b)** Histograms depict the localization density profiles along the nanofibre backbone (bin size 25 nm).

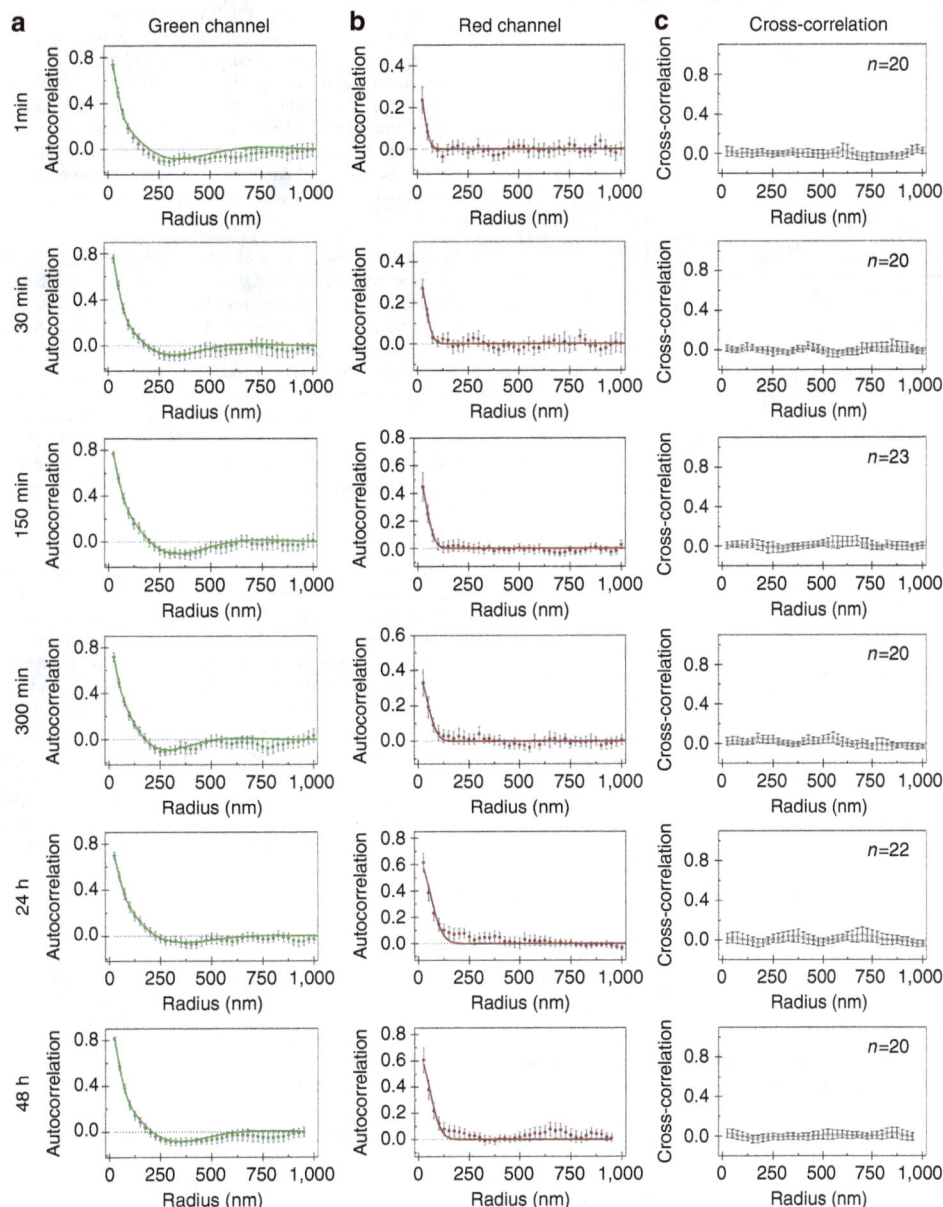

Figure 6 | Correlation of cyanine dyes along PA nanofibres initially only labelled with Cy3 during molecular exchange. Averaged autocorrelation plots for (**a**) Cy3 and (**b**) Cy5-labelled PAs and (**c**) averaged cross-correlation between both dyes. Solid lines correspond to model fittings, namely equation (2) for (**a**) green channel and equation (1) for (**b**) red channel (error bars represent 95% confidence level, $n \geq 20$).

Previous molecular dynamics simulations of a similar PA provide a theoretical framework for this structural diversity[27]. In these simulations, a broad distribution of secondary structures was found in the equilibrated fibre. The heterogeneous molecular exchange pattern observed in our study might stem from this conformational diversity. The key feature that led to the discovery of PA supramolecular nanofibres in the Stupp laboratory was the use of a β-sheet peptide domain to drive one-dimensional self-assembly. Using electron paramagnetic resonance, the β-sheet domain has been recently shown to give rise to locally solid-like behaviour in the interior of nanofibre PA assemblies[32]. On the other hand, the surface moieties in these nanofibres have been identified by the electron paramagnetic resonance experiments as regions where liquid-like dynamics prevail. β-Sheets within the nanofibres are highly cohesive assemblies and therefore the dynamic exchange among supramolecular nanofibres should give rise to a large diversity of supramolecular environments. This is in contrast to supramolecular systems with weaker internal cohesion

that could exchange molecules to produce completely homogeneous environments[19]. The formulation of supramolecular systems with high levels of internal order and cohesion such as the PA nanofibres will therefore open new avenues to generate structural diversity for functional purposes. One example will be their ability to adapt structurally to bind important bioactive targets.

Discussion

We have studied the dynamics of PA nanofibres using FRET and super-resolution localization microscopy. While ensemble FRET measurements prove that PAs exchange between different fibres, two-colour STORM imaging reveals a mechanism on the basis of the transfer of monomers and small clusters. Remarkably, the coexistence of fully dynamic and kinetically inactive areas in the aggregate architecture was observed, demonstrating the existence of structural diversity in PA nanofibres. This intriguing dynamic

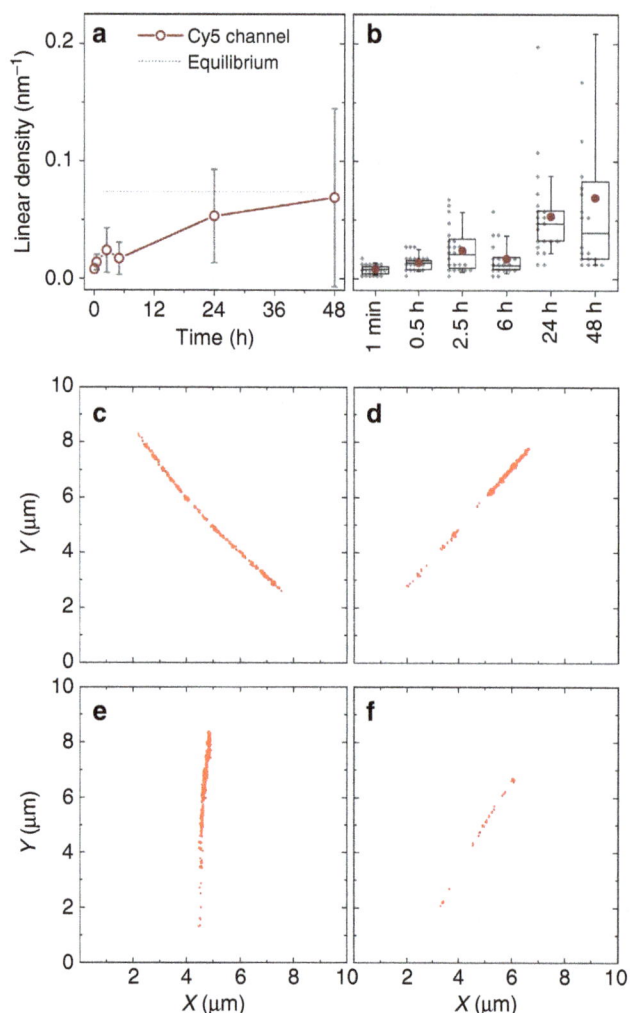

Figure 7 | PA molecular exchange kinetics estimated using the linear density of localizations calculated from STORM data. (a) Time course-averaged density of red dye (Cy5) on fibres that were initially only labelled with green dyes (Cy3) ($n \geq 20$, error bars correspond to standard deviation). The equilibrium guideline is estimated from the localization density of original single-colour red fibres. (b) Box chart of data represented on **a** depicting average (circle), median (box middle line), quartiles (box edges), 5 and 95% percentiles (error bars) and full data set points (diamond symbols, bin size $5 \mu m^{-1}$); (**c–f**) Examples of localization plots (after background cleaning) of individual fibres at 48 h.

behaviour is foreseen to have important implications in the biological performance of supramolecular systems.

Methods

Peptide synthesis and purification. PAs were synthesized using standard Fmoc solid-phase peptide synthesis. MBHA rink amide resin, Fmoc-protected amino acids and other solid-phase peptide synthesis reagents were purchased from Novabiochem (USA). Cyanine dyes were supplied by Cyandye (USA). ACS-grade solvents were used for synthesis. Water was purified on an EMD Milipore Milli-Q Integral Water Purification System. For each amino acid coupling, 4.1 equiv of Fmoc-protected amino acid was activated for 1 min with 4 equiv of HBTU (O-benzotriazole-N,N,N',N'-tetramethyluroniumhexafluorophosphate) and 6 equiv DIPEA (N,N-diisopropylethylamine) in 20 ml DMF. The coupling cocktail was then added to 0.5 mmol of MBHA rink resin and reacted for 1 h. Fmoc removal was performed with 30% piperidine in DMF. Fmoc-Glu(OtBu)-OH (N-α-Fmoc-L-glutamic acid γ-t-butyl ester) (3 ×), and Fmoc-Ala-OH (N-α-Fmoc-L-alanine) (6 ×) were successively coupled to the resin. Finally, the palmitic acid tail (8.1 equiv) was coupled using 8 equiv of HBTU and 12 equiv DIPEA in DMF/DCM (50:50). PAs were cleaved from the resin using a cleavage solution of 95% TFA, 2.5% TIPS (triisopropylsilane) and 2.5% H_2O. Cleavage solution was concentrated

by rotary evaporation and the PA precipitated in cold diethyl ether. The crude solid was dissolved in a diluted NH_4OH aqueous solution (~ 10 mg ml^{-1}) and purified by preparative-scale reversed-phase HPLC on a Varian Prostar 210 HPLC system, using acetonitrile/water gradient containing 0.1% NH_4OH. All eluents and additives were HPLC grade. Separation was achieved using a Phenomenex C18 Gemini NX column (150×30 mm) with $5 \mu m$ particle size and 110 Å pore size. Product-containing fractions were confirmed by ESI mass spectrometry (Agilent 6510 Q-TOF LC/MS) and combined. ACN was removed by rotary evaporation, lyophilized to yield a white solid product and stored at $-30\,^\circ C$.

Cyanine dye coupling. PAs labelled with cyanine dyes were synthesized first coupling Fmoc-Lys(Mtt)-OH (N-α-Fmoc-N-ε-4-methyltrityl-L-lysine) to the resin. The rest of the synthetic procedure proceeded as described above. After coupling the palmitic acid tail, Lys(Mtt) was selectively deprotected with 3% TFA, 5% TIPS and 92% DCM for 5 min (3 ×), to expose the side chain amine moiety. Disulfo-cyanine carboxylic acid dyes (Cy3 and Cy5) were dissolved in DMF and stock solutions were stored at -30 °C. (Benzotriazol-1-yloxy)tripyrrolidinophosphonium hexafluorophosphate (PyBOP) was also dissolved in DMF. Cyanine dyes (1 equiv) were activated with PyBOP (1 equiv) and DIPEA (2 equiv) for 1 min and coupled to the lysine-free amine. Cy3- and Cy5-labelled PAs were cleaved and purified as described above. Lyophilization yielded a red powder for Cy3 and blue for Cy5.

Purity. Purity was determined by reversed-phase analytical HPLC (see Supplementary Figs 7–9) using a Phenomenex Gemini C18 column (100×4.6 mm), with $5 \mu m$ particle size and 110 Å pore size, connected to a HPLC system equipped with an autosampler (Shimadzu SIL-20A XR), degasser (Shimadzu DGU-20A3) and a high-pressure gradient system comprising two LC-20AD XR pumps (Shimadzu). Separation was performed at a flow of 1 ml min^{-1} and using a 20-60% acetonitrile linear gradient in water containing 0.1% NH_4OH. All eluents and additives were HPLC grade. Peptides, cyanine dye-coupled peptides and eventual contaminants were detected using a PDA (Shimadzu SPD-M20A), acquiring a full ultraviolet-visible spectrum between 200 and 750 nm at any time point. Main chromatogram peaks were collected and product mass confirmed by ESI mass spectrometry using a LCQ Deca XP Max (Thermo Finnigan) ion-trap mass spectrometer. For that, samples were manually injected bypassing the column at a flow rate of 0.20 ml min^{-1} and positive ion mass spectra were acquired in standard enhanced mode.

Sample preparation. Stock solutions of non-labelled PA (10 mM), Cy3-labelled PA (1 mM) and Cy5-labelled PA (1 mM) were prepared in Milli-Q water, aliquoted and stored at $-30\,^\circ C$. Similar to other anionic PA systems, these formed hydrogels in the presence of $CaCl_2$. The degree of labelling of the fibres was accurately controlled by simply mixing PA stock solutions at the desired ratio. Solutions of non-labelled PA were mixed with aqueous solutions of either one or both labels at certain mole ratios, flash-frozen in liquid nitrogen and lyophilized. The obtained PA mixture was molecularly dissolved in TFA and readily evaporated under vacuum. The obtained material was redissolved in NH_4OH (aq.), flash-frozen and further lyophilized. These two last steps resembled the cleavage and purification conditions, respectively. These steps were undertaken to provide molecular mixing, as well as to reset PA self-assembly history. Non-labelled PA was treated using the same method before measurements where the fluorescence-labelled molecules were absent. The obtained powders were reconstituted in appropriate buffers before measurements. After the harsh acidic conditions (dissolution in TFA) used to create a homogenous molecular mixture, PAs were checked by analytical HPLC to rule out potential degradation of the original molecules. Analytical HPLC was performed as described above. Supplementary Fig. 10 shows that the PA does not degrade during the harsh procedure of molecular mixing in TFA. Fluorescence measurements showed pronounced FRET for nanofibres labelled with both Cy3 and Cy5. The FRET intensity increased with concentration of acceptor. This indicates that the sample preparation is suitable to assure co-assembly of the three PAs in a single supramolecular object.

Circular dichroism spectroscopy. Circular dichroism spectra were recorded using Jasco Circular Dichroism Spectrometer (model J-715). PA was dissolved in water at 10 mM. Before each measurement, PA stock solution was diluted in appropriate buffers to yield a final PA concentration of 0.1 mM. A quartz cuvette of 1 mm path length was used for the measurements. Each trace represents the average of five scans.

Cryogenic transmission electron microscopy. PA nanofibres were separately labelled with either PA-Cy3 (5%) or PA-Cy5 (5%). Both labelled and non-labelled PA nanofibres were dissolved at 1 mg ml^{-1} in 10 mM HEPES buffer (pH 7.5, NaCl 150 mM). CryoTEM was performed using a JEOL 1230 TEM at an accelerating voltage of 100 kV. Using a Vitrobot Mark IV (FEI) vitrification instrument at 25 °C with 100% humidity, a 6.5-μl drop of the sample was deposited on a 300-mesh copper grid with lacey carbon support (Electron Microscopy Sciences, EMS), blotted twice, plunge-frozen in liquid ethane, then stored in liquid N_2. For imaging, the sample was transferred to a Gatan 626 cryo-holder under liquid N_2, and images were obtained with a Gatan 831 CCD camera.

Critical micelle concentration by Nile Red assay. Nile Red is a hydrophobic solvatochromic dye (water soluble at low concentration) with high affinity to the hydrophobic pocket of micelles. In aqueous environment, the molecule is weakly fluorescent. In hydrophobic environments, fluorescence intensity is around 300 times higher and a pronounced blue shift is observed. Therefore, Nile Red is commonly used to probe the existence of hydrophobic pockets formed by surfactants in aqueous environment[38]. Briefly, a Nile Red 2 mM stock solution was prepared in MeOH. Non-labelled PA was dissolved at 10 mM in 10 mM HEPES buffer (pH 7.5, NaCl 150 mM). PA solution was diluted serially to obtain concentrations in the range 200 nM-10 mM. Nile Red stock solution was diluted in the same buffer to obtain a final concentration of 1 μM. Equal volumes of Nile Red aqueous diluted solution and PA solutions were mixed for each PA concentration. Solutions were aged for 24 h to assure full nanofibre disassembly at concentrations below the critical micelle concentration. Fluorescence-emission spectra (excitation 550 nm) were recorded for an emission range between 570 and 700 nm. The maximum intensity and respective wavelength at maximum intensity were both represented as a function of the logarithm of the PA concentration. At concentrations close to critical micelle concentration, it should be observed a sharp increase in fluorescence intensity and a hypsochromic effect.

Förster resonance energy transfer. Single-colour PA nanofibres labelled with either Cy3 or Cy5 at 1 mol% were assembled in 10 mM HEPES buffer (pH 7.5, NaCl 150 mM). Fluorescence spectra were recorded in a Varian Cary Eclipse Fluorimeter from Agilent Technologies. Emission spectra were acquired using excitation wavelengths of 520 or 615 nm for Cy3 or Cy5, respectively. Excitation spectra were collected using emission wavelengths of 600 or 710 nm for Cy3 or Cy5, respectively. Temperature was kept at 20 °C using the in-built peltier system. Supplementary Fig. 11a depicts the spectra of the individually labelled PA nanofibres, showing significant overlap between Cy3 emission (donor) and Cy5 excitation (acceptor). On an independent measurement, PA nanofibres labelled with both Cy3 (1 mol%) and Cy5 (0, 0.1, 0.2 and 0.5 mol%) were assembled in 10 mM HEPES buffer (pH 7.5, NaCl 150 mM) at a final PA concentration of 0.5 mM. Fluorescence-emission spectra of dual-labelled PAs were recorded at the excitation wavelength of Cy3 520 nm, at which direct excitation of Cy5 is negligible. Supplementary Fig. 11b shows considerable energy transfer by FRET, which increases monotonically with acceptor concentration, providing evidence for co-assembly of both Cy3- and Cy5-labelled PAs in the same nanofibres.

Ensemble molecular exchange kinetics. For molecular exchange kinetic experiments, two sets of PA nanofibres were pre-assembled from a mixture of non-labelled PAs and either Cy3 (0.5%) or Cy5 (0.5%). These independently labelled PA solutions were mixed at 1:1 ratio in a 50 μl quartz cuvette and fluorescence was followed over around 24 h (lag time ∼30 s). FRET ratio was determined by dividing Cy5 emission at 668 nm by Cy3 emission at 565 nm, at the Cy3 excitation wavelength (520 nm). The contribution of Cy3 emission at 668 nm was first subtracted from the Cy5 emission. At the beginning of the kinetics experiment, Cy3-PA and Cy5-PA are distributed in different nanofibres, being physically separated by distances much higher than the expected FRET radius[35]. Therefore, energy transfer at time zero should be negligible and, consequently, FRET ratio should be zero. A total PA concentration range of 0.5–5 mM was used. Kinetics was recorded at two different temperatures (20 or 37 °C).

Sample preparation for microscopy. Glass microscope coverslips were cleaned by successively immersing in acetone, isopropanol and Milli-Q water. Bath sonication was performed for 10 min with each solvent, followed by drying under N_2 flow. The glass coverslips were then etched with a fresh Piranha solution (3:1 v/v H_2SO_4 (98%):H_2O_2 (30%)) for 30 min. To finish the cleaning procedure, the slides were washed thoroughly with Milli-Q water and rinsed with acetone before drying under N_2 flow. To determine the labelled PA distribution along the nanofibres length, single fibres should be immobilized on a glass surface, and remain still during STORM acquisition. A flow chamber was assembled using a glass slide and a clean coverslip separated by double-sided tape. PA nanofibres were immobilized by adsorption onto the surface of the clean coverslip by flushing diluted PA solutions at high ionic strength (NaCl 1 M). After incubation for 1 min, the unbound nanofibres were washed out of the chamber by flushing with HEPES buffer (2 ×) followed by STORM buffer (2 ×). STORM buffer is composed by 50 mM HEPES buffer (pH = 8.5), 1 M NaCl, an oxygen-scavenging system (0.5 mg ml^{-1} glucose oxidase, 40 μg ml^{-1} catalase, 5 wt% glucose) and 200 mM 2-aminoethanethiol. The high ionic strength conditions were used to screen the electrostatic repulsion between highly charged PA nanofibres and the glass surface. However, repulsion between PA nanofibres is also reduced, causing fibre bundling. To avoid bundling, PA solutions were rapidly diluted at low ionic strength (1 μM). After a final (10 ×) dilution step in NaCl 1 M, nanofibres were immediately attached on the glass coverslip. At this concentration isolated nanofibres are mainly observed in the microscopic mounting.

Stochastic optical reconstruction microscopy. STORM images were acquired using a Nikon N-STORM system configured for total internal reflection fluorescence imaging. Excitation inclination was tuned to maximize the signal-to-noise ratio of the glass-absorbed fibres. Cy3- and Cy5-labelled samples were illuminated by 561 and 647 nm laser lines. No activation ultraviolet light was used. Fluorescence was collected by means of a Nikon 100 ×, 1.4NA oil immersion objective and passed through a quad-band pass dichroic filter (97335 Nikon). All time-lapses were recorded onto a 64 × 64 pixel region (pixel size 0.17 μm) of an EMCCD camera (ixon3, Andor). For each channel, 20,000 frames were sequentially acquired. STORM movies were analysed with the STORM module of the NIS Elements Nikon software.

Molecular imaging of fibres during exchange. PA nanofibres were separately labelled with either PA-Cy3 (5%) or PA-Cy5 (5%) at 0.5 mM in 10 mM HEPES buffer (pH 7.5, NaCl 150 mM). Subsequently, 250 μl-aliquots of each solution were mixed in a microcentrifuge tube and gently shaken at 37 °C for different amounts of time. At predefined time points, 2 μl of the solution was withdrawn, diluted and adsorbed on a glass surface as described above. We verified that once adsorbed on the slide, fibres are not able anymore to considerably exchange monomers with the solution in the time frame of the experiment. STORM imaging of Cy5 and Cy3 channels was performed sequentially. Fibres that are initially labelled only with Cy3 and incorporate Cy5 monomers over time were selected observing the fluorescence intensity in the low-resolution images. Original single-colour nanofibres were also observed before mixing as a control.

Image analysis. During STORM imaging, the Nikon software generates a list of localizations by 2D Gaussian fitting of blinking chromophores in the acquired movie of conventional microscopic images. This localizations list was subsequently analysed with our custom-made Matlab scripts, as described in detail elsewhere[19]. Briefly, a first script uses a density-based clustering algorithm to automatically identify the fibres in the image and to remove the background. Examples of output localization maps are shown in Figs 3a,b and 5a. A second script is run on the well-reconstructed fibres obtained in the Cy3-channel to find the fibre backbone. The lower localization density obtained in the Cy5-channel for shorter time periods in the molecular exchange experiment is not enough to fully reconstruct the nanofibres. The polymer backbone coordinates are used to obtain structural information. Next, histograms of localization density profiles along the backbone are generated for both channels, as shown in Figs 3c,d and 5b. The spatial autocorrelation (that is, the average correlation between the localization density at one backbone point in the fibre and the density at another backbone point, as a function of the distance between these points) was also computed using the Matlab-function xcov with unbiased normalization. The spatial cross-correlation of the two channels was calculated in a similar manner. Averaging over multiple fibres is required to obtain accurate correlation decay graphs, such as depicted in Figs 3e,f and 6. Investigation of the average correlation decays yields the distribution of dyes in a fibre, without interference from the stochastic fluctuations inherent to the technique.

References

1. Dominguez, R. & Holmes, K. C. Actin structure and function. *Annu. Rev. Biophys.* **40,** 169–186 (2011).
2. Janke, C. & Kneussel, M. Tubulin post-translational modifications: encoding functions on the neuronal microtubule cytoskeleton. *Trends Neurosci.* **33,** 362–372 (2010).
3. Pollard, T. D. & Borisy, G. G. Cellular motility driven by assembly and disassembly of actin filaments. *Cell* **112,** 453–465 (2003).
4. Pollard, T. D., Blanchoin, L. & Mullins, R. D. Molecular mechanisms controlling actin filament dynamics in nonmuscle cells. *Annu. Rev. Biophys. Biomol. Struct.* **29,** 545–576 (2000).
5. Stournaras, C., Gravanis, A., Margioris, A. N. & Lang, F. The actin cytoskeleton in rapid steroid hormone actions. *Cytoskeleton* **71,** 285–293 (2014).
6. Symons, M. & Rusk, N. Control of vesicular trafficking by Rho GTPases. *Curr. Biol.* **13,** R409–R418 (2003).
7. Albertazzi, L. *et al.* Spatiotemporal control and superselectivity in supramolecular polymers using multivalency. *Proc. Natl Acad. Sci. USA* **110,** 12203–12208 (2013).
8. Marchi-Artzner, V. *et al.* Selective adhesion, lipid exchange and membrane-fusion processes between vesicles of various sizes bearing complementary molecular recognition groups. *Chemphyschem Eur. J. Chem. Phys. Phys. Chem.* **2,** 367–376 (2001).
9. Ferrell, J. E., Lee, K. J. & Huestis, W. H. Lipid transfer between phosphatidylcholine vesicles and human-erythrocytes–exponential decrease in rate with increasing acyl chain-length. *Biochemistry* **24,** 2857–2864 (1985).
10. Lund, R., Willner, L., Richter, D. & Dormidontova, E. E. Equilibrium chain exchange kinetics of diblock copolymer micelles: tuning and logarithmic relaxation. *Macromolecules* **39,** 4566–4575 (2006).
11. Zinn, T., Willner, L., Lund, R., Pipich, V. & Richter, D. Equilibrium exchange kinetics in n-alkyl–PEO polymeric micelles: single exponential relaxation and chain length dependence. *Soft Matter* **8,** 623–626 (2011).

12. Fasting, C. *et al.* Multivalency as a chemical organization and action principle. *Angew. Chem. Int. Ed.* **51**, 10472–10498 (2012).

13. Helms, B. A. *et al.* High-affinity peptide-based collagen targeting using synthetic phage mimics: from phage display to dendrimer display. *J. Am. Chem. Soc.* **131**, 11683 (2009).

14. Silva, G. A. *et al.* Selective differentiation of neural progenitor cells by high-epitope density nanofibers. *Science* **303**, 1352–1355 (2004).

15. Baker, M. B. *et al.* Consequences of chirality on the dynamics of a water-soluble supramolecular polymer. *Nat. Commun.* **6**, 6234 (2015).

16. Bates, M., Huang, B., Dempsey, G. T. & Zhuang, X. Multicolor super-resolution imaging with photo-switchable fluorescent probes. *Science* **317**, 1749–1753 (2007).

17. Shim, S.-H. *et al.* Super-resolution fluorescence imaging of organelles in live cells with photoswitchable membrane probes. *Proc. Natl Acad. Sci. USA* **109**, 13978–13983 (2012).

18. Huang, B., Bates, M. & Zhuang, X. Super-resolution fluorescence microscopy. *Annu. Rev. Biochem.* **78**, 993–1016 (2009).

19. Albertazzi, L. *et al.* Probing exchange pathways in one-dimensional aggregates with super-resolution microscopy. *Science* **344**, 491–495 (2014).

20. Webber, M. J., Berns, E. J. & Stupp, S. I. Supramolecular nanofibers of peptide amphiphiles for medicine. *Isr. J. Chem.* **53**, 530–554 (2013).

21. Boekhoven, J. & Stupp, S. I. 25th anniversary article: supramolecular materials for regenerative medicine. *Adv. Mater.* **26**, 1642–1659 (2014).

22. Stupp, S. I. & Palmer, L. C. Supramolecular chemistry and self-assembly in organic materials design. *Chem. Mater.* **26**, 507–518 (2014).

23. Soukasene, S. *et al.* Antitumor activity of peptide amphiphile nanofiber-encapsulated camptothecin. *ACS Nano* **5**, 9113–9121 (2011).

24. Moyer, T. J. *et al.* Shape-dependent targeting of injured blood vessels by peptide amphiphile supramolecular nanostructures. *Small* **11**, 2750–2755 (2015).

25. Paramonov, S. E., Jun, H.-W. & Hartgerink, J. D. Self-assembly of peptide – amphiphile nanofibers: the roles of hydrogen bonding and amphiphilic packing. *J. Am. Chem. Soc.* **128**, 7291–7298 (2006).

26. Pashuck, E. T., Cui, H. G. & Stupp, S. I. Tuning supramolecular rigidity of peptide fibers through molecular structure. *J. Am. Chem. Soc.* **132**, 6041–6046 (2010).

27. Lee, O.-S., Stupp, S. I. & Schatz, G. C. Atomistic molecular dynamics simulations of peptide amphiphile self-assembly into cylindrical nanofibers. *J. Am. Chem. Soc.* **133**, 3677–3683 (2011).

28. Storrie, H. *et al.* Supramolecular crafting of cell adhesion. *Biomaterials* **28**, 4608–4618 (2007).

29. Stephanopoulos, N., Ortony, J. H. & Stupp, S. I. Self-assembly for the synthesis of functional biomaterials. *Acta Mater.* **61**, 912–930 (2013).

30. Greenfield, M. A., Hoffman, J. R., de la Cruz, M. O. & Stupp, S. I. Tunable mechanics of peptide nanofiber gels. *Langmuir* **26**, 3641–3647 (2010).

31. Newcomb, C. J. *et al.* Cell death versus cell survival instructed by supramolecular cohesion of nanostructuresl. *Nat. Commun.* **5**, 3321 (2014).

32. Korevaar, P. A., Newcomb, C. J., Meijer, E. W. & Stupp, S. I. Pathway selection in peptide amphiphile assembly. *J. Am. Chem. Soc.* **136**, 8540–8543 (2014).

33. Tantakitti, F. *et al.* Energy landscapes and functions of supramolecular systems. *Nat. Mater.* **15**, 469–476 (2016).

34. Levitt, M. Conformational preferences of amino acids in globular proteins. *Biochemistry* **17**, 4277–4285 (1978).

35. Bastiaens, P. I. & Jovin, T. M. Microspectroscopic imaging tracks the intracellular processing of a signal transduction protein: fluorescent-labelled protein kinase C beta I. *Proc. Natl Acad. Sci. USA* **93**, 8407–8412 (1996).

36. Rust, M. J., Bates, M. & Zhuang, X. Sub-diffraction-limit imaging by stochastic optical reconstruction microscopy (STORM). *Nat. Methods* **3**, 793–796 (2006).

37. Veatch, S. L. *et al.* Correlation functions quantify super-resolution images and estimate apparent clustering due to over-counting. *PLoS ONE* **7**, e31457 (2012).

38. Stuart, M. C. A., van de Pas, J. C. & Engberts, J. B. F. N. The use of Nile Red to monitor the aggregation behavior in ternary surfactant–water–organic solvent systems. *J. Phys. Org. Chem.* **18**, 929–934 (2005).

Acknowledgements

Molecular synthesis, purification and characterization were supported by the U.S. Department of Energy, Office of Science, Basic Energy Sciences, under Award # DE-FG02-00ER45810. The work in Eindhoven was supported by the Dutch Ministry of Education, Culture and Science (Gravity program 024.001.035) and the European Research Council (FP7/2007-2013, ERC Grant Agreement 246829). R.M.P.d.S. and his research activities were funded by the Marie Curie FP7-PEOPLE-2010-IOF program (ref. no. 273295). Compound preparation, purification and characterization were partially performed at the Peptide Synthesis Core at Simpson Querrey Institute. The Biological Imaging Facility (BIF) and Keck Biophysics facilities at Northwestern provided further instrumentation used in this work. Mark Seniw is acknowledged for the preparation of graphics.

Author contributions

R.M.P.d.S., L.A., D.v.d.Z., S.I.S. and E.W.M. conceived the project and contributed to the concept of the manuscript. R.M.P.d.S., L.A., D.v.d.Z. designed the experiments and analysed the results. R.M.P.d.S. performed peptide synthesis, spectroscopic and STORM experiments. S.S.L. performed cryoTEM imaging. R.M.P.d.S., L.A., D.v.d.Z., S.I.S. and E.W.M. wrote the manuscript and contributed to discussions on the project.

Additional information

Competing financial interests: The authors declare no competing financial interests.

19

Connecting two proteins using a fusion alpha helix stabilized by a chemical cross linker

Woo Hyeon Jeong[1,*], Haerim Lee[2,*], Dong Hyun Song[3], Jae-Hoon Eom[1], Sun Chang Kim[2], Hee-Seung Lee[1], Hayyoung Lee[4] & Jie-Oh Lee[1]

Building a sophisticated protein nano-assembly requires a method for linking protein components in a predictable and stable structure. Most of the cross linkers available have flexible spacers. Because of this, the linked hybrids have significant structural flexibility and the relative structure between their two components is largely unpredictable. Here we describe a method of connecting two proteins via a 'fusion α helix' formed by joining two pre-existing helices into a single extended helix. Because simple ligation of two helices does not guarantee the formation of a continuous helix, we used EY-CBS, a synthetic cross linker that has been shown to react selectively with cysteines in α-helices, to stabilize the connecting helix. Formation and stabilization of the fusion helix was confirmed by determining the crystal structures of the fusion proteins with and without bound EY-CBS. Our method should be widely applicable for linking protein building blocks to generate predictable structures.

[1] Department of Chemistry, KAIST, Daejeon 34141, Korea. [2] Department of Biological Sciences, KAIST, Daejeon 34141, Korea. [3] Agency for Defense Development, Daejeon 34186, Korea. [4] Institute of Biotechnology, Chungnam National University, Daejeon 34134, Korea. * These authors contributed equally to this work. Correspondence and requests for materials should be addressed to J.-O.L. (email: jieoh@kaist.ac.kr).

Linking two protein components to form a predictable and rigid structure is a prerequisite for generating complex protein assemblies in a pre-designed fashion[1]. Most of the chemical cross linkers available have long and flexible spacers to help them approach the reactive side chains of the target proteins. Because of this, the resulting hybrids have significant structural flexibility and the relative orientation and distance between their two components is largely unpredictable. This is the case even when the chemical cross linkers themselves have rigid structures since they are attached to flexible side chains such as cysteines or lysines. Recently several new methods have been proposed to assemble proteins in a pre-designed fashion.

Radford et al.[2–4] have shown that two proteins can be linked in a rigid and predictable way using a metal coordination method. Since the bound metal ion has additional coordination sites that are available for binding of another protein molecule, it can induce homodimerization of the protein. Because the metal coordination requires precise positioning of the coordinating amino acid side chains, the resulting protein dimer will have a predictable structure. The authors used a similar metal coordination method to produce a variety of protein assemblies[5–12]. King et al.[13] reported a computational method for designing cage-like structures by connecting two protein components. One of them was chosen from among natural proteins that can form stable homodimers and the other from among proteins that can form stable homotrimers. By optimizing the interface between these two protein components in the fusion proteins using computer calculations, they were able to produce several fusion proteins that could assemble into large cage-like structures.

Short alpha-helical linkers joining two protein components have been used to build large and highly symmetric assemblies with rigid and predictable structures. Because an α-helix is under strict structural restraint, the structure of the fusion protein ought to be predictable so long as a connecting helix is formed. Diverse architectures are possible, depending on the specific geometric arrangement between the protein components[14]. Using this method, Lai et al.[1,15,16] connected M1 matrix protein, which can form homodimers, and BPO protein, which can form stable homotrimers, with a short linker. By testing various linker sequences they were able to find one that could form an α-helix and supported the formation of cage-like structures. Recently, they applied a similar method to generate large and porous cube-like closed structures with high accuracy[17]. Rigidity of the helical linker is essential for the success of the design because apparently minor deviation of the structure from the intended ideal helix can lead to failure of protein assembly[18].

Here we propose a novel method for connecting the pre-existing α-helices of two proteins into a single extended helix using a chemical cross linker. To test our method, we chose as a model system an ankyrin family protein containing a C-terminal helix and a protein A fragment containing an N-terminal helix. We fused the two proteins by connecting the two terminal α-helices into a single continuous helix. In general, simple genetic ligation of the helices in the proteins does not lead to fusion of the helices. Therefore, we used a synthetic chemical that selectively reacts with α helices and stabilizes them. EY-CBS (3,3′-ethyne-1,2-diylbis-{6-[(chloroacetyl)amino]benzenesulfonic acid}) is a chemical cross linker that can react simultaneously with two cysteines in the i and i + 11 positions of the α-helix[19]. To test whether EY-CBS detects the formation of a helix, we chose two amino acid residues separated by 11 amino acids in the proposed fusion helix and mutated them to cysteines. The successful fusion helices showed high reactivity with EY-CBS, and cross linking with EY-CBS forced the structure of the fusion α-helix to resemble that of an ideal helix. Our method

does not require metal ions in the buffer, nor complicated computer simulation in the design process. It can be used to assemble protein components into asymmetrical and non-repeating structures.

Results

Design of the fusion helices. We chose an artificial ankyrin protein, named 'mbp3-16', that has been engineered to bind maltose-binding protein (MBP)[20], and the B4 domain of protein A as a model system to test our fusion helix method. We chose these proteins because both of them are composed entirely of α-helices and lack cysteine residues. In addition to this, they have been proven to be useful as building blocks of artificial protein assemblies because they can be engineered to bind other proteins[21,22]. We produced 17 fusion proteins by connecting the C-terminal helix of the ankyrin domain and the N-terminal helix of the protein A domain (Fig. 1 and Table 1). While designing connections between the two helices, obvious steric collision was avoided by using a molecular graphics program. To generate the EY-CBS reaction site, two amino acid residues separated by 11 amino acids, one from the ankyrin domain and the other from the protein A domain, were chosen and mutated to cysteines. Also, two additional amino acids, in positions i + 4 and i + 7, were mutated to alanines because long side chains may lead to steric collision with the EY-CBS molecules[19]. Using a similar

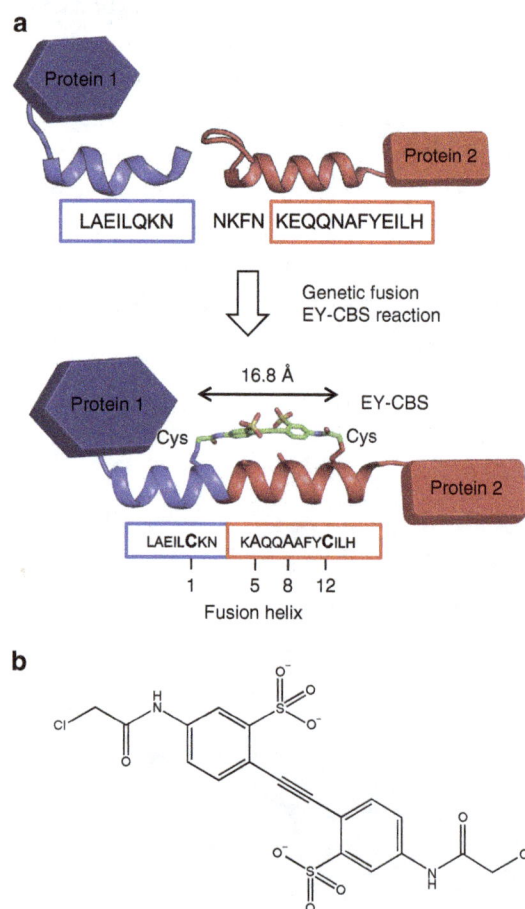

Figure 1 | Design of the fusion proteins. (a) The C-terminal helix of protein 1 and the N-terminal helix of protein 2 were genetically connected. Two amino acids separated by 11 amino acids were mutated to cysteines for reaction with EY-CBS. (b) Chemical structure of EY-CBS. EY-CBS has a rigid chemical structure with two α-chlorocarbonyl groups reactive with the thiol groups of cysteines.

Table 1 | Design of the fusion proteins.

Name	First protein	Second protein		Sequence of the fusion helix	
3,305	Protein A	mbp3-16	266*	QCAREAAREAAACD	13*
3,306	mbp3-16	Protein A	133	QCAREAAAREAACN	217
3,307	Protein A	mbp3-16	263	NCAQEAAAREAACD	13
3,308	mbp3-16	Protein A	129	ACILQAAAREAACN	217
3,309	Protein A	mbp3-16	266	QCAREAAARDLGCK	17
3,310	mbp3-16	Protein A	133	QCAREAAAREANCE	219
3,311	mbp3-16	Protein A	132	LCKNKAQQAAFYCI	227
3,312	mbp3-16	Protein A	125	NCRLAAILAKNKCQ	220
3,313	Protein A	mbp3-16	259	ACKLNAAQALGRCL	18
3,315	Protein A	mbp3-16	260	ACLNAAQAALGRCL	18
4,254	Protein A	mbp3-16	260	KCLNDALGAKLLCA	21
4,256	Protein A	mbp3-16	260	KCLNDAQAAKLLCA	21
4,257	Protein A	mbp3-16	259	ACKLNAAQALGRCL	18
4,258	Protein A	mbp3-16	259	ACKLNAAQAAGRCL	18
4,260	Protein A	mbp3-16	261	KCNDAAARALLECA	22
4,261	Protein A	mbp3-16	261	KCNDAALGAKLLCA	21
4,262	Protein A	mbp3-16	261	KCNDAAAGAKLLCA	21
6,758	Protein A	Calmodulin	260	KCLNDAEQIAEFCE	15
6,759	Protein A	Calmodulin	260	KCLNDEQIAEFKCA	16
6,760	Protein A	Calmodulin	259	ACKLNAQIAEFKCA	16
6,761	Protein A	Calmodulin	260	KCLNDAQAAAEECI	10

*Residue numbers are taken from the UniProt (protein A and calmodulin code number P38507 and P62158, respectively) and the GenBank (mbp3-16, AY326426) databases.

method, we designed four more fusion proteins with the protein A domain and the calmodulin N-terminal domain[23]. We intentionally positioned positively charged residues such as lysines and arginines in some of the fusion proteins to test whether the negatively charged EY-CBS had higher reactivity with positively charged targets (Fig. 1b). However, we found no obvious correlation between the location of the positively charged residues and the reactivity of the protein towards EY-CBS, and we did not artificially include positively charged amino acids in the later design. When necessary, we inserted alanine residues between the two proteins to control the relative orientation of the two protein components (Table 1). Note that insertion of one amino acid residue in an α-helix will rotate the helix by ~100 degrees. Therefore, one can control the relative orientation of the connected proteins by inserting precise number of amino acids into the fusion helix. Amino acid residues that may have roles in stabilizing the native helices were not changed.

Reaction with EY-CBS. The fusion proteins produced were purified and incubated with 1 mM EY-CBS (Fig. 2 and Supplementary Figs 1–3). Five of the 21 fusion proteins shifted substantially in the SDS–PAGE analysis. The molecular weight of EY-CBS is 521 daltons and if the two cysteines in the fusion helix react with EY-CBS, the shift in molecular weight should be by 521 daltons. In fact the change of molecular weight estimated from the shift of the protein bands was more than ~2 kDa (Fig. 3a). Therefore, we assumed the band shift was not only due to the additional molecular weight of the bound EY-CBS but to a change in the local structure of the fusion helix induced by circularization of the two cysteine side chains by EY-CBS. To analyse the situation further, we chose the 3,311 fusion protein because it had the highest reaction efficiency with EY-CBS (Fig. 3 and Supplementary Figs 4–6). To confirm that the 3,311 fusion indeed reacted with EY-CBS, the absorption spectrum of the reacted protein was measured after removing unreacted EY-CBS by gel filtration chromatography. The isolated EY-CBS molecule has an absorption peak at 340 nm due to the conjugated phenyl rings. As expected, the reacted fusion protein had a strong absorption peak at 340 nm, demonstrating that

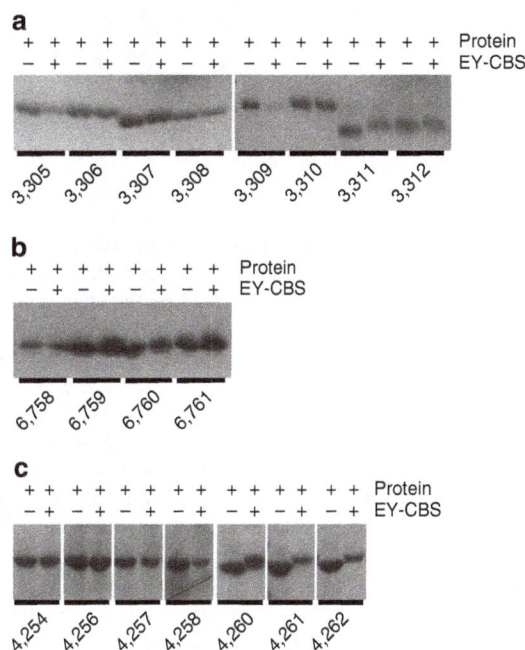

Figure 2 | Reactivity of the designed fusion proteins. The protein A-ankyrin (**a**, **c**) and the protein A-caldmodulin (**b**) fusion proteins were treated with 1mM EY-CBS. Constructs 3,311, 3,312, 4,260, 4,261 and 4,262 are substantially upshifted after the reaction.

EY-CBS was covalently linked to the protein (Fig. 3b). The ratio of the absorption peaks at 280 and 340 nm suggests that one EY-CBS molecule is bound per protein molecule.

To maximize the reaction efficiency of EY-CBS, we measured the reactivity of the 3,311 protein as a function of pH, anticipating that the deprotonated thiol groups of the cysteines would have higher reactivity at higher pH. As expected, the reaction was strongly dependent on pH and proceeded to near completion at pH 8 (Fig. 3c). At pH >9, the protein band and reaction

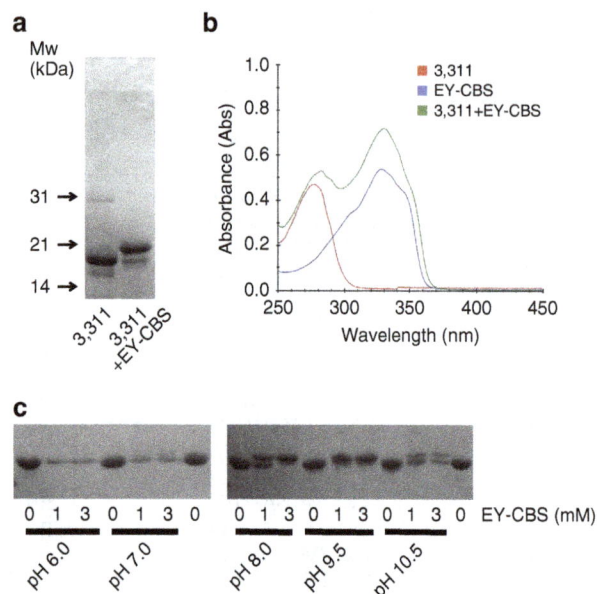

Figure 3 | Reactivity of the 3,311 fusion protein toward EY-CBS. (a) SDS–PAGE analysis of the 3,311 protein. Nearly 100% of the 3,311 protein is upshifted by reaction with EY-CBS. **(b)** Absorption spectrum of the 3,311 protein reacted with EY-CBS. Excess EY-CBS after reaction was removed by gel filtration chromatography. **(c)** pH dependence of the EY-CBS reaction. The 3,311 protein was reacted with EY-CBS at different pH and analysed by SDS–PAGE.

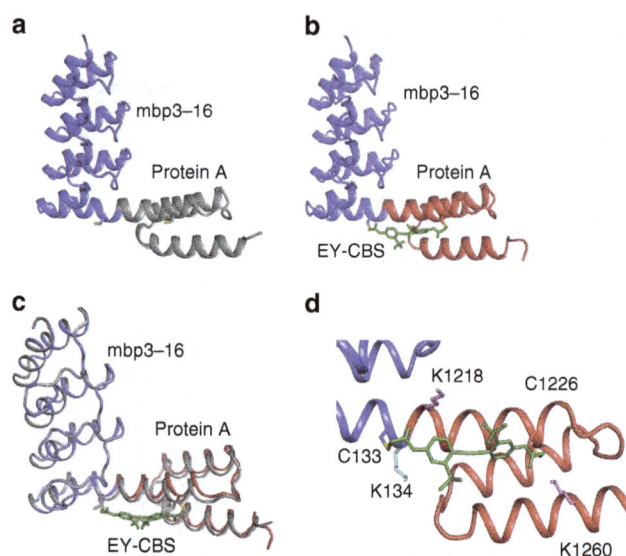

Figure 4 | Crystal structure of the 3,311 fusion protein after reaction with EY-CBS. (a) The intended structure of the 3,311 fusion protein. The ankyrin domain of mbp3-16 and the protein A domain are connected by an α-helix. **(b)** Crystal structure of 3,311 EY-CBS. EY-CBS is attached to the fusion helix and shown in green. For clarity the structure of the MBP bound to the ankyrin domain is not shown. **(c)** Superimposition of the intended and crystal structures of the 3,311 fusion. The intended and crystal structures are coloured in grey and blue/red, respectively. **(d)** Close-up view of the fusion helix. Positively charged amino acid side chains that interact with the sulfate groups of EY-CBS are shown.

efficiency declined presumably due to denaturation and precipitation of the fusion protein.

Crystal structure of the 3,311 fusion protein. To verify fusion of the helices, we crystallized the 3,311 fusion protein before and after reaction with EY-CBS and determined the structures of the two products (Figs 4 and 5). Since the ankyrin domain, mbp3–16, had been engineered to bind MBP, we were able to use MBP to facilitate crystallization[20]. In the crystal structure, one MBP molecule interacted exclusively with the mbp3–16 component of the hybrid protein and there was no interaction with either protein A or the fusion helix. The electron density map calculated using the refined protein structure showed a clear density corresponding to the EY-CBS, and one EY-CBS molecule could be modelled into the density (Supplementary Fig. 7). Both reactive ends of the EY-CBS are covalently linked to the cysteines in the fusion helix. In the crystal structure, the fusion helix adopts a nearly ideal α-helical structure and matches closely to the intended structure (Fig. 4c). The Cα atoms of the cysteines are separated by 16.8 angstrom and the mutated alanines are located under the phenyl rings of EY-CBS, as designed. The structures of the ankyrin and protein A domains are easily superimposable on those of the isolated proteins, which demonstrates that the helix fusion has no impact on the structure of the protein other than near the fusion site (Fig. 4b and c). The sulfate moieties of the EY-CBS groups form nonspecific ionic interactions with three lysine groups of the protein A and ankyrin domains (Fig. 4d). Other than that, EY-CBS does not have any noticeable interaction with the protein.

We determined the crystal structure of the unreacted 3,311 fusion protein at 2.8 angstrom resolution to understand why it reacts strongly with EY-CBS (Fig. 5). MBP was not included in the crystallized protein solution. The crystal structure showed that the protein A and ankyrin domains were connected by a slightly bent helix (Fig. 5a). When the structures of the fusion helix with and without the bound EY-CBS were superimposed, the fusion helix was seen to be bent by ∼20 degrees with respect to that of the EY-CBS-bound helix (Fig. 5b). The asymmetric unit of the crystal contains 12 fusion proteins that can be divided into two groups, each group containing 6 proteins (Supplementary Fig. 8). The proteins belonging to the same group are in a similar crystal environment and, as expected, have superimposable structures. However, the proteins belonging to different groups are in completely different crystal environments. Despite this, the proteins in the different groups have almost identical structures, with a Cα root mean square difference value of 0.32 angstrom. Furthermore, the simulated annealing omit map calculated using the refined protein structure showed a clear density near the fusion helix (Supplementary Fig. 9). Because of these, we conclude that formation of the fusion helix is unlikely to be a crystallization artifact.

Crystal structure of the unreacted 3,311 clearly explains why the 3,311 fusion protein is strongly reactive with EY-CBS: the fusion helix is already present before EY-CBS is bound; after the covalent reaction with EY-CBS the fusion helix resembles more closely the ideal helix because the distance between the two cysteines is forced to be 16.8 angstrom. In the crystal structure, Asn107 and Lys1218 are separated by 2.9 angstrom and are hydrogen bonded. In addition, Asn135 is connected to Gln1220 via a hydrogen bond and contributes to stability of the fusion helix (Fig. 5c). However, it is not clear whether these relatively weak interactions are enough to stabilize the helical connection between the two protein components. More sophisticated calculations are required to analyse this stabilization.

Crystal structure of a fusion protein that failed the EY-CBS test. Of the 21 fusion proteins, 16 failed to display any significant

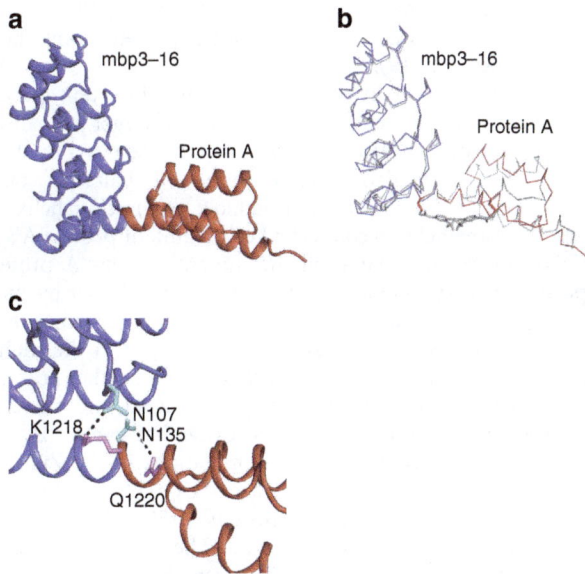

Figure 5 | Crystal structure of the 3,311 fusion protein not reacted with EY-CBS. (**a**) Crystal structure of the fusion protein. The fusion helix is bent by ∼20 degrees. (**b**) Superposition of the EY-CBS-bound and -unbound structures of 3,311 fusion protein. The 3,311 EY-CBS and 3,311 structures are coloured in grey and blue/red, respectively. (**c**) Close-up view of the fusion helix. Amino acid residues that may stabilize the fusion helix are labeled.

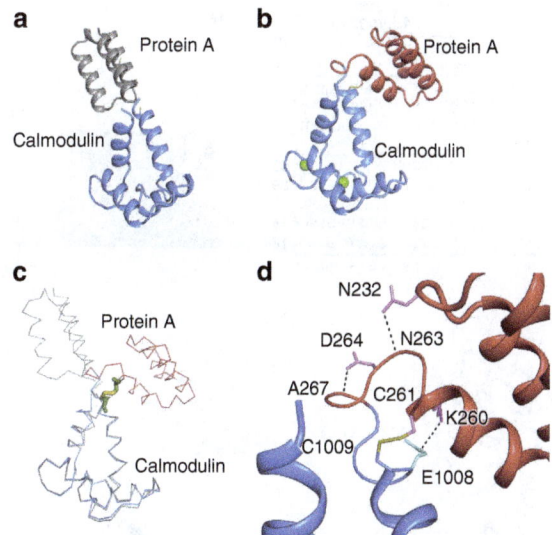

Figure 6 | Crystal structure of the 6,761 fusion protein containing the protein A domain and the calmodulin domain. (**a**) Intended structure of the 6,761 fusion protein. The two protein domains are connected by a continuous α-helix. (**b**) Crystal structure of the 6,761 fusion protein. An unexpected disulfide bridge that connects the mutated cysteines is shown. (**c**) Superposition of the intended and crystal structures of the 6,761 fusion protein. The intended structure is coloured in grey and the crystal structure is coloured in blue and red. (**d**) Close-up view of the fusion site. The cysteine 261 of the protein A domain and cysteine 1,009 of the calmodulin domain form a disulfide bridge. Amino acid residues that stabilize the interface of the two helices are drawn and labeled.

reactivity towards EY-CBS (Fig. 2). We were unable to crystallize any of these proteins except 6,761, presumably due to the structural instability of the interface between the fused protein components.

The crystallized 6,761 fusion protein is composed of an N-terminal protein A and a C-terminal calmodulin domain (Supplementary Fig. 10). Calmodulin is a calcium sensor composed of two homologous calcium-binding domains. Each calmodulin domain contains two EF hand motifs that bind calcium ions[24]. The crystal structure of the 6,761 fusion protein was determined at 2.7 angstrom resolution (Fig. 6). The structures of the two domains of the 6,761 fusion protein were easily superimposable with those of the separate protein A and calmodulin domains, and there were no noticeable structural changes except around the fusion site. In the crystal structure, it could be seen that the two helices designed to fuse had not formed a continuous helix, and the cysteines introduced to bind EY-CBS were close to one another and formed an unexpected disulfide bridge (Fig. 6c). The last and the first turns of the helices of the protein A and calmodulin domains, respectively, formed irregular loops and made sharp ∼120 degree turns. The crystal structure of the connecting loop and the disulfide bridge formed a distinct electron density demonstrating that the fusion site was structurally stable (Supplementary Fig. 11). It appears that the 6,761 fusion protein can be crystallized because the interface between the two protein components of the fusion protein is stabilized by the opportunistic disulfide bridge. Several strong interactions can be seen at the fusion site and these account for why the disulfide bridge is formed (Fig. 6d). Lys260 of the protein A domain and Glu1008 of the calmodulin domain are separated by only 3.8 angstrom and form a strong ionic interaction. In addition, hydrogen bonding of the backbone nitrogen of Ala267 and the carboxylic oxygen of Asp264 should also contribute to the stabilization of the disulfide bridge. Furthermore, the side chain of Asn232 and the backbone carbonyl oxygen of Asn263 are separated by 2.9 angstrom and form a strong hydrogen bond.

We conclude that that simple-minded ligation of two helices does not guarantee the formation of a single extended helix, and that high reactivity of the fusion helix with EY-CBS is a reliable indicator of helix formation.

Insertion of a binding adaptor protein into an internal loop. To confirm the validity of our EY-CBS method, we designed several fusion proteins containing the protein A domain inserted into loops preceding internal rather than terminal α-helices of T4 lysozyme. To identify the best insertion site, we tested four loop regions and inserted an artificial protein A analogue, Ztaq (Supplementary Fig. 12)[25]. Three of the four fusion proteins were produced as efficiently as wild-type lysozyme when expressed in *Escherichia coli*. The exception was the 8,133 fusion. All three fusion proteins eluted as monomers in size-exclusion chromatography and were resistant to limited subtilisin digestion, confirming their structural integrity. Of the three possible insertion sites, we chose the loop containing residue 37 and designed three helix-fusion proteins, 8,155, 8,157 and 8,158 containing the EY-CBS-binding sites (Supplementary Figs 13–15). Fusion protein 8,157 has two more amino acids in the fusion helix than protein 8,155 and therefore its protein A domain should be rotated by ∼200 degrees and translated by ∼7 angstroms. Fusion proteins 8,157 and 8,158 have the same EY-CBS sites except that the positions of the reactive cysteines are shifted by four amino acids.

All three helix-fusion proteins were produced in *E. coli* and purified to homogeneity. Unexpectedly, reaction with EY-CBS did not result in clearly visible upshifts of the protein bands on SDS–PAGE presumably because size of the fusion proteins are too big and resolution of the SDS–PAGE analysis is not good enough to detect small changes in the structure (Fig. 7b and Supplementary Figs 16 and 17). However, we believe that all three

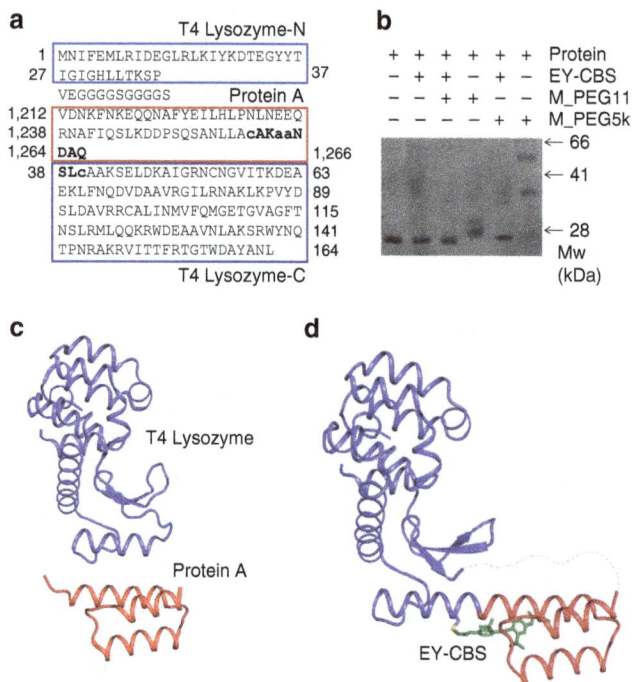

a

T4 Lysozyme-N

```
1    MNIFEMLRIDEGLRLKIYKDTEGYYT
27   IGIGHLLTKSP                37
     VEGGGGSGGGGS    Protein A
1,212 VDNKFNKEQQNAFYEILHLPNLNEEQ
1,238 RNAFIQSLKDDPSQSANLLAcAKaaN
1,264 DAQ                        1,266
38   SLcAAKSELDKAIGRNCNGVITKDEA  63
     EKLFNQDVDAAVRGILRNAKLKPVYD  89
     SLDAVRRCALINMVFQMGETGVAGFT  115
     NSLRMLQQKRWDEAAVNLAKSRWYNQ  141
     TPNRAKRVITTFRTGTWDAYANL     164
```

T4 Lysozyme-C

b

```
+ + + + + +  Protein
- + + - + -  EY-CBS
- - + + - -  M_PEG11
- - - - + +  M_PEG5k
                ← 66
                ← 41
                ← 28
              Mw
              (kDa)
```

c

T4 Lysozyme

Protein A

d

EY-CBS

Figure 7 | Insertion of the protein A domain into an internal loop of T4 lysozyme. (**a**) The amino acid sequence of fusion protein 8,155. (**b**) Reactivity of fusion protein 8,155 with EY-CBS. After one hour of the reaction, PEG-Maleimide reagent Maleimide-PEG11-Biotin (M_PEG11) or PEG-Maleimide 5000 (M_PEG5k) was added where appropriate. (**c**) Previously reported crystal structures of T4 lysozyme and protein A. (**d**) Crystal structure of fusion protein 8,155. The T4 lysozyme and the protein A parts of the fusion protein are coloured in blue and red, respectively. The 'GGGGS' linker and the N-terminal seven amino acids of protein A that are disordered in the crystal structure are drawn as a broken line.

proteins had reacted with EY-CBS with high efficiency because their cysteines became resistant to two maleimide-containing reagents, Maleimide-PEG11-Biotin and PEG-Maleimide 5000. These reagents are highly reactive with the free thiol groups of cysteines, and their reactivity is easier to detect after SDS–PAGE because they have high-molecular weights, 1.1 and 5 kDa, respectively. As shown in Fig. 7b and Supplementary Figs 16 and 17, the fusion proteins were resistant to the PEG-maleimide reagents after EY-CBS treatment, presumably because they had already formed covalent bonds with EY-CBS, whereas the SDS–PAGE bands formed by the same fusion proteins not reacted with EY-CBS were clearly shifted upwards, showing that their cysteines were free to react with the PEG-maleimide reagents.

To confirm fusion of the α-helices connecting protein A and lysozyme, we crystallized 8,155 after reaction with EY-CBS and determined its crystal structure. The purified and reacted 8,157 and 8,158 fusions were also crystallized, but we have not tried to optimize the crystallization conditions nor determined their structures. The 8,155 crystals diffracted X-rays towards 2.7 angstrom resolution. In the crystal structure, one EY-CBS molecule is covalently connected to the two cysteines in the fusion helix as expected (Supplementary Fig. 18). The distance between the Cα atoms of the reacted cysteines is 16.7 angstrom, which is shorter by only 0.1 angstrom than that of 3,311 treated with EY-CBS. The fusion helix thus adopts a nearly ideal α-helical structure and closely matches the intended structure. The lysozyme and protein A parts of the structure can be superimposed with the structures of the individual proteins,

demonstrating that fusion of the two helices had little impact on the overall structure of the individual protein components (Supplementary Fig. 19).

We chose the protein A domain as the insertion partner because it can be mutated to bind a variety of target proteins, as shown previously[26]. Because of this, it can be used as a universal adaptor protein mediating dimerization of pairs of target proteins. Provided we identify a suitable connecting helix, we can use the same helix to connect all other mutant protein A's for the following reasons. First, all the mutant protein A proteins adopt an essentially identical conformation, as shown by many crystal and NMR structures. Second, the C-terminal helix where our EY-CBS site is located is not changed in the mutant proteins because the mutations are limited to the first two α-helices. Among the known mutants, the Ztaq and anti-Ztaq proteins were selected for our study because they can form stable heterodimers[25]. To confirm that the mutations in the Ztaq and anti-Ztaq proteins do not affect the EY-CBS reaction, we replaced the protein A regions of fusions 8,155, 8,157 and 8,158 with the Ztaq or anti-Ztaq domain. The substituted fusion proteins retained similar reactivity with EY-CBS, as shown in Supplementary Fig. 16. Because the structure of the Ztaq-anti-Ztaq heterodimer was already known[25], we could predict the structures of the homo- and heterocomplexes 8,155–8,155, 8,157–8,157 and 8,155–8,157 with considerable confidence (Supplementary Fig. 20).

Thus, we have shown that universal adaptor proteins such as protein A or a coiled-coil domain can be inserted into a target protein to generate a predictable homo- or heterodimeric structure using our EY-CBS method.

Discussion

In this research, we have developed a novel method for connecting two proteins using a rigid fusion helix that is stabilized by a chemical cross linker. Using this method, we successfully fused the C-terminal helix of an ankyrin family protein to the N-terminal helix of a protein A fragment. We also succeeded inserting a protein A domain inside T4 lysozyme with predictable structure. Subsequent reaction with EY-CBS stabilized the structure of the fusion helix to that of an ideal helix. Our method, in principle, can be applied to majority of proteins as long as they contain at least one helix in their structure. In our method, EY-CBS reacts with thiol groups of cysteines. Therefore, proteins with natural cysteines may have unwanted reactions with EY-CBS. In such cases all the cysteines exposed to the surface need to be mutated before applying the method. Fortunately, most exposed cysteines found in proteins are not important functionally or structurally and can be converted into other amino acids without deleterious effects. In the PDB database of more than 100,000 protein structures, the majority have at least one α-helix and could therefore be used as building blocks for assembly of complex nanostructures by the helix-fusion method. Furthermore, application area of the helix-fusion method can be expanded by combining it with other methods like the metal chelation or the interface design method.

Several chemicals besides EY-CBS have been reported to stabilize the α-helical content of peptides[27]. Here are some notable examples of these helix stabilizers that can be applied to full size proteins. Two bisarylmethylene bromides named Bph and Bpy can selectively cross-link cysteines at the i and i + 7 positions in the α-helices[28]; the cross-linked peptides show increased helicity and cell permeability. Also spermine contains positively charged amines at 4.6 angstrom and 6.0 angstrom spacings, and has been shown to stabilize a peptide with four negatively charged glutamate or aspartate residues separated by

3–4 residues[29]. The stabilizing effect is highly dependent on pH, suggesting that ionic interactions play the major role in the helix stabilization. Similarly a diguanidium-containing compound has been shown to enhance the helical content of a peptide containing aspartate groups in the i and i + 3 positions[30]. These compounds may be able to replace EY-CBS in the helix-fusion method. Unlike EY-CBS, some of these helix stabilizing compounds do not use cysteines for cross linking. Therefore, they may be useful for connecting proteins with reactive cysteines.

Crystallization chaperones are proteins that can be easily produced and crystallized[31,32]. Fusing them to a target protein has proven successful in improving the crystallization property of the target protein. For example, T4 lysozyme has been inserted into a flexible loop of the β2-adrenergic receptor[33,34]. The authors generated a panel of fusion proteins and chose one where the linker between lysozyme and the receptor was rigid enough for crystallization. A similar method using cytochrome cb562 instead of T4 lysozyme has been essential for crystallizing several G-protein-coupled receptor receptors and determining their structures[35–38]. The success of this method relies on structural rigidity of the fusion site, because proteins with flexible structures cannot readily be crystallized. Since it is generally impossible to predict the structure of fusion sites, a large number of fusion proteins have to be screened and the linker sequence optimized for crystallization. Our method can be used to test whether a fusion protein has a rigid fusion site before extensive and time-consuming crystallization trials. For example, one can insert a helical chaperone protein like cytochrome cb562 into a loop preceding or following an α-helix in the target protein and connect the α-helices of the target and the chaperone protein using EY-CBS as shown in the crystal structure of 8,155 (Fig. 7). The resulting fusion protein will have a rigid helical connection and there will be more chance of successful crystallization.

In this study, we showed that reaction with EY-CBS could force the structure of a fusion helix into that of an ideal helix. The crystal structure demonstrated that the fusion helix was slightly bent at the fusion site before reaction with EY-CBS and was converted to a near-ideal helix after reaction with EY-CBS. For applications that involve crystallization of the fusion protein, the rigidity of the fusion protein is critical and reaction with EY-CBS would be desirable. However, in other applications small deviations from the ideal helical structure and some structural flexibility may be allowed. In such cases the EY-CBS reaction could be useful as a probe to test whether the two helices are indeed fused and a hybrid helix has been formed.

In conclusion, we have developed a new method that can convert two pre-existing α-helices into one long extended helix. This method can be used to link proteins to form a desired structure and may be useful in constructing complex protein nanoassemblies for a variety of purposes.

Methods

Design of fusion proteins. The fusion helices between pairs of proteins were designed using the molecular graphics program, COOT[39], assuming that their structures would not be significantly perturbed by the fusion. The coordinate files of the individual protein components were obtained from the PDB database[20,23,40,41]. The fusion helix was modelled by aligning the helices from the individual proteins with that of an ideal α-helix. Possible clashes between the two proteins were visually checked and avoided by addition or deletion of amino acid residues. Amino acid positions that were open to the solvent for easy approach of EY-CBS were chosen visually and mutated to cysteines for the EY-CBS reaction. Computer programs predicting protein structure were not used in our modelling.

Protein expression and purification. The genes of the fusion proteins were cloned into pET28a vector and over-expressed using E. coli strain BL21(DE3). Protein production was induced by adding 0.5 mM IPTG when OD 600 of the culture reached 0.7 and incubating the cells for an additional 20 h at 25 °C.

For proteins 3,305–3,315, glutathione-S-transferase was used as the purification tag. The cells were harvested by centrifugation, resuspended in lysis buffer containing 50 mM Tris pH 8.0, 200 mM NaCl and 10 mM β-mercaptoethanol and homogenized using a microfluidizer, model M-110L (Microfluidics). The supernatant after centrifugation was injected into a Glutathione-Sepharose (GE Healthcare) column equilibrated with a buffer containing 20 mM Tris pH 8.0, 200 mM NaCl and 1 mM dithiothreitol (DTT). The glutathione-S-transferase-tagged protein was eluted in a buffer containing 50 mM Tris pH 8.0, 100 mM NaCl, 10 mM glutathione and 1 mM DTT. The protein was cleaved overnight by thrombin to remove the purification tag and purified by a Q-Sepharose (GE Healthcare) anion exchange column equilibrated with a buffer containing 20 mM Tris 8.0 and 1 mM DTT. The bound protein was eluted by linear gradient of 1 M NaCl. The fractions containing the target protein were pooled and purified using a Supderdex 200 (GE Healthcare) gel filtration column equilibrated with a buffer containing 20 mM Tris pH 8.0, 200 mM NaCl and 1 mM TCEP.

For proteins 4,254–4,262, chitin-binding domain was used as the purification tag[42]. The harvested cells were homogenized in lysis buffer containing 50 mM Tris pH 8.0, 200 mM NaCl and 1 mM DTT. The supernatant after centrifugation was loaded onto a chitin affinity column (New England Biolab) equilibrated with 20 mM Tris 8.0, 200 mM NaCl and 1 mM DTT. The protein was eluted by thrombin cleavage and purified by a Q-Sepharose anion exchange column equilibrated with a buffer containing 20 mM Tris 8.0 and 1 mM DTT. The bound protein was eluted by linear gradient of 1 M NaCl. The fractions containing the target protein were pooled and purified using a Supderdex 200 gel filtration column equilibrated with a buffer containing 20 mM Tris pH 8.0, 200 mM NaCl and 1 mM TCEP.

Hexa-histidine was used as the purification tag for constructs 6,758–6,761. The harvested cells were homogenized in lysis buffer containing 50 mM Tris pH 8.0, 200 mM NaCl and 1 mM DTT. The supernatant was loaded onto a Ni-NTA affinity column equilibrated with 20 mM Tris 8.0, 200 mM NaCl and 1 mM CaCl$_2$. One millimolar DTT was added to the protein immediately after elution using 500 mM imidazole. The protein was cleaved overnight by thrombin to remove the purification tag. The cleaved protein was purified by Q-Sepharose anion exchange chromatography equilibrated with a buffer containing 20 mM Tris 8.0, 1 mM CaCl$_2$ and 1 mM DTT. The bound protein was eluted by linear gradient of 1 M NaCl. The fractions containing the target protein were pooled and purified using a Supderdex 200 gel filtration column equilibrated with a buffer containing 20 mM Tris pH 8.0, 200 mM NaCl, 1 mM CaCl$_2$ and 1 mM TCEP. The purified proteins were concentrated with an ultrafiltration kit (Amicon) and used for the EY-CBS reaction.

For proteins 8,132–8,136 and 8,155–8,158, hexa-histidine was used as the purification tag. The harvested cells were homogenized in lysis buffer containing 20 mM Tris pH 8.0, 200 mM NaCl, 0.1 mM phenylmethyl sulphonyl fluorideand 10 mM β-mercaptoethanol. The supernatant after centrifugation was loaded onto a Ni-NTA affinity column equilibrated with 20 mM Tris pH 8.0 and 200 mM NaCl. One millimolar TCEP was added immediately to the protein eluted using 300 mM imidazole. The eluted protein was reacted with EY-CBS according to the EY-CBS reaction protocol. The EY-CBS reacted protein was purified using a SP-Sepharose (GE Healthcare) cation exchange column equilibrated with a buffer containing 20 mM MES pH 5.5. The bound protein was eluted by linear gradient of 1 M NaCl. The fractions containing the target protein were pooled and cleaved overnight by thrombin to remove the purification tag. The cleaved protein was purified using a Supderdex 200 gel filtration column equilibrated with a buffer containing 20 mM Tris pH 8.0 and 200 mM NaCl.

EY-CBS reaction. EY-CBS was synthesized as described by Zhang et al.[19] The EY-CBS reaction, typically with 1 mg ml^{-1} fusion protein, was carried out at 23 °C for 1 h in 20 mM Tris HCl buffer at pH 8.0 and 200 mM NaCl. One millimolar TCEP was added to the reaction buffer to prevent oxidation of the cysteine during the reaction. Progress of the reaction was monitored by SDS–PAGE. The reaction was stopped and unreacted EY-CBS was removed by Superdex 200 gel filtration or SP-Sepharose ion exchange chromatography. The reacted and purified proteins were concentrated with an ultrafiltration kit and used for crystallization if necessary. The absorption spectrum of the protein after the EY-CBS reaction was measured with a JASCO UV-530 spectrophotometer.

The PEG-Maleimide reaction. For the PEG-Maleimide reaction, 10 mM Maleimide-PEG$_{11}$-Biotin (Thermo Scientific) or PEG-Maleimide 5000 (Nanocs) was added to fusion proteins that either had, or had not, been reacted with EY-CBS. The progress of the reaction was monitored by SDS–PAGE and the reaction was stopped by adding SDS–PAGE sample buffer containing 570 mM β-mercaptoethanol.

Crystallization. Purified and concentrated proteins were used for crystallization. Initial crystallization conditions were screened using a crystallization robot (Mosquito, TTP Labtech). Unreacted 3,311 fusion protein was crystallized in a buffer containing 0.1 M Bis-Tris pH 6.5, 45% v/v PEG 400 and 10 mM hexamine cobalt (III) chloride. The crystals were frozen in liquid nitrogen. Addition of cryoprotectant was not necessary. After EY-CBS reaction and addition of excess amount of MBP, the 3,311-EY-CBS–MBP complex protein was purified by Superdex 200 gel filtration chromatography and crystallized in a buffer containing 0.1 M MOPS pH 6.5, 0.2 M magnesium chloride and 36% w/v PEG 2000. Crystals were frozen in a buffer

Table 2 | Data collection and refinement statistics.

	3,311 + EY-CBS	3,311	6,761	8,155 + EY-CBS
Data collection				
Space group	$P2_12_12_1$	$P2_1$	$I4_1$	$C2$
Resolution (Å)	2.30–50	2.80–50	2.70–50	2.60–50
Coordinations::				
a, b, c (Å)	64.3, 75.8, 111.4	85.4, 220.0, 85.4	98.8, 98.8, 31.7	89.5, 149.9, 55.8
α, β, γ (°)	90.0, 90.0, 90.0	90.0, 60.0, 90.0	90.0, 90.0, 90.0	90.0, 122.7, 90.0
R_{sym}	0.063 (0.513)	0.051 (0.152)	0.114 (0.762)	0.079 (0.352)
$I/\sigma I$	31.9	25.3	17.2	18.3
Completeness (%)	99.7 (99.2)	92.9 (80.0)	98.7 (97.0)	99.1 (98.5)
Redundancy	6.8	3.2	6.5	3.5
Search probes (PDB code)	1SVX, 1DEE	1SVX, 1DEE	2SPZ, 1CLL	1LYD, 1DEE
Refinement				
No. of reflections	24,741	62,287	4,460	35,397
R_{work}/R_{free}	0.194/0.262	0.207/0.250	0.206/0.259	0.200/0.259
Twin				
Operator	NA	h, -k, h-l	NA	NA
Fraction	NA	0.46	NA	NA
No. of protein molecules	1 (MBP) +1 (3311)	12	1	2
No. of atoms	4,330	16,233	978	3,451
Protein	4,188	16,116	966	3,408
Ligands	30	0	0	60
Water	112	117	12	43
Average B factor (Å^2)	51.2	29.6	65.4	57.4
R.m.s. deviations				
Bond length (Å)	0.008	0.010	0.009	0.009
Bond angles (°)	1.080	1.081	1.105	1.451

EY-CBS, 3,3′-ethyne-1,2-diylbis-{6-[(chloroacetyl)amino]benzenesulfonic acid}; NA, not applicable; r.m.s., root mean square.
Highest resolution shell is shown in parenthesis.

containing the crystallization solution plus 14% PEG 2000 as the cryoprotectant. The 6,761 fusion protein was crystallized in a solution containing 10% w/v PEG 1000 and 10% w/v PEG 8000. Crystals were frozen in a buffer containing the crystallization solution plus 30% ethylene glycol as the cryoprotectant. The 8,155 protein reacted with EY-CBS was crystallized in a solution containing 1.84 M NaK phosphate pH 7.5. Crystals harvested with the crystallization buffer were flash frozen in a drop of paraffin oil (Hampton Research) using liquid nitrogen.

X-ray diffraction data and structure refinement. Diffraction data were collected at Pohang Accelerator Laboratory, beam lines 5C and 7A. Diffraction images were integrated and scaled with HKL2000. All the structures were solved by the molecular replacement method using the PHASER software[43]. Initial phasing data were then refined using PHENIX[44]. The program COOT was used for manual correction of the structure. The crystallographic data are summarized in Table 2. The 3,311 crystals were merohedrally twinned with a twin fraction of 0.46. Initially we could not solve the structure by the molecular replacement technique because the diffraction data were incorrectly indexed for the P6$_3$22 space group. Later we found that the crystals were almost perfectly twinned (Supplementary Fig. 21). Therefore, the data were re-indexed for space groups with lower symmetry. Promising solutions could be found for both the P6$_3$ and P2$_1$ space groups in molecular replacement calculations. However, refinement could not reduce the free R factor below 40% when the data were indexed for the P6$_3$ space group. Therefore, they were finally indexed for the P2$_1$ space group, and the structure was refined. The free R factor dropped from 0.369 to 0.261 after incorporation of the twin operator in the refinement protocol.

References

1. Lai, Y. T., King, N. P. & Yeates, T. O. Principles for designing ordered protein assemblies. *Trends Cell Biol.* **22**, 653–661 (2012).
2. Radford, R. J., Nguyen, P. C., Ditri, T. B., Figueroa, J. S. & Tezcan, F. A. Controlled protein dimerization through hybrid coordination motifs. *Inorg. Chem.* **49**, 4362–4369 (2010).
3. Radford, R. J., Nguyen, P. C. & Tezcan, F. A. Modular and versatile hybrid coordination motifs on alpha-helical protein surfaces. *Inorg. Chem.* **49**, 7106–7115 (2010).
4. Radford, R. J. & Tezcan, F. A. A superprotein triangle driven by nickel(II) coordination: exploiting non-natural metal ligands in protein self-assembly. *J. Am. Chem. Soc.* **131**, 9136–9137 (2009).
5. Brodin, J. D. *et al.* Metal-directed, chemically tunable assembly of one-, two- and three-dimensional crystalline protein arrays. *Nat. Chem.* **4**, 375–382 (2012).
6. Brodin, J. D. *et al.* Evolution of metal selectivity in templated protein interfaces. *J. Am. Chem. Soc.* **132**, 8610–8617 (2010).
7. Medina-Morales, A., Perez, A., Brodin, J. D. & Tezcan, F. A. *In vitro* and cellular self-assembly of a Zn-binding protein cryptand via templated disulfide bonds. *J. Am. Chem. Soc.* **135**, 12013–12022 (2013).
8. Salgado, E. N. *et al.* Metal templated design of protein interfaces. *Proc. Natl Acad. Sci. USA* **107**, 1827–1832 (2010).
9. Salgado, E. N., Faraone-Mennella, J. & Tezcan, F. A. Controlling protein-protein interactions through metal coordination: assembly of a 16-helix bundle protein. *J. Am. Chem. Soc.* **129**, 13374–13375 (2007).
10. Salgado, E. N., Lewis, R. A., Faraone-Mennella, J. & Tezcan, F. A. Metal-mediated self-assembly of protein superstructures: influence of secondary interactions on protein oligomerization and aggregation. *J. Am. Chem. Soc.* **130**, 6082–6084 (2008).
11. Song, W. J., Sontz, P. A., Ambroggio, X. I. & Tezcan, F. A. Metals in protein-protein interfaces. *Annu. Rev. Biophys.* **43**, 409–431 (2014).
12. Song, W. J. & Tezcan, F. A. A designed supramolecular protein assembly with in vivo enzymatic activity. *Science* **346**, 1525–1528 (2014).
13. King, N. P. *et al.* Computational design of self-assembling protein nanomaterials with atomic level accuracy. *Science* **336**, 1171–1174 (2012).
14. Padilla, J. E., Colovos, C. & Yeates, T. O. Nanohedra: using symmetry to design self assembling protein cages, layers, crystals, and filaments. *Proc. Natl Acad. Sci. USA* **98**, 2217–2221 (2001).
15. Lai, Y. T., Tsai, K. L., Sawaya, M. R., Asturias, F. J. & Yeates, T. O. Structure and flexibility of nanoscale protein cages designed by symmetric self-assembly. *J. Am. Chem. Soc.* **135**, 7738–7743 (2013).
16. Lai, Y. T., Cascio, D. & Yeates, T. O. Structure of a 16-nm cage designed by using protein oligomers. *Science* **336**, 1129 (2012).
17. Lai, Y. T. *et al.* Structure of a designed protein cage that self-assembles into a highly porous cube. *Nat. Chem.* **6**, 1065–1071 (2014).
18. Lai, Y. T., Jiang, L., Chen, W. & Yeates, T. O. On the predictability of the orientation of protein domains joined by a spanning alpha-helical linker. *Protein Eng. Des. Sel.* **28**, 491–500 (2015).
19. Zhang, F., Sadovski, O., Xin, S. J. & Woolley, G. A. Stabilization of folded peptide and protein structures via distance matching with a long, rigid cross-linker. *J. Am. Chem. Soc.* **129**, 14154–14155 (2007).

20. Binz, H. K. *et al.* High-affinity binders selected from designed ankyrin repeat protein libraries. *Nat. Biotech.* **22**, 575–582 (2004).
21. Nygren, P. A. Alternative binding proteins: affibody binding proteins developed from a small three-helix bundle scaffold. *FEBS J.* **275**, 2668–2676 (2008).
22. Pluckthun, A. Designed ankyrin repeat proteins (DARPins): binding proteins for research, diagnostics, and therapy. *Annu. Rev. Pharmacol. Toxicol.* **55**, 489–511 (2015).
23. Chattopadhyaya, R., Meador, W. E., Means, A. R. & Quiocho, F. A. Calmodulin structure refined at 1.7A resolution. *J. Mol. Biol.* **228**, 1177–1192 (1992).
24. Meador, W. E., Means, A. R. & Quiocho, F. A. Target enzyme recognition by calmodulin: 2.4A structure of a calmodulin-peptide complex. *Science* **257**, 1251–1255 (1992).
25. Lendel, C., Dogan, J. & Hard, T. Structural basis for molecular recognition in an affibody:affibody complex. *J. Mol. Biol.* **359**, 1293–1304 (2006).
26. Feldwisch, J. & Tolmachev, V. Engineering of affibody molecules for therapy and diagnostics. *Methods Mol. Biol.* **899**, 103–126 (2012).
27. Henchey, L. K., Jochim, A. L. & Arora, P. S. Contemporary strategies for the stabilization of peptides in the alpha-helical conformation. *Curr. Opin. Chem. Biol.* **12**, 692–697 (2008).
28. Muppidi, A., Wang, Z., Li, X., Chen, J. & Lin, Q. Achieving cell penetration with distance-matching cysteine cross-linkers: a facile route to cell-permeable peptide dual inhibitors of Mdm2/Mdmx. *Chem. Commun.* **47**, 9396–9398 (2011).
29. Tabet, M., Labroo, V., Sheppard, P. & Sasaki, T. Spermine-Induced Conformational-Changes of a Synthetic Peptide. *J. Am. Chem. Soc.* **115**, 3866–3868 (1993).
30. Albert, J. S., Goodman, M. S. & Hamilton, A. D. Molecular recognition of proteins—sequence-selective binding of aspartate pairs in helical peptides. *J. Am. Chem. Soc.* **117**, 1143–1144 (1995).
31. Koide, S. Engineering of recombinant crystallization chaperones. *Curr. Opin. Struct. Biol.* **19**, 449–457 (2009).
32. Lieberman, R. L., Culver, J. A., Entzminger, K. C., Pai, J. C. & Maynard, J. A. Crystallization chaperone strategies for membrane proteins. *Methods* **55**, 293–302 (2011).
33. Cherezov, V. *et al.* High-resolution crystal structure of an engineered human beta2-adrenergic G protein-coupled receptor. *Science* **318**, 1258–1265 (2007).
34. Rosenbaum, D. M. *et al.* GPCR engineering yields high-resolution structural insights into beta2-adrenergic receptor function. *Science* **318**, 1266–1273 (2007).
35. Liu, W. *et al.* Structural basis for allosteric regulation of GPCRs by sodium ions. *Science* **337**, 232–236 (2012).
36. Thompson, A. A. *et al.* Structure of the nociceptin/orphanin FQ receptor in complex with a peptide mimetic. *Nature* **485**, 395–399 (2012).
37. Fenalti, G. *et al.* Molecular control of delta-opioid receptor signalling. *Nature* **506**, 191–196 (2014).
38. Wu, H. *et al.* Structure of a class C GPCR metabotropic glutamate receptor 1 bound to an allosteric modulator. *Science* **344**, 58–64 (2014).
39. Emsley, P., Lohkamp, B., Scott, W. G. & Cowtan, K. Features and development of Coot. *Acta Crystallogr. D Biol. Crystallogr.* **66**, 486–501 (2010).
40. Graille, M. *et al.* Crystal structure of a Staphylococcus aureus protein A domain complexed with the Fab fragment of a human IgM antibody: Structural basis for recognition of B-cell receptors and superantigen activity. *Proc. Natl Acad. Sci. USA* **97**, 5399–5404 (2000).
41. Mooers, B. H. *et al.* Repacking the Core of T4 lysozyme by automated design. *J. Mol. Biol.* **332**, 741–756 (2003).
42. Mitchell, S. F. & Lorsch, J. R. Protein affinity purification using intein/chitin binding protein tags. *Methods Enzymol.* **559**, 111–125 (2015).
43. McCoy, A. J. *et al.* Phaser crystallographic software. *J. Appl. Crystallogr.* **40**, 658–674 (2007).
44. Adams, P. D. *et al.* PHENIX: a comprehensive Python-based system for macromolecular structure solution. *Acta Crystallogr. D Biol. Crystallogr.* **66**, 213–221 (2010).

Acknowledgements

We thank the staff of the beam lines 5C and 7A, Pohang Accelerator Laboratory for help with data collection, and Dr Julian Gross for critical reading of the manuscript. This research was supported by the Intelligent Synthetic Biology Center of Global Frontier Project (2011-0031955) and the National Research Foundation (NRF-2014R1A2A1A10050436) funded by the Ministry of Science, ICT and Future Planning of Korea.

Author contributions

W.H.J., H.L., D.H.S. and J.-H.E. performed experiments and data analysis. W.H.J., H.L., D.H.S., S.C.K., H.-S.L., H.L. and J.-O.L. designed experiments. W.H.J., H.L., S.C.K., H.-S. L., H.L. and J.-O.L. contributed to writing the paper. J.-O.L. supervised the project.

Additional information

Accession codes: Atomic coordinates and diffraction data have been deposited in the Protein Data Bank with code numbers 5CBN, 5CBO, 5COC and 5EWX for the 3311 + EY-CBS, 3311, 6761 and 8155 + EY-CBS structures, respectively.

Competing financial interests: The authors declare no competing financial interests.

20

Microelectromechanical reprogrammable logic device

M.A.A. Hafiz[1], L. Kosuru[1] & M.I. Younis[1]

In modern computing, the Boolean logic operations are set by interconnect schemes between the transistors. As the miniaturization in the component level to enhance the computational power is rapidly approaching physical limits, alternative computing methods are vigorously pursued. One of the desired aspects in the future computing approaches is the provision for hardware reconfigurability at run time to allow enhanced functionality. Here we demonstrate a reprogrammable logic device based on the electrothermal frequency modulation scheme of a single microelectromechanical resonator, capable of performing all the fundamental 2-bit logic functions as well as n-bit logic operations. Logic functions are performed by actively tuning the linear resonance frequency of the resonator operated at room temperature and under modest vacuum conditions, reprogrammable by the a.c.-driving frequency. The device is fabricated using complementary metal oxide semiconductor compatible mass fabrication process, suitable for on-chip integration, and promises an alternative electromechanical computing scheme.

[1] Physical Sciences and Engineering Division, King Abdullah University of Science and Technology, Thuwal 23955-6900, Saudi Arabia. Correspondence and requests for materials should be addressed to M.I.Y. (email: mohammad.younis@kaust.edu.sa).

The quest for mechanical computation is a century old and can be traced back to at least 1822 when Babbage presented his concept of difference engine[1]. Although the interest remained within the research community, the subsequent development in the fields of electronic transistor[2] and magnetic storage[3,4] outperformed the mechanical approach in computation both in terms of speed of operation and data density. However, recent advancements in micro-/nano-fabrication and measurement techniques have renewed the interest in the field of mechanical computation in the last decade[5–21].

The key to any computing machine are logic elements. The first demonstrated dynamic mechanical XOR logic gate was based on a piezoelectric nanoelectromechanical system (NEMS) structure where the presence (absence) of high-amplitude vibration in the linear regime denotes a logical high (low) state[7]. Later, OR/NOR and AND/NAND logic gates have been demonstrated utilizing the bistability of a nonlinearly resonating NEMS resonator mediated by the noise floor[12]. A universal logic device capable of performing AND, OR and XOR logic gates as well as multibit logic circuits has been implemented by parametrically exciting a single electromechanical resonator[15]. Same research group also demonstrated XOR and OR logic gates in an electromechanical membrane resonator under high vacuum and at room temperature condition[16]. On the basis of feedback control, a memory and OR logic operation have been demonstrated on a single microelectromechanical system (MEMS) resonator working in the nonlinear regime[20]. Recently, an unconventional and reversible logic gate (Fredkin gate) has been presented based on four coupled linearly resonating NEMS resonators[21] where AND, OR, NOT and FANOUT gate operations have been demonstrated. Note that room temperature and atmospheric operations are desirable prerequisites for any practical device implementation.

Here we demonstrate a reprogrammable logic device, capable of performing 2-bit AND, NAND, OR, NOR, XOR, XNOR and NOT logic operations using a single microelectromechanical resonator operating in the linear regime. The logic operations are performed by electrothermal modulation of the linear resonance frequency of the resonator, where two separate d.c. voltage sources represent logic inputs. The device can be programmed to perform any of these logic operations by simply tuning the a.c.-driving frequency. Also, we use this scheme of electrothermal frequency tuning to demonstrate 3-bit AND, NAND, OR and NOR logic gates on a single MEMS resonator. This can be extended to n-bit logic operations by adding a single d.c. voltage source per bit. This device works under room temperature and modest vacuum conditions and is fabricated using standard complementary metal oxide semiconductor-based fabrication techniques suitable for mass fabrication and on-chip integrated system development.

Results

Device fabrication and experimental set-up. The resonator is fabricated on a highly conductive Si device layer of silicon on insulator wafer by a two-mask process using standard photolithography, electron beam evaporation for metal layer deposition for actuating pad, deep reactive ion etch for silicon device layer etching and vapour hydrofluoric acid etch to remove the oxide layer underneath the resonating structure. It consists of a clamped–clamped arch-shaped microbeam with two adjacent electrodes to electrostatically induce the vibration and detect the generated a.c. output current due to the in-plane motion of the microbeam. The dimensions of the curved beam are 500 μm in length, 3 μm in width and 30 μm in thickness. The gap between the actuating electrode and the resonating beam is 8 μm at the

fixed anchors and 11 μm at the midpoint of the microbeam due to its 3-μm initial curvature.

Figure 1a shows the schematic of the arch microbeam and the two-port electrical transmission measurement configuration for electrostatic actuation and sensing that includes the parasitic current compensation circuit for enhanced transmission signal measurements[22]. The drive electrode is provided with an a.c. actuation signal from one of the outputs of a single-to-differential driver (AD8131), and the beam electrode is biased with a d.c. voltage source. The output current induced at the sense electrode is coupled with the variable compensation capacitor, C_{comp}, and followed by a low-noise amplifier whose output is coupled to the network analyser input port. Two logic inputs are provided with two d.c. voltage sources, V_A and V_B, connected in parallel across the microbeam with series resistors, R_A and R_B, and switches, A and B, respectively. The electrical wiring scheme for the logic inputs is depicted in red to differentiate it from the rest of the electrical connections. The binary logic input 1(0) is represented by connecting (disconnecting) V_A and V_B from the electrical network by the two switches, A and B, respectively. Hereafter, switch ON (OFF) condition for switches A and B corresponds to

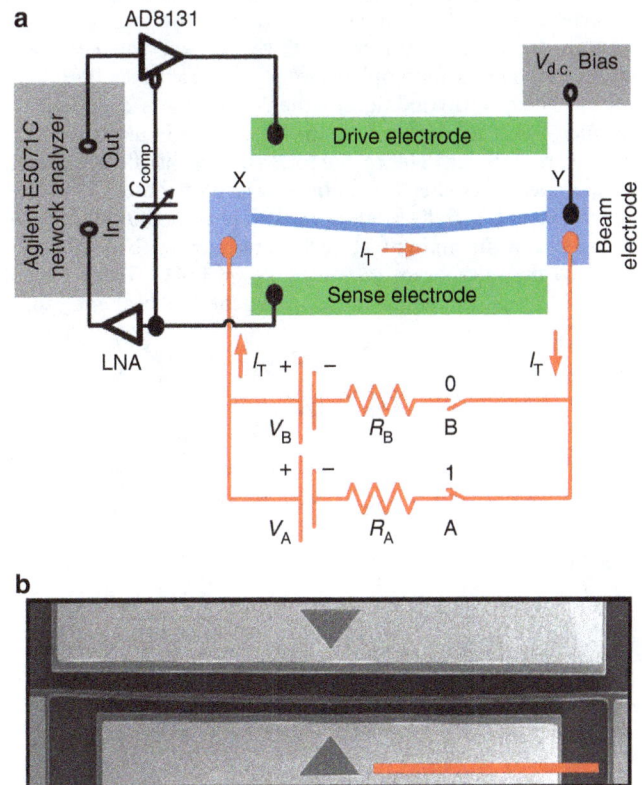

Figure 1 | Clamped–clamped arch resonator. (a) Schematic of the arch beam resonator and the two-port electrical transmission measurement configuration together with a parasitic current compensation circuitry using single-to-differential driver (AD8131) and a variable compensation capacitor, C_{comp}. The drive electrode is provided with an a.c. signal from one of the outputs from AD8131 and the beam electrode is biased with a d.c. voltage source. The output current induced at the sense-electrode is coupled with the compensation capacitor and followed by a low-noise amplifier (LNA) whose output is coupled to the network analyser input port. Two voltage sources, V_A and V_B and switches, A and B are connected in parallel across the beam to perform logic operations by electrothermal tuning of the resonance frequency. The arrow in the red represents the current flowing through the beam, responsible for electrothermal frequency modulation. **(b)** An SEM image of the microbeam resonator. Scale bar, 200 μm.

the binary logic input 1(0). The sensing electrode is used to obtain the logic output, where a relative high (low) S_{21} transmission signal corresponds to the logic output 1(0). Figure 1b shows an SEM image of the arch microbeam resonator.

Electrothermal frequency modulation. Electrothermal frequency modulation has an essential role in the execution of the logic functions in this architecture. Figure 2 shows four different electrical circuit configurations between nodes X and Y, shown in Fig. 1a. All the four logic input conditions, (0,0), (0,1), (1,0) and (1,1) are shown in Fig. 2a–d, respectively. For the case of (0,0) logic input condition, the total current flowing through the microbeam is $I_T = 0$ as depicted in the electrical circuit in Fig. 2a. In this case, the resonator exhibits series resonance peak and parallel resonance dip (anti-resonance) at 117.663 and 117.361 kHz, respectively, with an a.c. actuation voltage of 2 dBm (0.28 V_{rms}) and $V_{d.c.}$ of 45 V at 1 torr pressure and at room temperature (see Supplementary Note 1 and Supplementary Fig. 1a,b). The corresponding frequency response is plotted in black in Fig. 3. Note that due to over compensation of the feed through by the parallel variable compensation capacitance, C_{comp}, the parallel resonance appears earlier than the series resonance[22]. However, this does not put any limitation on the successful logic operation by the device. Moreover, we use both the series and parallel resonances for implementing the logic gates. For logic input (0,1) or (1,0) conditions, either V_B or V_A is connected to the microbeam as depicted in the electrical circuits shown in Fig. 2b,c, respectively. Hence, the total current that flows through the microbeam is either $I_T = I_B$ or $I_T = I_A$. We chose $V_A = 0.4$ V, $V_B = 0.7$ V, and $R_A = R_B = 50\,\Omega$ so that it satisfies the condition of the same current amount at each case; $I_A = I_B$. Note that we measured the microbeam resistance $R_{MB} = 114\,\Omega$. The electrical current flowing through the microbeam generates heat and causes

thermal expansion, which induces compressive axial force. This compressive force causes an increase in the microbeam curvature[23-25] and increases its stiffness. Hence, the series resonance frequency increases to 121.431 kHz for either (0,1) or (1,0) logic input conditions. The frequency responses due to the logic input (0,1) and (1,0) conditions are plotted as red and blue, respectively, in Fig. 3. For logic input condition (1,1), both the voltage sources V_A and V_B are connected to the microbeam as depicted in the electrical circuit shown in Fig. 2d. The total current generated in this case is $I_T = I'_A + I'_B > I_A$ or I_B. Hence, the series resonance frequency further increases to 128.969 kHz as depicted in green in Fig. 3. Thus, one can modulate the resonance frequencies (series and parallel) of the microbeam through the electrothermal effect by controlling the amount of current flow in the microbeam. Towards this, we build different logic gates by properly choosing the a.c.-driving frequency. We identify three regions in the frequency response plot of Fig. 3 to build all the six logic gates. Region I corresponds to frequency of operation for logic gates OR/NOR, region II corresponds to logic gates XOR/XNOR and finally, region III corresponds to logic gates AND/NAND. NOT logic operation can be built on any of these frequencies by proper conditioning of one of the inputs. The detail execution of the logic gates will be discussed in the following sections.

NOR/OR. The frequency responses of the resonator for different logic input conditions are shown in Fig. 4a, which lies in the region I of Fig. 3. To demonstrate NOR gate operation, the frequency of 117.663 kHz is chosen as it shows high S_{21} transmission signal denoted as the logic output 1 (in black) for (0,0) logic input

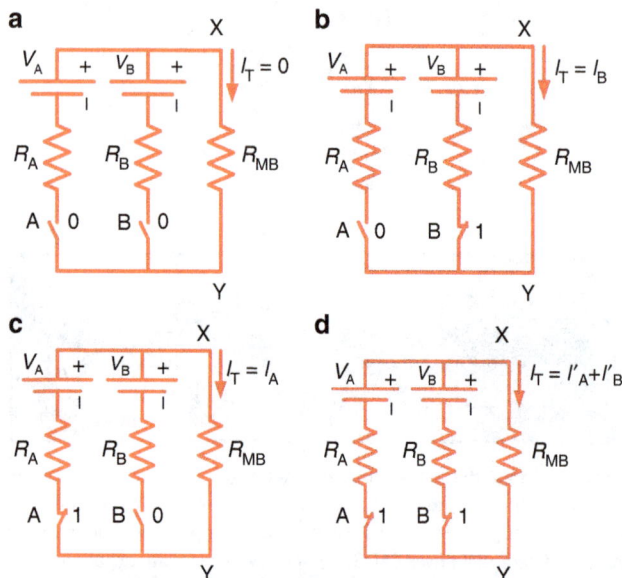

Figure 2 | Electrical circuit configuration of the logic input conditions.
(**a**) The electrical circuit represents the (0,0) logic input condition where the total current I_T through the beam R_{MB} is zero. (**b**) The circuit represents the (0,1) logic input condition corresponds to switch A, OFF and switch B, ON where the total current I_T flowing through the beam R_{MB} is I_B.
(**c**) The circuit represents the (1,0) logic input condition corresponds to switch A, ON and switch B, OFF where the total current I_T flowing through the beam R_{MB} is I_A. (**d**) The circuit represents the (1,1) logic input condition corresponds to switch A, ON and switch B, ON where the total current I_T flowing through the beam R_{MB} is $I'_A + I'_B$.

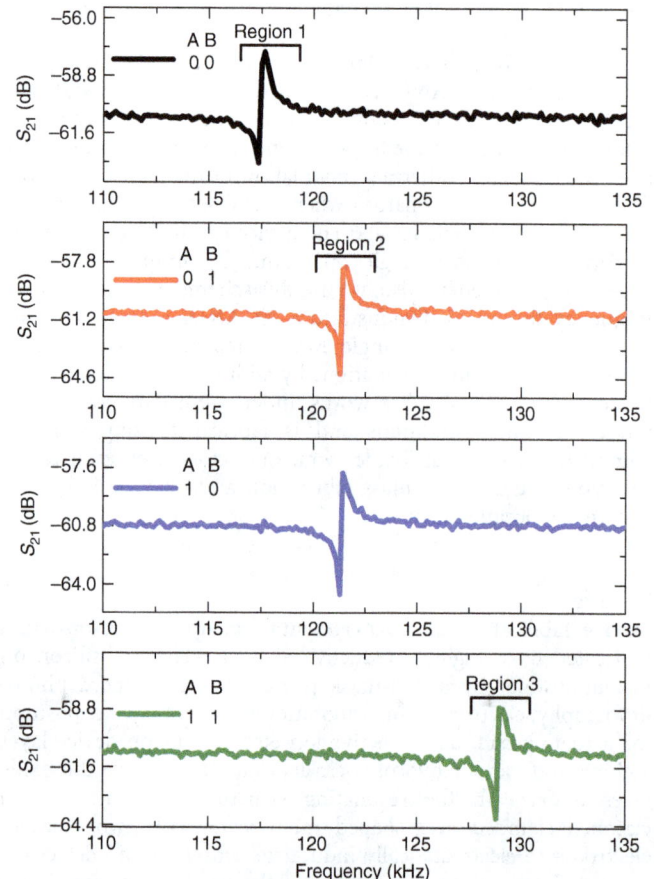

Figure 3 | Electrothermal frequency modulation. Frequency responses of the resonator for different logic input conditions, (0,0), (0,1), (1,0) and (1,1), shown in black, red, blue and green, respectively.

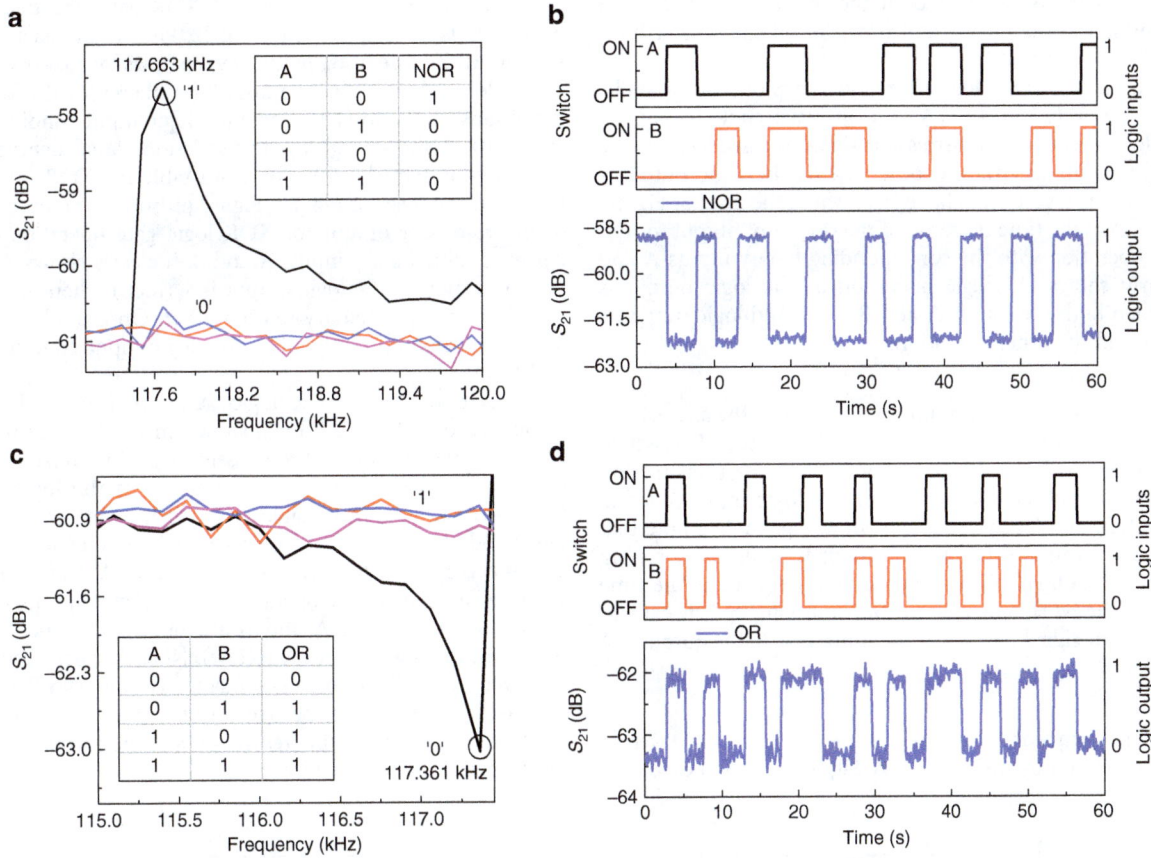

Figure 4 | Demonstration of 2-bit NOR and OR logic gates. (**a**) Frequency responses of the resonator for different logic input conditions where (0,0) logic input condition, shown in black has high S_{21} transmission signal at 117.663 kHz and others have low S_{21} transmission signal represented by 1 and 0, respectively. Truth table of NOR logic output is shown in the inset. (**b**) Demonstration of NOR logic operation when the frequency of the a.c. input signal is chosen as 117.663 kHz. Two input signals A and B are shown in black and red, respectively, where the switch OFF/ON corresponds to 0/1 logic input conditions. The S_{21} transmission signal in blue corresponds to the logic output and fulfills the NOR truth table. (**c**) Frequency responses of the resonator for different logic input conditions, where (0,0) logic input condition shown in black has low S_{21} transmission signal at 117.361 kHz and others have high S_{21} transmission signal, represented by 0 and 1, respectively. Truth table for OR logic output shown in the inset. (**d**) Demonstration of OR logic operation when the a.c. input signal frequency is chosen as 117.361 kHz. Two input signals, A and B are shown in black and red, respectively, and the switch OFF/ON corresponds to 0/1 logic input conditions. The S_{21} transmission signal in blue corresponds to the logic output that fulfills the OR truth table.

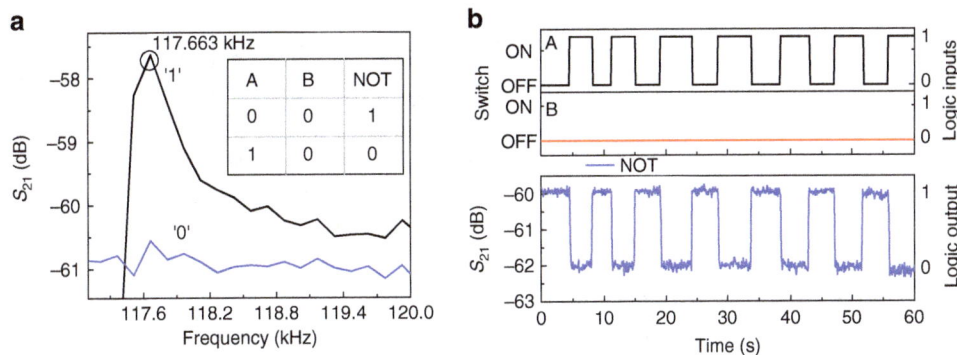

Figure 5 | Demonstration of NOT gate. (**a**) Frequency responses of the resonator for different logic input conditions, where (0,0) logic input condition shown in black has high S_{21} transmission signal at 117.663 kHz and others have low S_{21} transmission signal represented by 1 and 0, respectively. Truth table of NOT logic gate is shown in the inset. (**b**) Demonstration of NOT logic operation when the frequency of the a.c. input signal is chosen as 117.663 kHz. Two input signals A and B are shown in black and red, respectively, where the switch OFF/ON corresponds to 0/1 logic input conditions. S_{21} transmission signal in blue corresponds to the logic output and fulfills the NOT truth table.

condition only. The resonator is tuned away from its series resonance frequency of 117.663 kHz by other logic input conditions, (0,1), (1,0) and (1,1), respectively. Hence, shows low S_{21} transmission signal denoted as logic output 0 (in black) at the

frequency of 117.663 kHz. The NOR gate truth table is shown in the inset of Fig. 4a. The time response of the resonator showing binary inputs A and B and the corresponding logic output is depicted in Fig. 4b. It clearly shows NOR logic operation as the output is 1 (high) only when both the inputs A and B are 0 (switch OFF), and the output is 0 (low) for all the other conditions, (0,1), (1,0) and (1,1).

To demonstrate OR logic gate, we exploit the parallel resonance dip at 117.361 kHz, shown in black circle in Fig. 4c. Here the low level of S_{21} transmission signal is considered as the logic output 0 (in green), and otherwise as the logic output 1 (in green). The OR gate truth table is shown in the inset of Fig. 4c. Figure 4d shows the time response of the resonator output for OR logic gate operation with the corresponding binary inputs A and B. It clearly shows OR logic operation as the logic output is 0 (low) when both the inputs A and B are 0, and logic output is 1 (high) for all the other conditions.

NOT. To perform NOT operation on the input A, the a.c.-driving frequency is set to be at 117.663 kHz and the input B is set to 0 (switch OFF). For this set condition, a high S_{21} transmission signal (logic output 1) is achieved for the logic input A set at 0 (switch OFF) and vice versa as shown in Fig. 5a. We note that NOT operation can also be built on input B by properly setting input A (switch OFF/ON) and a.c.-driving frequency. The time response for the NOT operation is shown in Fig. 5b. It is evident from the output signal that when the input A is 0, the output is 1 and vice versa.

XOR/XNOR. Frequency responses of the resonator for different logic input conditions are shown in Fig. 6a, which lies in the

region II of Fig. 3. To implement XOR gate, the frequency of operation is chosen as 121.431 kHz, shown in black circle in Fig. 6a. At this operating frequency, it shows low S_{21} transmission signal denoted as the logic output 0 (in black) for the logic input conditions (0,0) and (1,1). For other logic input conditions, (0,1) and (1,0), it shows high S_{21} transmission signal denoted as the logic output 1 (in black). The truth table for XOR logic gate is shown in the inset of Fig. 6a. Figure 6b shows the time response of the resonator output for XOR logic gate operation with the corresponding binary inputs A and B. It clearly shows XOR logic gate operation as the logic output is 1 (high) when the inputs A and B are complementary to each other. On the other hand, the logic output is 0 (low) for the same logic input conditions, (0,0) and (1,1).

To demonstrate XNOR logic gate, we exploit the parallel resonance dip at 121.281 kHz, shown in black circle in Fig. 6c. Here the low level of S_{21} transmission signal is considered as the logic output 0 (in green), and otherwise as the logic output 1 (in green). XNOR truth table is shown in the inset of Fig. 6c. Figure 6d shows the time response of XNOR logic gate output and the corresponding binary logic inputs A and B. It clearly shows XNOR logic gate operation as the logic output is 1 (high) when both the inputs A and B are same, (0,0) and (1,1), and otherwise the logic output is 0 (low). Note that occasional spikes observed in the S_{21} transmission signal (in blue) in Fig. 6b,d are due to the switching between (0,1) and (1,0) logic input conditions. However, the resonator still performs the desired logic operations successfully.

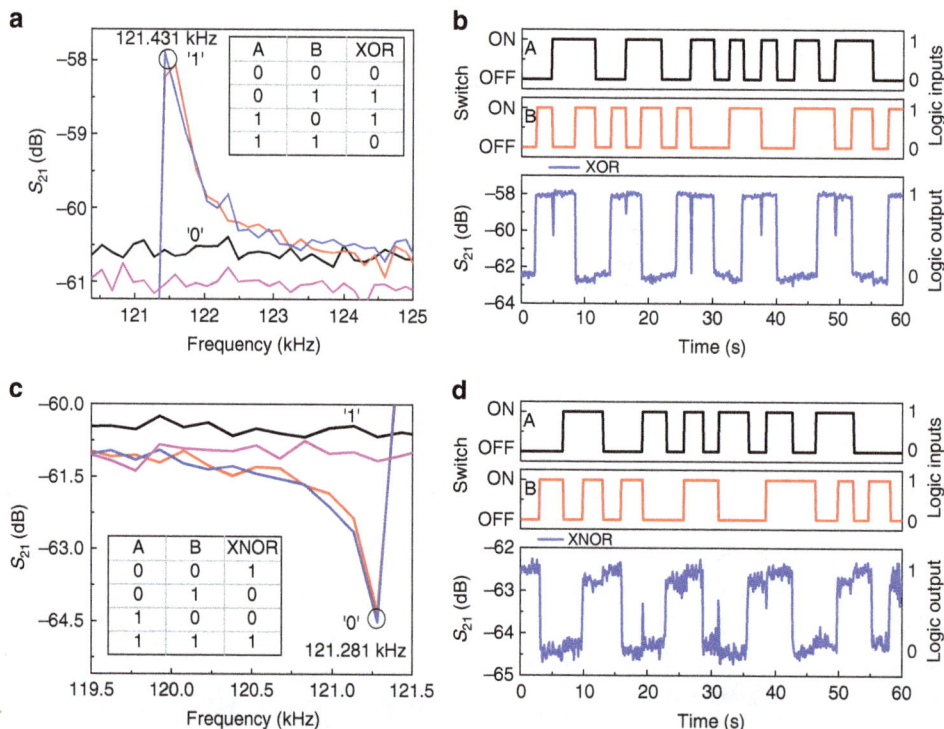

Figure 6 | Demonstration of 2-bit XOR and XNOR logic gates. (**a**) Frequency responses of the resonator for different logic input conditions, where (0,1) and (1,0) logic input condition shown in red and blue has high S_{21} transmission signal at 121.43 kHz and others have low S_{21} transmission signal represented by 1 and 0, respectively. Truth table of XOR logic gate is shown in the inset. (**b**) Demonstration of XOR logic operation when the operation frequency is chosen as 121.43 kHz. Two input signals, A and B are shown in black and red, respectively, where the switch OFF/ON corresponds to 0/1 logic input conditions. S_{21} transmission signal in blue corresponds to the logic output that fulfills the XOR truth table. (**c**) Frequency responses of the resonator for different logic input conditions, where (0,0) and (1,1) logic input conditions shown in red and blue, respectively, has low S_{21} transmission signal at 121.281 kHz and others have high S_{21} transmission signal represented by 0 and 1, respectively. Truth table of XNOR logic output is shown in the inset. (**d**) Demonstration of XNOR logic operation when the operating frequency is fixed at 121.281 kHz. Two input signals, A and B are shown in black and red, respectively, where the switch OFF/ON corresponds to 0/1 logic input conditions. S_{21} transmission signal in blue corresponds to the logic output and fulfills the XNOR truth table.

AND/NAND. Frequency responses of the resonator for different logic input conditions are shown in Fig. 7a, which falls in the region III of Fig. 3. To demonstrate AND gate operation, the frequency of 128.969 kHz is chosen, which is shown in black circle in Fig. 7a. When both the inputs A and B are 1 (switch ON), the high S_{21} transmission signal is observed at this operating frequency and denoted as the logic output 1 (in black). For other logic input conditions, (0,1), (1,0) and (0,0), it shows the low S_{21} transmission signal, which is denoted as the logic output 0 (in black). This is expressed in a truth table in the inset of Fig. 7a. The time response of the resonator for AND gate operation and the corresponding binary logic inputs A and B are shown in Fig. 7b. It clearly shows AND gate operation as the output is 1 (high) only when both the inputs A and B are 1, otherwise 0 (low).

To demonstrate NAND gate, the frequency of operation is chosen at 128.819 kHz, shown in black circle in Fig. 7c. Here the low level of S_{21} transmission signal of the parallel resonance dip is considered as the logic output 0 (in green), and otherwise as the logic output 1 (in green). NAND gate truth table is shown in the inset of Fig. 7c. Figure 7d shows the time response of NAND logic gate output and the corresponding binary logic inputs A and B. It shows NAND logic operation as the logic output is 0 (low) only when both the inputs A and B are 1 (switch ON).

Three-bit logic gates. We also implemented 3-bit logic gates by adding a third voltage source V_C (0.44 V) with series resistor R_C (50 Ω) and switch C, connected in parallel with the other two voltage sources, V_A (0.4 V) and V_B (0.7 V), in the electrical circuit

shown in Fig. 1a. Figure 8 shows the frequency responses of the resonator for different logic input conditions with an a.c. actuation voltage of 2 dBm (0.28 V_{rms}) and $V_{d.c.}$ of 40 V at 1 torr pressure and at room temperature. Three-bit NOR gate is realized by choosing the a.c.-driving frequency at 119.022 kHz marked in black circle as shown in Fig. 8. For (0,0,0) logic input condition, the frequency response shows high S_{21} transmission signal corresponds to the logic output (1). For all the other logic input conditions, the response shows low S_{21} transmission signal at this frequency, which corresponds to the logic output (0). Similar to the 2-bit OR logic operation, a 3-bit OR logic function can be realized by selecting the frequency of the anti-resonance dip as the a.c.-driving frequency. Next, a 3-bit AND gate is realized by choosing the frequency of operation at 132.105 kHz marked in black circle in Fig. 8, where only (1,1,1) logic input condition shows high S_{21} transmission signal corresponds to the logic output (1). For all the other logic input conditions shows low S_{21} transmission signal corresponds to the logic output (0). By selecting the corresponding anti-resonance frequency as the a.c.-driving frequency, a 3-bit NAND gate can be realized. Figure 9a shows the time response of the 3-bit NOR logic gate at the operation frequency of 119.022 kHz. Three logic input signals, A, B and C are shown in black, red and blue, respectively, where the switch OFF/ON corresponds to 0/1 logic input conditions. S_{21} transmission signal in green corresponds to the logic output, and fulfills the NOR truth table. Figure 9b shows the demonstration of a 3-bit AND logic function at the a.c.-driving frequency of 132.105 kHz. The output response shown in green fulfills the AND truth table.

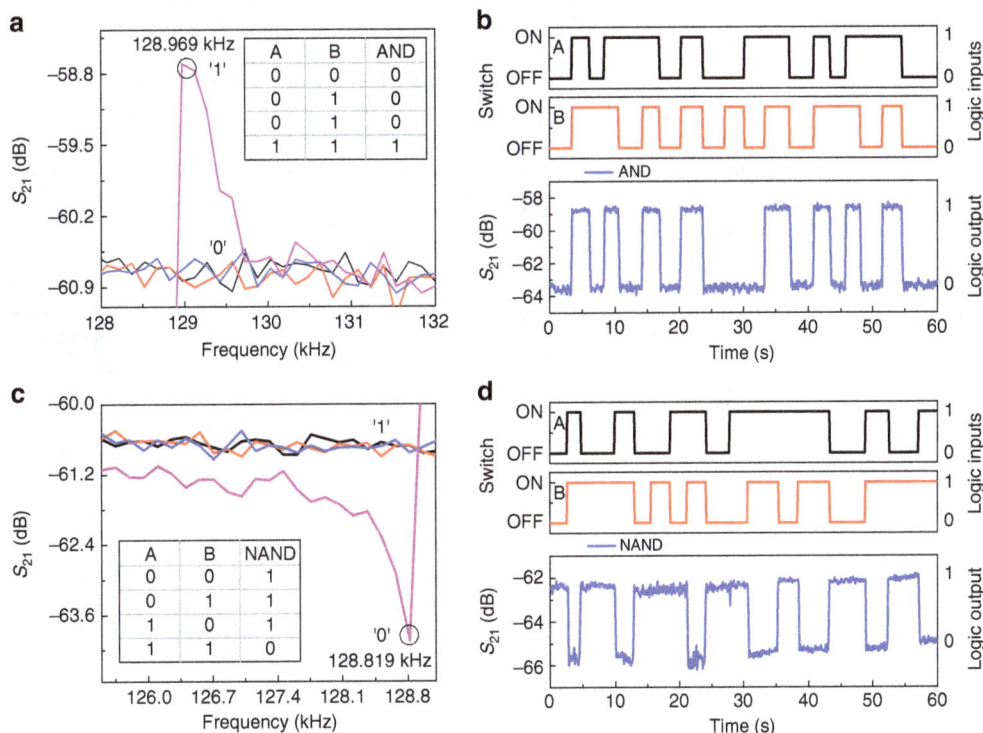

Figure 7 | Demonstration of 2-bit AND and NAND logic gates. (**a**) Frequency responses of the resonator for different logic input conditions, where (1,1) logic input condition shown in magenta has high S_{21} transmission signal at 128.969 kHz and others have low signal represented by 1 and 0, respectively. Truth table of AND logic output is shown in the inset. (**b**) Demonstration of AND logic operation when the operation of frequency is chosen as 128.969 kHz. Two input signals, A and B are shown in black and red, respectively, where the switch OFF/ON corresponds to 0/1 logic input conditions. S_{21} transmission signal in blue corresponds to the logic output and fulfills the AND truth table. (**c**) Frequency responses of the resonator for different logic input conditions, where (1,1) logic input condition has low S_{21} transmission signal at 128.819 kHz and others have high S_{21} transmission signal represented by 0 and 1, respectively. Truth table of NAND logic output is shown in the inset. (**d**) Demonstration of NAND logic operation when the operation of frequency is chosen as 128.819 kHz. Two input signals, A and B are shown in black and red, respectively, where the switch OFF/ON corresponds to 0/1 logic input conditions. S_{21} transmission signal in blue corresponds to the logic output and fulfills the NAND truth table.

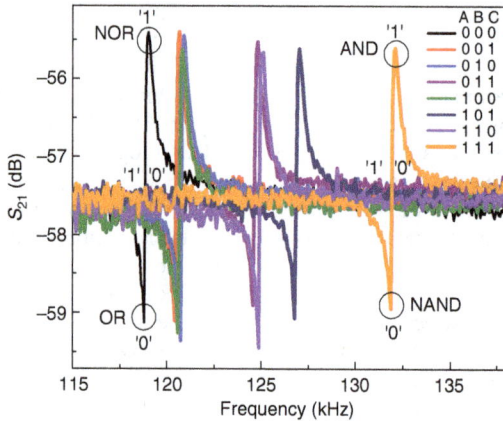

Figure 8 | Realization of 3-bit logic gates. Frequency responses of the resonator for three different input logic conditions. NOR gate is realized by choosing the frequency of operation at 119.022 kHz, where (0,0,0) logic input condition has high S_{21} transmission signal and all the others have low S_{21} transmission signal. By choosing the corresponding anti resonance dip frequency, 3-bit OR gate can be realized. A 3-bit AND gate is realized by choosing the frequency of operation at 132.105 kHz, where (1,1,1) logic input condition has high S_{21} transmission signal and all others have low S_{21} transmission signal. By choosing the corresponding anti resonance dip frequency, a 3-bit NAND gate can be realized.

One final remark is regarding the chosen d.c. bias voltage of this study. The demonstrated logic gates can be also operated at lower d.c. bias voltage. For example, we demonstrated a 2-bit NOR logic gate with a 20 V d.c. bias voltage (see Supplementary Note 2 and Supplementary Fig. 2a,b).

Discussion

An important feature of a logic gate is the operation speed. The speed of operation of the proposed logic device is governed by the speed of the electrothermal frequency modulation and the resonator switching speed. The characteristic time associated with electrothermal heating and cooling is typically much longer than the period of free vibrations of MEMS/NEMS structures[26,27]. Hence, electrothermal actuators have been mainly explored for static or low-frequency operations[26,27]. It is possible to calculate the thermal time constant of the microbeam[28,29] using the equation $\tau = \left[\frac{\pi^2 K_{Si}}{c \rho l^2} + \frac{F_s K_{air}}{g t c \rho} \right]^{-1}$, where l is the length of the microbeam, g is the gap between the beam and the substrate, t is the thickness of the beam, ρ is the density of silicon, c is the heat capacity of silicon, K_{Si} and K_{air} are the thermal conductivity of silicon and air, respectively. The beam shape factor, F_s, is a correction term that depends on the geometry of the beam. This correction term is necessary because the heat is conducted to the substrate not only through the bottom surface of the beam but also from the sides and the top surface. The formula for the shape factor[30] is given by $F_s = \frac{t}{w} \left(\frac{2g}{t} + 1 \right) + 1$, where w is the width of the beam. F_s for the studied microbeam is calculated to be 12.33. The calculated thermal time constant for the microbeam used in this study is 152 μs, which indicates an electrothermal switching speed of 6.5 kHz. The theoretical open-loop switching speed of the MEMS resonator is estimated to be, $f/Q \sim 238$ Hz. Thus, it can be inferred that the maximum operating speed of the proposed logic device is limited by the ring-up or ring-down time of the resonator rather than the thermal time constant. It is worth to note that by scaling the device dimensions to nanoscale, both the mechanical response time and the thermal time constant will be improved significantly. As an example, we have estimated the thermal time constant[28,29] to be in the order of 10^{-6} s for a

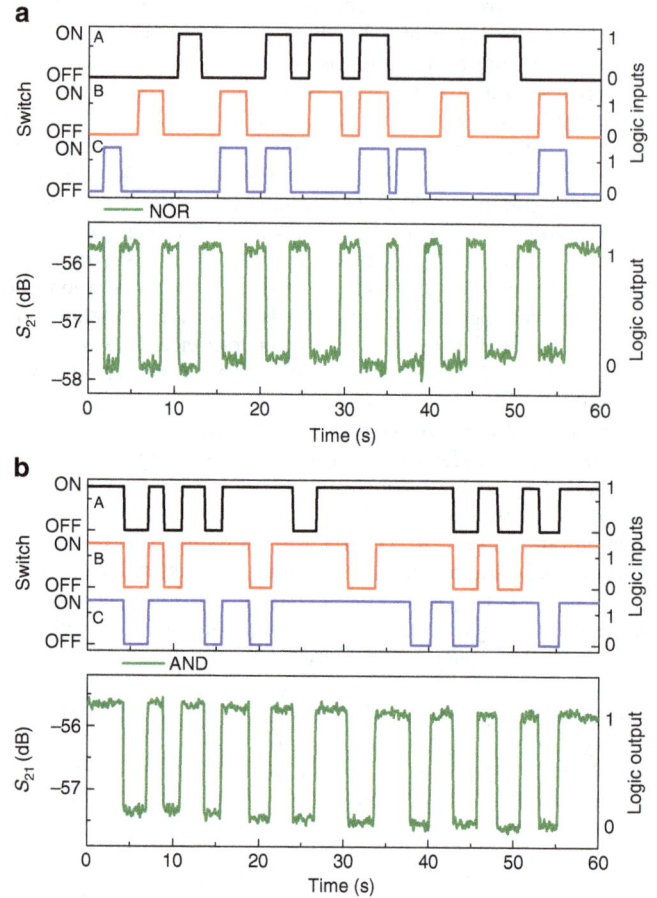

Figure 9 | Demonstration of 3-bit logic gates. (a) Demonstration of 3-bit NOR logic operation when the operating frequency is chosen at 119.022 kHz. Three input signals, A, B, and C are shown in black, red, and blue, respectively, where the switch OFF/ON corresponds to 0/1 logic input conditions. S_{21} transmission signal in green corresponds to the logic output and fulfills the NOR truth table. **(b)** Demonstration of 3-bit AND logic operation when the operation frequency is chosen at 132.105 kHz. Three input signals, A, B, and C are shown in black, red, and blue, respectively, where the switch OFF/ON corresponds to 0/1 logic input conditions. S_{21} transmission signal in green corresponds to the logic output and fulfills the AND truth table.

clamped–clamped beam resonator with a length of 20 μm, width of 300 nm and thickness of 500nm (ref. 12). This translates into a maximum electrothermal modulation speed in the order of 10^6 Hz. For the same resonator, the reported open-loop operation speed was around 48 kHz (ref. 12). It implies that the operation speed of logic devices built in these dimensions will be defined by the mechanical response time rather than the thermal response time. By considering a length of 600 μm and width of 50 μm for our device (includes electrodes and anchors), an integration density in the order of 10^4 per cm² can be achieved. Moreover, we note that the use of nanomechanical resonators would significantly increase the integration density. For a resonator with a length and width of 1 μm and 100 nm (resonance frequency around 1 GHz)[10], respectively, an integration density as high as 10^8 devices per cm² is plausible.

Another important aspect of a logic gate is the switching energy necessary to perform the desired logic operation. In this proposed scheme, the energy provided for the necessary switching events for the logic operation is in the form of resistive heating of the microbeam using the electrothermal circuit consisting of R_A, R_B and R_{MB}. While only a fraction of the total energy provided to the system is used by the microbeam for the state change during the logic operations, most of the energy is lost in the form of heat dissipation to the environment through R_A, R_B and R_{MB}. We estimated the maximum power cost for performing a single-logic operation as $P_{diss} \approx \frac{V_{RA}^2}{R_A} + \frac{V_{RB}^2}{R_B} + \frac{V_{RMB}^2}{R_{MB}} \approx 10^{-2}$ W. One can note that this energy cost is relatively high compared with other reported energy cost in performing a single-logic operation on nanomechanical resonator-based systems, such as in the work of Guerra et al.[12] and Wenzler et al.[21], which is based on electrostatic actuation. As traditionally well-known, thermal actuation, which is the base of this work, is considered less-energy efficient compared with other actuation methods. Nevertheless, the same principle demonstrated in this work applies when using other actuation techniques, as long as they can actively tune the stiffness of the resonating structure. It is also expected that the energy cost can be further reduced by orders of magnitude by optimizing device geometry.

The sensitivity of the proposed device to temperature variation is another important factor that needs to be addressed. The bandwidth, Δf, of the resonator of this study is estimated to be around 240 Hz ($\Delta f = f/Q$) at 1 torr. It implies that for the resonance frequency chosen as the operating frequency, the device will perform the desired logic operation successfully as long as the frequency shift due to the change in the ambient temperature lies within ± 120 Hz. We estimated the frequency shift due to temperature change according to ref. 31 $f(T) = f_0(1 + TC_f(T - T_0))$, where $TC_f = -30$ p.p.m. per °C, is the temperature coefficient of frequency for silicon resonators[32]. For the ambient temperature change between -10 °C and $+60$ °C from room temperature at 25 °C, the frequency shift is estimated to be $f_{shift} = \pm 120$ Hz, which is within the bandwidth of the resonator. Hence, the device will perform the desired logic operations successfully by selecting the resonance frequency as the driving frequency within this range of temperature variations. Additional temperature compensation scheme would be necessary to perform successful logic operation beyond this temperature range for the current device. Apart from this, the variation of resonance frequency due to phase noise is estimated to be around 105 Hz (see Supplementary Note 3 and Supplementary Fig. 3). Hence the device can still perform the desired logic operation successfully at a given operating conditions since the bandwidth is larger than the noise related frequency shift.

With regards to the potential interference between series and parallel resonances while selecting the a.c. operating frequency, it is noted that lowering down the compensation capacitance will broaden the separation between the series and parallel resonances. Also, improving the bandwidth will help to choose proper operating frequencies with lower margin of error.

A note is worth to be mentioned regarding the survivability of the resonators to mechanical shock. As was demonstrated[33–36] theoretically and experimentally, microstructures similar to the studied resonator shows excellent shock resilience up to 30,000–50,000 g. Downscaling the dimensions of the resonators will further improve the shock resilience.

The flexibility to cascade multiple gates is of paramount importance for realizing complex logic circuits. For the proposed scheme it is limited by two current challenges that warrant more future research. First, the strength of the output a.c. signal, which requires a transimpedence amplifier. Second is the fact that the signal waveforms as logic inputs and logic outputs are of different form. The output signal, a.c., needs to be converted into a d.c. signal. The d.c. output signal can be then used as an input to the next logic element, and hence enables sequencing. Also, the d.c. current can be split into various branches or pass through

multiple in-series resonators. If a single operating frequency is desired to be used throughout the grid of logic resonators, then one possibility is to fabricate several devices to have slightly different resonance frequencies, such that all can be driven at the same frequency. Also, the devices can be individually tuned by a separate d.c. biasing mechanism for each.

In summary, we demonstrated a reprogrammable logic device based on electrothermal tuning of the resonanance frequency, capable of performing all the fundamental 2-bit logic operations; AND, NAND, OR, NOR, XOR, XNOR and NOT, at room temperature and at modest vacuum conditions. We also demonstrated a single MEMS resonator-based reprogrammable 3-bit AND, NAND, OR and NOR logic gates. This device can be easily modified to perform n-bit OR/NOR and AND/NAND logic operations by simply adding one voltage source per bit in parallel in the electrical network responsible for the electrothermal frequency modulation. We program the device to perform a desired logic operation by simply choosing appropriate a.c.-driving frequency. This logic device operates in the linear regime of the resonator, and hence, may further reduce the voltage load if operated under low damping conditions. Although we have used an arch-shaped microbeam resonator, the same principle of electrothermal frequency modulation is equally applicable for a straight clamped–clamped MEMS/NEMS resonator. In fact, the demonstrated principle applies on any MEMS/NEMS resonator devices working in the linear frequency regime with a proper frequency tuning mechanism that can alter the stiffness property of the resonator, and hence, its linear resonance frequency. Future directions in this research can be targeted to simplify the bulky S_{21} parameter measurement set-up used in this paper. This complexity can be minimized by integrating necessary complementary metal oxide semiconductor devices, such as transimpedance amplifier, on-chip. This practical demonstration of essential elements of computation using MEMS resonators provide fundamental building blocks for alternative computing scheme in the electromechanical domain.

References

1. Babbage, H. P. Babbage's Calculating Engines Vol. 2 (The MIT Press, 1984).
2. Davis, M. The Universal Computer (Norton, 2000).
3. Theis, T. N. & Horn, P. M. Basic research in the information technology industry. Phys. Today 56, 44–49 (2003).
4. Swade, D. D. Redeeming Charles Babbage's mechanical computer. Sci. Am. 268, 86–91 (1993).
5. Halg, B. On a micro-electro-mechanical nonvolatile memory cell. IEEE Trans. Electron Devices 37, 2230–2236 (1990).
6. Badzey, R. L., Zolfagharkhani, G., Gaidarzhy, A. & Mohanty, P. A controllable nanomechanical memory element. Appl. Phys. Lett. 85, 3587–3589 (2004).
7. Masmanidis, S. C. et al. Multifunctional nanomechanical systems via tunably coupled piezoelectric actuation. Science 317, 780–783 (2007).
8. Charlot, B., Sun, W., Yamashita, K., Fujita, H. & Toshiyo, H. Bistable nanowire for micromechanical memory. J. Micromech. Microeng. 18, 045005 (2008).
9. Guerra, D. N., Imboden, M. & Mohanty, P. Electrostatically actuated silicon-based nanomechanical switch at room temperature. Appl. Phys. Lett. 93, 033515 (2008).
10. Mahboob, I. & Yamaguchi, H. Bit storage and bit flip operations in an electromechanical oscillator. Nat. Nanotechnol. 3, 275–279 (2008).
11. Roodenburg, D., Spronck, J. W., Van der Zant, H. S. J. & Venstra, W. J. Buckling beam micromechanical memory with on-chip readout. Appl. Phys. Lett. 94, 183501 (2009).
12. Guerra, D. N. et al. A noise assisted reprogrammable nanomechanical logic gate. Nano Lett. 10, 1168–1171 (2010).
13. Noh, H., Shim, S. B., Jung, M., Khim, Z. G. & Kim, J. A mechanical memory with a dc modulation of nonlinear resonance. Appl. Phys. Lett. 97, 033116 (2010).
14. Venstra, W. J., Westra, H. J. R. & van der Zant, H. S. J. Mechanical stiffening, bistability, and bit operations in a microcantilever. Appl. Phys. Lett. 97, 193107 (2010).
15. Mahboob, I., Flurin, E., Nishiguchi, K., Fujiwara, A. & Yamaguchi, H. Interconnect-free parallel logic circuits in a single mechanical resonator. Nat. Commun. 2, 198 (2011).

16. Hatanaka, D., Mahboob, I., Okamoto, H., Onomitsu, K. & Yamaguchi, H. An electromechanical membrane resonator. *Appl. Phys. Lett.* **101**, 063102 (2012).

17. Uranga, A. *et al.* Exploitation of non-linearities in CMOS-NEMS electrostatic resonators for mechanical memories. *Sen. Actuators A Phys.* **197**, 88–95 (2013).

18. Mahboob, I., Mounaix, M., Nishiguchi, K., Fujiwara, A. & Yamaguchi, H. A multimode electromechanical parametric resonator array. *Sci. Rep.* **4**, 4448 (2014).

19. Yao, A. & Hikihara, T. Counter operation in nonlinear micro-electro-mechanical resonators. *Phys. Lett. A* **377**, 2551–2555 (2013).

20. Yao, A. & Hikihara, T. Logic-memory device of a mechanical resonator. *Appl. Phys. Lett.* **105**, 123104 (2014).

21. Wenzler, J. S., Dunn, T., Toffoli, T. & Mohanty, P. A nanomechanical Fredkin gate. *Nano Lett.* **14**, 89 (2014).

22. Lee, J. E. Y. & Seshia, A. A. Parasitic feedthrough cancellation techniques for enhanced electrical characterization of electrostatic microresonators. *Sens. Actuators A Phys.* **156**, 36 (2009).

23. Younis, M. I., Ouakad, H., Alsaleem, F. M., Miles, R. & Cui, W. Nonlinear dynamics of MEMS arches under harmonic electrostatic actuation. *J. Microelectromech. Syst.* **19**, 647–656 (2010).

24. Alkharabsheh, S. & Younis, M. I. Dynamics of MEMS arches of flexible supports. *J. Microelectromech. Syst.* **22**, 216–224 (2013).

25. Alkharabsheh, S. & Younis, M. I. Statics and dynamics of MEMS arches under axial forces. *J. Vib. Acoust.* **135**, 021007 (2013).

26. Pelesko, J. A. & Bernstein, D. H. *Modeling of MEMS and NEMS* (CRC Press, 2002).

27. Kaajakari, V. *Practical MEMS* (Small Gear Publishing, 2009).

28. Schreiber, D. S., Cheng, W. J., Maloney, J. M. & DeVoe, D. L. in *Proceedings of 2001 ASME International Mechanical Engineering Congress and Exposition* 141–147 (New York, NY, USA, 2001).

29. Wang, Y., Zhihong, L., Daniel, M. T. & Norman, T. C. A low-voltage lateral MEMS switch with high RF performance. *J. Microelectromech. Syst.* **13**, 902 (2004).

30. Lin, L. & Chiao, M. Electrothermal responses of line shape microstructures. *Sens. Actuators A Phys.* **55**, 35–41 (1996).

31. Zhua, H., Tua, C., Shanb, G. & Lee, J. E. Y. Dependence of temperature coefficient of frequency (TCf) on crystallography and eigenmode in N-doped silicon contour mode micromechanical resonators. *Sen. Actuators A Phys.* **215**, 189–196 (2014).

32. Jeong, J. H., Chung, S. H., Lee, Se-Ho. & Kwon, D. Evaluation of elastic properties and temperature effects in Si thin films using an electrostatic microresonator. *J. Microelectromech. Syst.* **12**, 524–530 (2003).

33. Younis, M. I. *MEMS Linear and Nonlinear Statics and Dynamics* (Springer, 2011).

34. Ouakad, H., Younis, M. I. & Alsaleem, F. Dynamic response of an electrostatically actuated microbeam to drop-table test. *J. Micromech. Microeng.* **22**, 095003 (2012).

35. Younis, M. I., Jordy, D. & Pitarresi, J. Computationally efficient approaches to characterize the dynamic response of microstructures under mechanical shock. *J. Microelectromech. Syst.* **16**, 628–638 (2007).

36. Younis, M. I., Miles, R. & Jordy, D. Investigation of the response of microstructures under the combined effect of mechanical shock and electrostatic forces. *J. Micromech. Microeng.* **16**, 2463–2474 (2006).

Acknowledgements

We acknowledge Ulrich Buttner, EMPIRe Lab at KAUST for helping with laser cutting the chips. This research has been funded by KAUST.

Author contributions

M.A.H. conceived the idea, designed and fabricated the MEMS resonator. M.A.H. and L.K. performed the measurements and analysed the data. All authors discussed the results and wrote the paper. M.I.Y supervised the project.

Additional information

Competing financial interests: The authors declare no competing financial interests.

21

Rational design of efficient electrode–electrolyte interfaces for solid-state energy storage using ion soft landing

Venkateshkumar Prabhakaran[1], B. Layla Mehdi[1], Jeffrey J. Ditto[2], Mark H. Engelhard[3], Bingbing Wang[3,†], K. Don D. Gunaratne[1], David C. Johnson[2], Nigel D. Browning[1], Grant E. Johnson[1] & Julia Laskin[1]

The rational design of improved electrode–electrolyte interfaces (EEI) for energy storage is critically dependent on a molecular-level understanding of ionic interactions and nanoscale phenomena. The presence of non-redox active species at EEI has been shown to strongly influence Faradaic efficiency and long-term operational stability during energy storage processes. Herein, we achieve substantially higher performance and long-term stability of EEI prepared with highly dispersed discrete redox-active cluster anions (50 ng of pure ~ 0.75 nm size molybdenum polyoxometalate (POM) anions on 25 µg (~ 0.2 wt%) carbon nanotube (CNT) electrodes) by complete elimination of strongly coordinating non-redox species through ion soft landing (SL). Electron microscopy provides atomically resolved images of a uniform distribution of individual POM species soft landed directly on complex technologically relevant CNT electrodes. In this context, SL is established as a versatile approach for the controlled design of novel surfaces for both fundamental and applied research in energy storage.

[1] Physical Sciences Division, Pacific Northwest National Laboratory, PO Box 999, MSIN K8-88, Richland, Washington 99352, USA. [2] Department of Chemistry, University of Oregon, Eugene, Oregon 97403, USA. [3] Environmental Molecular Sciences Laboratory, Pacific Northwest National Laboratory, Richland, Washington 99352, USA. † Present address: State Key Laboratory of Marine and Environmental Science and College of Ocean and Earth Sciences, Xiamen University, Xiamen 361102, China. Correspondence and requests for materials should be addressed to J.L. (email: Julia.Laskin@pnnl.gov).

The development of improved materials for efficient energy storage is at the forefront of both fundamental and applied research in energy technology. Substantial progress both in understanding underlying phenomena and device fabrication is required to increase the charge-discharge capacity, rate and life cycle of energy storage devices while decreasing the cost associated with their manufacture[1,2]. Hence, the race to develop high performance, ecofriendly, low-cost energy storage devices is progressing along various parallel tracks. One familiar track is the development and optimization of the hybrid battery which is usually comprised of lithium ion batteries (Li-ion) coupled with supercapacitors[1,3-5]. The main challenge with Li-ion batteries is to provide improved power density and fast charge rates at a reduced price per $kW h^{-1}$, whereas supercapacitors need improvement in their energy densities so that the latter can respond to the coupled battery during combined operation[5-9]. In the past decade, both the power and energy density of supercapacitors have been significantly improved by combining non-Faradaic electrochemical double-layer capacitance and Faradaic pseudocapacitance[10,11] through a combination of high surface area carbon and metal oxides integrated at a molecular level[12,13]. Redox-supercapacitor electrodes have been fabricated by the direct transfer of redox pseudocapacitive materials from the solution phase onto electrodes using direct painting[14], ambient air spray[15], chemical vapour deposition[16], atomic layer deposition[17] and electrodeposition[18]. Electrodes prepared via such methods typically contain both electrochemically active components and inactive counter ions at the electrode–electrolyte interface (EEI). Polyoxometalates (POMs) are attractive redox species because of their stability, multi-electron redox activity[19,20] and strong interactions with electrodes[21-23]. Here, we demonstrate that fabrication of an energy storage device using uniformly distributed discrete POM anions without aggregate formation is an important breakthrough towards achieving superior performance not possible with conventional preparation techniques.

We employ soft landing (SL) of mass-selected ions for the preparation of well-defined electrode surfaces with an unprecedented level of control. SL enables the uniform deposition of ions with specific composition, charge state and kinetic energy[24-29] which is critical to gaining fundamental understanding of interfacial phenomena relevant to catalysis and materials science[27,30-33]. Previously, SL and reactive landing have been used to functionalize carbon nanotubes (CNTs), graphene, Si and Au surfaces to study electrochemical properties of well-defined deposited ions[34-38]. Notably, SL allows us to fabricate electrode surfaces for energy storage devices that are difficult to prepare using conventional solution- and vacuum-based deposition

techniques. This provides a unique opportunity to design efficient and stable EEI by (i) studying the inherent activity and efficacy of precisely defined electro-active species; (ii) understanding the effect of electrochemically inactive counter-ions at the EEI on overall device performance; and (iii) identifying pathways to enhance the overall efficacy and selectivity of reactions of interest.

We demonstrate that SL of only small amounts of redox-active species results in dramatically enhanced total specific capacitance and superior long-term stability of carbon-based electrodes. We also show that aggregation is largely responsible for the reduced performance of POM-based supercapacitors fabricated from solution.

Results

Role of ion soft landing in electrode fabrication. Using a specially designed high-flux ion SL instrument (see Supplementary Fig. 1)[39], we fabricated macroscopic energy storage devices and investigated the role of anion–cation interactions and aggregate formation on the performance of POM-based supercapacitors. The SL instrument is equipped with a dual ion funnel interface, which increases the deposition rate of ions (refer to Methods section for a detailed description of our SL instrument). Specifically, redox-supercapacitor devices were fabricated by immobilizing mass-selected $PMo_{12}O_{40}^{3-}$ onto CNT-coated carbon electrodes (CNT electrodes) and compared with a non-Faradaic CNT-based device. SL can be performed either in vacuum or at ambient conditions on a benchtop[29]. In this study, vacuum-based SL was used to achieve selective deposition of $PMo_{12}O_{40}^{3-}$ anions while excluding other charge states of POM (2- and 1-) produced by electrospray ionization (ESI) (see Supplementary Fig. 2). Predetermined amounts of POM were deposited onto CNT electrodes either using SL or ambient electrospray deposition (ESD) of $Na_3[PMo_{12}O_{40}]$, $(NH_4)_3[PMo_{12}O_{40}]$ or $H_3[PMo_{12}O_{40}]$ solutions (Fig. 1). In ESD, micron size charged droplets of solvent containing POM anions and counter-cations, are deposited onto electrodes with a set deposition rate controlled by the flow of POM solution. Therefore, the ions deposited using ESD are neither mass-selected nor charge-selected. Ambient ESD along with its analogue, ambient air spray, that does not use high voltage are commonly used fabrication methods for thin film deposition on surfaces and electrodes[40-45]. It has been demonstrated that deposition of small microdroplets by ESD produces more uniform films in comparison with bulk solution drop casting making it a preferred deposition method. Furthermore, the ability to control the amount of deposited material by ESD enables a quantitative comparison with SL.

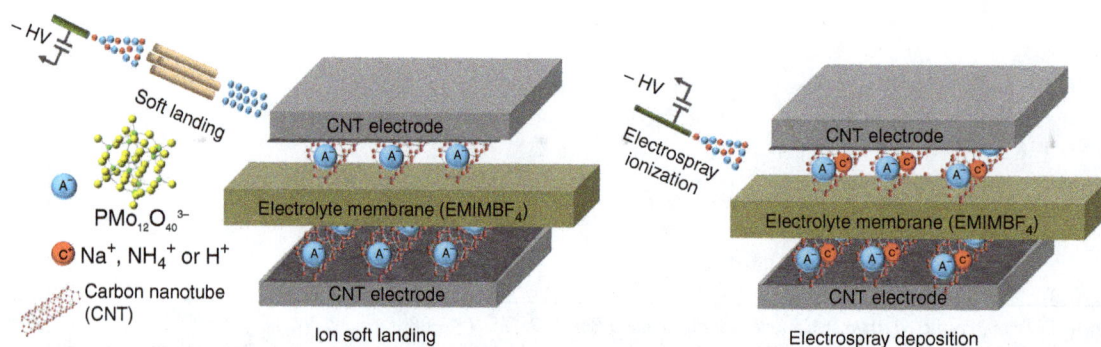

Figure 1 | Electrode-electrolyte interfaces of fabricated supercapacitor devices. Schematic representation of EEI of redox-supercapacitors fabricated using SL and ambient ESD. Note the absence of countercations in the device prepared by SL.

In addition, we fabricated supercapacitors by drop casting Na₃[PMo₁₂O₄₀] solution (DRP-NaPOM-CNT) at different POM loadings and compared the total specific capacitance of these electrodes with ones prepared by ESD. Consistent with the literature[46–48], the results demonstrated the superior performance of ESD (see Supplementary Fig. 3) making it a suitable benchmarking technique for comparison with SL.

A total of five different electrode configurations were employed in this study: (i) pristine CNT electrode (pCNT); (ii) SL electrode containing a known amount of charge- and mass-selected $PMo_{12}O_{40}^{3-}$ (SL-CNT); (iii) ESD electrode with Na₃[PMo₁₂O₄₀] (ESD-NaPOM-CNT); (iv) ESD electrode with (NH₄)₃[PMo₁₂O₄₀] (ESD-NH₄POM-CNT); and (v) ESD electrode with H₃[PMo₁₂O₄₀] (ESD-HPOM-CNT).

This study demonstrates that SL-CNT containing only a small amount of $PMo_{12}O_{40}^{3-}$ (~50 ng POM on ~25 µg of the CNT material) have remarkably higher total specific capacitance and superior long-term stability compared with ESD electrodes. High-angle annular dark-field scanning transmission electron microscopy (HAADF-STEM) demonstrates that SL electrodes exhibit an extremely uniform and narrow distribution of discrete 0.75 nm diameter [PMo₁₂O₄₀] clusters without any agglomeration. In contrast, the formation of aggregates and agglomerated POM in the presence of counter cations and solvent is observed on ESD electrodes. On the basis of these observations, we propose that aggregation is largely responsible for the reduced performance of POM-based supercapacitors fabricated from solution. Our results demonstrate that higher performance and longer stability can be achieved in redox-supercapacitors by uniform deposition of discrete redox-active species without strongly coordinating yet inactive counter ions or solvent at the EEI.

Electrochemical performance of fabricated devices. The supercapacitors were fabricated as described in the methods section using a 1-ethyl-3-methylimidazolium tetra fluoroborate (EMIMBF₄) ionic liquid membrane as a separator. The effect of the POM deposition technique and loading on total specific capacitance and stability of the fabricated devices was assessed. Our previous study showed that soft-landed $PMo_{12}O_{40}^{3-}$ remain intact and redox active on surfaces[49]. The redox activity of $PMo_{12}O_{40}^{3-}$ in the EMIMBF₄ ionic liquid was confirmed by examining the cyclic voltammograms (CVs) of different POM salts on glassy carbon (GC) electrodes (see Supplementary Fig. 4). A rectangular CV over different potential ranges was observed for supercapacitors fabricated using pCNT (see Supplementary Fig. 5) confirming the stability of the electrolyte membrane and the fabricated supercapacitor over a technologically relevant potential range[50–52].

The CVs of supercapacitors fabricated with pCNT, SL-CNT, ESD-NaPOM-CNT and ESD-NH₄POM-CNT (Fig. 2a) are rectangular in shape indicating that they exhibit an ideal capacitive-like behaviour. The respective galvanostatic charge-discharge (GCD) curves (Fig. 2b) are almost symmetrical triangles without any significant voltage drop related to internal resistance during the changing of polarity, suggesting fast transmission of ions at the EEI. However, ESD-HPOM-CNT showed neither rectangular CV (Fig. 2a) nor triangular GCD curves (Fig. 2b) which may be attributed to the presence of additional Faradaic reactions. The increase in current response closer to 1 V in the CV of ESD-HPOM-CNT indicates the presence of a side reaction such as oxygen evolution[53]. The GCD measurements were performed in triplicates and a total specific capacitance of 112 ± 12, 153 ± 8, 120 ± 16, 76 ± 16 and 128 ± 13 F g⁻¹ and specific energy densities of 15.0 ± 1.7,

21.3 ± 1.1, 16.7 ± 2.2, 10.6 ± 2.2 and 17.75 ± 1.5 Wh kg⁻¹ were obtained for the pCNT, SL-CNT, ESD-NaPOM-CNT, ESD-NH₄POM-CNT and ESD-HPOM-CNT, respectively, using equations 1 and 2. Similar values were also calculated from CV data using equation 3 indicating that the supercapacitors perform well both in constant potential (CV) and constant current mode (GCD). Of note, the total specific capacitance and energy density of pCNT are in agreement with literature values[54], which confirms the reliability of the baseline pCNT and the device fabrication method adopted in this study. The lack of dependence of the specific capacitance on the scan rates (see Supplementary Figs 6–9 and 11) indicates that, except for ESD-HPOM-CNT, the mobility of the ions and surface pseudocapacitance of unmodified and modified electrodes remain constant and stable over a wide range of charge-discharge rates of interest to energy storage applications. The ESD-HPOM-CNT showed an unstable capacitance over different scan rates (see Supplementary Figs 10 and 11), which may be attributed to the presence of additional Faradaic capacitance as discussed earlier.

The total specific capacitance and energy density of the SL-CNT (Fig. 2c) is ~36, 27, 101 and 20% higher than the values obtained for pCNT, ESD-NaPOM-CNT, ESD-NH₄POM-CNT and ESD-HPOM-CNT, respectively. Therefore, the efficient participation of $SL-PMo_{12}O_{40}^{3-}$ during redox reaction in the absence of counter cations contributes to the enhanced capacitance of SL-CNT. The 30% decrease and the 7 and 14% increase in the total specific capacitance of ESD-NH₄POM-CNT, ESD-NaPOM-CNT and ESD-HPOM-CNT, respectively, in comparison with pCNT clearly indicate the important role of counter cations that are absent in SL-CNT. For example, NH₄⁺ is known to strongly adsorb on carbon[55,56], which may decrease the electrochemically active surface area and reduce the non-Faradaic capacitance of the CNT surface along with the Faradaic capacitance of POM. Similar specific power densities of ~4 kW kg⁻¹ were obtained for both pristine and modified CNT electrodes using equation 4. However, the SL-CNT with a similar specific power density is characterized by a higher specific energy density demonstrating the improved performance of this electrode.

Effect of POM loading on specific capacitance. GCD measurements were also carried out with the SL-CNT, ESD-NaPOM-CNT and ESD-HPOM-CNT containing different loadings of POM (Fig. 2d). The ESD-NH₄POM-CNT was excluded in this further study as it showed lower capacitance than the pCNT. The SL-CNT achieved its maximum total specific capacitance of ~160 F g⁻¹ with 1.75 × 10¹³ POM ions. Meanwhile, the ESD-NaPOM-CNT required almost twice the amount of POM (3 × 10¹³ ions) and the ESD-HPOM-CNT required 2 × 10¹³ ions compared with the SL-CNT to achieve a similar maximum total specific capacitance. This demonstrates the higher efficacy of SL-CNT. For both SL and ESD electrodes, further addition of POM above the required maximum amount resulted in a decrease in the total specific capacitance.

It is remarkable that the maximum total capacitance was obtained only with 1.75 × 10¹³ $SL-PMo_{12}O_{40}^{3-}$ (~50 ng POM on 25 µg of the active CNT material) demonstrating the unexpectedly high contribution of nanogram quantities of pure $PMo_{12}O_{40}^{3-}$ to the total capacitance. The Faradaic component of the total specific capacitance (Fig. 2d) was calculated using equation 5. It is observed that the specific Faradaic capacitance of SL-CNT is greater than that of ESD-NaPOM-CNT and ESD-HPOM-CNT at all POM loadings. The observed decrease in capacitance at higher coverage may be attributed to formation

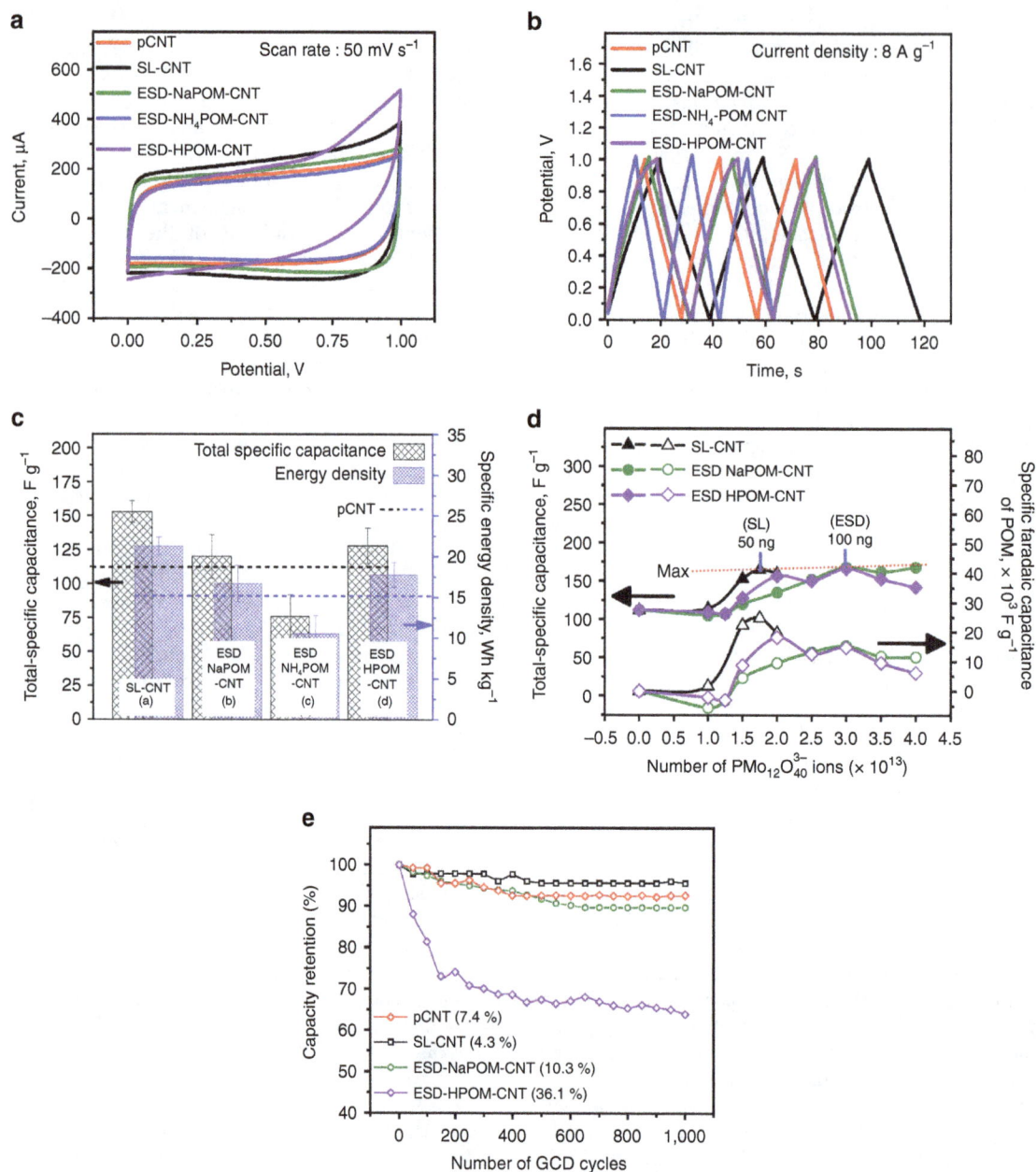

Figure 2 | Superior performance of SL supercapacitors. Characterization of supercapacitors fabricated with pCNT, SL-CNT, ESD-NaPOM-CNT, ESD-NH₄POM-CNT and ESD-HPOM-CNT and an EMIMBF₄ based membrane as a separator: (**a**) representative CV and (**b**) galvanostatic charge-discharge (GCD) curves. (**c**) Comparison of the total specific capacitance and specific energy density. Approximately 1.5×10^{13} $PMo_{12}O_{40}^{3-}$ were deposited in each case. (**d**) Effect of $PMo_{12}O_{40}^{3-}$ loading on the total specific capacitance (black) and specific Faradaic capacitance (green). (**e**) Capacity retention as a function of the number of GCD cycles. Current density: $8 \, A \, g^{-1}$.

of additional POM layers in which underlying layers do not participate in redox activity.

The capacity retention (stability) of the pCNT, SL-CNT, ESD-NaPOM-CNT and ESD-HPOM-CNT was evaluated by performing 1,000 GCD cycles. After cycling (Fig. 2e), the capacitance of ESD-NaPOM-CNT decreased by 10.3 %, whereas only a 4.3% decrease was observed with SL-CNT indicating that SL-CNT have almost twice the lifetime of the ESD-NaPOM-CNT. The capacitance of pCNT was decreased by 7.4 % after cycling, which further confirmed the higher stability of SL-CNT. Interestingly, the ESD-HPOM-CNT showed ~36% decrease in capacitance after 1,000 GCD cycles. The lower stability of ESD-HPOM-CNT was attributed to degradation of EEI due to

the presence of side reactions as evidenced in the CVs (See Fig. 2a and Supplementary Fig. 10) or chemical degradation of EMIMBF₄ in presence of free H^+ ions[57]. However, the exact mechanism responsible for the lower stability of ESD-HPOM-CNT is not clear. Further characterization of the instability of ESD-HPOM-CNT is beyond the scope of this work. The poor capacity retention of ESD-HPOM-CNT made them unsuitable to compare with SL-CNT.

Physical and chemical properties of fabricated electrodes. The exceptional performance and superior long-term stability of SL-CNT may reflect the absence of interactions between

$PMo_{12}O_{40}^{3-}$ ions and their strongly coordinating counterions, which facilitates the uniform distribution of discrete $PMo_{12}O_{40}^{3-}$ on SL-CNT without agglomeration, or differences in the oxidation state of POM. To evaluate these possibilities, the distribution and chemical state of POM on CNT were analysed using HAADF-STEM, scanning electron microscopy (SEM) and X-ray photoelectron spectroscopy (XPS).

Considering the fact that the ESD-NaPOM-CNT showed a similar ideal GCD behaviour and redox activity of POM anions with respect to SL-CNT, it was selected as a reference to compare with SL-CNT. Therefore, HAADF-STEM was used to examine the morphology of pCNT, SL-CNT and ESD-NaPOM-CNT at the nanometre scale (See Fig. 3a–f and Supplementary Figs 12 and 13). It is remarkable that a uniform distribution of soft-landed POM was observed directly on the complex commercially relevant CNT electrodes. A uniform and narrow size distribution of POM was observed on SL-CNT (Fig. 3b,d). The diameter of individual clusters was measured to be $\sim 0.74 \pm 0.04$ nm (see Fig. 3f and Supplementary Figs 14 and 15) which matches the theoretical size calculated for $PMo_{12}O_{40}^{3-}$ (ref. 20). STEM images indicate that $PMo_{12}O_{40}^{3-}$ on SL-CNT are predominantly preserved as discrete ions without any significant agglomeration and the estimated uniform cluster coverage is 2.1×10^5 POM clusters per μm^2. Fig. 3f also shows the presence of a minor fraction of individual Mo atoms that are challenging to group as part of POM clusters (see Supplementary Discussion). In contrast, STEM images of ESD electrodes (see Fig. 3c,e,f and Supplementary Fig. 13) show features in the size range of 3–

10 nm with an average size of 3.66 ± 2.11 nm, indicating agglomeration of POM during ESD. Such aggregate formation on ESD electrodes may be attributed to the agglomeration of POM anions in the presence of counter cations (Na^+, NH_4^+ or H^+) as the solvent evaporates from the surface. The presence of nanoscopic aggregates consisting of low-conductivity cation-POM assemblies at the EEI may decrease the overall interfacial surface area and impart additional contact resistance, which would decrease the overall capacitance of the supercapacitor. Thus, STEM images clearly show the agglomeration of POM in the ESD electrodes and a uniform distribution of individual discrete $[PMo_{12}O_{40}]$ clusters in the SL electrodes, which explains both the lower performance of ESD-NaPOM-CNT and higher performance of SL-CNT.

In addition to STEM, we also used SEM to characterize the agglomeration of POM in ESD electrodes at the microscopic scale. SEM images of pCNT, SL-CNT, ESD-NaPOM-CNT, ESD-NH4POM-CNT and ESD-HPOM-CNT are presented in Fig. 4. The presence of CNTs on top of the bulk carbon fibres of the base carbon electrode is clearly visible in Fig. 4a. CNT deposition helps increase the overall electrode surface area and also lowers the contact resistance at the EEI. The diameter of the pristine CNTs is found to be ~ 15 nm. The SL-CNT shows similar morphology to the pristine CNT electrode and does not exhibit any characteristic presence of POM aggregates (see Fig. 4b). Given the extremely small size of individual POM clusters (diameter ~ 0.75 nm), it is not possible to image them using conventional SEM. In contrast, ESD-NaPOM-CNT, ESD-NH4POM-CNT and

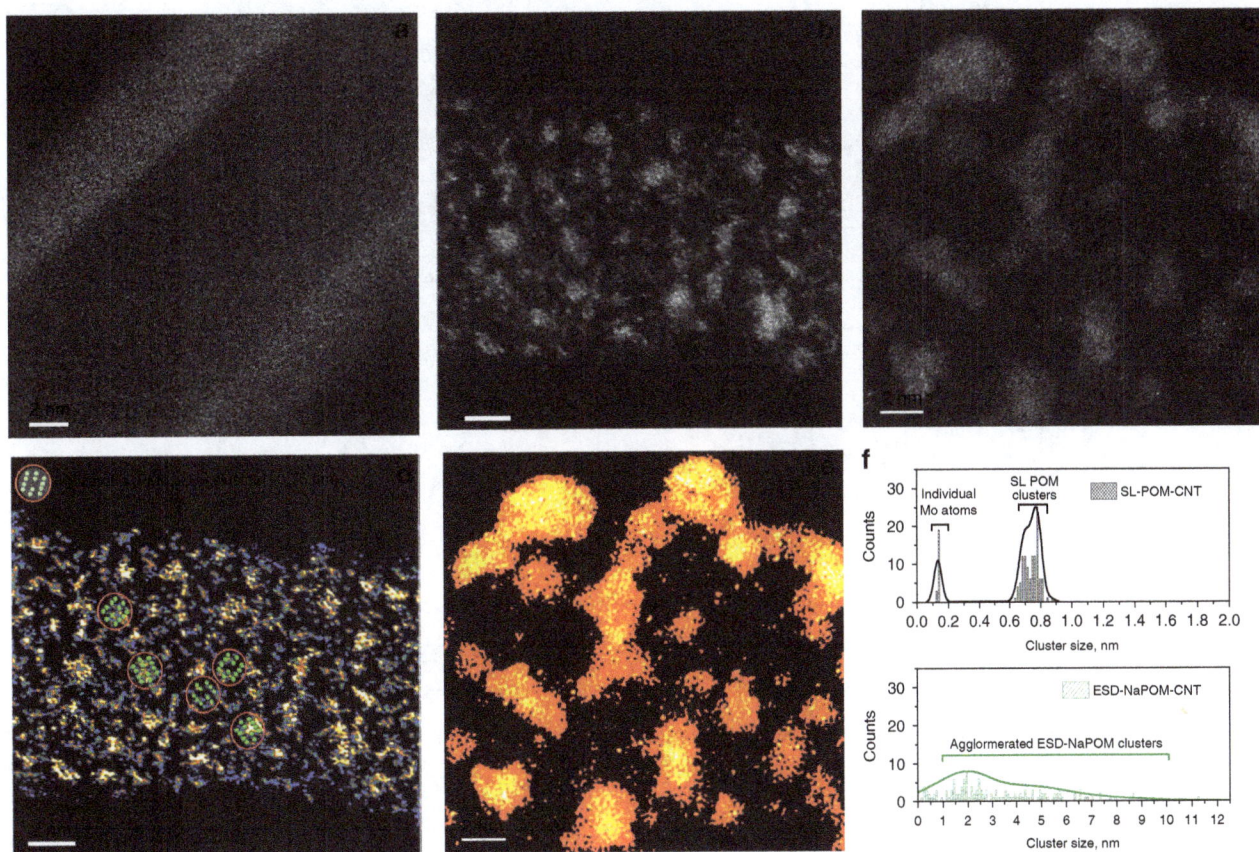

Figure 3 | Atomically resolved STEM images of monodisperse SL POM. (**a–c**) HAADF-STEM images of pCNT, SL-CNT and ESD-NaPOM-CNT, respectively. (**d,e**) The corresponding HAADF STEM images of SL-CNT and ESD-NaPOM-CNT, respectively, (**f**) Histogram of cluster size distribution in SL-CNT and ESD-NaPOM-CNT, respectively. Approximately 1.5×10^{13} $PMo_{12}O_{40}$ ions were loaded on the modified CNT electrodes. Selected examples of intact SL POM clusters were mapped in **d**—red circles. Note: The raw STEM images were processed using the Gatan Microscopy Suite DigitalMicrograph software to generate colour-enhanced Z-contrast images.

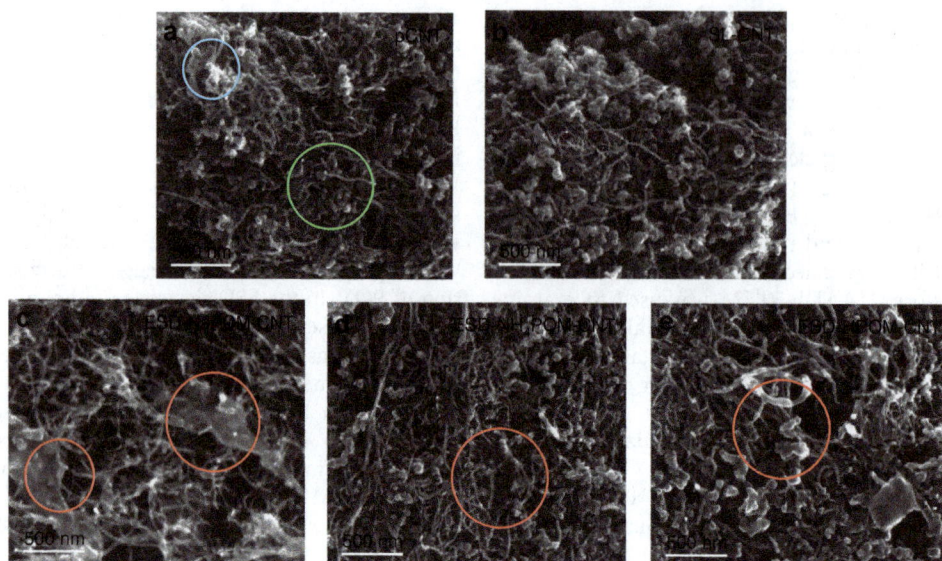

Figure 4 | POM aggregate formation at the microscopic scale. Scanning electron micrographs of (**a**) pCNT (**b**) SL-CNT (**c**) ESD-NaPOM-CNT (**d**) ESD-NH₄POM-CNT and (**e**) ESD-HPOM-CNT. Approximately 1.5×10^{13} PMo₁₂O₄₀ ions were loaded in the modified CNT electrodes. Green and blue circles in **a** represent the regions of CNT and carbon fibres, respectively. Red circles in **c–e** highlight microscopic aggregates observed on three ESD electrodes.

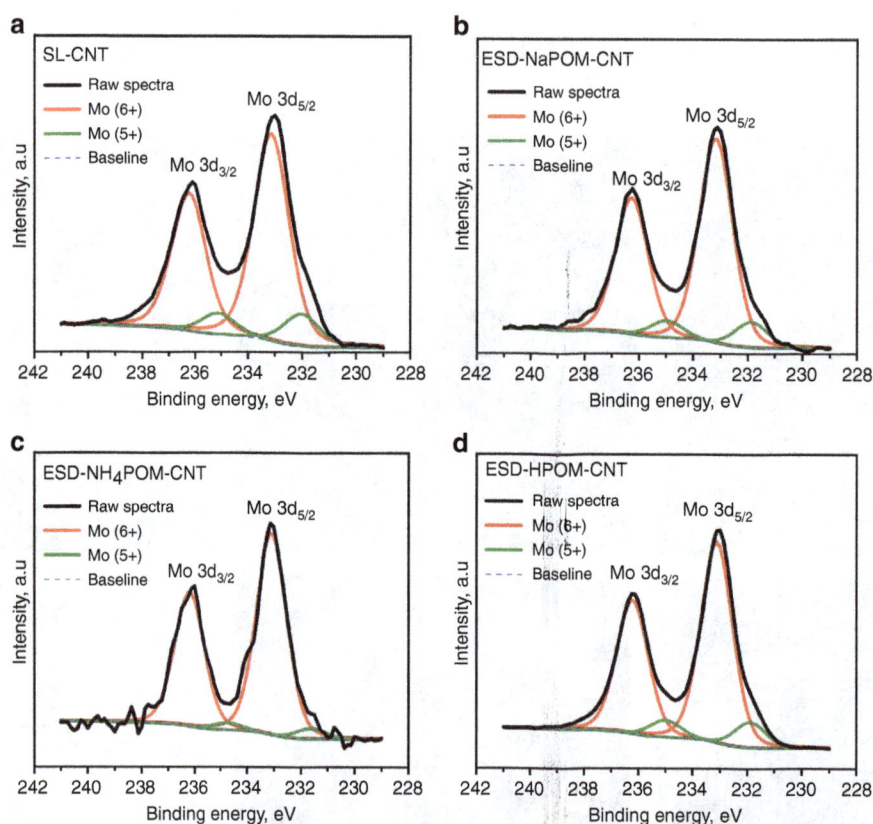

Figure 5 | Chemical state of SL and ESD electrodes. Mo $3d$ spectra of (**a**) SL-CNT (**b**) ESD-NaPOM-CNT (**c**) ESD-NH₄POM-CNT and (**d**) ESD-HPOM-CNT. Approximately 1.5×10^{13} POM ions were loaded in each case.

ESD-HPOM-CNT show the presence of distinct aggregates on the top layer. The diameter of the CNTs present in the ESD-NH₄POM-CNT and ESD-HPOM-CNT electrode are increased to ~50 nm from ~15 nm, indicating that the aggregates are preferentially formed around the outside walls of the CNTs. Formation of aggregates of POM clusters in the ESD electrodes observed with

SEM analysis is consistent with the agglomeration of POM seen in the STEM analysis. Again, such aggregate formation on the electrode surface post ESD may be attributed to the agglomeration of POM in the presence of its counter cations (Na⁺, NH₄⁺ and H⁺) as the solvent evaporates. Aggregates were not observed on the SL electrode at similar coverage. Assuming that the aggregates

Table 1 | XPS analysis results of Mo 3d spectra obtained on the CNT-carbon electrodes modified with POM anions.

Binding energy (eV)	Oxidation state		Area %			
			SL-CNT	ESD-NaPOM-CNT	ESD-NH₄POM-CNT	ESD-HPOM-CNT
232.0	Mo 3d$_{5/2}$	5+	8.1	6.7	3.0	6.8
233.1		6+	52.0	53.4	57.0	53.4
235.1	Mo 3d$_{3/2}$	5+	5.4	4.4	2.0	4.5
236.2		6+	34.6	35.6	38.0	35.5
	FWHM (eV)		1.6 eV	1.4 eV	1.3 eV	1.4 eV

CNT, carbon nanotube; ESD, electrospray deposition; FWHM, full-width at half-maximum; POM, polyoxometalate; SL, soft landing; XPS, X-ray photoelectron spectroscopy.

are composed of POM salt with lower electrical conductivity[58,59], the presence of the microscopic aggregates at the EEI may significantly decrease the overall interfacial surface area and impart additional contact resistance, which would decrease the overall capacitance of the device.

We also used XPS to determine the oxidation state of Mo in SL and ESD electrodes. Figure 5a–c shows Mo 3d XPS spectra of SL- CNT, ESD-NaPOM-CNT, ESD-NH₄POM-CNT and ESD-HPOM-CNT. All Mo 3d XPS spectra exhibit the characteristic 3d$_{5/2}$ and 3d$_{3/2}$ doublet caused by spin–orbit coupling of the Mo 3d orbitals. Deconvolution of all Mo 3d XPS spectra with a fixed intensity area ratio of 2:3 (corresponding to d orbital[60,61]) reveals two peaks for each Mo 3d$_{5/2}$ and Mo 3d$_{3/2}$ spin–orbit coupling. The peaks located at 232.0 eV/235.1 eV and 233.1 eV/236.2 eV in Mo (3d$_{5/2}$/3d$_{3/2}$) spectra correspond to Mo^{5+} and Mo^{6+}, respectively, which is in agreement with literature values[62–64]. In addition, the characteristic binding energy separation (Δ Mo 3d) between the Mo 3d$_{5/2}$ and 3d$_{3/2}$ doublet of Mo^{5+} and Mo^{6+} is ~3.1 eV as observed previously in the literature for Mo^{5+} and Mo^{6+} ions[65]. These XPS observations show that similar oxidation states of POM clusters on CNT are present following SL and ESD. The calculated area % of different Mo chemical states post XPS analysis of all Mo 3d spectra are reported in Table 1. The area % of Mo^{5+} and Mo^{6+} in both Mo 3d$_{5/2}$ and 3d$_{3/2}$ of the SL-CNT and ESD-NaPOM-CNT are similar indicating that the charge state of Mo is not affected in both cases. However, a slightly higher abundance of Mo^{6+} seen in the ESD-NH₄POM-CNT may be due to charge (e$^-$) transfer from POM to NH₄$^+$ (NH₄$^+$ is a Lewis acid). In addition, the C 1s spectra of different CNT electrodes were also examined, but no significant changes were observed (see Supplementary Fig. 16).

The preceding binding energy and area calculations do not reveal any substantial changes in the oxidation state of Mo in SL and ESD electrodes; however, the widths of the peaks observed on ESD and SL electrodes are noticeably different. Specifically, the full-width at half-maximum (FWHM) of peaks assigned to the 5+ and 6+ oxidation states of Mo in SL-CNT, ESD-NaPOM-CNT, ESD-NH₄POM-CNT and ESD-HPOM-CNT are 1.6, 1.4, 1.3 and 1.4 eV, respectively. The Mo 3d$_{5/2}$ and 3d$_{3/2}$ lines originating from SL POM ions are broader than those observed on both ESD electrodes. It is reasonable to attribute the increased broadening in SL electrodes to the presence of electronic interactions between [PMo₁₂O₄₀] clusters and the CNT support for the isolated SL clusters and lower crystallinity of the material prepared by SL compared with ESD[66–68]. In other words, the XPS analysis (see Fig. 5 and Supplementary Fig. 15) of Mo 3d and C 1s peaks reveals no significant change in the oxidation state of Mo and C between SL and ESD electrodes but indicates a higher crystallinity of POM on ESD electrodes which may be caused by agglomeration. This provides further evidence, in addition to the STEM and SEM images, for the surface immobilization of discrete POM anions without counter ions or solvent using SL.

Discussion

Collectively, evidence obtained using HAADF-STEM, SEM and XPS indicates that the uniform distribution of POM, lack of agglomeration and absence of counter ions on the electrode are responsible for the exceptionally high performance of SL-CNT. Such uniform distribution of active material on the support enables efficient access to redox active species during GCD operation and improves the overall efficacy and stability of the redox-supercapacitors. In contrast, ESD results in formation of a less active, agglomerated phase that shows reduced redox activity in comparison with isolated clusters prepared by SL. Once the agglomerated phase has formed, repartitioning of individual clusters into the ionic phase in a solid state electrolyte is likely inhibited. Our results demonstrate that higher performance and longer stability can be achieved in redox-supercapacitors by uniform deposition of discrete redox active species at the EEI.

In summary, ion SL enabled the first fabrication of high-density CNT electrodes containing monodisperse POM anions and direct atomically resolved imaging of the uniform distribution of individual redox-active species on complex commercially relevant electrodes using STEM. We present evidence that agglomeration of active species due to interaction with strongly coordinating counterions is one of the key factors affecting the performance and stability of solid-state electrochemical energy storage devices such as batteries and supercapacitors. This work clearly demonstrates, for the first time, that elimination of these interactions through mass-selection and uniform deposition of small amounts of intact active species on CNT electrodes substantially improves device performance. In addition to energy storage, SL deposition can be extended to design efficient energy conversion devices where improved charge transfer across active layers[69] is desired. In this context, SL is established as a breakthrough deposition technology that enables the rational design of efficient EEI for energy storage applications through fundamental understanding of the key limiting factors.

Methods

Electrode preparation. The CNT-coated carbon fibre paper was obtained from SGL carbon GmbH (Meitingen, Germany) (SIGRACET) and cut into 1 cm² pieces that were directly used in this study. Hereafter, the 1 cm² pieces of carbon fibre paper will be referred to as CNT electrode. The thickness of CNT electrode is ~300 μm and the surface loading of CNT on the electrode is ~25 μg cm^{-2}. It is assumed that only the CNT present on the top layer contribute to the non-Faradaic double layer capacitance of the surface supercapacitor fabricated in this work and the minor contribution from the layer behind the electrode interface is disregarded. Therefore, the combined weight of the CNT loading (~25 μg cm^{-2}) and immobilized [PMo₁₂O₄₀] were used as a total weight of the active material to calculate the total specific capacitance of the supercapacitor.

Soft landing of mass-selected PMo₁₂O₄₀$^{3-}$ *anions.* The PMo₁₂O₄₀$^{3-}$ anions were soft-landed onto CNT electrodes using a custom-designed ion SL instrument described in detail in previous studies[39,49]. See Supplementary Fig. 1. The ion SL instrument is equipped with an ESI source, a dual ion funnel, an RF-only collision quadrupole, a quadrupole mass filter (Extrel CMS, Pittsburgh, PA), and three einzel lenses that focus the ion beam onto a deposition target. A 150 μM

$Na_3[PMo_{12}O_{40}] \times H_2O$ solution in methanol is introduced at a flow rate of $65\,\mu l\,h^{-1}$ to the ESI emitter. The charged droplets produced using ESI are transferred into the vacuum system using a heated stainless steel inlet and two electrodynamic ion funnels. The desolvated ions from the funnel are then collimated by colliding with the background gas in an RF-only quadrupole (collisional quadrupole). Subsequently, the ions are mass-filtered to allow through only $PMo_{12}O_{40}^{3-}$ anions ($m/z = 607$; m = mass and z = ionic charge) using a resolving quadrupole. The collisional quadrupole and resolving quadrupole are maintained at pressures of 2×10^{-2} and at 8×10^{-5} Torr, respectively. The mass-selected ions are then refocused with three einzel lenses in series before SL on the CNT electrode, which is mounted inside the vacuum deposition system. The collision energy of the soft-landed ions, determined by the difference between the CQ DC potential and the surface, was in a range of $30-35\,eV$ per charge translating to an ion kinetic energy of $\sim 90\,eV$ for $PMo_{12}O_{40}^{3-}$. An ion current of $\sim 2\,nA$ on the surface was measured with a picoammeter (model 9103, RBD Instruments, Bend, OR) and remained steady throughout the deposition. The total number of ions deposited was calculated using the measured ion current over time. The ion beam produced in the ion SL instrument is circular in shape and $\sim 3\,mm$ in diameter. Therefore, the total deposition area is $\sim 7\,mm^2$ on $100\,mm^2$ CNT electrode.

Ambient electrospray of $PMo_{12}O_{40}$. Electrodes were also prepared by direct ambient ESD of $150\,\mu M$ solutions of three different POMs ($Na_3[PMo_{12}O_{40}]$, $(NH_4)_3[PMo_{12}O_{40}]$ and $H_3[PMo_{12}O_{40}]$) with a fixed flow rate of $20\,\mu l\,h^{-1}$ for a specific time period to deliver a predetermined amount of POM onto the CNT electrodes. POM solution was delivered to the ESD emitter through a fused silica capillary (inner diameter (ID): $100\,\mu m$, outer diameter (OD): $360\,\mu m$, Polymicro Technologies, Phoenix, AZ) using a syringe pump (KD Scientific, Holliston, MA). The emitter was produced by pulling a $500\,\mu m$ OD four bore capillary (VitroCom, NJ, USA) to a final OD of $130\,\mu m$ using a micropipette puller (P-2000, Sutter Instrument Company, Novato, CA). The ESD emitter was connected to the fused silica capillary using a microtight union (Upchurch Scientific). A voltage of $\sim -2.5\,kV$ was applied to the emitter to generate charged droplets. The electrode was positioned 3 mm away from the ESD emitter. It should be noted that ESD electrodes contain both anions ($[PMo_{12}O_{40}]^{n-}$, $n = 2,3$), counter cations (Na^+, NH_4^+, H^+) and solvent molecules while the SL electrodes contain only mass-selected $[PMo_{12}O_{40}]^{3-}$ anions. See Fig. 1. The diameter of the deposition area is $\sim 3\,mm$. We found that the performance of the supercapacitor does not change with respect to the deposition spot size. However, we maintained the same deposition area for both SL and ESD during electrode preparation. The total number of ions deposited over time during ESD was calculated based on the flow rate and concentration of the POM solution.

Preparation of electrolyte membrane. An ionic liquid-based electrolyte membrane was prepared using $EMIMBF_4$ and copolymer poly(vinylidene fluoride-co-hexafluoropropylene) (PVDF-HFP). A typical preparation method is as follows: 2 g of PVDF-HFP was dissolved in 13 ml of DMF and stirred overnight at room temperature to make a homogenous solution. Two millilitre of $EMIMBF_4$ ionic liquid was then added to the PVDF-HFP/DMF solution and stirred continuously for 4 h. Finally, 15 ml of $EMIMBF_4$/PVDF-HFP solution was cast onto a 9 cm diameter petri dish and dried it in the oven at $70\,°C$ for 12 h. The resulting membrane was peeled off and used as-prepared. See Supplementary Methods for detailed specifications of all chemicals used in this study.

Fabrication and testing of supercapacitor devices. The supercapacitor devices were fabricated by placing an $EMIMBF_4$/PVDF-HFP-based electrolyte membrane between two similar $1\,cm \times 1\,cm$ as-prepared electrodes. The electrode assembly was gently pressed at 100 psi using a mechanical press (International Crystal Laboratories, Garfield, NJ) for 5 min and assembled in a two electrode testing cell specially designed for performing electrochemical measurements such as CV and GCD.

The GCD characteristics and stability of the supercapacitor devices were evaluated using a VersaSTAT 3 potentiostat/galvanostat (Princeton Applied Research, Oak Ridge, TN). The GCD experiments were carried out in the voltage range of 0–1 V at a current density of $8\,A\,g^{-1}$—typical operating conditions used to study the performance of supercapacitors. Subsequently, the stability test was carried out by running the GCD experiment for 1,000 charge-discharge cycles.

Before performing GCD experiments, the intrinsic redox activity of $PMo_{12}O_{40}$ anions in $EMIMBF_4$ was evaluated using CV to make sure that $PMo_{12}O_{40}$ anions are redox active in $EMIMBF_4$ electrolyte. A CV was obtained on a GC working electrode containing 1×10^{14} $Na_3[PMo_{12}O_{40}]$ clusters deposited using ESD. Pristine $EMIMBF_4$ was used as an electrolyte; platinum and silver wires were used as a counter and pseudo-reference electrodes, respectively.

The total specific capacitance (that is, the sum of capacitance contributions from non-faradaic reactions on carbon and faradaic reactions using the redox active material) and energy density of the supercapacitor were calculated using the following equations:

$$C_{sp(GCD),t} = \frac{i}{-\left(\frac{\Delta V}{\Delta t}\right).m} \tag{1}$$

$$E = \frac{C_{sp(GCD),t}.\Delta V^2}{2.\left(10^{-3}\frac{kg}{g}\right).(3600\,s/h)} \tag{2}$$

$$C_{sp(CV),t} = \frac{\int I dV}{v.m.V} \tag{3}$$

$$P = \frac{E}{\Delta t} \tag{4}$$

where, i—current density used in the GCD experiment ($8\,A\,g^{-1}$); $\Delta V/\Delta t$—slope of the discharge curve ($V\,s^{-1}$); $C_{sp(GCD),t}$ and $C_{sp(CV),t}$- total specific capacitance of the supercapacitor calculated from GCD and CV experiments respectively; ΔV—voltage difference between the voltage at the beginning of discharging and the voltage at the end of discharge in GCD measurement; I and V—response current (A) and potential window (V) in CV measurement, respectively; v—scan rate in the CV measurement; m—total mass of the active material on the electrode surface; E—energy density of the supercapacitor ($Wh\,kg^{-1}$), P—Power density of the supercapacitor ($kW\,kg^{-1}$), Δt—discharge time (s).

To calculate the specific Faradaic capacitance of POM-based supercapacitors, it is assumed that only CNT present on the top layer contributes to the non-Faradaic double layer capacitance and the minor contribution from the layer behind the electrode interface is disregarded. Therefore, the combined weight of the CNT loading ($\sim 25\,\mu g\,cm^{-2}$) and the weight of immobilized $[PMo_{12}O_{40}]$ were used as a total weight of the active material to calculate the total specific capacitance of the supercapacitor using equation 5.

$$C_{sp,[PMo12]} = \frac{[(C_{sp,t} * W_{t,CNT + PMo12}) - (C_{sp,CNT} * W_{CNT})]}{W_{PMo12}} \tag{5}$$

$C_{sp,t}$—total specific capacitance (F/g) of modified CNT electrodes calculated using equation 1. $W_{t,CNT + PMo12}$—combined weight (g) of Faradaic and non-Faradaic active material (CNT loading $\sim 25\,\mu g$ and loading of $[PMo_{12}O_{40}]$ clusters in each case); $C_{sp,CNT}$—total specific capacitance (F/g) of pristine CNT (only CNT as an active material); W_{CNT}—weight (g) of CNT loading in a pristine CNT ($\sim 25\,\mu g$); $W_{PMo12}W_{PMo12}$—mass of $[PMo_{12}O_{40}]$ clusters at each loadings; $C_{sp,[PMo12]}$—specific Faradaic capacitance contributed from $[PMo_{12}O_{40}]$ clusters alone in the modified electrodes.

The oxidation state and surface characteristics of $[PMo_{12}O_{40}]$ clusters on the different CNT electrodes were examined using XPS, SEM and STEM so that structural characteristics could be correlated with the electrochemical activity.

Electrode characterization. *Scanning transmission electron microscopy.* The size and distribution of $[PMo_{12}O_{40}]$ clusters on CNT electrodes were determined using STEM. In a typical sample preparation, a lift-out procedure was employed using a focused ion beam workstation, in which a portion of the CNT electrode was removed and transferred onto a copper TEM grid (Ted Pella, Inc., Redding, CA)[70]. STEM micrographs were then obtained using a FEI TITAN 80–300 eV TEM/STEM operated at 300 kV. The microscope is fitted with a spherical aberration corrector for the probe forming lens, enabling sub-Angstrom resolution in the STEM imaging mode. STEM analysis provides insight into the size and distribution of SL and ESD $[PMo_{12}O_{40}]$ clusters on CNT electrodes through Z-contrast imaging in the HAADF mode. Approximately 1.5×10^{13} $[PMo_{12}O_{40}]$ clusters were deposited using SL and ESD on CNT electrode and used for STEM analysis.

The STEM images were analysed as follows: the histograms of the cluster size distribution for SL and ESD samples were prepared with the Gatan Digital Micrograph software, which allows one to establish contrast threshold difference for Mo atoms. To analyse the STEM images of the SL sample (Fig. 3b), intact POM clusters and individual Mo atoms were identified and counted towards the generation of a histogram. A similar method was adopted for analysing images of ESD samples in Fig. 3c. These images were analysed by manually defining the features corresponding to both agglomerated POM clusters and individual Mo atoms and calculating the areas of these features. The average size of individual features shown in Fig. 3f was obtained assuming they have circular shapes.

Scanning electron microscopy. The surface morphology of different CNT electrodes was examined using a scanning electron microscope (Quanta 3D model, FEI, Inc.) operated at 10 kV accelerating voltage. Approximately 1.5×10^{13} $[PMo_{12}O_{40}]$ clusters were deposited using SL and ESD on CNT electrodes for SEM analysis.

X-ray photoelectron spectroscopy. The oxidation states of Mo on different CNT electrodes were evaluated using Mo $3d$ XPS spectra. XPS measurements were performed with a Physical Electronics Quantera Scanning X-ray Microprobe. This system uses a focused monochromatic Al $K\alpha$ X-ray (1,486.7 eV) source for excitation and a spherical section analyser. The instrument has a 32 element multichannel detection system. A 100 W X-ray beam focused to $100\,\mu m$ diameter was rastered over a $1.2\,mm \times 0.1\,mm$ rectangular region on the sample. The X-ray beam is incident normal to the sample and the photoelectron detector is at 45° off-normal. High energy resolution spectra were collected using a pass-energy of

69.0 eV with a step size of 0.125 eV. For the Ag $3d_{5/2}$ line, these conditions produced a FWHM of 0.91 eV. Approximately 1.5×10^{13} [PMo$_{12}$O$_{40}$] clusters were deposited using SL and ESD on CNT electrode and used for XPS analysis. The XPS spectra were fitted using Multipak Spectrum software and the following peak fitting parameters were used. The intensity ratio of Mo $3d_{5/2}$ and Mo $3d_{3/2}$ peaks arising from spin–orbit coupling was fixed at 2:3 corresponding to d orbital and the spit–orbit splitting was fixed at 3.1 eV. The Shirley background subtraction method with 80% Gaussian, 20% Lorentz line shapes was employed to fit both Mo $3d$ and C 1s XPS spectra.

References

1. Goodenough, J. B. Electrochemical energy storage in a sustainable modern society. *Energy Environ. Sci.* 7, 14–18 (2014).
2. Lavine, M., Szuromi, P. & Coontz, R. Electricity now and when. *Science* 334, 921–921 (2011).
3. Suárez-Guevara, J., Ruiz, V. & Gomez-Romero, P. Hybrid energy storage: high voltage aqueous supercapacitors based on activated carbon–phosphotungstate hybrid materials. *J. Mater. Chem. A* 2, 1014–1014 (2014).
4. Vlad, a. *et al.* Hybrid supercapacitor-battery materials for fast electrochemical charge storage. *Sci. Rep.* 4, 4315–4315 (2014).
5. Choi, N. S. *et al.* Challenges facing lithium batteries and electrical double-layer capacitors. *Angew. Chem. Int. Ed. Engl.* 51, 9994–10024 (2012).
6. Simon, P. & Gogotsi, Y. Materials for electrochemical capacitors. *Nat. Mater.* 7, 845–854 (2008).
7. Armand, M. & Tarascon, J. M. Building better batteries. *Nature* 451, 652–657 (2008).
8. Tarascon, J. M. & Armand, M. Issues and challenges facing rechargeable lithium batteries. *Nature* 414, 359–367 (2001).
9. Goodenough, J. B. & Park, K. S. The Li-ion rechargeable battery: a perspective. *J. Am. Chem. Soc.* 135, 1167–1176 (2013).
10. Wang, G., Zhang, L. & Zhang, J. A review of electrode materials for electrochemical supercapacitors. *Chem. Soc. Rev.* 41, 797–828 (2012).
11. Simon, P., Gogotsi, Y. & Dunn, B. Where do batteries end and supercapacitors begin? *Science* 343, 1210–1211 (2014).
12. Sathiya, M., Prakash, A. S., Ramesha, K., Tarascon, J. M. & Shukla, A. K. V2O5-anchored carbon nanotubes for enhanced electrochemical energy storage. *J. Am. Chem. Soc.* 133, 16291–16299 (2011).
13. Gomez-Romero, P. Hybrid organic–inorganic materials—in search of synergic activity. *Adv. Mater.* 13, 163–174 (2001).
14. Liu, Q., Nayfeh, M. H. & Yau, S. T. Brushed-on flexible supercapacitor sheets using a nanocomposite of polyaniline and carbon nanotubes. *J. Power Sources* 195, 7480–7483 (2010).
15. Huang, C. & Grant, P. S. One-step spray processing of high power all-solid-state supercapacitors. *Sci. Rep.* 3, 2393–2393 (2013).
16. Wang, X. *et al.* Chemical vapor deposition growth of crystalline monolayer MoSe2. *ACS Nano* 8, 5125–5131 (2014).
17. Boukhalfa, S., Evanoff, K. & Yushin, G. Atomic layer deposition of vanadium oxide on carbon nanotubes for high-power supercapacitor electrodes. *Energy Environ. Sci.* 5, 6872–6872 (2012).
18. Yang, J., Lian, L., Ruan, H., Xie, F. & Wei, M. Nanostructured porous MnO2 on Ni foam substrate with a high mass loading via a CV electrodeposition route for supercapacitor application. *Electrochim. Acta* 136, 189–194 (2014).
19. Mattes, R. Heteropoly and isopoly oxometalates. Von M. T. Pope. Springer-Verlag, Berlin 1983. XIII, 180 S., geb. DM 124.00. *Angew. Chem. Int. Ed. Engl.* 96, 730–730 (1984).
20. Wang, H. *et al.* In operando X-ray absorption fine structure studies of polyoxometalate molecular cluster batteries: polyoxometalates as electron sponges. *J. Am. Chem. Soc.* 134, 4918–4924 (2012).
21. Kulesza, P. J. *et al.* Fabrication of network films of conducting polymer-linked polyoxometallate-stabilized carbon nanostructures. *Electrochim. Acta* 51, 2373–2379 (2006).
22. Rong, C. & Anson, F. C. Spontaneous adsorption of heteropolytungstates and heteropolymolybdates on the surfaces of solid electrodes and the electrocatalytic activity of the adsorbed anions. *Inorg. Chim. Acta* 242, 11–16 (1996).
23. Rausch, B., Symes, M. D., Chisholm, G. & Cronin, L. Decoupled catalytic hydrogen evolution from a molecular metal oxide redox mediator in water splitting. *Science* 345, 1326–1330 (2014).
24. Franchetti, V., Solka, B. H., Baitinger, W. E., Amy, J. W. & Cooks, R. G. Soft landing of ions as a means of surface modification. *Int. J. Mass Spectrom. Ion Processes* 23, 29–35 (1977).
25. Johnson, G. E. & Laskin, J. Soft landing of mass-selected gold clusters: Influence of ion and ligand on charge retention and reactivity. *Int. J. Mass Spectrom.* 377, 205–213 (2015).
26. Johnson, G. E., Hu, Q. & Laskin, J. Soft landing of complex molecules on surfaces. *Annu. Rev. Anal. Chem.* 4, 83–104 (2011).
27. Johnson, G. E., Gunaratne, D. & Laskin, J. Soft- and reactive landing of ions onto surfaces: concepts and applications. *Mass Spectrom. Rev.* 35, 439–479 (2016).
28. Rauschenbach, S. *et al.* Electrospray ion beam deposition of clusters and biomolecules. *Small* 2, 540–547 (2006).
29. Badu-Tawiah, A. K., Wu, C. & Cooks, R. G. Ambient ion soft landing. *Anal. Chem.* 83, 2648–2654 (2011).
30. Verbeck, G., Hoffmann, W. & Walton, B. Soft-landing preparative mass spectrometry. *Analyst* 137, 4393–4407 (2012).
31. Gologan, B., Green, J. R., Alvarez, J., Laskin, J. & Graham Cooks, R. Ion/surface reactions and ion soft-landing. *Phys. Chem. Chem. Phys.* 7, 1490–1500 (2005).
32. Li, A. *et al.* Using ambient ion beams to write nanostructured patterns for surface enhanced raman spectroscopy. *Angew. Chem. Int. Ed.* 53, 12528–12531 (2014).
33. Krásný, L. *et al.* In-situ enrichment of phosphopeptides on MALDI plates modified by ambient ion landing. *J. Mass Spectrom.* 47, 1294–1302 (2012).
34. Evans, C. *et al.* Surface modification and patterning using low-energy ion beams: Si − O bond formation at the vacuum/adsorbate interface. *Anal. Chem.* 74, 317–323 (2002).
35. Pepi, F. *et al.* Chemically modified multiwalled carbon nanotubes electrodes with ferrocene derivatives through reactive landing. *J. Phys. Chem. C* 115, 4863–4871 (2011).
36. Mazzei, F., Favero, G., Frasconi, M., Tata, A. & Pepi, F. Electron-transfer kinetics of microperoxidase-11 covalently immobilised onto the surface of multi-walled carbon nanotubes by reactive landing of mass-selected ions. *Chemistry* 15, 7359–7367 (2009).
37. Dubey, G. *et al.* Chemical modification of graphene via hyperthermal molecular reaction. *J. Am. Chem. Soc.* 136, 13482–13485 (2014).
38. He, Q. *et al.* In Situ bioconjugation and ambient surface modification using reactive charged droplets. *Anal. Chem.* 87, 3144–3148 (2015).
39. Gunaratne, K. D. D. *et al.* Design and performance of a high-flux electrospray ionization source for ion soft landing. *Analyst* 140, 2957–2963 (2015).
40. Jaworek, A. Electrospray droplet sources for thin film deposition. *J. Mater. Sci.* 42, 266–297 (2007).
41. Jaworek, A. & Sobczyk, A. T. Electrospraying route to nanotechnology: an overview. *J. Electrostat.* 66, 197–219 (2008).
42. Morozov, V. N. & Morozova, T. Y. Electrospray deposition as a method to fabricate functionally active protein films. *Anal. Chem.* 71, 1415–1420 (1999).
43. Morozov, V. N. & Morozova, T. Y. Electrospray deposition as a method for mass fabrication of mono- and multicomponent microarrays of biological and biologically active substances. *Analytical Chemistry* 71, 3110–3117 (1999).
44. James, N. O. S. *et al.* Electrospray deposition of carbon nanotubes in vacuum. *Nanotechnology* 18, 035707 (2007).
45. Saywell, A. *et al.* Electrospray deposition of C60 on a hydrogen-bonded supramolecular network. *J. Phys. Chem. C* 112, 7706–7709 (2008).
46. Chaparro, A. M., Gallardo, B., Folgado, M. A., Martín, A. J. & Daza, L. PEMFC electrode preparation by electrospray: optimization of catalyst load and ionomer content. *Catal. Today* 143, 237–241 (2009).
47. Beidaghi, M. & Gogotsi, Y. Capacitive energy storage in micro-scale devices: recent advances in design and fabrication of micro-supercapacitors. *Energy Environ. Sci.* 7, 867–884 (2014).
48. Ervin, M. H., Miller, B. S., Hanrahan, B., Mailly, B. & Palacios, T. A comparison of single-wall carbon nanotube electrochemical capacitor electrode fabrication methods. *Electrochim. Acta* 65, 37–43 (2012).
49. Gunaratne, K. D. D. *et al.* Controlling the charge state and redox properties of supported polyoxometalates via soft landing of mass-selected ions. *J. Phys. Chem. C* 118, 27611–27622 (2014).
50. Conway, B. E. Transition from "supercapacitor" to "battery" behavior in electrochemical energy storage. *J. Electrochem. Soc.* 138, 1539–1539 (1991).
51. Kowsari, E. *High-Performance Supercapacitors Based on Ionic Liquids and a Graphene Nanostructure, Ionic Liquids - Current State of the Art* (InTech, 2015).
52. Armand, M., Endres, F., MacFarlane, D. R., Ohno, H. & Scrosati, B. Ionic-liquid materials for the electrochemical challenges of the future. *Nat Mater* 8, 621–629 (2009).
53. Bard, A. J. & Faulkner, L. R. *Electrochemical Methods: Fundamentals and Applications*, 2nd edn (Wiley, 2000).
54. Yang, X. *et al.* A high-performance graphene oxide-doped ion gel as gel polymer electrolyte for all-solid-state supercapacitor applications. *Adv. Funct. Mater.* 23, 3353–3360 (2013).
55. Long, X.-L. *et al.* Adsorption of ammonia on activated carbon from aqueous solutions. *Environ. Prog.* 27, 225–233 (2008).
56. Moradi, O. & Zare, K. Adsorption of ammonium ion by multi-walled carbon nanotube: kinetics and thermodynamic studies. *Fullerenes Nanotubes Carbon Nanostruct.* 21, 449–459 (2012).
57. De Vos, N., Maton, C. & Stevens, C. V. Electrochemical stability of ionic liquids: general influences and degradation mechanisms. *ChemElectroChem* 1, 1258–1270 (2014).
58. Coronado, E. & Gómez-García, C. J. Polyoxometalate-based molecular materials. *Chem. Rev.* 98, 273–296 (1998).
59. Bellitto, C. *et al.* BEDT-TTF salts with.alpha.-Keggin poly(oxometallates): electrical, magnetic, and optical properties of (BEDT-TTF)8[PMo12O40] and (BEDT-TTF)8[SiW12O40] and X-ray crystal structure of (BEDT-TTF)8

[PMo12O40].cntdot.{(CH3CN.cntdot.H2O)2}. *Chem. Mater.* **7,** 1475–1484 (1995).

60. Moulder, J. F. & Chastain, J. *Handbook of X-ray photoelectron spectroscopy: a reference book of standard spectra for identification and interpretation of XPS data.* 6th edn (Physical Electronics Division, Perkin-Elmer Corp, 1992).

61. Taylor, A. in *Practical Surface Analysis* 2nd edn., Vol. I (eds Briggs, D. & Seah, M. P.) (John Wiley, New York, 1990).

62. Quincy, R. B., Houalla, M., Proctor, A. & Hercules, D. M. Distribution of molybdenum oxidation states in reduced molybdenum/titania catalysts: correlation with benzene hydrogenation activity. *J. Phys. Chem.* **94,** 1520–1526 (1990).

63. Nakayama, M., Ii, T. & Ogura, K. Spectrosopic and paramagnetic properties of molybdenum oxyhydroxide films electrochemically formed from the Keggin-type (PMo12O40)3 − complex. *J. Mater. Res.* **18,** 2509–2514 (2003).

64. Spevack, P. A. & McIntyre, N. S. A Raman and XPS investigation of supported molybdenum oxide thin films. 2. Reactions with hydrogen sulfide. *J. Phys. Chem.* **97,** 11031–11036 (1993).

65. Belanger, D. & Laperriere, G. Electrochromic molybdenum trioxide thin film preparation and characterization. *Chem. Mater.* **2,** 484–486 (1990).

66. Anwar, M., Hogarth, C. A. & Bulpett, R. Effect of substrate temperature and film thickness on the surface structure of some thin amorphous films of MoO3 studied by X-ray photoelectron spectroscopy (ESCA). *J. Mater. Sci.* **24,** 3087–3090 (1989).

67. Black, L., Garbev, K., Beuchle, G., Stemmermann, P. & Schild, D. X-ray photoelectron spectroscopic investigation of nanocrystalline calcium silicate hydrates synthesised by reactive milling. *Cement Concrete Res.* **36,** 1023–1031 (2006).

68. Li, Z., Gao, L. & Zheng, S. SEM, XPS, and FTIR studies of MoO3 dispersion on mesoporous silicate MCM-41 by calcination. *Mater. Lett.* **57,** 4605–4610 (2003).

69. Vasilopoulou, M., Douvas, A. M., Palilis, L. C., Kennou, S. & Argitis, P. Old metal oxide clusters in new applications: spontaneous reduction of Keggin and Dawson polyoxometalate layers by a metallic electrode for Improving efficiency in organic optoelectronics. *J Am. Chem. Soc.* **137,** 6844–6856 (2015).

70. Giannuzzi, L. A. & Stevie, F. A. A review of focused ion beam milling techniques for TEM specimen preparation. *Micron* **30,** 197–204 (1999).

Acknowledgements

This work was supported by the U.S. Department of Energy's (DOE) Office of Basic Energy Sciences, Division of Chemical Sciences, Geosciences & Biosciences and performed in EMSL, a national scientific user facility located at Pacific Northwest National Laboratory (PNNL). TEM work was supported by the Joint Center for Energy Storage Research (JCESR), an Energy Innovation Hub. PNNL is operated by Battelle for DOE. We thank Sigracet SGL carbon GmbH for providing the CNT substrates used in this study.

Author contributions

V.P., D.D.G., G.E.J. and J.L. conceived and designed the experiments. V.P. performed all the device fabrication and electrochemical experiments. B.L.M., J.J.D., D.C.J., N.D.B. performed STEM measurements; B.L.M. and V.P. analysed the STEM data. M.H.E. performed XPS measurements; V.P. and M.H.E. analysed the XPS data. B.W. performed SEM measurements; V.P. and B.W. analysed the SEM data. V.P., G.E.J. and J.L. co-wrote the paper. All authors contributed to the interpretation of the results and assisted in writing the paper.

Additional information

22

Lithium-coated polymeric matrix as a minimum volume-change and dendrite-free lithium metal anode

Yayuan Liu[1,*], Dingchang Lin[1,*], Zheng Liang[1], Jie Zhao[1], Kai Yan[1] & Yi Cui[1,2]

Lithium metal is the ideal anode for the next generation of high-energy-density batteries. Nevertheless, dendrite growth, side reactions and infinite relative volume change have prevented it from practical applications. Here, we demonstrate a promising metallic lithium anode design by infusing molten lithium into a polymeric matrix. The electrospun polyimide employed is stable against highly reactive molten lithium and, via a conformal layer of zinc oxide coating to render the surface lithiophilic, molten lithium can be drawn into the matrix, affording a nano-porous lithium electrode. Importantly, the polymeric backbone enables uniform lithium stripping/plating, which successfully confines lithium within the matrix, realizing minimum volume change and effective dendrite suppression. The porous electrode reduces the effective current density; thus, flat voltage profiles and stable cycling of more than 100 cycles is achieved even at a high current density of $5\,mA\,cm^{-2}$ in both carbonate and ether electrolyte. The advantages of the porous, polymeric matrix provide important insights into the design principles of lithium metal anodes.

[1] Department of Materials Science and Engineering, Stanford University, Stanford, California 94305, USA. [2] Stanford Institute for Materials and Energy Sciences, SLAC National Accelerator Laboratory, 2575 Sand Hill Road, Menlo Park, California 94025, USA. * These authors contributed equally to this work. Correspondence and requests for materials should be addressed to Y.C. (email: yicui@stanford.edu).

The ever-increasing demand for high-energy-density storage systems for transportation (electric vehicles), portable electronics and other applications has stimulated intensive research on rechargeable batteries that go beyond the conventional lithium (Li) ion chemistry[1]. Among all the possible options[2,3], Li metal is the most ideal anode material due to its high theoretical capacity (3,860 mAh g^{-1}) as well as its low electrochemical potential (-3.040 V versus standard hydrogen electrode)[4,5]. Despite the appealing properties, Li metal electrode has been plagued for decades with the problem of ramified growth during repeated stripping/plating and the associated electrolyte decomposition, which lead to serious safety concerns and poor battery cycling efficiency[6,7].

It is well known that Li is highly reactive such that in liquid electrolyte it reacts spontaneously with solvent molecules and salt anions to form an insoluble layer of solid-electrolyte interphase (SEI)[8]. When SEI becomes stabilized to block electron transfer, this passivating film can slow down or, ideally, prevent the electrolyte from further decomposition. Nevertheless, as a 'hostless' electrode, the Li metal anode has a virtually infinite relative volume change during stripping/plating, resulting in the mechanical instability of the SEI layer and the formation of cracks. The cracks expose fresh Li underneath and locally enhance the Li ion flux, leading to non-homogeneous Li growth (dendrite, filament, etc.) that can induce internal short circuit and thermal runaway with potential safety hazards[9,10]. Moreover, the large-surface-area, dendritic Li growth brings about a continuous loss of both working Li and electrolyte (recurrent SEI formation), which gives rise to low Coulombic efficiency (CE) and rapid capacity decay.

For the past four decades, continuous research on Li metal stripping/plating has deepened our understanding of the process, but has not helped in solving the above-mentioned problems in an effective manner[11-14]. On one hand, the use of solid electrolytes to suppress dendrite propagation remains premature in the current stage[15-17], for they often fall short of meeting the high-power requirement at ambient temperature due to limited ionic conductivity[18,19], together with issues such as large interfacial impedance[20,21]. On the other hand, the most common approaches to dendrite mitigation in liquid electrolyte focus on the stabilization of SEI via adjusting the electrolyte composition and additives[22-26]. Though proven to be effective, most additives will be continuously consumed during battery cycling so that the suppression effect is not fully sustainable[27-29]. Alternatively, the application of a mechanically stable artificial SEI coating such as polymer or solid-state blocking layers has been proposed[30-33]. For example, a promising nanoscale interfacial engineering approach has been demonstrated recently based on interconnected hollow carbon nanospheres, ultrathin two-dimensional boron nitride or oxidized polyacrylonitrile fibres to control the dendrite growth and improve the cycling CE[34-36]. Nevertheless, all these studies adopted the galvanostatic Li plating/stripping approach on a current collector, which is still unable to address the issue of infinite volume change since the electrodes expand during Li plating and shrink during stripping. In addition, contrary to Li-ion batteries where Li ions are stored in prelithiated cathodes, many of the intensively studied high-energy-density battery chemistries (e.g., Li-air and Li-S) involve cathodes in the non-lithiated form. Therefore, it is apparent that a metallic Li anode design with no volume change at the whole-electrode scale and long-term cycling stability in liquid electrolyte is of paramount research importance[3].

Herein, we demonstrate a rational design of metallic Li anode that successfully achieves minimum volume change at the whole-electrode level and stable, dendrite-free Li cycling. Several important design principles are employed. Firstly, in order to

realize negligible volume change, a chemically as well as electrochemically stable matrix is required to sustain a constant electrode volume during cycling. In addition, complete confinement of Li within the matrix is necessary to preserve a constant electrode dimension; therefore, the direct nucleation of Li on the top surface of the matrix should be prevented, as reported in previous studies[31,34]. Moreover, a porous electrode is desirable since a reduced local current density is beneficial to alleviate dendrite propagation[37,38]. Following the aforementioned rationale, a Li-coated polyimide (PI) matrix design for metallic Li anode is proposed. The electrospun polymeric fibres guarantee a chemically and electrochemically inert matrix, which is favourable to confine the stripping/plating of Li solely within the matrix. Notably, the choice of PI is rather unique for it is one of the only few high-performance polymers that exhibit excellent chemical stability, heat resistance and mechanical strength above the melting point of Li (180 °C)[39]. However, the wetting of molten Li on PI polymer is poor. By applying a layer of ZnO coating via atomic layer deposition (ALD) on the PI fibres, we discover that molten Li can react with ZnO and subsequently infuse into the PI matrix, resulting in a free-standing, current collector-free Li electrode. More importantly, we separate the conducting function from the matrix, where infused metallic Li itself serves as the only electron transport media. As a result, the electrically insulating surface after Li stripping effectively prevents the direct plating of Li on the top surface of the matrix in the subsequent cycle, bringing about a well-confined, dendrite-free Li stripping/plating behaviour that successfully addresses the problem of infinite volume change present in all the previous designs. Moreover, the obtained electrode is highly porous so that the reduced effective current density results in flat voltage profiles and stable cycling of at least 100 cycles in both carbonate- and ether-based electrolytes even at a high current density of 5 mA cm^{-2}, which stands in stark contrast to the fluctuated and unstable cycling profile of bare Li foil electrodes.

Results

Fabrication of the Li-coated PI electrode. Figure 1 illustrates the fabrication process of the Li-coated PI matrix electrode. We employed a facile electrospinning method to obtain the PI fibre matrix. Thermogravimetric analysis (Supplementary Fig. 1) confirmed that the electrospun PI fibre is stable up to 450 °C, which is well above the melting point of Li (180 °C). Such high heat resistance ensures that the matrix can withstand the temperature of molten Li in order to fabricate the metallic Li anode. Nevertheless, molten Li cannot wet the bare PI matrix (Supplementary Fig. 2). Due to the high surface tension of molten Li on PI fibre, a large driving force is needed for Li to infuse into the matrix. Rather than physical absorption, a surface chemical reaction that can afford much higher driving force is necessary. Through universal screening on materials that can undergo conversion reaction with Li, we found that a layer of conformal ZnO coating applied to the matrix via ALD can render the matrix wet by molten Li, or 'lithiophilic'. Subsequently, when the core-shell PI-ZnO matrix was put into contact with molten Li, ZnO reacted with molten Li and, interestingly, extra Li can be drawn into the matrix, affording a Li-coated PI electrode.

Characterization of the Li-coated PI electrode. The morphology of the electrospun PI fibres was characterized using scanning electron microscopy (SEM). As can be seen from Fig. 2a, the fibres were continuous and uniform in general with a diameter of ~400 nm. After ALD coating, the surface of the fibres roughened due to the accumulation of a conformal layer of

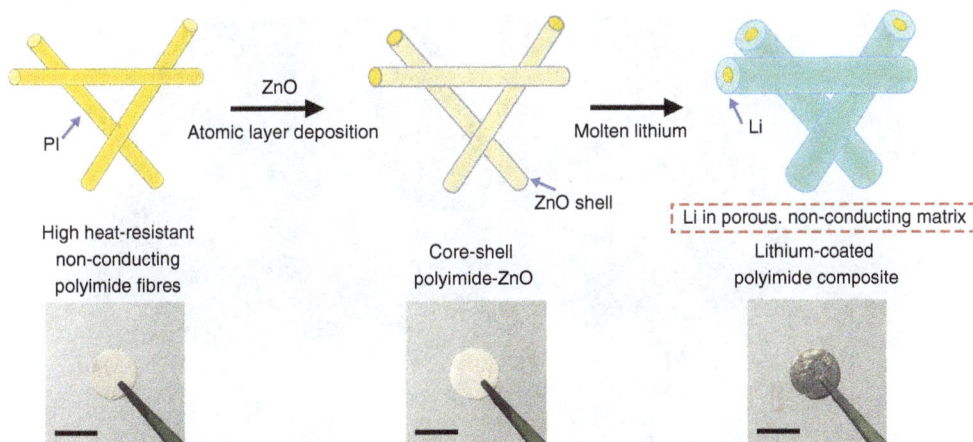

Figure 1 | Schematic of the fabrication of the Li-coated PI matrix. Electrospun PI was coated with a layer of ZnO via ALD to form core-shell PI-ZnO. The existence of ZnO coating renders the matrix 'lithiophilic' such that molten Li can steadily infuse into the matrix. The final structure of the electrode is Li coated onto a porous, non-conducting polymeric matrix. Scale bar, 1 cm.

ZnO nanoparticles (Fig. 2b). Evident PI-ZnO core-shell structure can be observed from the cross-sectional SEM image (Fig. 2c), where the PI core appeared darker in colour and the thickness of the ZnO shell was measured to be ~ 30 nm (Supplementary Fig. 3). Scanning transmission electron microscopy energy-dispersive X-ray elemental mapping (Fig. 2d) as well as line scan (Supplementary Fig. 4) resolved the distribution of C (from the PI backbone) and Zn (from the ZnO coating), further confirming the core-shell structure of the fibre matrix after ALD. Figure 2e shows the SEM top view of the PI matrix after Li coating. It appears that Li was drawn into the matrix preferentially along the fibres, and hence the matrix was not densely coated with Li. The porous nature of the resulting matrix can be further revealed from the cross-sectional SEM image (Supplementary Fig. 5), where obvious pores can be observed. X-ray diffraction (XRD) was employed to understand the compositional evolution of the matrix (Fig. 2f). As can be seen clearly from the XRD spectra, the ZnO layer reacted with molten Li to form LiZn alloy and Li_2O during the Li coating process[40]. Since the electron percolation pathway of the alloy particles within the non-conducting Li_2O matrix is generally limited it is justified to believe that the matrix remained low in electrical conductivity compared to metallic Li after Li coating[41,42]. Such a porous and electrically insulating matrix renders favourable electrochemical features to the resulting Li electrode, which will be elaborated in detail in later sections.

It is important to ensure the stability of the polymeric backbone in contact with the highly reductive molten Li. Therefore, Fourier transform infrared spectroscopy was employed (Fig. 2g), where the existence of PI can be identified from three characteristic peaks corresponding to asymmetric C=O stretching, symmetric C=O stretching and C-N stretching[43]. After ALD and the subsequent contact with molten Li, the transmittance intensity was reduced due to the coating layers, while the three characteristic peaks remained, confirming the intact polymeric matrix. In addition, neither anodic nor cathodic decomposition peaks can be observed from the cyclic voltammetry of the pristine PI (Supplementary Fig. 6), indicating the stability of the polymeric matrix towards electrochemical cycling.

The capacity of the Li electrode was determined to be above 2,000 mAh g^{-1} based on the weight of the whole composite electrode via Li stripping (Supplementary Fig. 7). Thus, the existence of the matrix did not seriously compromise the high specific capacity of the Li anode. Noticeably, by adjusting the thickness of the electrospun matrix, which can be done easily by changing the electrospinning time, the thickness of the final Li electrode can be tuned accordingly with great ease to match the capacity of the battery cathode (Supplementary Fig. 8), manifesting the facileness of our proposed method for real applications.

Well-confined Li stripping/plating within the matrix. We investigated the stripping/plating process of the Li-coated PI matrix using a two-electrode symmetric cell configuration assembled in 2,032 coin cells with carbonate-based electrolyte (1 M lithium hexafluorophosphate ($LiPF_6$) in 1:1 ethylene carbonate (EC)/diethyl carbonate (DEC), BASF). Interestingly, the electrode exhibited a well-confined stripping/plating behaviour (Fig. 3). Top fibres of the matrix were exposed after stripping away 5 mAh cm^{-2} Li at a current density of 1 mA cm^{-2} (Fig. 3a), which indicates that the top Li layers were dissolved more favourably during stripping. Subsequently, when 3 mAh cm^{-2} Li was plated, Li was observed to be deposited into the matrix and partially fill the space between the fibres (Fig. 3b). Finally, when all the stripped Li was plated back (Fig. 3c), the top surface of the matrix was covered again by Li (similar to Fig. 2e) with no discernable dendrites. The well-confined plating behaviour can be rationalized by the removal of the conductive Li component in the prior stripping process, and thus the exposure of the electrically insulating PI surface. Since Li plating only occurs where electrons meet Li ions, the exposed insulating surface was rendered unfavourable for Li nucleation. Instead, the metallic Li confined within the matrix served as the only electron conductor such that the deposition of Li occurred majorly on the underlying reserved Li. In addition, the much larger effective surface area (Supplementary Fig. 9) lowered the overall deposition barrier, thus preventing the formation of 'hot spot'. As a result, uneven Li deposition can be suppressed. On the contrary, if the electrons could be efficiently transported to the electrolyte-facing top surface or the electrodes exhibited limited surface area, undesirable Li stripping/plating behaviour may occur after recurrent cycles (Fig. 3d), as discussed in previous studies[31,34]. Direct Li nucleation on the top surface might be easier due to the high availability of both electrons and Li ions, which provides favourable sites for dendrite growth while leaving the interior voids empty.

Dendrite-free cycling with minimum volume change. The morphology of the top surface of the Li-coated PI matrix was studied after 10 cycles of galvanostatic stripping/plating in

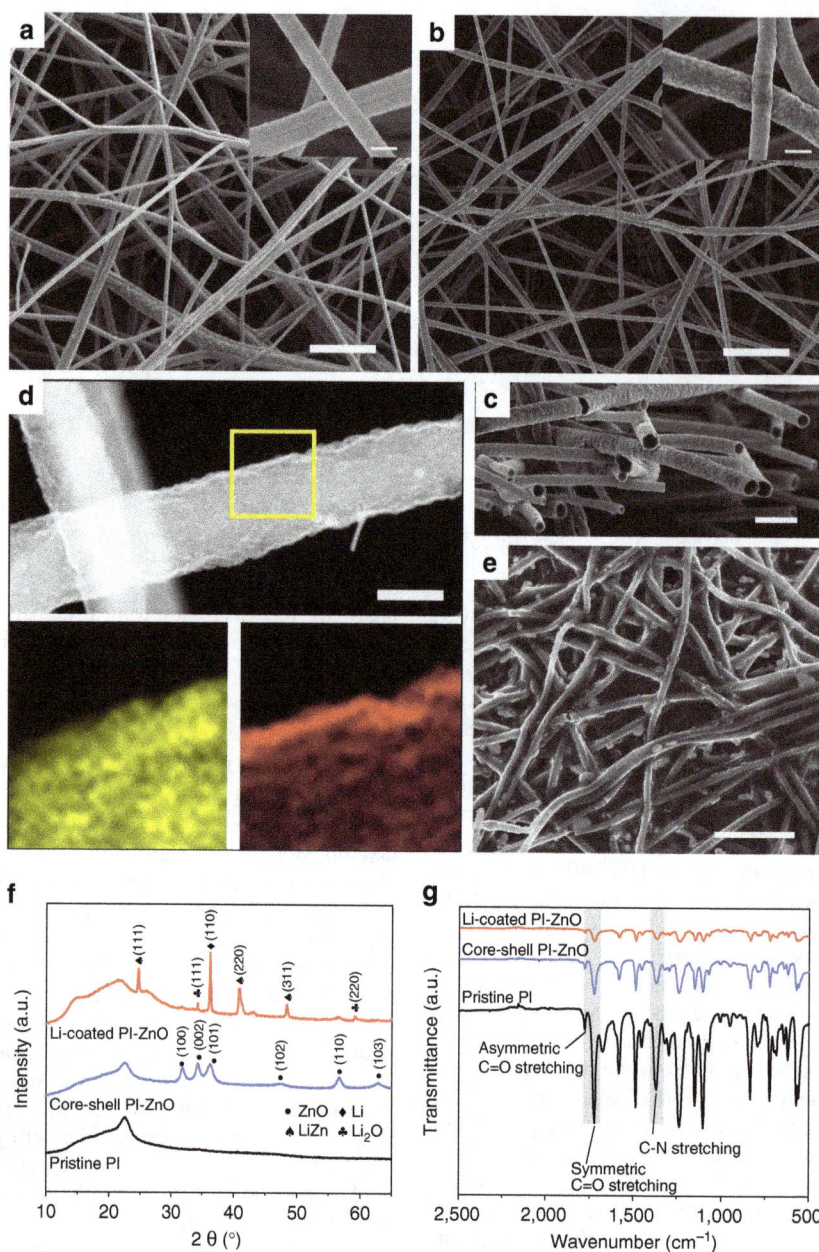

Figure 2 | Characterization of the Li-coated PI electrode. SEM images of the electrospun PI fibres (**a**) before and (**b**) after ZnO coating. Scale bars, 5 μm; insets scale bars, 500 nm. (**c**) Cross-sectional SEM image of the core-shell PI-ZnO, where the conformal ZnO coating can be observed clearly from the contrast of the fibre cross-sections. Scale bar, 1 μm. (**d**) Scanning transmission electron microscopy (STEM) image of a single core-shell PI-ZnO fibre and the corresponding energy-dispersive X-ray (EDX) elemental mapping of C and Zn distribution. Scale bar, 200 nm. (**e**) SEM image of the Li-coated PI matrix, showing the porous nature of the Li electrode. Scale bar, 5 μm. (**f**) XRD spectra of the pristine PI, the core-shell PI-ZnO and the Li-coated PI-ZnO matrix, where the Li-coated PI-ZnO exhibited the signals of LiZn alloy, Li$_2$O and metallic Li. (**g**) Fourier transform infrared spectra of the pristine PI, the core-shell PI-ZnO and the Li-coated PI-ZnO matrix (Li was scraped away to expose the underlying matrix in order to obtain the signal). The characteristic peaks of PI remained after ZnO and Li coating, indicating the stability of the polymeric matrix.

EC/DEC (Fig. 4a–c, Supplementary Fig. 10). Due to the abovementioned well-confined Li cycling behaviour, the surface of the Li-coated PI electrode remained consistently flat even at a high current density of 5 mA cm^{-2} (note that the uniform fibrous features in Fig. 4a–c are the fibre matrix and shall not be mistaken as Li dendrites). Moreover, no excessive dendrite formation can be observed after long-term cycling of 100 cycles (Supplementary Fig. 11). On the contrary, for the bare Li electrodes, rough surface and excessive mossy Li growth can be observed after 10 cycles even at a relatively low current density of 1 mA cm^{-2}. Such drastically different result further demonstrates the merit of the PI matrix on dendrite suppression.

Moreover, due to the existence of the host matrix, the issue of infinite volume change associated with the 'hostless' Li stripping/plating can now be solved. Even with the complete stripping of Li, the change in electrode thickness was minimal. For example, as shown in Fig. 4f,g, the electrode size was on average ~253 μm before stripping and remained at ~247 μm after complete stripping, which was merely ~2.4% of change (a relatively thick electrode was chosen for the more precise determination of thickness variation). However, for bare Li foil, 1 mAh cm^{-2} capacity represents ~4.85 μm thickness of Li (see Supplementary Method). Therefore, at least tens of microns of electrode thickness fluctuation can be expected for merely a

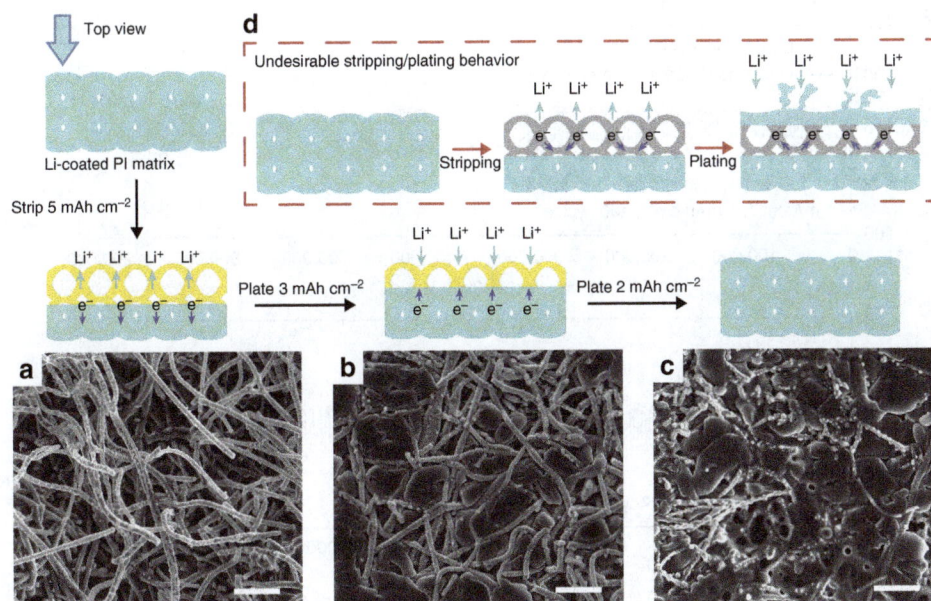

Figure 3 | Well-confined stripping/plating behaviour of the Li-coated PI matrix. Top-view SEM images of (**a**) the exposed top fibres of the Li-coated PI electrode after stripping away 5 mAh cm^{-2} Li; (**b**) exposed top fibres partially filled with Li when plating 3 mAh cm^{-2} Li back and (**c**) completely filled PI matrix after plating an additional 2 mAh cm^{-2} Li back (current density 1 mA cm^{-2}, in EC/DEC). The polymeric matrix ensures that Li is dissolved and deposited from the underlying conductive Li substrate and, as a result, Li is effectively confined into the matrix. (**d**) Schematic illustrating the alternative undesirable Li stripping/plating behaviour where, after stripping, Li nucleate on the top surface, leading to volume change and dendrites shooting out of the matrix. Scale bars, 5 μm.

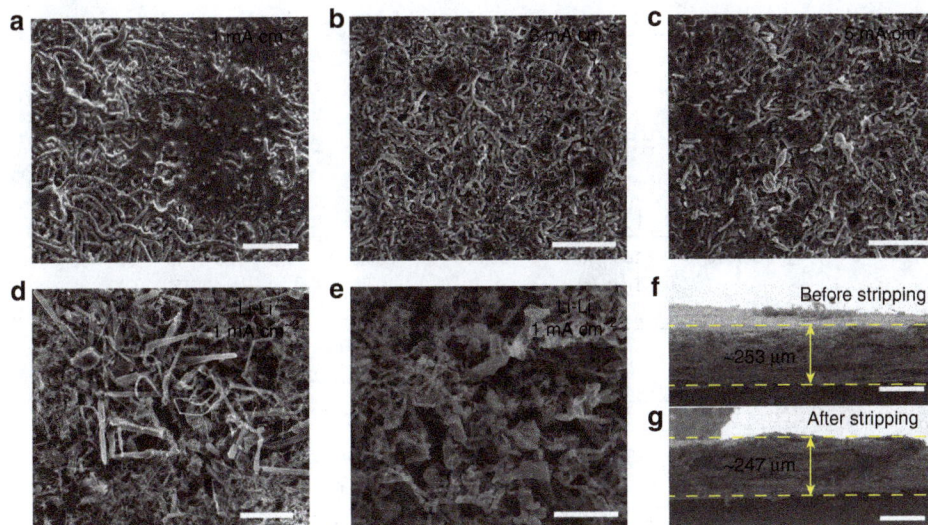

Figure 4 | Morphology of the Li electrodes after cycling at different current densities. Top-view SEM images of the Li-coated PI matrix after 10 cycles of stripping/plating in EC/DEC at a current density of (**a**) 1 mA cm^{-2}, (**b**) 3 mA cm^{-2} and (**c**) 5 mA cm^{-2}. (**d,e**) Top-view SEM images of the bare Li electrode after 10 cycles of stripping/plating in EC/DEC at a current density of 1 mA cm^{-2} with large amounts of mossy Li dendrites. Cross-sectional SEM images of the Li-coated PI matrix (**f**) before and (**g**) after complete Li stripping, from which no significant volume change can be seen. (Note that the uniform fibrous features in **a**–**c** are not dendrites but the fibrous matrix, which are distinctly different from the non-uniform, random-sized mossy Li dendrites in **d**,**e**). Scale bars, (**a**–**d**) 10 μm, (**e**) 5 μm, (**f**,**g**) 200 μm.

single layer of Li electrode in a commercial cell. Considering the conventionally applied stacking or rolling battery configuration with multiple layers, the accumulated dimension fluctuation can be tremendous. It is noted that for later cycles, due to the formation of dendritic Li and thus a porous electrode, the dimension fluctuation can be even larger. Thus, it is apparent that the existence of a stable matrix and the well-confined Li cycling behaviour are essential to alleviate the

electrode-level volume change, addressing the potential safety concerns.

Electrochemical cycling stability. The galvanostatic cycling performance of the Li-coated PI matrix was studied in both carbonate (EC/DEC) and ether (1 M lithium bis(tri-fluoromethanesulfonyl)imide in 1:1 w/w 1,3-dioxolane (DOL)/

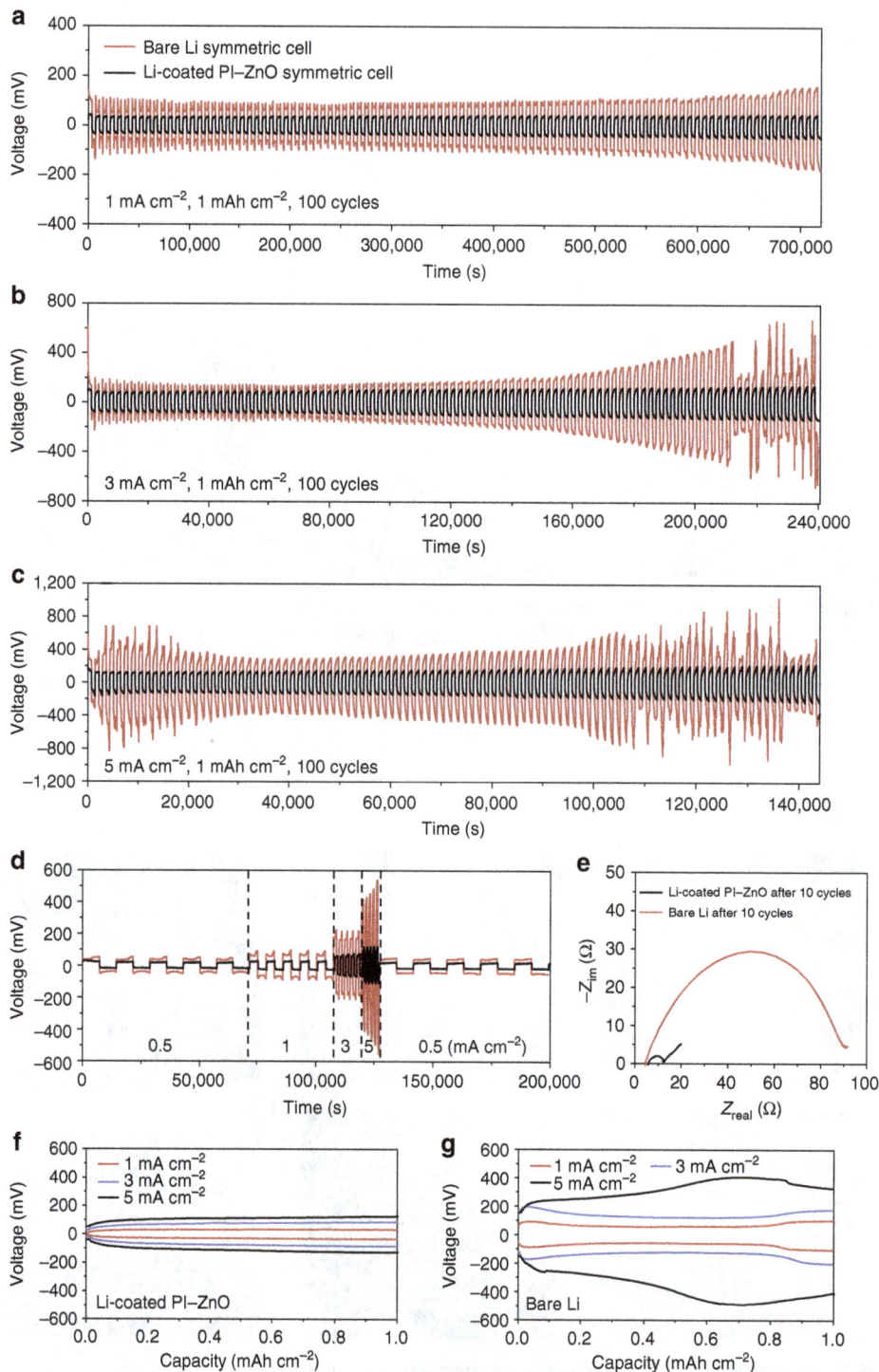

Figure 5 | Electrochemical characterization in EC/DEC electrolyte. Comparison of the cycling stability of the Li-coated PI matrix and the bare Li electrode at a current density of (**a**) $1 \, \text{mA cm}^{-2}$, (**b**) $3 \, \text{mA cm}^{-2}$ and (**c**) $5 \, \text{mA cm}^{-2}$. (**d**) Rate performance of the Li-coated PI matrix and the bare Li electrode. (**e**) Nyquist plot of the impedance spectra of the symmetrical Li-coated PI matrix and the bare Li cell after 10 cycles at a current density of $1 \, \text{mA cm}^{-2}$. Voltage profiles of (**f**) Li-coated PI matrix and (**g**) bare Li electrode at different current densities after 10 cycles. The amount of Li cycled was $1 \, \text{mAh cm}^{-2}$ in all cases.

1,2-dimethoxyethane (DME) with 1 wt% lithium nitrate) based electrolyte and compared with bare Li electrode (Fig. 5, Supplementary Figs 12–18). At a current density of $1 \, \text{mA cm}^{-2}$ in EC/DEC (Fig. 5a), the symmetrical cell of bare Li exhibited a large Li stripping/plating overpotential ($>100 \, \text{mV}$ versus Li^+/Li), which increased considerably within the first 100 cycles ($>170 \, \text{mV}$ in the 100th cycle). In contrast, the Li-coated PI

matrix not only showed a much lower overpotential ($\sim 35 \, \text{mV}$ in the initial cycle) but also achieved very stable cycling for at least 100 cycles ($\sim 40 \, \text{mV}$ overpotential in the 100th cycle). The difference in cycling stability (Fig. 5b,c) and overpotential (Fig. 5d) became increasingly pronounced at higher current densities. The stripping/plating overpotential for Li-coated PI matrix was ~ 70 and $\sim 110 \, \text{mV}$ at a current densities of 3 and

$5\,\mathrm{mA\,cm^{-2}}$, respectively, and the values remained constant within 100 cycles. However, the bare Li electrode succumbed to substantial voltage fluctuation at only 88 cycles at a current density of $3\,\mathrm{mA\,cm^{-2}}$ and 75 cycles at a current density of $5\,\mathrm{mA\,cm^{-2}}$, which might be attributed to possible dendrite-induced soft short circuit. Greatly improved cycling stability was also observed as we increased the cycling capacity to $3\,\mathrm{mAh\,cm^{-2}}$ (Supplementary Fig. 12). Similarly, in DOL/DME electrolyte, the Li-coated PI matrix again outperformed the bare Li electrode (Supplementary Fig. 13). Noticeably, although it is generally recognized that DOL can improve the cycling life of Li metal anodes due to the formation of a relatively flexible oligomer SEI[44], the bare Li electrode still exhibited a necking behaviour (the overpotential first decreases, then increases) during cycling, which is a characteristic sign for dendrite formation in early stage and SEI accumulation later[37]. Nevertheless, the Li-coated PI matrix maintained flat, constant cycling profiles and reduced overpotential at all current densities. Such exceptional long-term cycling performance is a good indicator of the superior CE and more uniform Li deposition/dissolution of the Li-coated PI electrode. Finally, since it is technically challenging to directly determine the CE of electrodes with pre-stored Li through Li plating/stripping as in previous studies[27,34], we paired our electrode with high-areal-capacity $\mathrm{Li_4Ti_5O_{12}}$ (LTO, $\sim 3\,\mathrm{mAh\,cm^{-2}}$) electrode as an indirect method to study the CE of the electrode (Supplementary Fig. 14). Li-coated PI electrode ($\sim 10\,\mathrm{mAh\,cm^{-2}}$) exhibited quite stable cycling comparable to the highly oversized Li metal counterpart (750 μm Li), while bare 50 μm Li foil ($\sim 10\,\mathrm{mAh\,cm^{-2}}$) started to decay at ~ 20 cycles. The results gave rise to semiquantitative CE values and illustrate the superior CE of the Li-coated PI electrode (see Supplementary Note 1).

Discussion

Two key factors contributed to the excellent electrochemical performance of the Li-coated PI matrix, namely, the porous nature of the electrode and the non-conducting nature of the exposed matrix surface after Li stripping. As mentioned previously, molten Li was drawn into the matrix preferentially along the fibres during the Li coating step, resulting in a porous Li electrode. The high porosity can be further confirmed by the fast electrolyte uptake during cell assembly (Supplementary Fig. 15). Such high porosity increased the surface area of the electrode, which can in turn significantly reduce the effective current density during cycling. As a result, the Li stripping/plating overpotential was much smaller for the Li-coated PI matrix, especially at high current densities (Fig. 5d). Correspondingly, the electrochemical impedance spectroscopy revealed a much reduced interfacial charge transfer resistance for the Li-coated PI matrix compared to bare Li (~ 10 times lower in EC/DEC, Fig. 5e and Supplementary Fig. 16). More importantly, the reduced effective current density rendered the Li-coated PI matrix a flat stripping/plating voltage profile (Fig. 5f,g and Supplementary Fig. 17). For the bare Li electrode, large 'overpotential bumps' at the beginning and the end of each stripping or plating process can be observed, especially for the early cycles (Supplementary Fig. 18). This phenomenon can be explained by the high specific kinetic hindrance for non-uniform Li dissolution/deposition at high current density, which has been investigated in great detail in previous reports[37,45]. The fluctuation attenuated in later cycles due to the formation of mossy Li, which increased the surface area of the electrode, reducing the effective current density. However, when compared to the fluctuated profile of the bare Li electrode, the voltage profile of Li-coated PI was flat at all current densities throughout the

cycling, clearly demonstrating the advantages of the porous electrode. In addition, it is well known that the effective current density during Li stripping/plating has a crucial impact on the dendrite formation and growth[37,46]. Lower effective current density results in reduced electrolyte decomposition and related SEI formation during cycling so as to suppress the dendrite growth[37]. Therefore, the porous electrode structure, in addition to the non-conducting polymeric matrix, which led to the well-confined Li dissolution/deposition behaviour, effectively ensured the dendrite-free cycling of the Li-coated PI electrode, giving rise to stable long-term performance.

In conclusion, our work demonstrates the method of obtaining a free-standing, porous metallic Li anode by infusing molten Li into a core-shell PI-ZnO matrix. The excellent heat resistance and chemical stability of the PI fibres guaranteed the structural integrity of the matrix during Li coating and the subsequent battery cycling, while the conformal ALD ZnO coating provided the driving force for the molten Li infusion. Noticeably, the exposed non-conducting fibres after Li stripping prevented the direct plating of Li on the top surface of the electrode, which effectively confined Li within the matrix. In this manner, dendrite-free and minimum-volume-change Li stripping/plating can be successfully achieved, addressing the biggest concerns regarding Li metal anode. Remarkably, different from the dense bare Li electrode, the large porosity of the Li-coated PI electrode considerably decreased the effective current density during cycling. As a result, flat voltage profiles and long-term cycling stability can be realized even at a high current density of $5\,\mathrm{mA\,cm^{-2}}$. The benefits of a non-conducting polymeric matrix and electrode porosity shed new light on the design principles of metallic Li anodes and open up new opportunities to the realization of the next-generation high-energy-density battery systems based on Li metal chemistries.

Methods

Electrospinning of the PI matrix. 15 wt% PI powder (DuPont CP-0650) was dissolved in N-methyl-2-pyrrolidone and stirred at 750 r.p.m. in a 60 °C oil bath overnight to afford a homogeneous solution. Subsequently, the solution was loaded into a glass syringe with a stainless steel needle tip, which is connected to a voltage supply (ES30P-5 W, Gamma High Voltage Research). The applied potential on the needle was 15 kV, the distance between the needle tip and the graphite paper collector was 15 cm and the pumping rate was $10\,\mu\mathrm{l\,min^{-1}}$. In addition, a negative voltage of 1 kV was applied at the collector to improve the homogeneity of the electrospun film. The thickness of the electrospun PI matrix can be tuned easily by adjusting the electrospinning time.

Atomic layer deposition. The conformal ZnO layer was coated on the PI fibres via ALD using the Cambridge Nanotech Savannah S100 at 80 °C with diethyl zinc (DEZn) and DI water as precursors. The pulse times for DEZn and DI water were 15 ms each, with 60 s waiting between each pulse. Approximately 300 cycles were needed to obtain a ZnO coating with desirable Li wetting property.

Li coating of the core-shell PI-ZnO matrix. The Li coating process was carried out in an argon-filled glovebox with sub-p.p.m. O_2 level. In a typical process, freshly scraped Li foil (99.9%, Alfa Aesar) was put into a stainless-steel crucible and heated to melt on a hotplate (VWR). Subsequently, the edge of the core-shell PI-ZnO matrix (punched into 1-cm² discs) was put into contact with the molten Li. Li can steadily climb up and wet the whole matrix, affording the final Li electrode.

Electrochemical measurements. The processes of Li stripping and plating were studied using a symmetric cell configuration by assembling the electrodes into 2,032-type coin cells. The electrolytes employed were either 1 M lithium hexa-fluorophosphate (LiPF₆) in 1:1 EC/DEC (BASF Selectilyte LP40) or 1 M lithium bis(trifluoromethanesulfonyl)imide in 1:1 w/w DOL/DME with 1 wt% lithium nitrate additive. The separator used was Celgard 2325 (25 μm PP/PE/PP). The control bare Li cells were assembled using freshly scraped Li foil. Galvanostatic cycling was conducted on a standard eight-channel battery tester (Wuhan LAND Electronics Co., Ltd.). A constant current was applied to the electrodes during repeated stripping/plating while the potential was recorded over time. The impedance measurements were carried out using a Biologic VMP3 multichannel system. For the half-cell study, the LTO electrodes were prepared by mixing LTO (MTI), polyvinylidene fluoride (MTI) and carbon black (TIMCAL) in the ratio of 8:1:1 with N-methyl-2-pyrrolidone as the solvent. The areal mass loading of the LTO electrodes was ~ 20 mg cm^{-2}.

Characterization. SEM images were taken using a FEI XL30 Sirion scanning electron microscope at an acceleration voltage of 5 kV. In order to observe the surface morphology of Li after cycling, the electrodes were disassembled from the coil cell in the glovebox followed by gentle rinse in DOL. Scanning transmission electron microscopy images, the corresponding energy-dispersive X-ray elemental mapping and line scan were obtained on a FEI Tecnai G2 F20 X-TWIN. XRD patterns were recorded on a PANalytical X'Pert instrument. Noticeably, the Li electrode was loaded on a glass slide and covered with Kapton tape during XRD measurements to avoid direct contact with air. Fourier transform infrared spectra were recorded on a Nicolet iS50 FT/IR Spectrometer (Thermo Scientific). Thermogravimetric analysis was performed on a TA Instrument Q500 TGA in air at a heating rate of 5 °C min^{-1}. N$_2$ sorption studies were performed using a Micromeritics ASAP 2,020 adsorption apparatus at 77 K and at pressure up to 1 bar after the samples were first degassed at 180 °C overnight. The Brunauer-Emmett-Teller surface area was calculated using the adsorption data in a relative pressure ranging from 0.1 to 0.3.

References

1. Armand, M. & Tarascon, J. M. Building better batteries. *Nature* **451**, 652–657 (2008).
2. Chan, C. K. *et al.* High-performance lithium battery anodes using silicon nanowires. *Nat. Nanotechnol.* **3**, 31–35 (2008).
3. Bruce, P. G., Freunberger, S. A., Hardwick, L. J. & Tarascon, J. M. Li-O$_2$ and Li-S batteries with high energy storage. *Nat. Mater.* **11**, 19–29 (2012).
4. Xu, W. *et al.* Lithium metal anodes for rechargeable batteries. *Energy Environ. Sci.* **7**, 513–537 (2014).
5. Kim, H. *et al.* Metallic anodes for next generation secondary batteries. *Chem. Soc. Rev.* **42**, 9011–9034 (2013).
6. Aurbach, D., Zinigrad, E., Cohen, Y. & Teller, H. A short review of failure mechanisms of lithium metal and lithiated graphite anodes in liquid electrolyte solutions. *Solid State Ion.* **148**, 405–416 (2002).
7. Brissot, C., Rosso, M., Chazalviel, J. N. & Lascaud, S. Dendritic growth mechanisms in lithium/polymer cells. *J. Power Sources* **81**, 925–929 (1999).
8. Peled, E. The electrochemical-behavior of alkali and alkaline-earth metals in non-aqueous battery systems—the solid electrolyte interphase model. *J. Electrochem. Soc.* **126**, 2047–2051 (1979).
9. Cohen, Y. S., Cohen, Y. & Aurbach, D. Micromorphological studies of lithium electrodes in alkyl carbonate solutions using in situ atomic force microscopy. *J. Phys. Chem. B* **104**, 12282–12291 (2000).
10. Aurbach, D., Gofer, Y. & Langzam, J. The correlation between surface-chemistry, surface-morphology, and cycling efficiency of lithium electrodes in a few polar aprotic systems. *J. Electrochem. Soc.* **136**, 3198–3205 (1989).
11. Gireaud, L., Grugeon, S., Laruelle, S., Yrieix, B. & Tarascon, J. M. Lithium metal stripping/plating mechanisms studies: a metallurgical approach. *Electrochem. Commun.* **8**, 1639–1649 (2006).
12. Bhattacharyya, R. *et al.* In situ NMR observation of the formation of metallic lithium microstructures in lithium batteries. *Nat. Mater.* **9**, 504–510 (2010).
13. Harry, K. J., Hallinan, D. T., Parkinson, D. Y., MacDowell, A. A. & Balsara, N. P. Detection of subsurface structures underneath dendrites formed on cycled lithium metal electrodes. *Nat. Mater.* **13**, 69–73 (2014).
14. Chandrashekar, S. *et al.* Li-7 MRI of Li batteries reveals location of microstructural lithium. *Nat. Mater.* **11**, 311–315 (2012).
15. Kamaya, N. *et al.* A lithium superionic conductor. *Nat. Mater.* **10**, 682–686 (2011).
16. Murugan, R., Thangadurai, V. & Weppner, W. Fast lithium ion conduction in garnet-type Li$_7$La$_3$Zr$_2$O$_{12}$. *Angew. Chem. Int. Ed.* **46**, 7778–7781 (2007).
17. Li, J. C., Ma, C., Chi, M. F., Liang, C. D. & Dudney, N. J. Solid electrolyte: the key for high-voltage lithium batteries. *Adv. Energy Mater.* **5** doi:10.1002/aenm.201401408 (2015).
18. Croce, F., Persi, L., Ronci, F. & Scrosati, B. Nanocomposite polymer electrolytes and their impact on the lithium battery technology. *Solid State Ion.* **135**, 47–52 (2000).
19. Bouchet, R. *et al.* Single-ion BAB triblock copolymers as highly efficient electrolytes for lithium-metal batteries. *Nat. Mater.* **12**, 452–457 (2013).
20. Kim, K. H. *et al.* Characterization of the interface between LiCoO$_2$ and Li$_7$La$_3$Zr$_2$O$_{12}$ in an all-solid-state rechargeable lithium battery. *J. Power Sources* **196**, 764–767 (2011).
21. Zhang, T. *et al.* A novel high energy density rechargeable lithium/air battery. *Chem. Commun.* **46**, 1661–1663 (2010).
22. Xu, K. Electrolytes and interphases in Li-ion batteries and beyond. *Chem. Rev.* **114**, 11503–11618 (2014).
23. Li, W. *et al.* The synergetic effect of lithium polysulfide and lithium nitrate to prevent lithium dendrite growth. *Nat. Commun.* **6**, 7436 (2015).
24. Lu, Y. Y., Tu, Z. Y. & Archer, L. A. Stable lithium electrodeposition in liquid and nanoporous solid electrolytes. *Nat. Mater.* **13**, 961–969 (2014).
25. Qian, J. *et al.* High rate and stable cycling of lithium metal anode. *Nat. Commun.* **6**, 6362 (2015).
26. Suo, L. M., Hu, Y. S., Li, H., Armand, M. & Chen, L. Q. A new class of solvent-in-salt electrolyte for high-energy rechargeable metallic lithium batteries. *Nat. Commun.* **4**, 1481 (2013).
27. Ding, F. *et al.* Dendrite-free lithium deposition via self-healing electrostatic shield mechanism. *J. Am. Chem. Soc.* **135**, 4450–4456 (2013).
28. Umeda, G. A. *et al.* Protection of lithium metal surfaces using tetraethoxysilane. *J. Mater. Chem.* **21**, 1593–1599 (2011).
29. Ishikawa, M., Morita, M. & Matsuda, Y. In situ scanning vibrating electrode technique for lithium metal anodes. *J. Power Sources* **68**, 501–505 (1997).
30. Kang, I. S., Lee, Y. S. & Kim, D. W. Improved cycling stability of lithium electrodes in rechargeable lithium batteries. *J. Electrochem. Soc.* **161**, A53–A57 (2014).
31. Ji, X. L. *et al.* Spatially heterogeneous carbon-fiber papers as surface dendrite-free current collectors for lithium deposition. *Nano Today* **7**, 10–20 (2012).
32. Ryou, M. H. *et al.* Excellent cycle life of lithium-metal anodes in lithium-ion batteries with mussel-inspired polydopamine-coated separators. *Adv. Energy Mater.* **2**, 645–650 (2012).
33. Kozen, A. C. *et al.* Next-generation lithium metal anode engineering via atomic layer deposition. *ACS Nano* **9**, 5884–5892 (2015).
34. Zheng, G. Y. *et al.* Interconnected hollow carbon nanospheres for stable lithium metal anodes. *Nat. Nanotechnol.* **9**, 618–623 (2014).
35. Yan, K. *et al.* Ultrathin two-dimensional atomic crystals as stable interfacial layer for improvement of lithium metal anode. *Nano Lett.* **14**, 6016–6022 (2014).
36. Liang, Z. *et al.* Polymer nanofiber-guided uniform lithium deposition for battery electrodes. *Nano Lett.* **15**, 2910–2916 (2015).
37. Heine, J. *et al.* Coated lithium powder (CLiP) electrodes for lithium-metal batteries. *Adv. Energy Mater.* **4** doi:10.1002/aenm.201300815 (2014).
38. Yang, C.-P., Yin, Y.-X., Zhang, S.-F., Li, N.-W. & Guo, Y.-G. Accommodating lithium into 3D current collectors with a submicron skeleton towards long-life lithium metal anodes. *Nat. Commun.* **6**, 8058 (2015).
39. Wright, W. W. & Hallden-Abberton, M. in *Ullmann's Encyclopedia of Industrial Chemistry* (Wiley-VCH Verlag GmbH & Co. KGaA, 2000).
40. Fu, Z. W., Huang, F., Zhang, Y., Chu, Y. & Qin, Q. Z. The electrochemical reaction of zinc oxide thin films with lithium. *J. Electrochem. Soc.* **150**, A714–A720 (2003).
41. Taberna, L., Mitra, S., Poizot, P., Simon, P. & Tarascon, J. M. High rate capabilities Fe$_3$O$_4$-based Cu nano-architectured electrodes for lithium-ion battery applications. *Nat. Mater.* **5**, 567–573 (2006).
42. Grugeon, S. *et al.* Particle size effects on the electrochemical performance of copper oxides toward lithium. *J. Electrochem. Soc.* **148**, A285–A292 (2001).
43. Mrsevic, M., Dusselberg, D. & Staudt, C. Synthesis and characterization of a novel carboxyl group containing (co)polyimide with sulfur in the polymer backbone. *Beilstein J. Org. Chem.* **8**, 776–786 (2012).
44. Aurbach, D., Youngman, O., Gofer, Y. & Meitav, A. The electrochemical behaviour of 1,3-dioxolane-LiClO$_4$ solutions-I. Uncontaminated solutions. *Electrochim. Acta* **35**, 625–638 (1990).
45. Bieker, G., Winter, M. & Bieker, P. Electrochemical in situ investigations of SEI and dendrite formation on the lithium metal anode. *Phys. Chem. Chem. Phys.* **17**, 8670–8679 (2015).
46. Kim, J. S., Baek, S. H. & Yoon, W. Y. Electrochemical behavior of compacted lithium powder electrode in Li/V$_2$O$_5$ rechargeable battery. *J. Electrochem. Soc.* **157**, A984–A987 (2010).

Acknowledgements

Y.C. acknowledges the support from the Assistant Secretary for Energy Efficiency and Renewable Energy, Office of Vehicle Technologies of the US Department of Energy under the Battery Materials Research (BMR) program.

Author contributions

Y.L., D.L. and Y.C. conceived the idea. Y.L. and D.L. carried out the materials synthesis and electrochemical tests. J.Z. carried out the TEM characterization. K.Y. and Z.L. provided important experimental insights. Y.L., D.L. and Y.C. co-wrote the paper. All the authors discussed the results and commented on the manuscript.

Additional information

23

Stimulus-responsive light-harvesting complexes based on the pillararene-induced co-assembly of β-carotene and chlorophyll

Yan Sun[1], Fang Guo[1], Tongfei Zuo[1], Jingjing Hua[1] & Guowang Diao[1]

The locations and arrangements of carotenoids at the subcellular level are responsible for their designated functions, which reinforces the necessity of developing methods for constructing carotenoid-based suprastructures beyond the molecular level. Because carotenoids lack the binding sites necessary for controlled interactions, functional structures based on carotenoids are not easily obtained. Here, we show that carotene-based suprastructures were formed via the induction of pillararene through a phase-transfer-mediated host–guest interaction. More importantly, similar to the main component in natural photosynthesis, complexes could be synthesized after chlorophyll was introduced into the carotene-based suprastructure assembly process. Remarkably, compared with molecular carotene or chlorophyll, this synthesized suprastructure exhibits some photocatalytic activity when exposed to light, which can be exploited for photocatalytic reaction studies of energy capture and solar conversion in living organisms.

[1] College of Chemistry and Chemical Engineering, Yangzhou University, Yangzhou, Jiangsu 225002, China. Correspondence and requests for materials should be addressed to G.D. (email: gwdiao@yzu.edu.cn).

Carotenoids are a diverse group of structurally related compounds[1] that perform various functions in living organisms[2]. Their conjugated double-bond systems determine their photochemical properties and their chemical reactivities[3], which, in turn, influence the properties of these subcellular structures[4]. They are relevant to human and animal processes related to nutrition[5], immune systems[6] and antioxidation[7]. Their roles in plants and microorganisms are ascribed mostly to acting as a stabilizing structure for light-harvesting complexes (LHCs)[8], photoprotection[9], absorbing light and transferring energy to chlorophyll pigments[10], which acts as a driving force for photosynthesis[11]. Furthermore, the molecular geometry and specific interactions with other molecules in the structures are critical for ensuring that the carotenoids fit into the suprastructures in the correct location and orientation[12]. Considerable effort has been devoted to investigating the biochemistry and molecular biology of carotenoids[13]. Nevertheless, in the fields of both nutrition utilization and suprastructure construction, suprastructure development has been limited by the hydrophobic nature of carotenoids[14].

Recently, in the carotenoid-related fields, this problem has been partially solved through the use of hydrophilic vectors[15]. For example, natural carotene has been wrapped into hydrophilic vectors by spinning disk processing to construct a carotene-loaded particle[16]. Notably, supramolecular studies have been focused primarily on the characterization of the natural photosynthesis machinary[17]. The design and synthesis of programmable entities based on carotenoids still faces many issues due to the low solubility of natural pigments in aqueous media. Thus, no carotenoid molecules have been introduced as building blocks to form suprastructures that mimic natural structures. To solve these problems, synthetic water-soluble analogues have often been used instead of natural pigments as building units[18]. Compared with the successful fabrication of many synthetic supramolecular systems based on hydrophilic biomolecules, including peptides[19], fatty acids[20], nucleotides[21] and porphyrins[22], the preparation of synthetic suprastructures based on hydrophobic biomolecules remains in the initial stages. Therefore, the development of effective strategies for assembling carotenoids into intricate and customizable large-scale systems that perform the same physiological functions as they perform *in vivo* is now essential[23].

Fortunately, supramolecular chemistry provides a new framework to tackle this difficult question. Thus far, supramolecular macrocycles, such as crown ethers[24-26], cyclodextrins[27] and calixarenes[28,29], have been used for various applications. Pillararenes[30,31], which are composed of hydroquinone units linked by methylene bridges at the para positions, are an especially interesting new class of macrocyclic host in supramolecular chemistry. Their unique, intrinsically rigid and symmetrical pillar architecture endows them with an outstanding ability to selectively bind various types of guest molecules[32]. This property has been used to construct various interesting supramolecular systems[33], including nanomaterials[34], sensors[35], ion channels[36] and drug-delivery systems[37].

In this article, β-carotene (β-CAR) was selected as a type of carotenoid to represent the carotenoid family. A synthetic water-soluble carboxyl-modified pillararene (WP5) was also chosen because its cavity dimensions are appropriate for the accommodation of β-CAR, which could further induce the construction of a β-CAR-based substructure. We envisioned that β-CAR might bind WP5 in water to achieve WP5 ⊃ β-CAR complexation (WCC) via the hydrophobic effect. Then, the resulting amphiphilic WCC might be able to form a β-CAR-containing suprastructure in water via self-assembly.

On the basis of the fascinating properties exhibited by natural chlorophyll/carotenoid complexes in photosynthesis, chlorophyll-b (Chl-b) was selected as the co-assembly factor participating in the preparation process with WCC, and a similar suprastructure of Chl-b-containing LHC was synthesized. These hydrophilic complexes possess a suite of unusual properties, including spontaneous growth, fusion, pH stimulus responsiveness and even some photocatalytic activity.

Results

Host–guest interaction between WP5 and β-CAR. The pillar[5]arene carboxyl derivative WP5 was prepared according to published procedures[38] and was identified by ^{1}H-NMR spectroscopy after being dissolved in D_2O. According to the results of Gaussian 09 calculations, the internal diameter of the cavity is 0.5 nm (ref. [32] and the length of the WP5 is 1.1 nm. Within the as-prepared synthetic LHC systems, the role of WP5 (Fig. 1b) is primarily to improve the solubility of β-CAR in aqueous environments and to further induce the hierarchical arrangement of β-CAR during assembly. β-CARs are naturally abundant pigments that are extremely hydrophobic and have low solubility in water. As shown in the energy-minimized structure of β-CAR (Fig. 1c), the length and width of β-CAR are ∼ 3.0 and 0.5 nm, respectively. A remarkable feature of β-CAR is that its molecular shape and size (a slight twist of conformation) fit well within the cavity of WP5, which provides the molecular basis for the interaction between WP5 and β-CAR. β-CAR units serve as the structural skeleton, providing photoprotective functions and binding sites for Chl-b (Fig. 1d). Chl-b molecules serve as the light-absorbing and energy-transferring components; they can attach to the hydrophobic domains by inserting their long alkyl chains into WCC. Given that the porphyrin head group of Chl-b exhibits photocatalytic activity, it is a rational choice for a functional component for integrating the photoprotective properties of β-CAR. The length of the hydrophobic tail of Chl-b (2.4 nm) matches the length of the hydrophobic moieties of WCC (2.3 nm) well. Thus, the hydrophobic tail of Chl-b could be used to anchor Chl-b to the hydrophobic segments of WCC and further facilitate the investigation of photosynthesis on the supramolecular level. In addition, WP5 as the carboxylic sodium salt and its precipitation from water as the corresponding carboxylic acid could be reversibly adjusted by changing the pH of the solution, which could endow the proposed suprastructure with pH responsiveness. The samples were prepared by a simple procedure. β-CAR powder was first dissolved in ethanol, and then, 2 ml of this β-CAR solution (150 μM) was added to 2 ml of a dilute sodium hydroxide aqueous solution of WP5 (150 μM) using a pipette, without stirring. The formation of WCC (Fig. 2a) is likely to be mainly driven by hydrophobic interactions between the hydrophobic cavity of WP5 and the lipophilic β-CAR. As shown in Fig. 2b, a 'tadpole-like' WCC with the bulky group at one side of β-CAR was threaded into the cavity of WP5, and the residue of β-CAR protruding from the WP5 cavity. The TCH (ten-carboxylic acid head) segment is the hydrophilic region with an extended length of 1.1 nm, and the SCT (single-carotene tail) is the segment (2.3 nm) that possesses a long hydrophobic tail that imparts amphiphilic character to WCC. The complexation between WP5 and β-CAR was first demonstrated by ^{1}H-NMR spectroscopy. As shown in Supplementary Figs 1, 2, the proton peaks H_3 of the methylene moieties at both rims were split into quartets from singlets, likely because of the inclusion of terpene in the cavity of WP5 restricted the swinging of the constituent units, which led to the loss of internal symmetry of H_3 protons. At the same time, downfield shifts of the aromatic protons H_1 and H_2 of

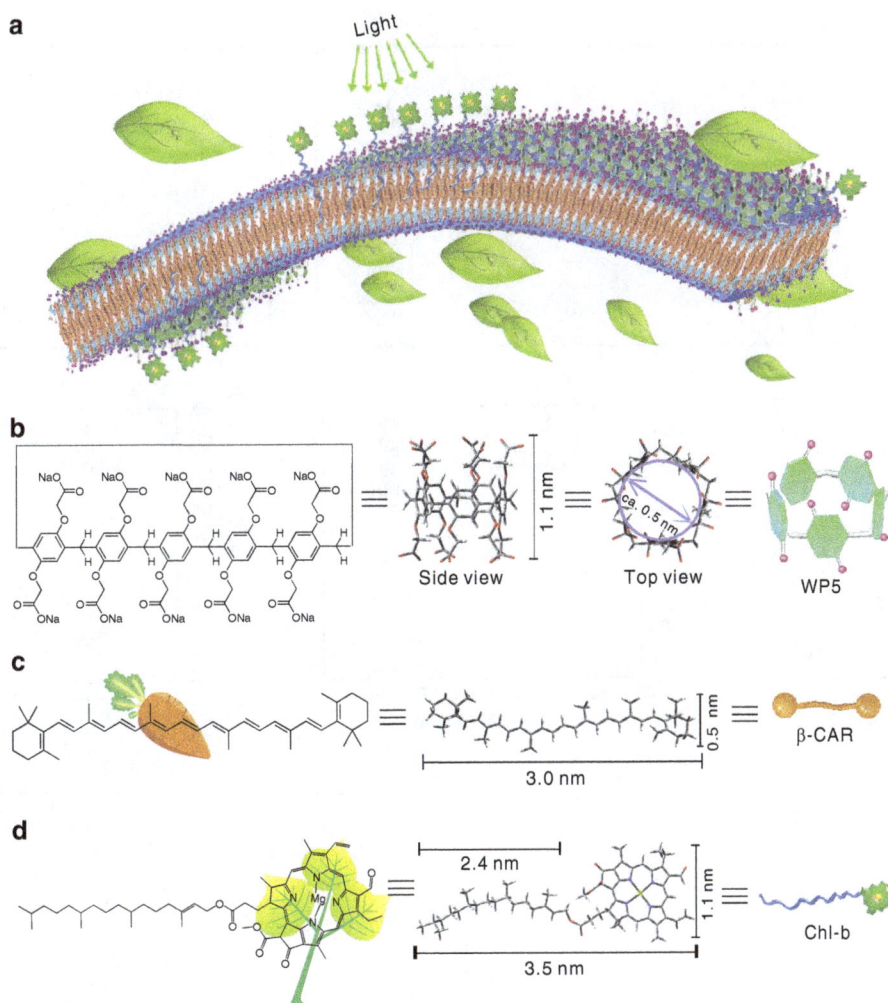

Figure 1 | Structural model of building blocks. (**a**) LHC containing Chl-b formed by (**b**) WP5, (**c**) β-CAR and (**d**) Chl-b. For clarity, the β-CAR-based hydrophobic interior layer is orange, and the WP5-based hydrophilic exterior layer is green.

WP5 were observed. In contrast, remarkable upfield chemical shifts of the methylene protons (H_b, H_c and H_d) and methyl protons (H_a and H_e) of the bulky terpene were observed ($\Delta\delta(H_b) = 0.228$ p.p.m., $\Delta\delta(H_c) = 0.259$ p.p.m., $\Delta\delta(H_d) = 0.354$ p.p.m., $\Delta\delta(H_a) = 0.160$ p.p.m. and $\Delta\delta(H_e) = 0.150$ p.p.m.) on the addition of WP5 due to the shielding effect of the electron-rich cavities of WP5 toward β-CAR, which clearly demonstrated the inclusion of the terpene section of β-CAR into the hydrophobic WP5 cavity. Moreover, the signals derived from protons H_g ($\Delta\delta(H_g) = -0.027$ p.p.m.) and methyl protons (H_f) on linear unsaturated carbon chain shifted from 6.1–6.8 p.p.m. to downfield 6.6–6.3 p.p.m. ($\Delta\delta(H_f) = -0.5$ p.p.m.). A similar phenomenon was also observed for the inclusion complexation between water-soluble pillar[5]arene and guest in previous reports[39,40]. The above results revealed that the cavity was fully threaded by β-CAR with the protons H_a, H_b, H_c, H_d and H_e in the hydrophobic WP5 cavity and that the other protons on a linear unsaturated carbon chain (H_f and H_g) protruded out of the cavity. The binding affinity for such host–guest inclusion might be mainly driven by the hydrophobic interactions. The generation of β-CAR-based WCC was then supported by Raman spectroscopy (Fig. 2c). The features of WCC in the 1,600–900 cm^{-1} region are similar to those of the pure β-CAR spectra[41]: the peak at 1,526 cm^{-1} originates from the stretching modes of the conjugated C=C bonds; the peak at 1,160 cm^{-1} arises from a mixture of C=C and C–C bond stretching modes

with C–H bending modes; and the peak at 1,009 cm^{-1} is attributed to the stretching modes of C–CH_3 bonds between the main chain and the side methyl carbons. These peaks indicate that β-CAR is present in the WCC. In addition, the ν_1 band assigned to C=C stretching (which is sensitive to host–guest interactions) blue-shifted (5 cm^{-1}) from 1,521 to 1,526 cm^{-1}, which not only provided further direct evidence of host–guest interactions between WP5 and β-CAR, but also further confirmed that the bulky groups at the ends of the β-CAR units were inserted into the cavity[42].

The encapsulation of a guest (G) in the hydrophobic cavity might also influence the polarizability of the host (H) cavity[43], which can be characterized by the fluorescence spectra. As shown in Fig. 2d, different (G)/(H) mole ratios ((H) = 150 µM) demonstrated that the confinement of β-CAR in the cavity strongly affects the fluorescence behaviour. The stoichiometry of the WCC between WP5 and β-CAR was further investigated based on the fluorescence spectra. The peak at 653 nm was selected for the investigation, which revealed a 1:1 binding stoichiometry between WP5 and β-CAR (Fig. 2e). Notably, the peak at 654 nm was selected because of the emission peak overlap of WP5 and β-CAR at 334 nm (Supplementary Fig. 3a–f). Furthermore, the fluorescence spectra of (G)/(H) at a higher host concentration ((H) = 200 µM) were also monitored, as also shown in (Supplementary Fig. 3g–i). The peak at 375 nm was selected for the investigation of the intensity change of β-CAR

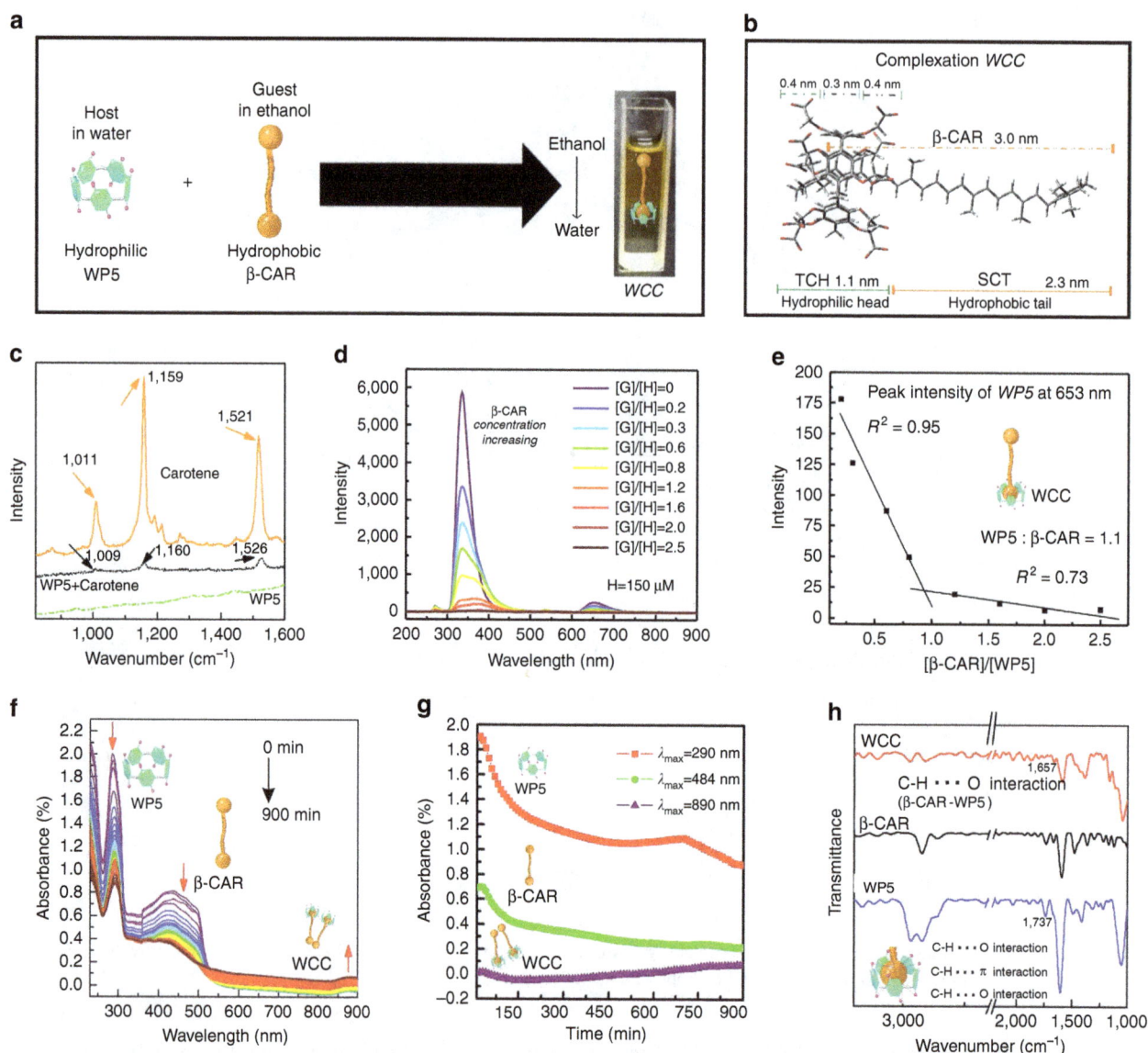

Figure 2 | Synthesis of the WCC. (**a**) A schematic displaying the synthetic route to the WCC. (**b**) Simulated structure of the WCC. (**c**) Raman spectra of β-CAR and the WCC. (**d**) Fluorescence spectra of WP5 (150 μM) on addition of β-CAR in ethanol/water (excited at 268 nm) at room temperature. (**e**) The fluorescence intensity changes of WP5 at 653 nm. (**f**) Time-dependent ultraviolet–visible spectra showing the formation of the WCC, as indicated by the changes in the intensities of the peaks at 290 and 484 nm (20 min each). (**g**) Plot of absorbance intensity as a function of time (20 min each). (**h**) FT-IR spectra of WP5, β-CAR and the WCC.

(this peak could not be observed when the WP5 concentration was 150 μM, which resulted in a relatively low concentration of β-CAR), and the peak at 653 nm was selected for the investigation of the intensity change of WP5. The results of both of these experiments further demonstrated that the (H)/(G)-binding stoichiometry is 1:1 (ref. 34). According to the fluorescence intensity change in Fig. 2d, the association constant (K_a) of WP5 ⊃ β-CAR was calculated to be $(2.34 \pm 2.06) \times 10^5 M^{-1}$ using a non-linear curve-fitting method (Supplementary Fig. 3j). Aside from the above fluorescence results, the stoichiometry of complexation for WP5 ⊃ β-CAR was further proven by Job's plot method using ultraviolet–visible spectroscopy, which confirmed the 1:1 binding stoichiometry for WP5 ⊃ β-CAR complexation (Supplementary Fig. 4)[44]. To gain further insight into the dynamic interaction between WP5 and β-CAR, we acquired 45 sequential ultraviolet–visible spectra at 20-min intervals for the same sample. As shown in Fig. 2f, the intensities of the characteristic peaks at ∼290 nm and 484 nm

(attributed to the absorbance of WP5 and β-CAR, respectively) decreased from 0 to 900 min. Meanwhile, the emergence and continued increase in intensity of a peak at 890 nm were observed, providing another piece of evidence for the formation of the WCC. Figure 2g clearly shows that the increase in intensity of the peak at 890 nm is associated with a concomitant decrease in the intensities of the features at 290 nm and 484 nm, which might be caused by the formation of microaggregates (Supplementary Fig. 5) and leads to the decrease of WP5 and β-CAR absorption intensity. Further evidence for the existence of the WCC was obtained by Fourier-transform infrared (FT-IR) spectroscopy, which shows that hydrogen bonds formed in the WCC (between the –COO⁻ of WP5 and the CH₃– of β-CAR). As shown in Fig. 2h, in the spectrum of the WP5 before self-assembly, the absorbance at 1,737 cm⁻¹ is attributed to a non-hydrogen-bonded –COOH group, which implies the presence of free carboxyl groups[45]. In contrast, this stretching vibration at 1,737 cm⁻¹ disappeared after the complexation.

These results show that strong and orderly hydrogen bonds were formed in the WCC.

Construction of HMS based on the WCCs.

WCCs were then utilized as building blocks in the construction of light-harvesting antenna complexes (LHCs). After the as-prepared solutions were aged for 7 days, orange aggregates appeared in the solutions. As shown in Fig. 3a (Supplementary Fig. 6), the aggregates were determined to be hollow microspheres (HMSs) based on optical microscopy (OM) observations, and the orange colour of the wall confirmed the homogeneous distribution of β-CAR across the HMSs. The DLS result showed that the WP5 ⊃ β-CAR-based HMS had an average diameter of 1,855 nm (Supplementary Fig. 5b). Autofluorescence is exhibited in Fig. 3b–d (Supplementary Fig. 7) and is attributed to the presence of the conjugated system in β-CAR and WP5. Scanning electron microscopy (SEM) micrographs further confirmed the micro-spherical morphology of the HMSs (Fig. 3e, Supplementary Fig. 8). The elemental mapping analysis demonstrated the homogeneous distribution of O and Na, which are characteristic elements in WP5 (Fig. 3f), across the HMSs, providing powerful evidence that the HMSs are composed of WP5. Interestingly, an HMS with a partially broken shell was observed (Fig. 3g) by SEM. The layer peeled off in the form of an arc, which indicated that the HMS might be composed of concentric multilayers (the observed wall thickness is ca. 40 nm). Furthermore, the marked red-shift of the ultraviolet–visible spectra suggested that J-aggregates had been formed (Fig. 3h)[46], indicating that the β-CARs are oriented in the WCC bilayer with the long axis almost perpendicular to the bilayer surface and with the two WP5 located in the hydrophilic regions on both sides of the bilayer. Strong anisotropic photoluminescence (birefringence) was observed when the dispersion was placed between crossed polarizers (Fig. 3i, Supplementary Fig. 9), providing further evidence for the presence of ordered arrays of WP5 and β-CAR within the HMS. Transmission electron microscopy (TEM) was used to characterize the fine structure of the HMSs. Figure 3j shows a bright-field TEM micrograph of one HMS. The intensity of the central part of the HMS is much higher than that of the edge, indicating that the sphere is hollow. The thickness of the microspherical wall was calculated to be ∼100 nm based on this TEM image. The extended geometries of the β-CAR and their 1:1 complexation with the WCC are depicted Fig. 3k. Given that the maximal length of the WCC-based bilayer calculated using Gaussian is ∼4.6 nm, this TEM observation suggests that the HMS may possess a multilayer structure. Notably, WP5s with ten carboxylic acids possess negative charges, which might result in the enhancement of electrostatic repulsion. However, similar to the multiwalled microtubule assemblies obtained in weakly acidic environments[47,48], the carboxylic acid groups in this study are partially dissociated in weakly basic environments. Hydrogen bonds thus dominate the interactions within the HMS and further weaken the electrostatic repulsion, facilitating the formation of multilayer microspheres. The packing information with respect to the WCC was obtained from small-angle X-ray scattering (SAXS) data. As shown in Fig. 3l, a broad Bragg reflection peak centred at 0.20 Å^{-1}, corresponding to the ordered structure with an interlayer spacing of ∼3.1 nm, was observed. Note that this broad peak covers the q region range from 0.15 to 0.25 Å^{-1}, corresponding to an interlayer spacing between 2.5 to 4.2 nm.

These distances were slightly shorter than the maximal calculated value (4.6 nm), possibly due to the conformational change or partial curling of the unsaturated carbon chain[49]. In addition, two slight peaks at lower q values (0.05 and 0.1 Å^{-1}) were observed in the profile, indicating that there are multiple orders of structure with an interlayer spacing of ∼3.1 nm in the HMS. However, these peaks are weak, possibly because of the partially inhomogeneous lamellar structure within the HMS. In addition, considering the SAXS results, we conjectured that an interlayer spacing centred at 3.1 nm may exist in the HMS. Furthermore, the zeta-potential of the HMS decreased from ∼0 mV (pure β-CAR) to −31.6 mV, which provides further evidence that the hydrophilic WP5 is located on the exterior surface of the HMS (Supplementary Fig. 10).

On the basis of the aforementioned results, we proposed the following mechanism for the formation of the WCC-based HMSs (Fig. 3m): The hydrophobic interactions between β-CAR and WP5 provide the driving force for the formation of host–guest complexation WCC, which results in the phase transfer of β-CAR from ethanol to water. Meanwhile, hydrogen bonding (CH—O) and the CH—π interaction[50] between β-CAR and WP5 render the WCC stable in water. Subsequently, hydrophobic interactions associated with aromatic stacking[51,52] drive the WCC to assemble into an HMS, in which the hydrophobic β-CARs are shielded from the water, while the hydrophilic WP5s are in contact with the aqueous environment.

Synthesis of HMS-based LHC containing Chl-b.

Additional experiments were carried out to investigate the possibility of reproducing similar 'tadpole-like' host–guest complexes by using Chl-b instead of β-CAR. When Chl-b solution was added to WP5, the proton peaks H$_3$ of the methylene moieties on both rims were split into quartets from singlets, likely because of the inclusion of Chl-b in the cavity of WP5, which affected the swinging of the constituent units and led to the loss of the internal symmetry of the H$_c$ protons. However, different from the remarkably chemical shift (β-CAR) before and after the complexation, the chemical shift of Chl-b changed slightly (Supplementary Figs 11–13). Furthermore, as shown in Supplementary Fig. 14a–d, according to the fluorescence intensity change of Chl-b (monitored at 450 nm), the K_a of WP5 ⊃ Chl-b (WCB) was calculated to be $(3.46 \pm 0.14) \times 10^3$ M^{-1} (Supplementary Fig. 14e). The above results indicate that although Chl-b could thread into the cavity, the size mismatch between alkyl and cavity facilitates the alkyl in protruding from the cavity freely and thus could not result in the occupation of the cavity. The possible aggregates based on the unstable complexation was investigated, and no notable hierarchical nanostructure was observed when Chl-b was added to the solution of WP5 by using a similar method for HMS construction (Supplementary Fig. 15). Thus, based on the different binding affinities between WP5 ⊃ β-CAR and WCB in binary solvents, Chl-b could play a role as a co-assembly component instead of a competitive guest during LHCs-b construction.

The route to synthesize the light-harvesting antenna is proposed in Fig. 4a. In the initial stage (step 1), the WCC is formed via host–guest recognition. According to the ^{1}H-NMR results, WP5 is incapable of fixing Chl-b into a firm and durable cavity. Thus, once the cavity was filled by β-CAR, Chl-b had no opportunity to form complexation with WP5. In contrast, β-CAR could form complexation with WP5 when Chl-b protruded out of the cavity. Thus, WCC and Chl-b act as the main assembly factors. In this case, hydrophobic interactions further drive the WCC and Chl-b to assemble into LHCs containing Chl-b, in which the hydrophobic β-CAR and Chl-b tail are shielded from the water, whereas the hydrophilic WP5s and porphyrin are in contact with the aqueous environment (step 2). As in the LHCs containing Chl-b synthesis method, the β-CAR and Chl-b powders were first dissolved in ethanol, and then, 2 ml of the

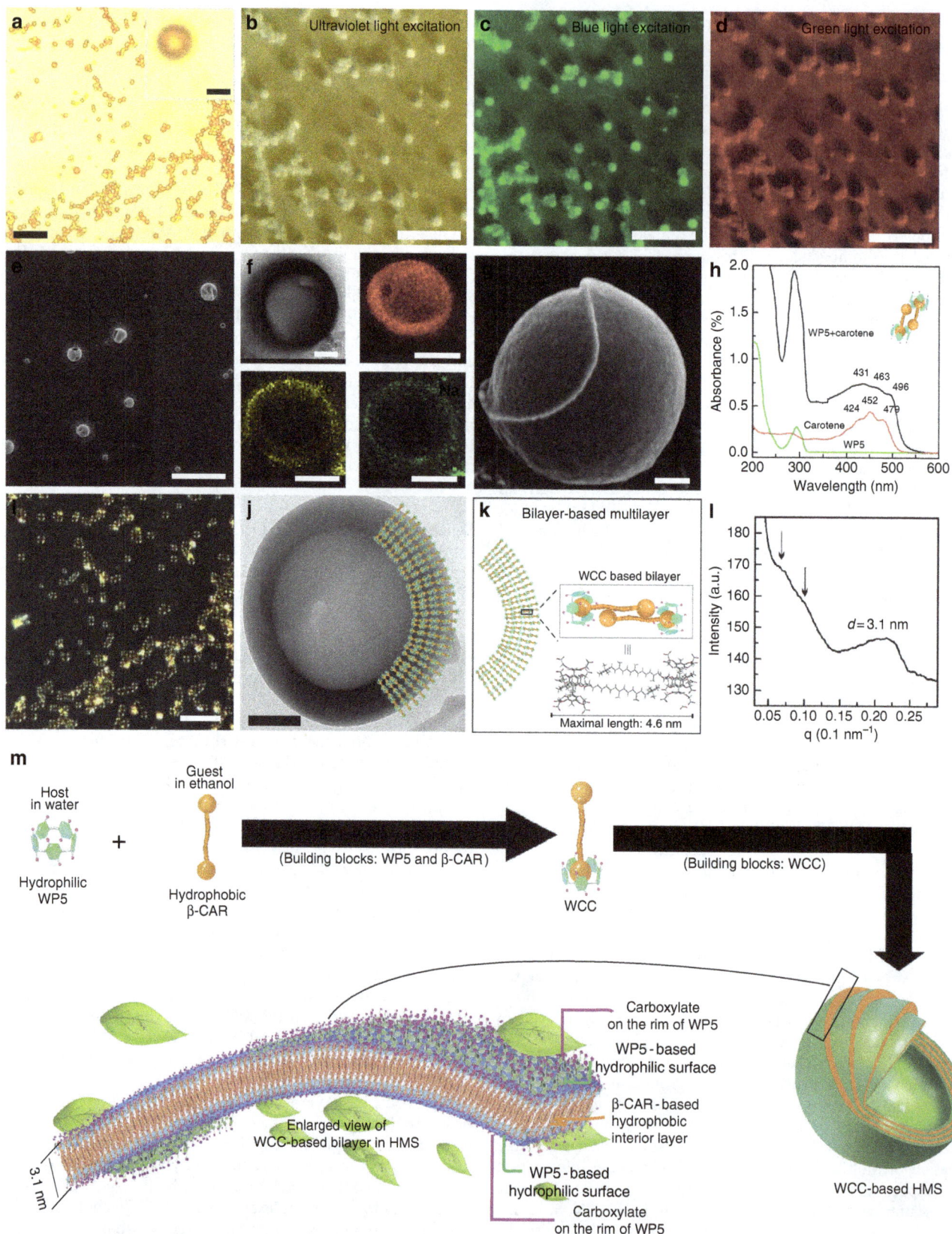

Figure 3 | Synthesis of WCC-based HMS. (**a**) OM image of HMSs over a large area. Scale bar, 10 μm (inset: enlarged image of an HMS; Scale bar, 1 μm). Large-area fluorescence microscopy images of HMSs under (**b**) Ultraviolet-light excitation, (**c**) blue-light excitation and (**d**) green-light excitation. (**e**) SEM image of HMSs. Scale bar, 10 μm. (**f**) Top left, TEM image of an enlarged WCC-based HMS; Scale bar, 500 nm. EDX mapping images of a WCC-based HMS: top right, distribution of element C; lower left, distribution of element O; lower right, distribution of element Na. Scale bar, 1 μm. (**g**) SEM image of an HMS with a broken shell. Scale bar, 200 nm. (**h**) ultraviolet–visible spectra of HMSs, β-CAR and WP5. (**i**) Polarized image of HMSs (the colour in this image is false). Scale bar, 10 μm. (**j**) TEM image of an HMS. Scale bar, 200 nm. (**k**) Cartoon of WCC-based multilayers in the HMSs. (**l**) SAXS profile of the HMS sample (the final concentration of both HMS and β-CAR is 75 μM) with an acquisition time of 5 h. (**m**) Schematic of the HMS formation process.

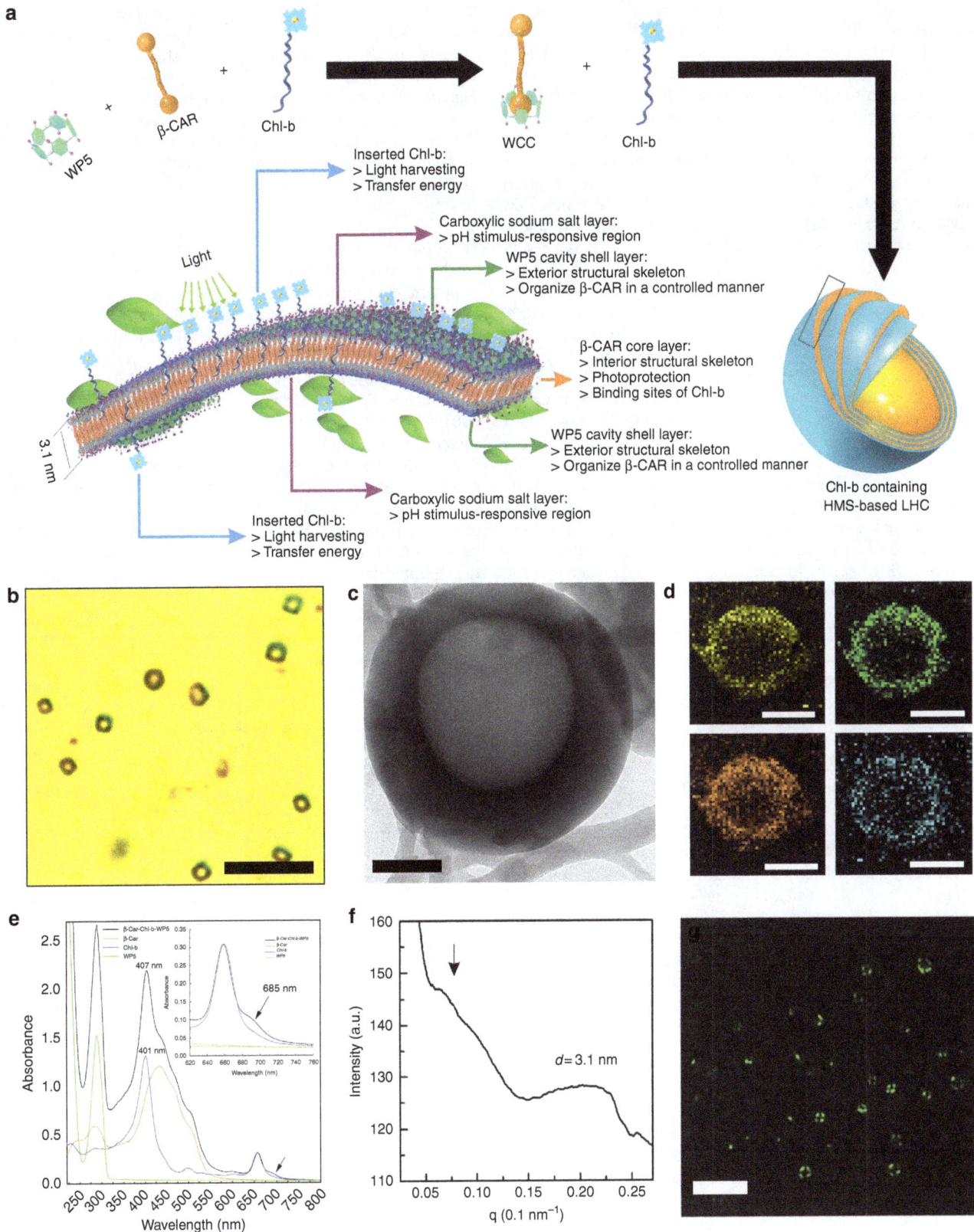

Figure 4 | Synthesis of HMS-based LHCs-b. (**a**) Synthesis of LHCs-b. (**b**) OM image showing LHCs-b. Scale bar, 10 μm. (**c**) TEM image of an LHCs-b. Scale bar, 200 nm. (**d**) EDX mapping image of an LHCs-b (the colour in this image is false). Scale bar, 500 nm. (**e**) Ultraviolet–visible spectra of β-CAR, Chl-b, WP5 and LHCs-b. (**f**) SAXS profile of an LHCs-b. (**g**) Polarized image of LHCs-b (the colour in this image is false). Scale bar, 10 μm.

β-CAR and 2 ml of the Chl-b solutions were added to 4 ml of an aqueous solution of WP5 using a pipette, without stirring. After the solution was aged for 7 days, pale-orange aggregates appeared.

Meanwhile, a clear Tyndall effect (Supplementary Fig. 16a) could be observed, indicating the formation of microaggregates. As shown in Fig. 4b, HMS-based LHCs containing Chl-b (LHCs-b) with orange–green-coloured walls were observed, confirming the homogeneous distribution of β-CAR and Chl-b throughout the LHCs-b. Compared with the narrow size distribution of HMSs (diameter range from 650 to 2,000 nm), the DLS results showed that the insertion of Chl-b causes the size distribution to become wider (Supplementary Fig. 16b). OM observation demonstrated that both LHCs-b with diameters <650 nm and >2,000 nm co-existed and also possessed hollow spherical structures (Supplementary Fig. 17). It is worth noting that some green liquid could be found in the cavities of LHCs-b through OM observation (blue arrows). It is speculated that a small amount of Chl-b did not participate in the wall construction and, as a result, was encapsulated in the cavities of the hollow spheres during the formation of the hollow spheres. Chl-b solution shows no photocatalytic activity and they were isolated by the cavities. Thus, no further experiment was designed to remove the non-aggregated Chl-b. The TEM image in Fig. 4c (Supplementary Figs 18,19) demonstrates that the hollow spherical structure was maintained after the insertion of Chl-b. The elemental mapping analysis demonstrates the homogeneous distribution of O and Na (the characteristic elements in the WP5), as well as the homogeneous distribution of N and Mg (the characteristic elements in Chl-b), across the entire LHC (Fig. 4d, Supplementary Fig. 19). The ultraviolet–visible spectra of LHCs-b showed a red-shifted shoulder (at 685 nm) absorption, which arose from alterations in the pigment–pigment interactions in LHCs-b (Fig. 4e).[53] In addition, visible CD spectroscopy is a sensitive technique to monitor excitonic pigment–pigment and pigment–protein interactions[54]. As shown in Supplementary Fig. 20, the CD bands at $(-)474$, $(-)488$, $(+)499$, $(-)653$, $(-)666$, $(+)702$ and $(+)724$ nm were observed in LHCs-b. In the Q region, the CD spectra of the LHCs-b showed a negative peaks at $(-)666$ nm, which is characteristic of Chl-b[55] and accompanied by negative peak at $(-)653$ nm. The positive peaks at $(+)499$ nm and $(-)474$ nm in the CD band were accompanied by a negative peak at $(-)488$ nm, originating from β-CAR[56]. Notably, the band at 649 nm (Chl-b in ethanol) was red-shifted to 666 nm with a shoulder at 653 nm on constitution into HMS. These results indicate that the Chl-b–Chl-b and/or Chl-b–β-CAR interactions were affected by constituting into HMS, which resulted from the changes in microenvironment of Chl-b in LHCs-b[57]. The bands at $(-)474$ and $(+)499$ nm as well as those at $(-)666$ nm and $(+)724$ nm may originate from excitonic interactions involving Chl-b and β-CAR molecules. The integrated areas of the positive and negative bands appear to be approximately equal[58]. These spectra suggest that in LHCs-b, a complicated set of excitonic interactions occurs between several Chl-b and β-CAR molecules[59]. Furthermore, the peak obtained from SAXS showed that the d-spacing remains at 3.1 nm (Fig. 4f), which provides further evidence that the ordered structure of the HMS was not destroyed after the co-assembly with Chl-b. In addition, the strong anisotropic photoluminescence shown in Fig. 4g is in good agreement with the SAXS data. Because the K_a of WP5 ⊃ β-CAR was higher than that of WCB by two orders of magnitude, it could be deduced that almost all of the β-CAR in the system was wrapped by WP5. Thus, the position of β-CAR mainly located in the wall of the LHCs-b. Considering that hierarchical aggregates could not be obtained by WCB, a large amount of Chl-b, but not WCB randomly inserted into the HMS due to non-specific interactions. Therefore, it is hard to obtain a constant proportion of Chl-b among all LHCs-b (caused by the randomly insertion manner). As a result, it is difficult to obtain a convincing data by Elementary Analysis or Inductive Coupled Plasma Emission Spectrometer (proportion of Chl-b in each LHCs-b is different). Fortunately, EDX spectra provide a fine-testing method to investigate the β-CAR/Chl-b ratio in individual LHC-b. Thus, as shown in Supplementary Fig. 19a–d, LHCs-b with different diameter were selected and analysed. On the basis of the obtained semi-quantitative data, rough range of the Chl-b proportion could be provided (Supplementary Fig. 19a–d inset). Then, the average β-CAR/Chl-b ratio of ~ 1.8 was obtained by calculation. Being confined to the accuracy of the semi-quantitative analysis (and interference of Chl-b solution in cavities), these data are provided only for providing a rough estimate.

HMS-based LHCs served as photocatalytic entities. The catalytic activity of LHCs-b was evaluated for the reduction of the pollutant 4-nitrophenol (4-NP) after the addition of $NaNO_2$. Figure 5a shows the ultraviolet–visible spectra of 4-NP in ethanol/water solutions containing distilled water (blank sample), β-CAR, Chl-b, HMS and LHCs-b; in this case, the photoirradiation time was 0 min. All of these spectra display a main absorption at 318 nm (4-NP) and a shoulder peak at 400 nm (4-nitrophenolate ions). After irradiation with a mercury lamp (500 W) for 10 min (Fig. 5b), the peak intensity at 318 nm decreased and the peak intensity at 400 nm increased for all of the samples due to the formation of 4-nitrophenolate ions[60]. Importantly, a new peak at 290 nm was observed in the spectrum of the sample of LHCs-b (blue line), which indicated the generation of 4-aminophenol (4-AP) and further demonstrated the photocatalytic activity of LHCs-b. Remarkably, after being irradiated with the mercury lamp for 20 min, the HMS sample also exhibited photocatalytic activity towards 4-NP; as shown in Fig. 5c, a new peak at 290 nm appeared in the HMS spectrum (green line).

These findings reveal that β-CAR suprastructures also exhibited catalytic activity. Reasonably, as the irradiation time was prolonged (Fig. 5d), a peak was also observed at 290 nm in the spectra of HMS and LHCs-b. As revealed by the literature, the carotene molecules form a layer between 0–40-mm deep under the insect's cuticle, putting them in the perfect position to capture the Sun's light. Similar to the literature, the hierarchical way in which the carotene molecules are arranged in our studies provides the possibility of photocatalytic activity by the HMS[61,62]. Control experiments using untreated samples were performed for the same samples at different time intervals in the absence of photoirradiation. As shown in Fig. 5e–h, no changes were observed in the ultraviolet–visible spectra compared with those of the experimental group (Fig. 5a–d). Subsequently, LHCs-b were irradiated for up to 120 min (in 5 min intervals, from 0–120 min). As shown in Fig. 5i, when a mixture of 4-NP and LHCs-b was irradiated, the intensity of the peak at 290 nm increased due to the generation of 4-AP, which further confirmed that the LHCs-b can be used as a photocatalyst for the reduction of organic species.

Excitingly, when an aqueous solution of K_2PtCl_4, ascorbic acid (as an electron donor) and LHCs-b was irradiated for 2 h, LHC microspheres decorated with platinum nanoparticles were observed (Fig. 5j). Similarly, photoreduction of an Ag^I salt occurred within 2 h of irradiation and produced AgNP-loaded spheres, as revealed by TEM (Supplementary Fig. 21). Palladium nanoparticles were similarly obtained (Fig. 5k). The photocatalytic reaction is believed to occur through two steps. First, Chl-b captures light to produce photoexcited states that are rapidly reduced by an electron donor. Then, the resulting Chl-b radical anion can be used as a catalyst to reduce various metal salts to their metallic state through successive light-harvesting and photochemical cycles (Fig. 5l)[63]. Microspheres with stimuli responsiveness are extremely important for potential applications. A decrease in pH induces marked damage in HMSs (Fig. 5m),

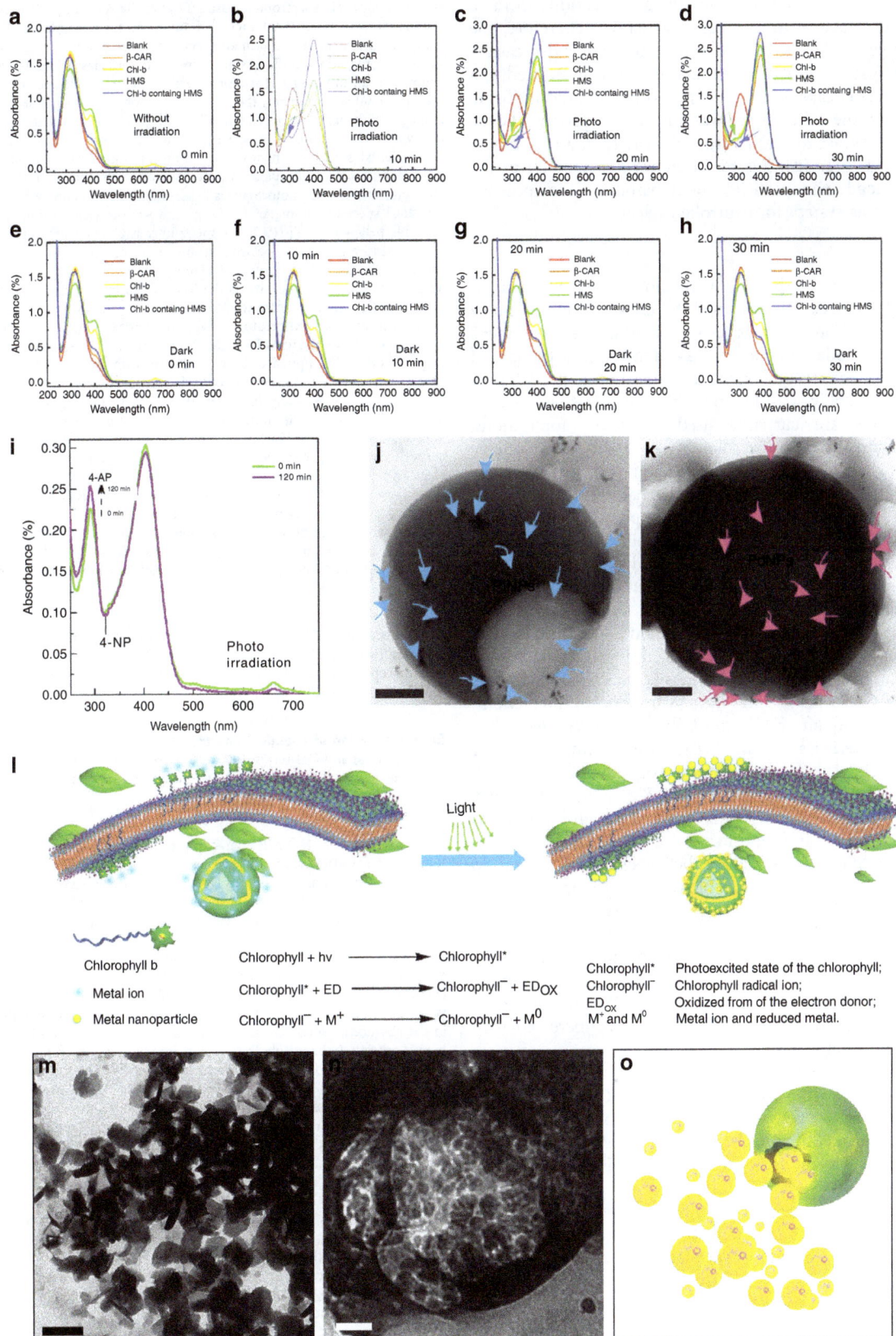

Figure 5 | HMS-based LHCs-b served as photocatalytic entities. Ultraviolet–visible spectra of the transformation of 4-NP into 4-AP under photocatalysis by β-CAR, Chl-b, HMS, and LHCs-b for (**a**) 0 min, (**b**) 10 min, (**c**) 20 min and (**d**) 30 min. (**e–h**) are the corresponding untreated control groups for (**a–d**). (**i**) Ultraviolet–visible spectra of the transformation of 4-NP into 4-AP under photocatalysis by LHCs-b. TEM images of **j**, a PtNPs-loaded LHCs-b (the blue arrows indicate the PtNPs); Scale bar, 100 nm, and (**k**) a PdNPs-loaded LHCs-b (the pink arrows indicate the PdNPs); Scale bar, 200 nm. (**l**) Possible photosynthesis mechanism of LHCs-b. (**m**) TEM micrograph of a self-broken HMS during the HCl-triggered degradation. Scale bar, 500 nm. (**n**) TEM micrograph of an HMS containing AgNPs during the HCl-triggered degradation. Scale bar, 100 nm. (**o**) Schematic showing the release of nanoparticles.

which appears to be attributable to a transition from a microspherical structure to a micellar structure. Therefore, the loading and release of guest molecules in response to a decrease in pH was investigated. Here, WP5-functionalized AgNPs used as a model guest were encapsulated in the microsphere interior during assembly. When the solution pH was adjusted to produce acidic conditions, the release of AgNPs from the cavity of an HMS was clearly observed (Fig. 5n). Figure 5o shows a schematic of the release of loaded nanoparticles, demonstrating the possible application of this system for controlled release.

Discussion

In previous reports of β-CAR with host β-cyclodextrin or calixarenes, β-cyclodextrin was used to encapsulate the bulky group of β-CAR[64,65] and calixarenes were used as nanocontainers. The natural β-cyclodextrins are concave macrocycles capable of expressing molecular recognition via the formation of noncovalent inclusions with suitably sized hydrophobic guests in their cavities. These complexes are currently used in many applications, including enhancing the water solubility of hydrophobic guests, controlling the release of volatile guests and protecting labile guests from degradation promoted by external agents. As for the calixarenes, β-CAR molecules were simply loaded in the hydrophilic vector by spinning disk processing. In this sense, a study of the association between β-CAR and other hosts is very important because it could provide better binding sites for carotenoids, allowing improved structural control via higher binding specificity (Supplementary Discussion). The formation of the WCC was supported by the Raman and ^{1}H-NMR spectra, which demonstrated that the binding site was in the terpene section of β-CAR based on the hydrophobic interactions. The generation of an HMS must be associated with the structural characteristics of the WCC; when either WP5 or β-CAR was removed from the water or ethanol, no HMS formation was observed (Supplementary Fig. 22), convincingly demonstrating that the WCC formed by WP5 and β-CAR is undoubtedly the critical factor for the construction of HMSs. Interestingly, HMSs can provide a better chemical system in terms of stability without the loss of bioavailability (Supplementary Figs 23,24). The HMS was observed to display characteristics that differ from those of traditional synthetic molecule-based suprastructures (Supplementary Fig. 25). Furthermore, the peak in the SAXS profile is weak as a consequence of the low concentration of HMSs. As shown in Supplementary Fig. 26, the intensity of the peak increased with increasing acquisition time.

As shown in Supplementary Fig. 27, LHC containing Chl-a could also be obtained based on WCC-based HMS. LHCs-b exhibit photocatalytic activity, as was demonstrated through a series of experiments. Interestingly, the HMSs also exhibit photoactivity, which differs from that of LHCs-b in that it exhibits slower kinetics.

The successful creation of HMS provides sophisticated strategies to generate carotenoids-based hierarchical model, which will result in innovative approaches to generate controllable supramolecular structures composed of carotenoids. The introduction of various biomolecules, could lead to the development of this supramolecular platform into diverse artificial biological cells for mimicking and optimizing photosynthetic systems, which would further provide a photosynthetic model and important tools for the investigation of the origins of the bioenergy system in living organisms.

Methods

General considerations. All chemical reagents were purchased from Aladdin and used without further purification. ^{1}H-NMR spectra were recorded on a Bruker

Advance 600 MHz spectrometer using D_2O as the solvent. β-CAR and Chl-b were purchased from EKEAR (Shanghai). Raman spectra were measured in solution using an In Renishaw Via Raman spectrometer ($\lambda_{exc} = 532$ nm). For Raman measurements, aqueous dispersions were dried onto glass slides; the focus was then centred on an individual HMS for detection. Fluorescence spectra were recorded on an F-4500 spectrometer; the spectra were corrected against a photomultiplier and against the lamp intensity. The slit width of both monochromators was 5.0 nm. ultraviolet–visible absorption spectra of the photocatalytic experiment (the experimental group and the untreated control group) were recorded on a UV-2550 spectrophotometer. Ultraviolet–visible absorption spectra for assembly characterization and photocatalysis experiments (0–120 min) were recorded on a UV-2501 spectrophotometer. FT-IR spectra were collected using potassium bromide pellets on a TENSOR 27 spectrometer. Circular dichroism spectra were measured in a Jasco-815 spectropolarimeter. To compare the shift of each peak more clearly, the intensity of the 649 nm peak in the LHCs containing Chl-b was adjusted to the same intensity as the base of Chl-b. Dynamic light scattering and zeta-potential measurements were carried out on a Malvern Nanosizer S instrument at room temperature. OM and polarized OM images were obtained using a Leica Microsystems DM LM/P instrument. Fluorescence OM images were acquired on an Olympus IX 73 W. D. 27 mm with an inverted configuration. For OM, polarized OM and fluorescence OM, aqueous dispersions of orange aggregates were dried onto glass slides for observation. The TEM images were obtained using a Philips TECNAI-12 instrument with an accelerating voltage of 120 kV. For TEM, aqueous dispersions of HMSs and Chl-b-containing LHCs were dried onto carbon-coated copper support grids. A Hitachi S-4800 field-emission scanning electron microscope was used to investigate the HMS. For SEM, aqueous dispersions of HMSs were dried onto silicon wafers. HRTEM and elemental mapping images were obtained using a Tecnai G2 F30 S-TWIN instrument. SAXS experiments were performed at room temperature on a NanoSTAR, Bruker-AXS, 30 W. For SAXS, dispersions of WCC-based HMSs and Chl-b-containing LHCs were loaded into special glass X-ray capillary tubes with an internal diameter of 1.5 mm.

Synthesis of pillar[5]arene WP5. Carboxyl-modified pillar[5]arene WP5 was prepared by a modified literature procedure[38] and was identified by ^{1}H-NMR spectroscopy in D_2O. ^{1}H-NMR (D_2O, 600 MHz, p.p.m.): δ 6.7 (s, 10 H), 4.28 (d, 20 H), 3.90 (s, 10 H).

Characterization of chemical and biomolecular materials. β-CAR (purity ≥ 97%) and Chl-b (purity ≥ 95%) were purchased from EKEAR (Shanghai). WP5 was prepared by a modified literature procedure and was identified by ^{1}H-NMR spectroscopy in D_2O.

WCC studies. For ^{1}H-NMR characterization, β-CAR was dissolved in ethanol-d^6 to produce a 150 μM solution (curve a in Supplementary Figs 1, 2, solvent: ethanol-d^6). WP5 was dissolved in D_2O to produce a 300 μM aqueous solution, and then, an equal volume of ethanol-d^6 was added to the WP5 solution to produce a 150 μM binary solvent solution (curve c in Supplementary Figs 1, 2, solvent: ethanol-d^6: $D_2O = 1:1$). The D_2O aqueous solution of WP5 added to the ethanol-d^6 solution to produce a 150 μM WP5 and β-CAR complexation solution (curve b in Supplementary Figs 1, 2, solvent: ethanol-d^6: $D_2O = 1:1$). Clear ^{1}H-NMR spectra could not be achieved when β-CAR ethanol-d^6 solution was added by an equal volume of D_2O due to the poor solubility and low concentration of β-CAR in water in the absence of WP5. Thus, to obtain clear spectra, ethanol-d^6 instead of binary solvents (ethanol-d^6: $D_2O = 1:1$) were used for the β-CAR ^{1}H-NMR characterization and for further comparison. In addition, the comparison between β-CAR (ethanol-d^6) and WCC (ethanol-d^6: $D_2O = 1:1$) is closer to the solvent environment changes in the assembly process conditions. For fluorescence characterization, various concentrations of β-CAR in ethanol solution were added to an aqueous solution of WP5 to produce β-CAR/WP5 solutions with different molar ratios of 0, 0.2, 0.3, 0.6, 0.8, 1.2, 1.6, 2.0 and 2.5. The emission intensities of the WP5 solutions at various molar ratios were monitored by fluorescence spectroscopy ($\lambda_{ex} = 268$ nm). Using a non-linear curve-fitting method, the K_a of WP5 ⊃ β-CAR was estimated. The non-linear curve fitting was based on the equation $\Delta F = (\Delta F_\infty/[H]_0)$ $(0.5[G]_0 + 0.5([H]_0 + 1/K_a) - (0.5 ([G]_0{}^2 + (2[G]_0(1/K_a - [H]_0)) + (1/K_a + [H]_0)^2)^{0.5}))$ (1) (refs 37,66). On addition of β-CAR, the fluorescence intensity of WP5 (monitored at 653 nm) was gradually quenched, and the association constant of WP5 ⊃ β-CAR was calculated. ΔF is the fluorescence intensity changes at 653 nm at $[G]_0$; ΔF_∞ is the fluorescence intensity changes at 653 nm when WP5 is completely complexed, $[G]_0$ is the initial concentration of β-CAR, $[H]_0$ is the fixed initial concentration of WP5.

WP5 and Chl-b interaction studies. For ^{1}H-NMR characterization, Chl-b was dissolved in ethanol-d^6 to produce a 150 μM solution (curve a in Supplementary Figs 11–13, solvent: ethanol-d^6). WP5 was dissolved in D_2O to produce a 300 μM aqueous solution, and then, an equal volume of ethanol-d^6 was added to WP5 solution to produce a 150 μM binary solvent solution (curve c in Supplementary Figs 11–13, solvent: ethanol-d^6: $D_2O = 1:1$). The D_2O aqueous solution of WP5 added to the ethanol-d^6 solution to produce a 150 μM WP5 and Chl-b complexation solution (curve b in Supplementary Figs 11–13, solvent: ethanol-d^6: D$_2$O = 1:1). In accordance with the method used in previous experiments (WCC ^{1}H-NMR characterization), ethanol-d^6 was used for the Chl-b ^{1}H-NMR characterization and for further comparison instead of binary solvents (ethanol-d^6: D$_2$O = 1:1).

To determine the K_a for the complexation between WP5 and Chl-b, fluorescence titration experiments were carried out in solutions that had a constant concentration of Chl-b and varying concentrations of WP5. Various concentrations of WP5 solutions were added to an ethanol solution of Chl-b at the concentration of 100 μM (1,000 μl) to produce WP5/Chl-b solutions with different molar ratios (0, 0.2, 0.4, 0.6, 0.8, 1.0, 1.2, 1.4, 1.6, 2.2, 2.6 and 2.8). In a typical experiment, 1,000 μl of WP5 at the concentration of 20 μM were added to the Chl-b solution to produce WP5/Chl-b solutions with a molar fraction of 0.2. Then, 1,000 μl of WP5 at the concentration of 40 μM were added to the Chl-b solution to produce WP5/Chl-b solutions with a molar fraction of 0.4. Subsequently, 1,000 μl of WP5 at the concentration of 60 μM were added to the Chl-b solution to produce WP5/Chl-b solutions with a molar fraction of 0.6. The solutions with the other molar ratios (0.8, 1.0, 1.2, 1.4, 1.6, 2.2, 2.6 and 2.8) were prepared by a similar method. The emission intensities of the WP5 solutions at various molar ratios were monitored by fluorescence spectroscopy (λ_{ex} = 268 nm). The non-linear curve fitting was based on the equation $\Delta F = (\Delta F_\infty/[G]_0) (0.5[H]_0 + 0.5([G]_0 + 1/Ka) - (0.5 ([H]_0{}^2 + (2[H]_0(1/K_a - [G]_0)) + (1/K_a + [G]_0)^2)^{0.5}))$ (2) (refs 37,66), which is was obtained by a modified literature equation (fix the concentration of guest and increase the concentration of host). According to the different emission peaks of WP5 and Chl-b in the range from 300 to 750 nm, the peak at 450 nm was selected to monitor the intensity change of Chl-b. On the basis of the fluorescence intensity change of Chl-b (monitored at 450 nm), the association constant of WCB was calculated. ΔF is the fluorescence intensity change at 450 nm at $[H]_0$, ΔF_∞ is the fluorescence intensity change at 450 nm when WP5 is completely complexed; $[G]_0$ is the fixed initial concentration of Chl-b; $[H]_0$ is the initial concentration of WP5.

Synthesis of HMSs. WP5 was first basified with sodium hydroxide to produce an aqueous solution. Then, β-CAR was dissolved in ethanol to produce an ethanol solution of β-CAR. Subsequently, β-CAR-based HMSs were prepared by pipetting an ethanol solution of β-CAR into an aqueous solution of WP5 and then aging. The final concentration of host and guest could be controlled by adjusting their initial concentration. In a typical experiment, HMSs were prepared by pipetting 2 ml of an ethanol solution of β-CAR (150 μM) into 2 ml of an aqueous solution of WP5 (150 μM) and then aging the solution for 7 days at room temperature.

Synthesis of LHCs-b. WP5 was first basified with sodium hydroxide to produce an aqueous solution. Then, β-CAR was dissolved in ethanol to produce an ethanol solution of β-CAR. Chl-b was dissolved in ethanol to produce an ethanol solution of Chl-b. Simlar to the synthesis method of HMSs, LHCs-b were prepared at room temperature by pipetting an ethanol solution of β-CAR and Chl-b into an aqueous solution of WP5. The final concentration of host and guest could be controlled by adjusting their initial concentration. In a typical experiment, LHCs-b were prepared at room temperature by pipetting 2 ml of an ethanol solution of β-CAR (150 μM) and 2 ml of an ethanol solution of Chl-b (150 μM) into 4 ml of an aqueous solution of WP5. Mixing the solution resulted in a pale-orange turbid suspension that was subsequently aged for 7 days.

Photocatalytic synthesis of metal nanoparticles. In a typical experiment, 0.50 ml of a suspension of LHCs-b, 20 μl of aqueous K$_2$PtCl$_4$ (10 mM) and 25 μl of aqueous ascorbic acid (0.2 M) were sequentially added to a 2-ml glass vial and then irradiated for 2 h with visible light from a 300-W Xe lamp. Pd and Ag nanoparticles were prepared using similar methods.

Photocatalytic reduction of 4-NP. *Experimental group.* Blank sample: 1 ml of 4-NP stock solution (2 mM), 100 μl of sodium nitrite stock solution (0.1 M) and 300 μl of distilled water were mixed and homogenized in a glass vial and then irradiated with visible light from a 500-W mercury lamp. β-CAR sample: One millilitre of 4-NP stock solution (2 mM), 100 μl of sodium nitrite stock solution (0.1 M) and 300 μl of a dispersion of β-CAR (75 μM) were mixed and homogenized in a glass vial and then irradiated with visible light from a 500-W mercury lamp. Chl-b sample: 1 ml of 4-NP stock solution (2 mM), 100 μl of sodium nitrite stock solution (0.1 M) and 300 μl of a dispersion of Chl-b (75 μM) were mixed and homogenized in a glass vial and then irradiated with visible light from a 500-W mercury lamp. HMS sample: 1 ml of 4-NP stock solution (2 mM), 100 μl of sodium nitrite stock solution (0.1 M) and 300 μl of a dispersion of HMS (75 μM) were mixed and homogenized in a glass vial and then irradiated with visible light from a 500-W mercury lamp. LHCs-b sample: 1 ml of 4-NP stock solution (2 mM), 100 μl sodium nitrite stock solution (0.1 M) and 300 μl of a dispersion of LHCs-b (75 μM) were mixed and homogenized in a glass vial and then irradiated with visible light from a 500-W mercury lamp.

Untreated control group. Blank sample: 1 ml of 4-NP stock solution (2 mM), 100 μl sodium nitrite stock solution (0.1 M) and 300 μl of distilled water were

mixed and homogenized in a glass vial and then stored in the dark. β-CAR sample: 1 ml of 4-NP stock solution (2 mM), 100 μl sodium nitrite stock solution (0.1 M) and 300 μl of a dispersion of β-CAR (75 μM) were mixed and homogenized in a glass vial and then stored in the dark. Chl-b sample: 1 ml of 4-NP stock solution (2 mM), 100 μl sodium nitrite stock solution (0.1 M) and 300 μl of a dispersion of Chl-b (75 μM) were mixed and homogenized in a glass vial and then stored in the dark. HMS sample: 1 ml of 4-NP stock solution (2 mM), 100 μl sodium nitrite stock solution (0.1 M) and 300 μl of a dispersion of HMS (75 μM) were mixed and homogenized in a glass vial and then stored in the dark. LHCs-b: 1 ml of 4-NP stock solution (2 mM), 100 μl sodium nitrite stock solution (0.1 M) and 300 μl of a dispersion of LHCs-b (75 μM) were mixed and homogenized in a glass vial and then stored in the dark. Changes in the absorption peaks over time associated with the photocatalytic reduction of 4-NP to 4-AP (experimental group and untreated control group) were monitored by ultraviolet–visible spectroscopy (UV-2501).

For the photocatalytic activity investigation of the LHCs-b sample using a longer duration of irradiation, the procedure was as follows: A 4-NP stock solution (2 mM), a sodium nitrite stock solution (0.1 M) and a dispersion of LHCs-b were mixed and homogenized in a glass vial and then irradiated with visible light from a 300-W Xe lamp. Changes in the absorption peaks over time associated with the photocatalytic reduction of 4-NP to 4-AP (0–120 min) were monitored by ultraviolet–visible spectroscopy (UV-2501).

References

1. Álvarez, R., Vaz, B., Gronemeyer, H. & Lera, Á. R. Functions, therapeutic applications, and synthesis of retinoids and carotenoids. *Chem. Rev.* **114**, 1–125 (2014).
2. Britton, G. Structure and properties of carotenoids in relation to function. *FASEB J.* **9**, 1551–1558 (1995).
3. Chatgilialoglu, C., Ferreri, C., Melchiorre, M., Sansone, A. & Torreggiani, A. Lipid geometrical isomerism: from chemistry to biology and diagnostics. *Chem. Rev.* **114**, 255–284 (2014).
4. Vishnevetsky, M., Ovadis, M. & Vainstein, A. Carotenoid sequestration in plants: the role of carotenoid associated proteins. *Trends Plant Sci.* **4**, 232–235 (1999).
5. Edge, R., Land, E. J., McGarvey, D., Mulroy, L. & Truscott, T. G. Relative one-electron reduction potentials of carotenoid radical cations and the interactions of carotenoids with the vitamin E radical cation. *J. Am. Chem. Soc.* **120**, 4087–4090 (1998).
6. Blount, J. D., Metcalfe, N. B., Birkhead, T. R. & Surai, P. F. Carotenoid modulation of immune function and sexual attractiveness in zebra finches. *Science* **300**, 125–127 (2003).
7. El-Agamey, A. & McGarvey, D. Evidence for a lack of reactivity of carotenoid addition radicals towards oxygen: a laser flash photolysis study of the reactions of carotenoids with acylperoxyl radicals in polar and non-polar solvents. *J. Am. Chem. Soc.* **125**, 3330–3340 (2003).
8. Havaux, M. Carotenoids as membrane stabilizers in chloroplasts. *Trends Plant Sci.* **3**, 147–151 (1998).
9. Staleva, H. *et al.* Mechanism of photoprotection in the cyanobacterial ancestor of plant antenna proteins. *Nat. Chem. Biol.* **11**, 287–291 (2015).
10. Leverenz, R. L. *et al.* A 12 Å carotenoid translocation in a photoswitch associated with cyanobacterial photoprotection. *Science* **348**, 1463–1466 (2015).
11. Pascal, A. A. *et al.* Molecular basis of photoprotection and control of photosynthetic light-harvesting. *Nature* **436**, 134–137 (2005).
12. Lakshmi, K. V. *et al.* Pulsed high-frequency EPR study on the location of carotenoid and chlorophyll cation radicals in photosystem II. *J. Am. Chem. Soc.* **125**, 5005–5014 (2003).
13. Moise, A. R., Al-Babili, S. & Wurtzel, E. T. Mechanistic aspects of carotenoid biosynthesis. *Chem. Rev.* **114**, 164–193 (2014).
14. Foss, B. J., Nœss, S. N., Sliwka, H.-R. & Partali, V. Stable and highly water-dispersible, highly unsaturated carotenoid phospholipids-surface properties and aggregate size. *Angew. Chem. Int. Ed.* **42**, 5237–5240 (2003).
15. Auweter, H. *et al.* Supramolecular structure of precipitated nanosize β-carotene particles. *Angew. Chem. Int. Ed.* **15**, 2188–2191 (1999).
16. Anantachoke, N. *et al.* Fine tuning the production of nanosized β-carotene particles using spinning disk processing. *J. Am. Chem. Soc.* **128**, 13847–13853 (2006).
17. Zouni, A. *et al.* Crystal structure of photosystem II from *Synechococcus elongatus* at 3.8 Å resolution. *Nature* **409**, 739–743 (2001).
18. Qiu, Y., Chen, P. & Liu, M. Evolution of various porphyrin nanostructures via an oil/ aqueous medium: controlled self-assembly, further organization, and supramolecular chirality. *J. Am. Chem. Soc.* **132**, 9644–9652 (2010).

19. Rufo, C. M. *et al.* Short peptides self-assemble to produce catalytic amyloids. *Nat. Chem.* **6**, 303–309 (2014).

20. Tang, T-Y. D. *et al.* Fatty acid membrane assembly on coacervate microdroplets as a step towards a hybrid protocell model. *Nat. Chem.* **6**, 527–533 (2014).

21. Calabrese, C. M. *et al.* Biocompatible infinite-coordination-polymer nanoparticle–nucleic-acid conjugates for antisense gene regulation. *Angew. Chem. Int. Ed.* **54**, 476–480 (2015).

22. Liu, K. *et al.* Peptide-induced hierarchical long-range order and photocatalytic activity of porphyrin assemblies. *Angew. Chem. Int. Ed.* **54**, 500–505 (2015).

23. Scholes, G., Fleming, G., Olaye-Castro, A. & Grondelle, R. Lessons from nature about solar light harvesting. *Nat. Chem.* **3**, 763–774 (2011).

24. Han, Y., Meng, Z., Ma, Y.-X. & Chen, C.-F. Iptycene-derived crown ether hosts for molecular recognition and self-assembly. *Acc. Chem. Res.* **47**, 2026–2040 (2014).

25. Wang, F. *et al.* Self-sorting organization of two heteroditopic monomers to supramolecular alternating copolymers. *J. Am. Chem. Soc.* **130**, 11254–11255 (2008).

26. Yan, X. *et al.* A multiresponsive, shape-persistent, and elastic supramolecular polymer network gel constructed by orthogonal self-assembly. *Adv. Mater.* **24**, 362–369 (2012).

27. Crini, G. Review: a history of cyclodextrins. *Chem. Rev.* **114**, 10940–10975 (2014).

28. Zhang, H. & Zhao, Y. Pillararene-based assemblies: design principle, preparation and applications. *Chem. Eur. J* **19**, 16862–16879 (2013).

29. Shen, M., Sun, Y., Han, Y., Yao, R. & Yan, C. Strong deaggregating effect of a novel polyamino resorcinarene surfactant on gold nanoaggregates under microwave irradiation. *Langmuir* **24**, 13161–13167 (2008).

30. Xue, M., Yang, Y., Chi, X., Zhang, Z. & Huang, F. Pillararenes, a new class of macrocycles for supramolecular chemistry. *Acc. Chem. Res.* **45**, 1294–1308 (2012).

31. Yao, Y., Wang, Y. & Huang, F. Synthesis of various supramolecular hybrid nanostructures based on pillar[6]arene modified gold nanoparticles/nanorods and their application in pH- and NIR-triggered controlled release. *Chem. Sci.* **5**, 4312–4316 (2014).

32. Ogoshi, T. & Yamagishi, T. Pillar [5]-and pillar[6]arene-based supramolecular assemblies built by using their cavity-size-dependent host–guest interactions. *Chem. Commun.* **50**, 4776–4787 (2014).

33. Yu, G. *et al.* Pillar[6]arene/paraquat molecular recognition in water: high binding strength, ph-responsiveness, and application in controllable self-assembly, controlled release, and treatment of paraquat poisoning. *J. Am. Chem. Soc.* **134**, 19489–19497 (2012).

34. Yu, G. *et al.* A water-soluble pillar[6]arene: synthesis, host–guest chemistry, and its application in dispersion of multiwalled carbon nanotubes in water. *J. Am. Chem. Soc.* **134**, 13248–13251 (2012).

35. Yao, Y., Xue, M., Chen, J., Zhang, M. & Huang, F. An amphiphilic pillar[5]arene: synthesis, controllable self-assembly in water, and application in calcein release and TNT adsorption. *J. Am. Chem. Soc.* **134**, 15712–15715 (2012).

36. Chen, L. *et al.* Chiral selective transmembrane transport of amino acids through artificial channels. *J. Am. Chem. Soc.* **135**, 2152–2155 (2013).

37. Duan, Q. *et al.* pH-responsive supramolecular vesicles based on water-soluble pillar[6]arene and ferrocene derivative for drug delivery. *J. Am. Chem. Soc.* **135**, 10542–10549 (2013).

38. Ogoshi, T., Hashizume, M., Yamagishi, T. & Nakamoto, Y. Synthesis, conformational and host–guest properties of water-soluble pillar[5]arene. *Chem. Commun.* **46**, 3708–3710 (2010).

39. Cao, Y. *et al.* Supramolecular nanoparticles constructed by DOX-based prodrug with water-soluble Pillar[6]arene for self-catalyzed rapid drug release. *Chem. Mater.* **27**, 1110–1119 (2015).

40. Wu, X., Li, Y., Lin, C., Hu, X. & Wang, L. GSH- and pH-responsive drug delivery system constructed by water-soluble pillar[5]arene and lysine derivative for controllable drug release. *Chem. Commun.* **51**, 6832–6835 (2015).

41. Takaya, T. & Iwata, K. Relaxation mechanism of β-carotene from S_2 ($1B_u^+$) State to S_1 ($2Ag^-$) state: femtosecond time-resolved near-ir absorption and stimulated resonance raman studies in 900–1550 nm region. *J. Phys. Chem. A.* **118**, 4071–4078 (2014).

42. Oliveira, V. E. *et al.* Carotenoids and β-cyclodextrin inclusion complexes: Raman spectroscopy and theoretical investigation. *J. Phys. Chem. A.* **115**, 8511–8519 (2011).

43. Schneider, H.-J. Dispersive interactions in solution complexes. *Acc. Chem. Res.* **46**, 916–926 (2015).

44. Cao, Y. *et al.* Multistimuli-responsive supramolecular vesicles based on water-soluble pillar[6]arene and SAINT complexation for controllable drug release. *J. Am. Chem. Soc.* **136**, 10762–10769 (2014).

45. Matsuzawa, Y. *et al.* Assembly and photoinduced organization of mono- and oligopeptide molecules containing an azobenzene moiety. *Adv. Funct. Mater.* **17**, 1507–1514 (2007).

46. Spano, F. C. Analysis of the UV/Vis and CD spectral line shapes of carotenoid assemblies: spectral signatures of chiral *H*-aggregates. *J. Am. Chem. Soc.* **131**, 4267–4278 (2009).

47. Sun, Y., Yao, Y., Yan, C.-G., Han, Y. & Shen, M. Selective decoration of metal nanoparticles inside or outside of organic microstructures *via* self-assembly of resorcinarene. *ACS Nano* **4**, 2129–2141 (2010).

48. Sun, Y., Yan, C.-G., Yao, Y., Han, Y. & Shen, M. Self-assembly and metallization of resorcinarene microtubes in water. *Adv. Funct. Mater.* **18**, 3981–3990 (2008).

49. Soni, S. P., Ward, J. A., Sen, S. E., Feller, S. E. & Wassall, S. R. Effect of trans unsaturation on molecular organization in a phospholipid membrane. *Biochemistry* **48**, 11097–11107 (2009).

50. Li, C. *et al.* Self-assembly of [2] pseudorotaxanes based on pillar[5]arene and bis(imidazolium) cations. *Chem. Commun.* **46**, 9016–9018 (2010).

51. Yamamoto, Y. *et al.* Photoconductive coaxial nanotubes of molecularly connected electron donor and acceptor layers. *Science* **314**, 1761–1764 (2006).

52. Blau, W. J. & Fleming, A. J. Designer nanotubes by molecular self-assembly. *Science* **304**, 1457–1458 (2004).

53. Koehne, B., Elli, G., Jennings, R., Wilhelm, C. & Trissl, H.-W. Spectroscopic and molecular characterization of a long wavelength absorbing antenna of *Ostreobium* sp. *Biochim. Biophys. Acta.* **1412**, 94–107 (1999).

54. Hu, Z., Xu, Y., Gong, Y. & Kuang, T. Effects of heat treatment on the protein secondary structure and pigment microenvironment in photosystem I complex. *Photosynthetica* **43**, 529–534 (2005).

55. Yang, Z., Su, X., Wu, F., Gong, Y. & Kuang, T. Effect of phosphatidylglycerol on molecular organization of photosystem I. *Biophys. Chem.* **115**, 19–27 (2005).

56. Goss, R., Wilhelm, C. & Garab, G. Organization of the pigment molecules in the chlorophyll a/b/c containing alga *Mantoniella squamata* (Prasinophyceae) studied by means of absorption, circular and linear dichroism spectroscopy. *Biochim. Biophys. Acta.* **1457**, 190–199 (2000).

57. Dobrikova, A. *et al.* Structural rearrangements in chloroplast thylakoid membranes revealed by differential scanning calorimetry and circular dichroism spectroscopy. thermo-optic effect. *Biochemistry.* **42**, 11272–11280 (2003).

58. Agostiano, A., Cosma, P., Trotta, M., Monsu-Scolaro, L. & Micali, N. Chlorophyll a behavior in aqueous solvents: formation of nanoscale self-asembled complexes. *J. Phys. Chem. B* **106**, 12820–12829 (2002).

59. Kwa, S. *et al.* Steady-state and time-resolved polarized light spectroscopy of the green plant light-harvesting complex II. *Biochim. Biophys. Acta.* **1101**, 143–146 (1992).

60. Dauthal, P. & Mukhopadhyay, M. *Prunus domestica* fruit extract-mediated synthesis of gold nanoparticles and its catalytic activity for 4-nitrophenol reduction. *Ind. Eng. Chem. Res.* **51**, 13014–13020 (2012).

61. Moran, N. & Jarvik, T. Lateral transfer of genes from fungi underlies carotenoid production in aphids. *Science* **328**, 624–628 (2010).

62. Valmalette, J. *et al.* Light-induced electron transfer and ATP synthesis in a carotene synthesizing insect. *Sci. Rep.* **2**, 579 (2012).

63. Zou, Q. *et al.* Multifunctional porous microspheres based on peptide-porphyrin hierarchical co-assembly. *Angew. Chem. Int. Ed.* **53**, 2366–2370 (2014).

64. Mele, A., Mendichi, R., Selva, A., Molnar, P. & Toth, G. Non-covalent associations of cyclomaltooligosaccharides(cyclodextrins) with carotenoids in water. A study on the α-and β-cyclodextrin/ψ,ψ-carotene (lycopene) systems by light scattering, ionspray ionization and tandem mass spectrometry. *Carbohydr. Res.* **337**, 1129–1136 (2002).

65. Fernandez, J., Mellet, C. & Defaye, J. Glyconanocavities: cyclodextrins and beyond. *J. Inclusion Phenom. Macrocyclic Chem* **56**, 149–159 (2006).

66. Ashton, P. R. *et al.* Self-assembly, spectroscopic, and electrochemical properties of [n] rotaxanes1. *J. Am. Chem. Soc.* **118**, 4931–4951 (1996).

Acknowledgements

This work was financially supported by the China Postdoctoral Science Foundation (137070124), the National Natural Science Foundation of China (21273195, 21503185), we thank J.G. Zheng of the University of California (Irvine) for help in TEM discussions, as well as F.M Zhang and Z.F. Wang for the NMR and SAXS experiments.

Author contributions

Y.S. and G.D. designed the experiments and analysed the data. ^{1}H-NMR, TEM, SEM, HRTEM, EDX mapping, DLS and CD investigations were performed by Y.S. F.G. and Y.S. designed and performed the fluorescence characterization and analysed the corresponding fluorescence data. Y.S. and T.Z. performed the optical microscopy and polarized optical microscopy experiments. Y.S. and J.H. performed the fluorescene optical microscopy experiments, FT-IR, and zeta-potential experiments. Y.S., T.Z., and J.H. performed the ultraviolet–visible and Raman spectra experiments. Y.S. and G.D. conceived the experiments and co-wrote the paper.

Additional information

Competing financial interests: The authors declare no competing financial interests.

24

Growth of semiconducting single-wall carbon nanotubes with a narrow band-gap distribution

Feng Zhang[1,*], Peng-Xiang Hou[1,*], Chang Liu[1], Bing-Wei Wang[1], Hua Jiang[2], Mao-Lin Chen[1], Dong-Ming Sun[1], Jin-Cheng Li[1], Hong-Tao Cong[1], Esko I. Kauppinen[2] & Hui-Ming Cheng[1,3]

The growth of high-quality semiconducting single-wall carbon nanotubes with a narrow band-gap distribution is crucial for the fabrication of high-performance electronic devices. However, the single-wall carbon nanotubes grown from traditional metal catalysts usually have diversified structures and properties. Here we design and prepare an acorn-like, partially carbon-coated cobalt nanoparticle catalyst with a uniform size and structure by the thermal reduction of a $[Co(CN)_6]^{3-}$ precursor adsorbed on a self-assembled block copolymer nanodomain. The inner cobalt nanoparticle functions as active catalytic phase for carbon nanotube growth, whereas the outer carbon layer prevents the aggregation of cobalt nanoparticles and ensures a perpendicular growth mode. The grown single-wall carbon nanotubes have a very narrow diameter distribution centred at 1.7 nm and a high semiconducting content of >95%. These semiconducting single-wall carbon nanotubes have a very small band-gap difference of ~0.08 eV and show excellent thin-film transistor performance.

[1] Shenyang National Laboratory for Materials Science, Advanced Carbon Division, Institute of Metal Research, Chinese Academy of Sciences, 72 Wenhua Road, Shenyang 110016, China. [2] Nano Materials Group, Department of Applied Physics and Center for New Materials, School of Science, Aalto University, PO Box 15100, FI-00076 Aalto, Finland. [3] Faculty of Science, Chemistry Department, King Abdulaziz University, Jeddah 21589, Saudi Arabia.
* These authors contributed equally to this work. Correspondence and requests for materials should be addressed to C.L. (email: cliu@imr.ac.cn) or to H.M.C. (email: cheng@imr.ac.cn).

Single-wall carbon nanotubes (SWCNTs) have a small diameter, high carrier mobility, tunable band-gap and good stability[1–3]. Therefore, they are considered to be an ideal channel material for high-performance field effect transistors (FETs), integrated circuits and related electronic devices[4–6]. However, slight changes in the diameter and chiral angle of a SWCNT lead to the change of its type of electrical conductivity, that is, semiconducting to metallic or *vice versa*. As a result, as-prepared SWCNTs are usually a mixture of semiconducting and metallic nanotubes. In recent times, great efforts have been devoted to the selective preparation of pure semiconducting SWCNTs (s-SWCNTs) in large scale for their potential use in new-generation integrated circuits and notable progress has been made. One way is to separate SWCNTs by postsynthesis chemical and physical treatments[7–10]. Although high-purity s-SWCNTs can be obtained, this approach suffers from drawbacks that contamination and defects are inevitably introduced and the resultant SWCNTs are usually shortened. The other way is to synthesize s-SWCNTs directly by using metal nanoalloy as catalyst[11,12] or *in situ* selective etching during growth based on the principle that metallic SWCNTs (m-SWCNTs) are chemically more active than s-SWCNTs[13,14]. For example, Zhang and colleagues[15] grew s-SWCNT arrays by applying ultraviolet irradiation and subsequent photochemical reactions; Liu and colleagues[16] prepared SWCNTs by using a mixture of ethanol and methanol as both carbon source and oxidant precursor; Cheng and colleagues[17,18] synthesized s-SWCNTs by introducing oxygen or hydrogen as etchants; and Li and colleagues[19] grew s-SWCNTs from a metal catalyst loaded on CeO_2 supports that can release oxygen for selective etching. In these above studies, high-quality s-SWCNTs with purities higher than 90% were obtained. However, little attention has been paid to the band-gap uniformity of the s-SWCNTs obtained. In fact, the diameters of these s-SWCNTs are usually distributed over a wide range. According to theoretical and experimental studies, the band-gap of s-SWCNT is inversely proportional to their diameter[2]. It is only by using s-SWCNTs with a narrow band-gap range that devices with a uniform and stable performance can be fabricated. From this point of view, it is most desirable to grow SWCNTs with uniform chirality. Progress has been made on the controlled growth of SWCNTs by catalyst engineering[20–22]. Harutyunyan et al.[20] first achieved preferential growth of m-SWCNTs by modifying the morphology and coarsening behaviour of catalyst nanoparticles. Scott et al.[23] synthesized short (5, 5) CNTs from a corannulene seed by stepwise chemical method and by Diels–Alder reaction[24]. Theoretically, Ding et al.[25] proposed that the growth rate of an SWCNT is proportional to its chiral angle, whereas Artyukhov et al.[26] pointed out that the naturally enriched near-armchair SWCNTs can be attributed to a 'compromise' of the kinetic and thermodynamic growth trends for a rigid state catalyst. Very recently, high-purity single-chirality SWCNTs were grown from high-melting-point, W-containing alloy nanoparticles[21,22], as well as growth seeds derived from molecular precursors[27]. However, the enriched (12, 6)[21], (6, 6)[27] and so on SWCNTs are metallic, whereas the content of (16, 0) s-SWCNTs is ~80% (ref. 22).

To obtain s-SWCNTs with a narrow diameter distribution and a small range of band-gap, it is essential to use uniform catalyst nanoparticles. However, it was found that even with a similar initial catalyst nanoparticle size, the grown SWCNTs may have different diameters[28,29]. There are mainly two reasons for this phenomenon: the first is that catalyst nanoparticles tend to aggregate, to form larger ones during the growth of SWCNTs at high temperatures; second, SWCNTs may grow following either the 'tangential' mode (with diameter equal to the nanoparticle size)[28] or the 'perpendicular' mode (with diameter smaller than

the nanoparticle size)[28]. Therefore, to achieve well-controlled growth of SWCNTs, it is crucial to prevent the aggregation of catalyst nanoparticles and to control their mode of nucleation and growth. In this study, we design and prepare an acorn-like bicomponent catalyst, that is, Co nanoparticle with a partial carbon coating layer. The inner Co nanoparticle functions as active catalytic phase for SWCNT growth, whereas the outer carbon layer prevents the aggregation of Co nanoparticles and ensures the perpendicular growth of SWCNT from the catalyst. This acorn-like catalyst with uniform size and structure is prepared using a block copolymer (BCP) self-assembly technique[30]. The SWCNTs grow from the catalyst have a very narrow diameter distribution. Based on the principle that m-SWCNTs are chemically more reactive than s-SWCNTs with similar diameters due to the smaller ionization potential of the former[31,32], hydrogen is introduced as an etchant and m-SWCNTs are selectively removed *in situ*. As a result, high-purity (>95%) s-SWCNTs with a narrow range of band gaps (<0.08 eV) are obtained, which show an excellent thin-film transistor (TFT) performance.

Results

Preparation of Co nanoparticle partially coated with carbon. A schematic showing the preparation of the Co nanoparticles partially coated with carbon is shown in Fig. 1. Details of catalyst preparation are described in Methods. In brief, an asymmetric poly-(styrene-block-4-vinylpyridine) (PS-b-P4VP) film (Fig. 1a) was self-assembled into vertical P4VP nanocylinders by placing it in a toluene/tetrahydrofuran (THF) mixture vapor annealing for 24 h. The formed nanodomains, consisting of hydrophilic vertical P4VP nanocylinders enclosed by a hydrophobic PS matrix, were then immersed in an acidic aqueous solution of $K_3[Co(CN)_6]$ (Fig. 1b). The protonated pyridinic nitrogen sites in the P4VP nanocylinders attracted anionic Co complexes as $[Co(CN)_6]^{3-}$, after which the material was treated in an air plasma to oxidize the polymer. As the Co precursor was anchored at one end of the nanodomains, CoO nanoclusters partially encapsulated by a thin polymer layer were obtained (Fig. 1c). It is worth noting that we intentionally used a relatively weak plasma treatment to remove the polymer in a controlled manner. The PS-b-P4VP used, PS40000-b-P4VP5600, has a high PS to P4VP mass ratio (~7), which also makes it difficult to completely remove the polymer. Furthermore, owing to the similar size of the self-assembled nanodomains, the resulting CoO clusters have a uniform size and exposed surface area. In subsequent heating in a H_2 atmosphere, the CoO nanoclusters were reduced to Co and aggregated into Co nanoparticles, and the residual polymer was carbonized to produce Co catalysts that were partially coated with carbon; thus, the acorn-like bicomponent catalyst is obtained (Fig. 1d). SWCNTs with a uniform diameter (Fig. 1e) were grown from these catalyst particles by the chemical vapour deposition (CVD) of ethanol. When H_2 was introduced as an *in situ* growth etchant, m-SWCNTs with higher chemical reactivity were selectively removed and s-SWCNTs with a narrow range of band-gap were obtained (Fig. 1f).

Characterization of the partially coated Co nanoparticles. To understand the structure of the catalyst synthesized using this method, atomic force microscopy (AFM) and transmission electron microscopy (TEM) characterization were performed. An AFM image (Fig. 2a) shows that the catalyst particles are well-dispersed and uniformly distributed on the substrate, whereas the TEM image (Fig. 2b) verifies the uniform size of the monodispersed Co nanoparticles. These nanoparticles were

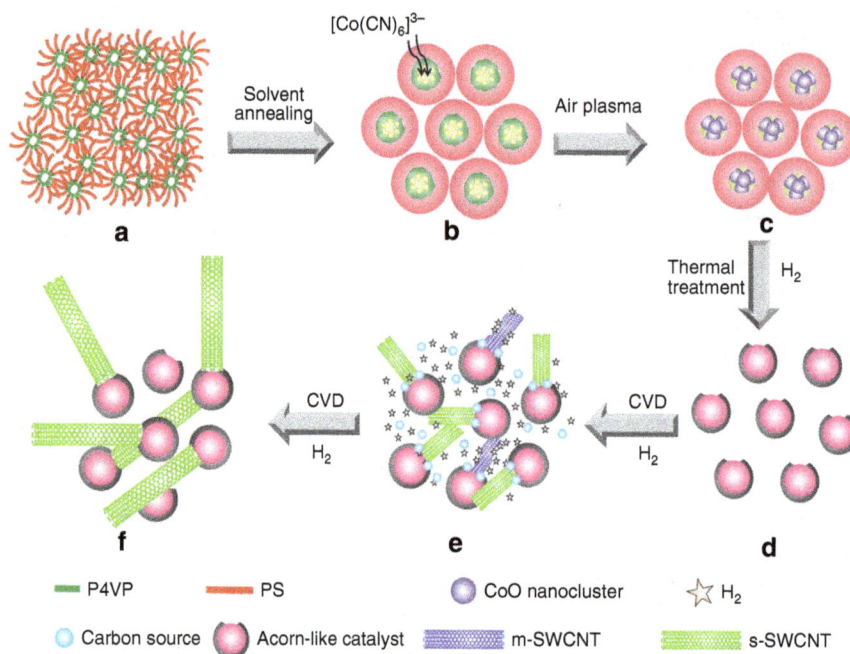

Figure 1 | Schematic showing the formation of carbon-coated Co nanoparticles and the growth of s-SWCNTs. (a) A self-assembled PS-b-P4VP film. **(b)** Formation of phase-separated nanodomains and adsorption of $[Co(CN)_6]^{3-}$ catalyst precursors. **(c)** CoO nanoclusters partially surrounded by residual polymer. **(d)** Co nanoparticles partially coated with carbon. **(e)** Nucleation and growth of SWCNTs with a narrow diameter distribution grown from the partially carbon-coated Co nanoparticles and *in situ* etching of m-SWCNTs. **(f)** s-SWCNTs with a narrow band-gap range.

observed to be partially coated with a carbon layer (indicated by red arrows in Fig. 2b). High-resolution TEM images (Supplementary Fig. 1a–c) clearly demonstrate the partially carbon-coated Co nanoparticle structure of the catalyst. More detailed characterizations of the catalysts are reported in Methods. The sizes of ∼130 nanoparticles were measured from TEM images and the resultant histogram of their diameters is shown in Fig. 2c. It can be seen that most of the nanoparticles (>90%) have diameters in the range of 2.5–4.5 nm, with a mean diameter of 3.1 nm. These uniform and monodispersed Co catalyst particles partially coated with carbon are critical for the growth of SWCNTs with a narrow diameter distribution.

Structure characterization of SWCNTs. SWCNTs were synthesized by CVD at 700 °C using the carbon-coated Co particles as catalyst, alcohol as carbon source and hydrogen as carrier gas. Details of the SWCNT synthesis procedure and characterization are described in Methods. Scanning electron microscope (SEM) image of Fig. 3a shows that carbon nanotubes with a length of 10 μm are randomly dispersed on the substrate. TEM observations show that the SWCNTs obtained are isolated and straight (Fig. 3b), suggesting that the catalyst has a good catalytic activity for SWCNT growth even at the relatively low temperature of 700 °C. A SWCNT grown from a partially carbon-coated catalyst nanoparticle was clearly observed (Fig. 3c). The diameters of 130 SWCNTs were measured from TEM images and the resultant histogram is shown in Fig. 3d. The diameters are narrowly distributed in the range of 1.6–1.9 nm with a mean diameter of 1.7 nm. This diameter distribution is much narrower than most previously reported results[17,33]. In addition, the SWCNTs are isolated rather than in bundles, which can be attributed to the monodispersed Co catalysts used.

Characterization by Raman and absorption spectroscopy. We further characterized the SWCNTs using Raman spectroscopy

with excitation laser wavelengths of 488, 532, 633 and 785 nm. Figure 4a–d show the Raman spectra, where the radial breathing mode (RBM) peaks originating from m- and s-SWCNTs are highlighted according to the Kataura plot[34]. In Fig. 4a, sharp peaks narrowly centred at ∼141 cm^{-1} are shown with an excitation wavelength of 532 nm (2.33 eV) and these peaks originate from s-SWCNTs. Figure 4b shows the Raman spectra excited with a laser wavelength of 633 nm (1.96 eV) and the RBM peaks originating from s-SWCNTs are located in a narrow range around ∼141 cm^{-1}. Raman spectra excited by a 785 nm (1.58 eV) laser (Fig. 4c) shows very few weak RBM peaks at ∼141 cm^{-1} originating from metallic tubes. No RBM Raman peak is detected with the 488-nm laser (2.54 eV) (Fig. 4d). All the excited Raman RBM signals under these four laser wavelengths are centred at ∼141 cm^{-1} and most can be assigned to s-SWCNTs, suggesting that the sample contains s-SWCNTs with a narrow diameter distribution. Using the inverse relationship between RBM peak frequency (ω_{RBM}) and tube diameter (d_t) ($d_t = 235.9/(\omega_{RBM} - 5.5))$[35], the calculated diameters of the SWCNTs are around 1.7 nm, which is in good agreement with the TEM observations. The content of s-SWCNTs in our sample is calculated to be ∼98% based on the number of the metallic and semiconducting RBM peaks averaged from 160 Raman spectra[19,36]. The G-band of the Raman spectra (Fig. 4e) show an obvious Lorentzian line shape, which is characteristic of s-SWCNTs[37] and this further verifies the effective enrichment of semiconducting nanotubes. As the diameters of the SWCNTs are mostly in the range of 1.6–1.9 nm, these s-SWCNTs have an extremely small band-gap difference of 0.08 eV (ref. 38). The ultraviolet/visible/near-infrared absorption spectrum of an aqueous SWCNT dispersion collected from 125 SWCNT samples (for more details, see Methods and Supplementary Fig. 2) is shown in Fig. 4f. One intense narrow peak and two weak peaks distributed in the range of 1,200–1,430 nm and 600–715 nm, respectively, were detected, corresponding to the second and third van Hove

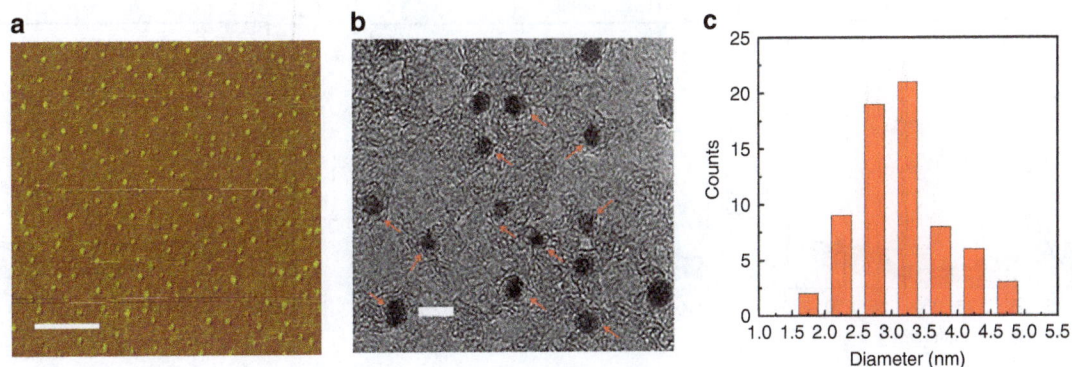

Figure 2 | Morphology and diameter distribution of the Co nanoparticle partially coated with carbon. Typical (**a**) AFM and (**b**) TEM images of the Co nanoparticles partially coated with carbon, showing that the nanoparticles are monodispersed and uniform in size. The red arrows indicate Co nanoparticles partially coated by a carbon layer. Scale bar, 400 nm (**a**) and 5 nm (**b**). (**c**) A histogram showing the diameter distribution of the nanoparticles based on TEM observations.

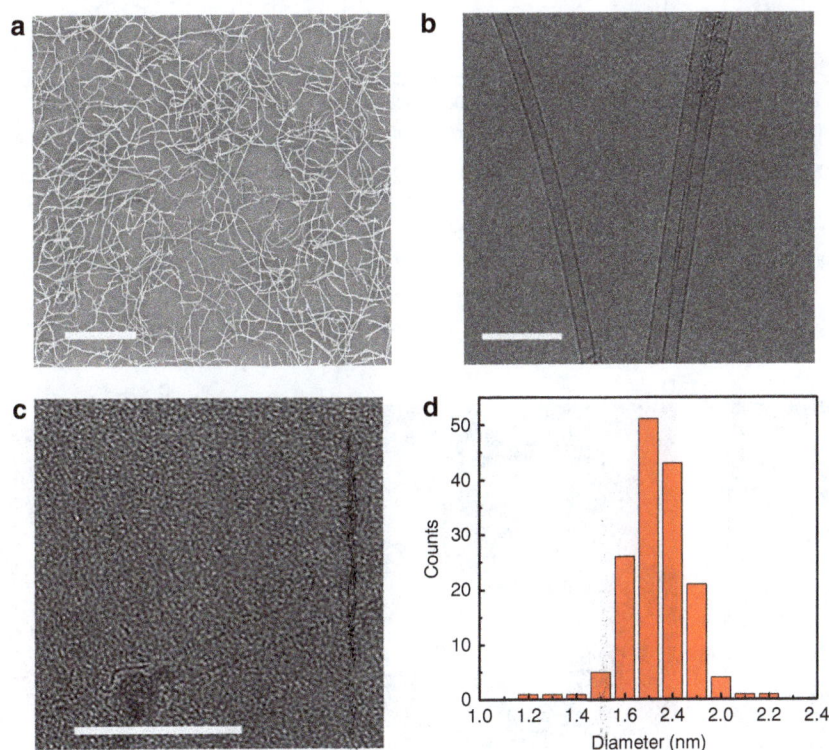

Figure 3 | SEM and TEM observations of the as-prepared SWCNTs. SEM (**a**) and TEM (**b,c**) image of the as-prepared SWCNTs. Scale bar, 10 μm (**a**) and 10 nm (**b,c**). (**d**) Diameter distribution of the SWCNTs based on TEM observations.

singularity transition of s-SWCNTs (S_{22} and S_{33}) with diameter distributions of 1.6–1.9 nm (refs 34,38). No obvious first van Hove singularity transition of m-SWCNTs (800–950 nm) was detected by the ultraviolet/visible/near-infrared, confirming the high purity of s-SWCNTs in our sample. Quantitatively, the content of s-SWCNTs in the sample was calculated to be ∼ 99% (ref. 39).

Conductivity characterization by electron diffraction. Electron diffraction analysis is an effective method for directly determining the chiral indices (n, m) of SWCNTs[40]. To further confirm the enrichment of s-SWCNTs in our sample, we transferred the SWCNTs from a Si substrate to a TEM grid (for details see Methods) and carried out electron diffraction using a TEM (JEOL

JEM-2200FS 2 × CS corrected) operated at 80 kV with a Gatan 794 multiscan charge-coupled device camera (1 k × 1 k). Figure 5a shows a freestanding SWCNT and its corresponding electron diffraction pattern is shown in Fig. 5b. In addition to the bright spot at the centre caused by the direct electron beam, the diffraction pattern mainly consists of a set of parallel diffracted layer lines, which are separated by certain distances from the equatorial layer line at the centre. By analysing the layer line distance, the chiral index of this SWCNT was determined to be (19, 5) with a tube diameter of 1.72 nm. The chiral indexes of total 95 SWCNTs were assigned by using this electron diffraction method (Supplementary Figs 3 and 4). The resultant numbers of s- and m-SWCNTs are shown in Fig. 5c. The s-SWCNT content reaches ∼ 96%, which verifies the enrichment of s-SWCNTs in our sample.

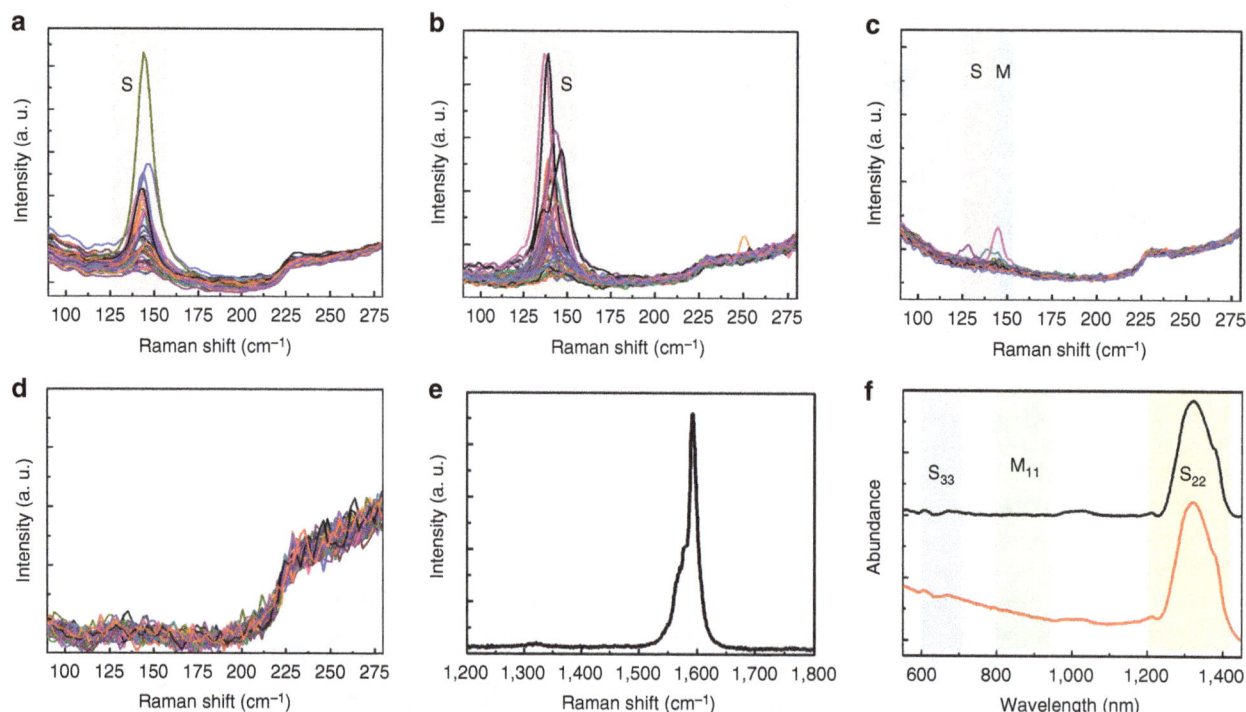

Figure 4 | Electrical type of the SWCNTs. Electrical type characterized by multi-wavelength Raman spectroscopy and ultraviolet/visible/near-infrared absorption spectroscopy. Raman spectra of the SWCNTs measured with (**a**) 532, (**b**) 633, (**c**) 785 and (**d**) 488 nm lasers. The regions corresponding to semiconducting and metallic transitions are labelled as S (pink) and M (blue), respectively. (**e**) D and G band excited with a 633-nm laser. (**f**) Ultraviolet/visible/near-infrared absorption spectrum of an aqueous dispersion collected from 125 SWCNT samples: as-collected curve (red line) and background-subtracted curve (black line).

Figure 5 | Electron diffraction of the SWCNTs. (**a**) Typical TEM image of an isolated SWCNT. Scale bar, 10 nm. (**b**) Electron diffraction pattern of the SWCNT shown in **a**, the chirality of which is assigned as (19, 5). (**c**) A pie chart showing the content of s- and m-SWCNTs based on electron diffraction measurements of a total of 95 SWCNTs.

Electrical performance of SWCNT TFTs. Electrical measurements can provide direct evidence of the transport properties of SWCNTs[41,42]. Thus, TFT devices based on the SWCNTs with a growth time of 25 min were fabricated (for more details, see Methods). Supplementary Fig. 5 shows a typical SEM image of a long channel bottom-gated SWCNT TFT. It can be seen that the SWCNTs are pure and randomly distributed. The channel width and length of the TFTs are 100 and 200 μm, respectively. Figure 6a shows typical output characteristics ($I_{ds} - V_{ds}$)of the TFTs, which indicates the formation of ohmic contacts between SWCNTs and metal eletrode. Figure 6b shows transfer

characteristics ($I_{ds} - V_{gs}$) of ten devices with the same channel geometry ($L_{ch} = 200$ μm, $W_{ch} = 100$ μm); simultaneously, high on/off ratios of 3.1×10^3–3.6×10^6 and high carrier mobilities of 36–143 cm^2 V^{-1} s^{-1}, as evaluated by the standard formula[6], were demonstrated. Remarkably, the mean carrier mobility is as high as 95.2 cm^2 V^{-1} s^{-1}. Furthermore, as shown in Fig. 6c, when compared with the previously reported SWCNT-based TFTs[6,8,18,43–49], including the TFTs fabricated using chirality-enriched nanotubes[46], our TFTs show obviously superior performance, in terms of both high current on/off ratio and high carrier mobility, which further verifies the

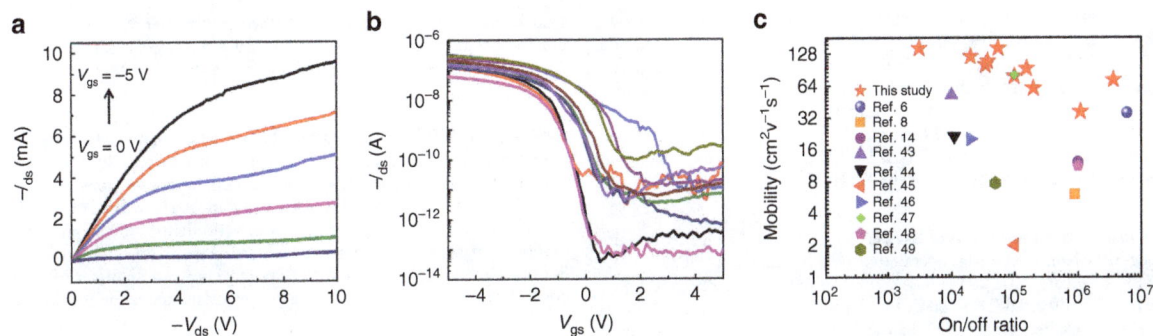

Figure 6 | Electrical performance of the SWCNT TFTs. (a) Output ($I_{ds} - V_{ds}$) characteristics of the same device measured at various V_{gs} from -5 to 0 V in 1 V step. $L_{ch} = 200\,\mu m$, $W_{ch} = 100\,\mu m$. **(b)** Transfer characteristics of 10 SWCNT TFTs. **(c)** A comparison of the mobility performance of our SWCNT TFTs with those previously reported.

high quality and high s-SWCNTs content of our sample. We further constructed a series of bottom-gate TFTs from the grown s-SWCNTs with L_{ch} ranging from 20 to 200 μm (Supplementary Fig. 6). It was found that on/off ratios higher than 10^5 were obtained for the TFTs with L_{ch} longer than 40 μm, and even for the devices with a L_{ch} value of 20 μm, the on/off ratios were still around 10^2. TFTs with a short channel length of 1.5 μm were also fabricated by electron-beam lithography (EBL) using the same sample. High on/off ratios and high carrier mobilities were demonstrated for the majority of the devices (Supplementary Fig. 7), although a few TFTs showing current on/off ratios <5 caused by the short circuiting of m-SWCNTs were also detected.

Discussion

We attribute this selective growth of s-SWCNTs to both novel catalyst and *in situ* growth selective etching. As mentioned above, we prepared an acorn-like, partially carbon-coated Co nanoparticle catalyst with a narrow diameter distribution by a BCP self-assembly method. The $[Co(CN)_6]^{3-}$ catalyst precursor was absorbed on the phase-separated P4VP nanodomains so that only part of the Co nanoparticle is exposed and serves as the catalyst for SWCNT growth following a perpendicular mode (see Fig. 1). As shown in Fig. 3c, SWCNTs were observed to grow from the partially carbon-coated Co nanoparticles via a perpendicular growth mode. As the Co nanoparticles prepared by the copolymer self-assembly method have a narrow range of diameters, the exposed Co active catalyst obtained after polymer wrapping and partial etching processes should have an even smaller size difference. As a consequence, the SWCNTs grown on the exposed Co regions have an extremely narrow diameter distribution. The hydrogen used as carrier gas during the CVD growth also serves as an etchant, that is, m-SWCNTs can be preferentially removed by H radicals decomposed from H_2 at 700 °C in the presence of Co[50]. The narrow diameter distribution of the SWCNTs guarantees efficient selective etching of m-SWCNTs, as the chemical stability of SWCNTs is determined by both their electrical type and diameter[14,16,17,31,32,51,52]. Therefore, high purity ($>95\%$) s-SWCNTs with a uniform diameter and band-gap are obtained.

To verify the above controlled growth mechanism, we further performed a series of comparative experiments. First, fully exposed Co catalyst particles were prepared by omitting the solvent annealing process and a thermal treatment at 750 °C in air for 5 min was added to completely remove the polymer (see Methods). These Co catalysts were used for growing SWCNTs. The Raman spectra (Supplementary Fig. 8) of the obtained sample show a broad RBM peak distribution, originating from both s- and m-SWCNTs according to the Kataura plot[34].

This result proves that the partial carbon coating of the Co nanoparticles plays a key role in growing SWCNTs with uniform diameters. Following this controlled growth mechanism, it may be feasible to tune the diameter of SWCNTs by controlling the exposed area of the Co nanoparticles. We therefore tried to enlarge the exposed area by performing heat treatment in a H_2 atmosphere at 800 °C for 10 min and then used the resulting material for SWCNT growth (Methods). Supplementary Fig. 9 shows the Raman spectra of the SWCNTs obtained. It can be seen that the RBM peaks are distributed in the range of 110–120 cm^{-1}, corresponding to nanotube diameters of 2.0–2.2 nm. Based on the Raman spectra from three laser wavelengths, the content of s-SWCNT was calculated to be $\sim 96\%$ (refs 19,34,36). This result further verifies the growth mechanism of s-SWCNTs from the Co catalyst partially coated with carbon.

In summary, acorn-like partially carbon-coated Co nanoparticles with a mean diameter of 3.1 nm were prepared by a BCP self-assembly and subsequent partial oxidation approach. The bicomponent nanoparticles were used as a catalyst for growing SWCNTs by CVD. SWCNTs with a narrow and tunable diameter distribution were synthesized in a hydrogen atmosphere following a perpendicular growth mode. It was found that a high content ($>95\%$) of s-SWCNTs with a narrow diameter distribution was obtained. The mean diameter of the s-SWCNTs could be tuned to be ~ 1.7 or 2.1 nm by controlling the carbon coverage of the Co catalyst. The band-gap range of the s-SWCNTs was <0.08 eV, due to their very narrow diameter distribution. The TFTs fabricated using these s-SWCNTs showed an excellent electrical performance. This combined catalyst design and *in situ* etching approach may pave the way for chirality-controlled growth of SWCNTs.

Methods

Preparation of the partially carbon-coated Co nanoparticles. A BCP self-assembly technique was used to prepare Co nanoparticles with a uniform diameter[53,54]. Briefly, 0.25 wt% PS-b-P4VPs (Mn: 40,000 g mol^{-1} for PS and 5,600 g mol^{-1} for P4VP, Polymer Source Inc.) were dissolved in a toluene/THF = 3/1 (wt ratio) solution and electromagnetically stirred for 2 h in a 70 °C water bath. The solution was then spin-coated (Spin Master100) onto a Si/SiO$_2$ (300 nm thick) substrate at 5,000 r.p.m. for 2 min. The substrate had previously been treated in a sulfuric acid/hydrogen peroxide (5/1 in wt ratio) solution and ultrasonicated in ethanol and acetone for 15 min. The BCP thin films were then annealed under a saturated vapour of toluene/THF (1/4 vol ratio) solution, after which they were immersed in a 1 mmol l^{-1} K$_3$[Co(CN)$_6$] (98%, J&K) acidic aqueous solution for 3 min, to allow the adsorption of [Co(CN)$_6$]$^{3-}$, followed by washing three times in de-ionized water. The BCP films were then dried at 60 °C for 20 min followed by exposure to air plasma for 10 min, to remove the polymer covered on the top of particles. The CoO nanoclusters were reduced under a hydrogen atmosphere with a flow rate of 200 s.c.c.m. at 500 °C for 5 min, subsequently heated up to 700 °C for 5 min, to obtain Co nanoparticles that were partially coated with carbon.

Synthesis of SWCNTs. For the growth of SWCNTs, a Si/SiO$_2$ substrate with catalyst dispersed on the surface was placed at the low-temperature zone (end) of a quartz tube reactor (25 mm in diameter) inserted into a horizontal tube furnace. When the furnace temperature reached 700 °C, the Si substrate was put into the centre of the reactor. At the same time, a 75 s.c.c.m. argon flow through an ethanol bubbler (in a 35 °C water bath) and a 200 s.c.c.m. hydrogen flow were introduced into the reactor for 10–25 min, to grow SWCNTs. The furnace was then cooled to room temperature under the protection of the Ar/H$_2$ flow.

Characterizations of the catalysts and SWCNTs. An AFM (Digital Instruments Multiple Mode SPM Nanoscope IIIa, operated at tapping-mode), an SEM (FEI XL30 S-FEG, operated at 1 kV), a TEM, (FEI Tecnai F20, 200 kV; JEOL JEM-2200FS 2× CS corrected TEM, 80 kV) and a micro-Raman spectroscope (Jobin Yvon HR800, excited by 488, 532, 633 and 785 nm He-Ne laser with laser spot size of ~1 µm^2, in line mapping mode) were used to characterize the SWCNTs. For Raman characterization, more than 40 locations in each sample were measured randomly by moving the laser spot in 10 µm steps. For TEM observation, we used the transfer procedure developed by Jiao et al.[55,56] with a slight modification. Briefly, a polymethylmethacrylate (PMMA) solution (AR-P 679, molecular weight = 950 K, 4 wt.% in ethyl lactate, Allresist) was first spin-coated (3,000 r.p.m. for 1 min) onto the substrate on which the SWCNTs were grown. The substrate was then baked at 80 °C for 4 h in an oven. The PMMA solidified and formed a thin film with SWCNTs embedded in it. It was then put into a freshly prepared NaOH aqueous solution (5 M) with a temperature of 80 °C and kept there for 100 min. The slight etching of the SiO$_2$ surface by the hot NaOH solution released the PMMA film from the substrate[57]. Finally, the PMMA film embedded with SWCNTs was peeled off and attached to a Cu grid for TEM characterization after the PMMA was removed in an acetone bath[58]. For ultraviolet/visible/near-infrared absorption characterization, total 125 Si substrates (10 mm × 10 mm) with SWCNTs grown were inmmersed into a 5-ml 1 wt% sodium benzenesulfonate (SDS) D$_2$O solution (one by one), followed by ultrasonic treatment for 5 min, to detach the SWCNTs from the substrate and to disperse in the D$_2$O solution.

Calculation of the s-SWCNT content from absorption spectrum. According to the tight-binding model[34], the band gaps of s-SWCNTs and m-SWCNTs are dependent on their diameters as follows[38]:

$$E_{11}^S = \frac{2a_0\gamma_0}{d_t}, \; E_{22}^S = \frac{4a_0\gamma_0}{d_t}, \; E_{33}^S = \frac{8a_0\gamma_0}{d_t}, \; E_{11}^M = \frac{6a_0\gamma_0}{d_t} \qquad (1)$$

a_0 is the C–C band distance and γ_0 is interaction energy between neighbouring C atoms.

In this study, the diameter of the SWCNTs is distributed in the range of 1.6–1.9 nm. Thus, their absorption spectra peaks of S_{11}, S_{22}, S_{33} and M_{11} are calculated to be at 2,408–2,860 nm, 1,204–1,430 nm, 602–715 nm and 803–953 nm, respectively. The absorption peaks of S_{11} cannot be detected due to the strong noise absorption from the solvent at corresponding wavelength. Therefore, the content of s-SWCNTs was calculated by using absorption peak areas of M_{11} and S_{22} with the following formula:

$$R_S = \frac{n_S}{n_M + n_S} = \frac{1}{1 + \frac{M_{11}}{S_{22}f}} \qquad (2)$$

Where R_S is the content of s-SWCNTs in the sample, n_M and n_S are the numbers of m- and s-SWCNTs, M_{11} and S_{22} are the areas of the M_{11} or S_{22} peaks measured, and f is the absorption coefficient, here 1.24 according to the SWCNT diameter distribution. After baseline subtraction, the peak areas of S_{22} and M_{11} were measured to be 3.81 and 0.0265, respectively. Thus, the content of s-SWCNTs in our samples was calculated to be ~99% (ref. 39).

Fabrication of SWCNT TFTs. Bottom-gate TFTs based on our as-grown SWCNT network were fabricated on Si substrates with a thermally grown SiO$_2$ layer (300 nm) as a gate dielectric. The bottom-gate electrode (Ti/Au: 10/50 nm) was deposited by electron-beam evaporation after the SiO$_2$ layer on the backside of the wafer was etched by reactive ion etching. Source and drain electrodes (Ti/Au: 10/50 nm) were fabricated by standard photolithography, electron-beam evaporation and lift-off processes. Subsequently, SWCNTs outside the channel area were removed by oxygen plasma. For the fabrication of TFTs with short channel lengths, the patterning of electrodes and channel was realized by EBL. For details, the source and drain electrodes (Ti/Au: 5/50 nm) on Si/SiO$_2$ substrates with grown SWCNT network were fabricated by EBL, electron-beam evaporation and lift-off processing. Next, the SWCNTs outside the channel area were removed by EBL and oxygen plasma etching. Electrical measurements were conducted using a semiconductor analyser (Agilent B1500A) at ambient conditions.

Preparation of exposed Co catalyst and growth of SWCNTs. For comparison, a parallel experiment was conducted by using completely exposed Co nanoparticles as the catalyst to grow SWCNTs. The catalyst was prepared by omitting the solvent annealing step to obtain a uniform BCP film without microphase separation. We

treated the material in air at 750 °C for 5 min to remove the carbon coating, to obtain uncoated Co nanoparticles. The fully exposed Co catalyst was used to grow SWCNTs with the same CVD conditions described in Methods.

Synthesis of SWCNTs with a tunable mean diameter. To tune the mean diameter of SWCNTs, we prepared carbon-coated Co catalyst with a larger area of exposed Co by changing the thermal reduction process from 500 °C for 5 min to 700 °C for 10 min and to 800 °C for 10 min under H$_2$ with a flow rate of 200 s.c.c.m. The catalyst was used to grow SWCNTs with identical CVD growth conditions as described in Methods. The result of Raman characterization is shown in Supplementary Fig. 9. These SWCNTs have a mean diameter of 2.1 nm and the content of semiconducting nanotubes was calculated to be higher than 96%.

References

1. Frank, S., Poncharal, P., Wang, Z. L. & de Heer, W. A. Carbon nanotube quantum resistors. *Science* **280**, 1744–1746 (1998).
2. Saito, R., Fujita, M., Dresselhaus, G. & Dresselhaus, M. S. Electronic-structure of chiral graphene tubules. *Appl. Phys. Lett.* **60**, 2204–2206 (1992).
3. Durkop, T., Getty, S. A., Cobas, E. & Fuhrer, M. S. Extraordinary mobility in semiconducting carbon nanotubes. *Nano Lett.* **4**, 35–39 (2004).
4. Tulevski, G. S. et al. Toward high-performance digital logic technology with carbon nanotubes. *ACS Nano* **8**, 8730–8745 (2014).
5. Cao, Q. & Rogers, J. A. Ultrathin films of single-walled carbon nanotubes for electronics and sensors: a review of fundamental and applied aspects. *Adv. Mater.* **21**, 29–53 (2009).
6. Sun, D. M. et al. Flexible high-performance carbon nanotube integrated circuits. *Nat. Nanotechnol.* **6**, 156–161 (2011).
7. Liu, H. P., Nishide, D., Tanaka, T. & Kataura, H. Large-scale single-chirality separation of single-wall carbon nanotubes by simple gel chromatography. *Nat. Commun.* **2**, 309 (2011).
8. Lee, H. W. et al. Selective dispersion of high purity semiconducting single-walled carbon nanotubes with regioregular poly(3-alkylthiophene)s. *Nat. Commun.* **2**, 541 (2011).
9. Tu, X. M., Manohar, S., Jagota, A. & Zheng, M. DNA sequence motifs for structure-specific recognition and separation of carbon nanotubes. *Nature* **460**, 250–253 (2009).
10. Zheng, M. et al. DNA-assisted dispersion and separation of carbon nanotubes. *Nat. Mater.* **2**, 338–342 (2003).
11. Chiang, W. H. & Sankaran, R.M. Linking catalyst composition to chirality distributions of as-grown single-walled carbonnanotubes by tuning Ni$_x$Fe$_{1-x}$ nanoparticles. *Nat. Mater.* **8**, 882–886 (2009).
12. Zhang, S. C., Tong, L. M., Hu, Y., Kang, L.X. & Zhang, J. Diameter-specific growth of semiconducting SWNT arrays using uniform Mo$_2$C solid catalyst. *J. Am. Chem. Soc.* **137**, 8904–8907 (2015).
13. Li, Y. M. et al. On the origin of preferential growth of semiconducting single-walled carbon nanotubes. *J. Phys. Chem. B* **109**, 6968–6971 (2005).
14. Zhang, G. Y. et al. Selective etching of metallic carbon nanotubes by gas-phase reaction. *Science 2006* **314**, 974–977 (2005).
15. Hong, G. et al. Direct growth of semiconducting single-walled carbon nanotube array. *J. Am. Chem. Soc.* **131**, 14642–14643 (2009).
16. Ding, L. et al. Selective growth of well-aligned semiconducting single-walled carbon nanotubes. *Nano Lett.* **9**, 800–805 (2009).
17. Yu, B. et al. Bulk synthesis of large diameter semiconducting single-walled carbon nanotubes by oxygen-assisted floating catalyst chemical vapor deposition. *J. Am. Chem. Soc.* **133**, 5232–5235 (2011).
18. Li, W. S. et al. High-quality, highly concentrated semiconducting single-wall carbon nanotubes for use in field effect transistors and biosensors. *ACS Nano* **7**, 6831–6839 (2013).
19. Qin, X. J. et al. Growth of semiconducting single-walled carbon nanotubes by using ceria as catalyst supports. *Nano Lett.* **14**, 512–517 (2014).
20. Harutyunyan, A. R. et al. Preferential growth of single-walled carbon nanotubes with metallic conductivity. *Science* **326**, 116–120 (2009).
21. Yang, F. et al. Chirality-specific growth of single-walled carbon nanotubes on solid alloy catalysts. *Nature* **510**, 522–524 (2014).
22. Yang, F. et al. Growing zigzag (16,0) carbon nanotubes with structure-defined catalysts. *J. Am. Chem. Soc.* **137**, 8688–8691 (2015).
23. Scott, L. T. et al. A short, rigid, structurally pure carbon nanotube by stepwise chemical synthesis. *J. Am. Chem. Soc.* **134**, 107–110 (2012).
24. Fort, E. H., Donovan, P. M. & Scott, L. T. Diels-Alder reactivity of polycyclic aromatic hydrocarbon bay regions: implications for metal-free growth of single-chirality carbon nanotubes. *J. Am. Chem. Soc.* **131**, 16006–16007 (2009).
25. Ding, F., Harutyunyan, A. R. & Yakobson, B. I. Dislocation theory of chirality-controlled nanotube growth. *Proc. Natl Acad. Sci. USA* **106**, 2506–2509 (2009).
26. Artyukhov, V. I., Penev, E. S. & Yakobson, B. I. Why nanotubes grow chiral. *Nat. Commun.* **5**, 4892 (2014).
27. Sanchez-Valencia, J. R. et al. Controlled synthesis of single-chirality carbon nanotubes. *Nature* **512**, 61–64 (2014).

28. Fiawoo, M. F. C. *et al.* Evidence of correlation between catalyst particles and the single-wall carbon nanotube diameter: a first step towards chirality control. *Phys. Rev. Lett.* **108**, 195503 (2012).

29. Yao, Y. G. *et al.* Temperature-mediated growth of single-walled carbon nanotube intramolecular junctions. *Nat. Mater.* **6**, 283–286 (2007).

30. Mai, Y. Y. & Eisenberg, A. Self-assembly of block copolymers. *Chem. Soc. Rev.* **41**, 5969–5985 (2012).

31. Lu, J. *et al.* Selective interaction of large or charge-transfer aromatic molecules with metallic single-wall carbon nanotubes: critical role of the molecular size and orientation. *J. Am. Chem. Soc.* **128**, 5114–5118 (2006).

32. Strano, M.S. *et al.* Electronic structure control of single-walled carbon nanotube functionalization. *Science* **301**, 1519–1522 (2003).

33. Zhu, Z., Jiang, H., Susi, T., Nasibulin, A. G. & Kauppinen, E. I. The use of NH_3 to promote the production of large-diameter single-walled carbon nanotubes with a narrow (n,m) distribution. *J. Am. Chem. Soc.* **133**, 1224–1227 (2011).

34. Kataura, H. *et al.* Optical properties of single-wall carbon nanotubes. *Synth. Met.* **103**, 2555–2558 (1999).

35. Zhang, D. Q. *et al.* (n,m) assignments and quantification for single-walled carbon nanotubes on SiO_2/Si substrates by resonant Raman spectroscopy. *Nanoscale* **7**, 10719–10727 (2015).

36. Kang, L. X. *et al.* Growth of close-packed semiconducting single-walled carbon nanotube arrays using oxygen-deficient TiO_2 nanoparticles as catalysts. *Nano Lett.* **15**, 403–409 (2015).

37. Jorio, A. *et al.* G-band resonant Raman study of 62 isolated single-wall carbon nanotubes. *Phys. Rev. B* **65**, 1554121 (2002).

38. Saito, R., Dresselhaus, G. & Dresselhaus, M. S. Trigonal warping effect of carbon nanotubes. *Phys. Rev. B* **61**, 2981–2990 (2000).

39. Huang, L. *et al.* A Generalized method for evaluating the metallic-to-semiconducting ratio of separated single-walled carbon nanotubes by UV-vis-NIR characterization. *J. Phys. Chem. C* **114**, 12095–12098 (2010).

40. Jiang, H., Nasibulin, A. G., Brown, D. P. & Kauppinen, E. I. Unambiguous atomic structural determination of single-walled carbon nanotubes by electron diffraction. *Carbon* **45**, 662–667 (2007).

41. Tans, S. J., Verschueren, A. R. M. & Dekker, C. Room-temperature transistor based on a single carbon nanotube. *Nature* **393**, 49–52 (1998).

42. Martel, R., Schmidt, T., Shea, H. R., Hertel, T. & Avouris, P. Single- and multi-wall carbon nanotube field-effect transistors. *Appl. Phys. Lett.* **73**, 2447–2449 (1998).

43. LeMieux, M. C. *et al.* Self-sorted, aligned nanotube networks for thin-film transistors. *Science* **321**, 101–104 (2008).

44. Wang, C. *et al.* Wafer-scale fabrication of separated carbon nanotube thin-film transistors for display applications. *Nano Lett.* **9**, 4285–4291 (2009).

45. LeMieux, M. C. *et al.* Solution assembly of organized carbon nanotube networks for thin-film transistors. *ACS Nano* **3**, 4089–4097 (2009).

46. Arnold, M. S., Green, A. A., Hulvat, J. F., Stupp, S. I. & Hersam, M. C. Sorting carbon nanotubes by electronic structure using density differentiation. *Nat. Nanotechnol.* **1**, 60–65 (2006).

47. Cao, Q. *et al.* Medium-scale carbon nanotube thin-film integrated circuits on flexible plastic substrates. *Nature* **454**, 495–500 (2008).

48. Cao, X. *et al.* Screen printing as a scalable and low-cost approach for rigid and flexible thin-film transistors using separated carbon nanotubes. *ACS Nano* **8**, 12769–12776 (2014).

49. Chen, H. T., Cao, Y., Zhang, J. L. & Zhou, C. W. Large-scale complementary macroelectronics using hybrid integration of carbon nanotubes and IGZO thin-film transistors. *Nat. Commun.* **5**, 4097 (2014).

50. Liu, W. W., Aziz, A., Chai, S. -P., Mohamed, A. R. & Hashim, U. Synthesis of single-walled carbon nanotubes: effects of active metals, catalyst supports, and metal loading percentage. *J. Nanomater.* **2013**, 592464 (2013).

51. Liu, B. L. *et al.* Chirality-dependent reactivity of individual single-walled carbon nanotubes. *Small* **9**, 1379–1386 (2013).

52. Li, J. H. *et al.* Importance of diameter control on selective synthesis of semiconducting single-walled carbon nanotubes. *ACS Nano* **8**, 8564–8572 (2014).

53. Mun, J. H. *et al.* Monodisperse pattern nanoalloying for synergistic intermetallic catalysis. *Nano Lett.* **13**, 5720–5726 (2013).

54. Park, S., Kim, B., Wang, J. Y. & Russell, T. P. Fabrication of highly ordered silicon oxide dots and stripes from block copolymer thin films. *Adv. Mater.* **20**, 681–685 (2008).

55. Jiao, L. Y. *et al.* Creation of nanostructures with poly(methylmethacrylate)-mediated nanotransfer printing. *J. Am. Chem. Soc.* **130**, 12612–12613 (2008).

56. Jiao, L. Y., Xian, X. J. & Liu, Z. F. Manipulation of ultralong single-walled carbon nanotubes at macroscale. *J. Phys. Chem. C* **112**, 9963–9965 (2008).

57. Reina, A. *et al.* Transferring and identification of single- and few-layer graphene on arbitrary substrates. *J. Phys. Chem. C* **112**, 17741–17744 (2008).

58. Wang, Y. *et al.* Optimizing single-walled carbon nanotube films for applications in electroluminescent devices. *Adv. Mater.* **20**, 4442–4449 (2008).

Acknowledgements

We thank Professor Xiaohui Wang for help in 488 nm Raman measurements. We thank Mr Chao Shi and Mr Jian Luan for their contributions in sample preparation. We thank Professor Xiuliang Ma and Mr Zhibo Liu for useful discussion and help in TEM characterization. This work was supported by the Ministry of Science and Technology of China (Grants 2011CB932601), National Natural Science Foundation of China (Grants 51532008, 51572264, 51272256, 61422406, 51272257, 61574143 and 51372254), Chinese Academy of Sciences (Grant KGZD-EW-T06), CAS/SAFEA International Partnership Program for Creative Research Teams, Liaoning BaiQianWan Talents Program and EU-JST joint project IRENA.

Author contributions

C.L., P.H., F.Z. and H.M.C. designed the experiments. F.Z. carried out the catalyst preparation, SWCNTs growth and most characterizations. P.H., F.Z., C.L., J.L. and H.M.C. performed data analysis. B.W., M.C. and D.S. performed electrical property measurement. H.J. and E.K. performed electron diffraction experiment. C.L., P.H., F.Z., H.M.C., B. W., D.S., H.C., H.J. and E.K. wrote and revised the manuscript.

Additional information

25

Array of nanosheets render ultrafast and high-capacity Na-ion storage by tunable pseudocapacitance

Dongliang Chao[1], Changrong Zhu[1], Peihua Yang[1], Xinhui Xia[2], Jilei Liu[1], Jin Wang[3], Xiaofeng Fan[4], Serguei V. Savilov[5], Jianyi Lin[3], Hong Jin Fan[1] & Ze Xiang Shen[1,3]

Sodium-ion batteries are a potentially low-cost and safe alternative to the prevailing lithium-ion battery technology. However, it is a great challenge to achieve fast charging and high power density for most sodium-ion electrodes because of the sluggish sodiation kinetics. Here we demonstrate a high-capacity and high-rate sodium-ion anode based on ultrathin layered tin(II) sulfide nanostructures, in which a maximized extrinsic pseudocapacitance contribution is identified and verified by kinetics analysis. The graphene foam supported tin(II) sulfide nanoarray anode delivers a high reversible capacity of $\sim$1,100 mAh g^{-1} at 30 mA g^{-1} and $\sim$420 mAh g^{-1} at 30 A g^{-1}, which even outperforms its lithium-ion storage performance. The surface-dominated redox reaction rendered by our tailored ultrathin tin(II) sulfide nanostructures may also work in other layered materials for high-performance sodium-ion storage.

[1] School of Physical and Mathematical Sciences, Nanyang Technological University, Singapore 637371, Singapore. [2] State Key Laboratory of Silicon Materials, Department of Materials Science and Engineering, Zhejiang University, Hangzhou 310027, China. [3] Energy Research Institute @ NTU, Nanyang Technological University, Singapore 639798, Singapore. [4] College of Materials Science and Engineering, Jilin University, Changchun 130012, China. [5] Department of Chemistry, Moscow State University, Moscow 119992, Russia. Correspondence and requests for materials should be addressed to Z.X.S. (email: zexiang@ntu.edu.sg).

While open framework materials have proven effective in constructing kinetically favourable sodium (Na)-ion channels as cathodes for sodium-ion batteries (SIBs)[1–7], the sluggish Na-ion transport and severe volume expansion currently limit the rate performance and stability in most anode materials. The typical alloying materials (for example, Sn, Ge, Pb, Sb) possess high capacities ($Na_{15}Sn_4$: 847 mAh g^{-1}, Na_3Ge: 1,108 mAh g^{-1}, $Na_{15}Pb_4$: 484 mAh g^{-1}, Na_3Sb: 660 mAh g^{-1}) for Na storage but have severe volume expansion/contraction during the Na alloying/dealloying ($\sim 360 - 420\%$) (refs 8–10). To address this pulverization issue, one effective approach is to design integrated electrodes in which nanosized active materials are grafted to a secondary matrix[8,10]. Compared with tin (Sn) metal anodes, tin-based oxides and chalcogenides can store Na$^+$ through a combined electrochemical conversion and alloying mechanisms, giving rise to higher theoretical capacities (SnO_2: 1,378 mAh g^{-1}, SnS_2: 1,136 mAh g^{-1}, SnS: 1,022 mAh g^{-1}) (ref. 11). Sulfides are typically more reversible than oxides due to relatively weaker $M - S$ ionic bonds compared with M–O bonds, resulting in kinetically more favourable and higher first-cycle efficiency of tin chalcogenides[11]. Moreover, with merits of high electrical conductivity ($0.193 - 0.0083$ S cm^{-1}), higher capacity and earth abundancy, tin(II) sulfide (SnS) is considered to be a very promising anode material for SIBs. SnS has a smaller lattice expansion (242%) in the sodiation/desodiation process than SnS_2 (324%) (ref. 11). This correlates to a two-structure phase reaction in SnS (from orthorhombic-SnS to cubic-Sn to orthorhombic-$Na_{15}Sn_4$) compared with the three-structure transformation in SnS_2 (from hexagonal-SnS_2 to tetragonal-Sn to orthorhombic-$Na_{15}Sn_4$ (ref. 11). Despite the high capacities in Sn-based alloying materials, high rate capability and fast-charging have not yet been reported, to the best of our knowledge.

Compared with the diffusion-controlled process (insertion, conversion and alloying) in conventional Li/Na-ion storage battery materials, capacitive charge storage has the advantage of rendering high charging rate and therefore high power. In particular, pseudocapacitance refers to underpotential deposition, faradaic charge-transfer reactions including surface or near-surface redox reactions and bulk fast ion intercalation[12–16]. Pseudocapacitance can be intrinsic or extrinsic to a material[12]; intrinsic ones (such as RuO_2, MnO_2 and Nb_2O_5) display the capacitive characteristics for a wide range of particle sizes and morphologies, whereas extrinsic ones (such as $LiCoO_2$, MoO_2 and V_2O_5) emerge only when they are made into nanoscale dimensions to maximize reaction sites on the surface[14–17]. So far, enhanced pseudocapacitive contributions have been realized in some insertion and conversion Li-ion battery (LIB) and SIB materials with high-rate performance (see summary in Supplementary Table 1)[12,18–20]. However, it has not yet been implemented in alloying-type materials where the challenge is to realize high-capacity materials that can accommodate fast kinetics.

Here we demonstrate and prove by quantitative kinetics analysis, the pseudocapacitive contribution to the high capacity of Na-ion storage in few-layered SnS nanosheet arrays directly grown on a graphene foam (GF) backbone (Fig. 1). To the best of our knowledge, our purposely engineered SnS nanohoneycomb structure exhibits the highest reversible capacity, rate capability compared with the reported carbon allotrope, metal/alloy and metal oxides/sulfides as SIB anodes. Excitingly, due to the maximized pseudocapacitive contribution, its rate performance in Na-ion storage even exceeds that for Li ion. Our result may bring a paradigm shift in SIB anode materials to layered metal sulfides, and also afford deeper understanding as well as other nanoscale engineering strategies to boost the performance of SIBs.

Results

Structure and growth mechanism.

Figure 1a–c illustrates the fabrication procedure of the flexible GF-supported SnS electrodes by a rapid one-step in situ hot bath route (details are described in the Methods section). To demonstrate the morphology-dependent property, we obtained three types of samples with gradient morphologies that are denoted as nanowall (NW), nanoflakes (NF) and nanohoneycomb (NH). Herein, it is important to control the precursor concentration and nucleation rates to achieve desirable SnS nanostructure. Our synthesis leads to a full and uniform coverage of the substrate by SnS nanostructures, for which the size distribution can also be tailored (Supplementary Fig. 1). The SnS presents average lateral sizes of 400–500, 400–500 and 50–70 nm and thickness of ~ 150, 10 and 5 nm for NW, NF and NH, respectively. The growth of the SnS nanostructures is also proved substrate friendly (Ni foam, carbon cloth and ITO, see Supplementary Fig. 2). A growth mechanism is proposed based on the crucial roles of the substrate and ethanol, which involves (i) the hydrolysis of thioacetamide and (ii) the in situ metathesis reactions, self-assembly and oriented crystallization processes (details are provided in Supplementary Fig. 3 and Supplementary Note 1)[5]. The structure evolution and purity are verified by X-ray diffraction , energy-dispersive spectroscopy and Raman measurements (Supplementary Figs 4 and 5). The surface chemical-bonding state of GF-SnS electrode is also detected by X-ray photoelectron spectroscopy (XPS) and presented in Supplementary Fig. 6. The existence of C–S bonds is confirmed by spectra of both S 2p (163.7 eV) and C 1s (285.7 eV). XPS results suggest that the SnS might be chemically bonded with the GF matrix besides physical deposition[19] (details are provided in Supplementary Note 2).

Transmission electron microscopy (TEM) and high-resolution transmission electron microscopy (HRTEM) images in Fig. 2 further confirm the crystallographic orientation and unique nanosheet-on-microstructure three-dimensional (3D) porous nanowall, nanoflake and nanohoneycomb architectures. For SnS NW (Fig. 2a,b), the lattice-resolved HRTEM image shows interplanar spacing of 2.9 Å for the (101) planes of SnS. Further, the inset fast Fourier transform spots also reveal the existence of (101), (002) and (100) facets in the [010] zone axis, demonstrating that the layers of SnS are stacked along the [010] direction (see illustration in Fig. 1e). For SnS NF (Fig. 2c,d), the interconnected nanoflakes are regular with a periodic stacking of fringes (~ 15 layers) enforced by van der Waals interactions along the [010] direction. The layer distance is measured as ~ 6.2 Å, which is slightly larger than the layer-to-layer spacing in reported work[21]. The lateral view of the nanoflake in Fig. 2d also illustrates an interplanar spacing of 2.9 Å for the (101) planes of SnS with a crystal grain size ~ 20 nm. The results also confirm the formation of nanosheets by stacking of (010) facets. Similarly, SnS NH also presents intrinsic corrugations and lamellar structure in the nanosheets with interlayer spacing ~ 6.2 Å, but with thinner thickness (~ 6 layers). Interestingly, numerous tiny nanoclusters (~ 5 nm) and nanocavities (3–5 nm) are also observed from the cross-section of the wrinkles in Fig. 2f. Meanwhile, HRTEM and fast Fourier transform pattern reveals clear lattices with spacings of 3.4 and 2.8 Å for (120) and (040) planes, respectively, under [001] zone axis. A schematic illustration of the SnS laminar structure seen from [001] zone axis can be seen from Fig. 1d.

Na-ion storage performance.

Our designed battery electrodes allow electron and sodium-ion transfer through the GF–SnS network without the necessity of extra binders, conductive additives or metal Cu current collectors, which are essential for

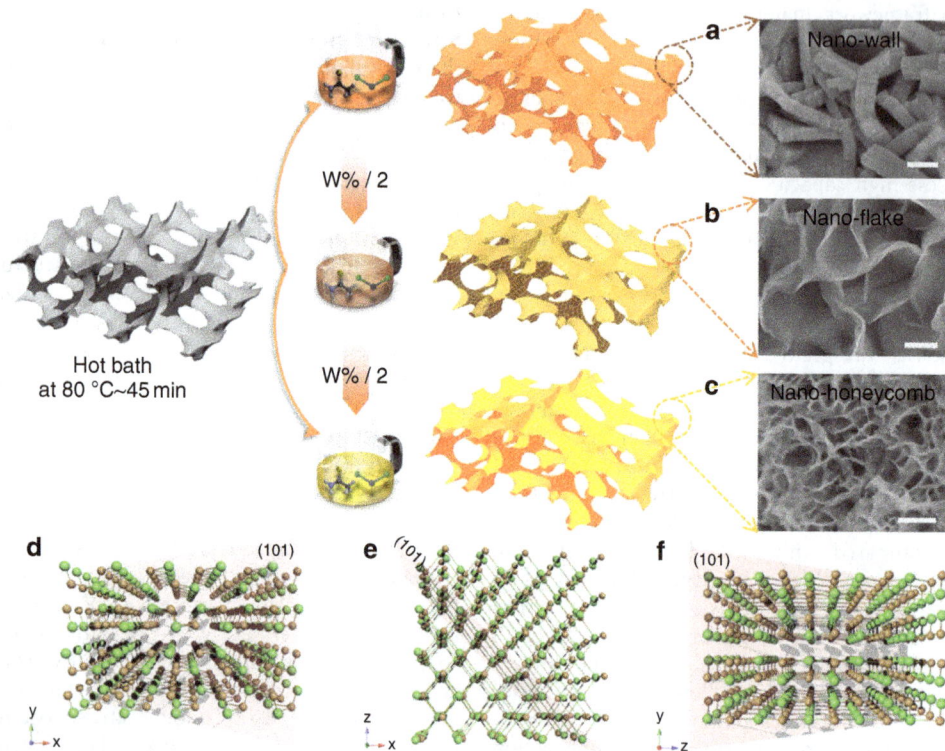

Figure 1 | Synthesis and structure of SnS nanostructures. (a–c) SnS nanostructures synthesized in different solution concentrations for (**a**) nano-wall; (**b**) nano-flake; (**c**) nano-honeycomb. Scale bar, 200 nm. (**d–f**) Schematic illustrations of the SnS laminar structure viewed along the [001], [010] and [100] zone axis with inserted (101) planes.

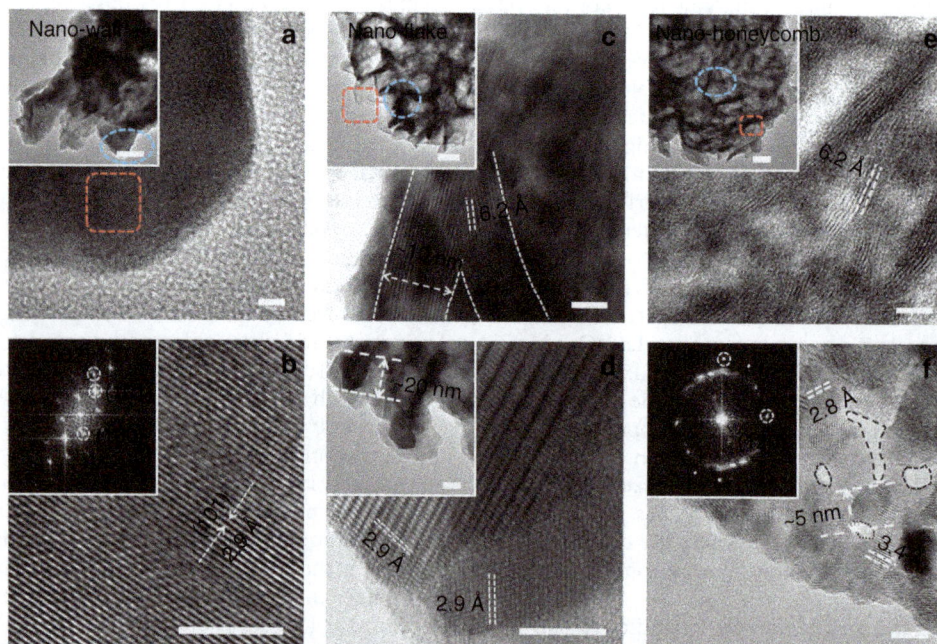

Figure 2 | TEM images of SnS nanostructures. (a,b) HRTEM images for the nanowall SnS structure. Scale bar, 5 nm. Inset a: low magnification TEM image (scale bar, 100 nm) and (**b**) fast Fourier transform (FFT) pattern along [010] zone axis. (**c,d**) HRTEM images for nanoflake SnS structure. Scale bar, 5 nm. Insets **c,d**: low magnification TEM images. Scale bars of 400 and 10 nm, respectively. (**e,f**) HRTEM images for nanohoneycomb SnS structure. Scale bar, 5 nm. Inset **e**: low magnification TEM image. Scale bar, 30 nm. Inset **f**: FFT pattern in the [001] zone axis. The dashed loops denote the mesopores.

exploring intrinsic sodium-storage properties of active material and increasing the energy/power densities of the full cell[5,22]. Herein, GF serves as both a lightweight 3D porous current collector (for electron transfer) and compressible/flexible backbone. The porous nanoarray feature prevents the

aggregation and expansion of SnS during charge/discharge cycles. During sodiation, electrolyte can enter the interval between nanoarrays on both outside and inner surface of GF, so that the Na ion and electrons can react with the SnS nanoarrays effectively.

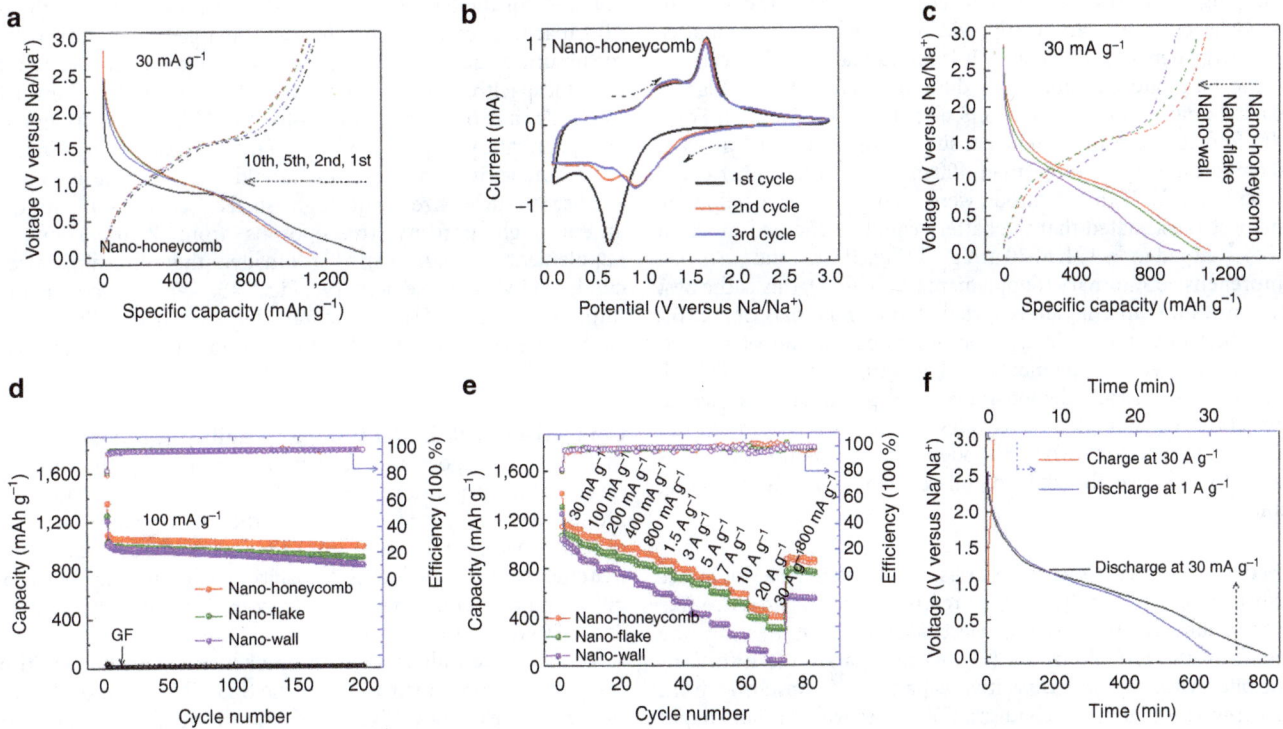

Figure 3 | Electrochemical Na⁺ storage performance of SnS electrodes. (a) Galvanostatic charge/discharge profiles during the first 10 cycles of the nanohoneycomb electrode. **(b)** CV curves of the first three cycles of the SnS nanohoneycomb electrode at a scan rate of $0.2\,mV\,s^{-1}$. **(c)** Galvanostatic charge/discharge profiles of SnS electrodes after five cycles activation. **(d)** Long-term cycling performances of SnS electrodes and the pure graphene foam electrode at a current density of $100\,mA\,g^{-1}$. **(e)** Rate performances of SnS electrodes at various current densities from 30 to $30,000\,mA\,g^{-1}$. **(f)** Fast charging (charge at $30\,A\,g^{-1}$ in 1 min, discharge at $30\,mA\,g^{-1}$ and $1\,A\,g^{-1}$ with $\sim 13\,h$ and 30 min, respectively) properties of the nanohoneycomb electrode.

The reaction of SnS with sodium undergoes two steps: conversion followed by alloying according to equations (1) and (2) (ref. 11):

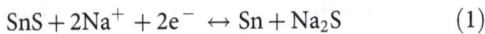

$$SnS + 2Na^+ + 2e^- \leftrightarrow Sn + Na_2S \qquad (1)$$

$$4Sn + 15Na^+ + 15e^- \leftrightarrow Na_{15}Sn_4. \qquad (2)$$

From the galvanostatic discharge–charge results (Fig. 3a and Supplementary Fig. 7), the first discharge presents a steep decrease and remains relatively flat in voltage. This may be due to the following two possible reasons: (1) deactivated surface and related high electrochemical polarization during the first sodiation; (2) transition from crystalline SnS to metal Sn nanograins and Na₂S (equation (1)) (refs 11,23). However, the subsequent curves after the first three cycles almost overlap, indicating a stable surface state, structure and electrochemical reversibility after the initial activation process. The detailed reaction processes could be disclosed by the cyclic voltammetric (CV) curves (Fig. 3b). In the first cathodic scan, two prominent peaks were observed at around $0.6-0.7$ and $0.01-0.1\,V$, respectively. The former is associated with both the conversion reaction (equation (1)) and the alloying reaction (equation (2), Na_xSn, $x \sim 0.75$) because it is difficult to distinguish the conversion and alloying peaks especially in the first discharge process[11,24]. Obviously this peak also includes the contribution from solid electrolyte interphase (SEI) formation since its intensity significantly decays in the subsequent cycles. The peak at $0.01-0.1\,V$ is regarded as the reaction between Na and Na_xSn ($x \sim 3.75$) alloy due to the multi-step Na–Sn alloying feature[11]. In the anodic scan, a serial of small peaks below $1.4\,V$ (~ 0.3, 0.7, $1.2-1.4\,V$) correspond to the multi-step dealloying reaction of Na_xSn (ref. 24); whereas the distinct peak at $\sim 1.7\,V$ could be attributed to the reversible conversion reaction from Sn to SnS, which was also observed in SnS@G[11], 3D porous

interconnected SnS (ref. 25), and proved by *ex situ* Raman results of the electrodes after charging and discharging processes (Supplementary Fig. 8).

The first sodiation and desodiation capacities of GF–SnS NH are 1,416 and $1,147\,mAh\,g^{-1}$, respectively, calculated based on active materials. A rather high first-cycle coulombic efficiency of $>80\%$ is observed. A reversible capacity more than $1,100\,mAh\,g^{-1}$ could be maintained in the following cycles. We note this value is higher than all reported ones from sodium alloys, oxides, sulfides and carbonaceous anodes so far (see comprehensive comparison in Supplementary Table 2), to the best of our knowledge. Note that the pure GF contributed negligibly to the capacity: $\sim 30\,mAh\,g^{-1}$ for the first discharge process and $<10\,mAh\,g^{-1}$ in the following cycles. Compared with NH electrode, NF presents a similar behaviour due to their similar microtopography, whereas the NW one exhibits a suppressed capacity and higher polarization. More importantly, the voltage profile in NW electrode shows more obvious plateaus, which is a typical feature of diffusion-controlled charge storage of battery materials[16]. The improvements in the coulombic efficiency and high reversible capacity of GF–SnS NH electrode should be correlated to the reversible formation/decomposition of the polymeric film on the surface of SnS NH (refs 11,26), its distinct disorder two-dimensional structure, ultrathin-layered mesoporous SnS nanocrystals, and as a result, its unique electrochemical mechanism (capacitive contribution), which is to be discussed below. The GF–SnS electrodes deliver excellent capacity retention from the third cycle onwards. After 200 cycles, the capacity retains at $1,010\,mAh\,g^{-1}$ for the NH electrode with well-preserved microstructure (Supplementary Fig. 9).

The rate capability is a crucial indicator for large scale application of batteries, such as regenerative braking and fast recharging of electric vehicles and cellphones. The drawback of low power becomes particularly evident in high capacity (that is, high energy density) materials[14]. It is found that the NH electrode has the best rate capability, in addition to consistently highest capacity, among the three GF–SnS electrodes (see Fig. 3e). For a 1,000-fold increase in current density (from 30 to 30 A g^{-1}), a discharge capacity of more than 400 mAh g^{-1} (in 1 min) could still be retained. If this electrode were used to power a cellphone system, it is estimated that the battery could be charged in 1 min and discharged in $\sim$13 h at 30 mA g^{-1} (Fig. 3f). On the basis of a comprehensive summary (Supplementary Table 2), this is the best rate capability among all reported anode materials for SIBs, to the best of our knowledge. The preliminary result of full-cell fabrication (see Supplementary Fig. 10, Na$_3$(VO)$_2$(PO$_4$)$_2$F cathode\\SnS anode) demonstrates the potential commercial application of our SnS electrodes to be considered as an anode material for SIBs, although further optimization is urgent to improve the skill in full-cell fabrication and cycling stability.

Kinetics and quantitative analysis. To explain the high-rate performance, we analysed the redox pseudocapacitance-like contribution in the GF–SnS electrodes by investigating the kinetics of the SnS electrodes (Fig. 4) to separate the diffusion-controlled capacity and capacitive capacity[15,27]. Resulting from the stepwise sodiation mechanism, CV curves with similar shapes at various scan rates from 0.2 to 0.8 mV s^{-1} (Supplementary Fig. 11) display two broad cathodic peaks as the scan rate increases. As cation intercalation reaction can be ruled out from our SnS electrode, we mainly consider the below three charge-storage mechanisms: the diffusion-controlled faradaic contribution from conversion and alloying reaction, the faradaic contribution from charge transfer with surface/subsurface atoms (that is, extrinsic pseudocapacitance effect), and the non-faradaic contribution from electrical double-layer effect[17,19,20].

The ratios of Na-ion capacitive contribution can be further quantitatively quantified by separating the current response i at a fixed potential V into capacitive effects (proportional to the scan rate v) and diffusion-controlled reactions ($k_2v^{1/2}$), according to Dunn[15,28]:

$$i(V) = k_1v + k_2v^{1/2} \qquad (3)$$

By determining both k_1 and k_2 constants, we can distinguish the fraction of the current from surface capacitance and Na$^+$ semi-infinite linear diffusion. Fig. 4a shows the typical voltage profile for the capacitive current (red region) in comparison with the total current. A dominating capacitive contribution ($\sim$84%) is

obtained for the NH electrode. As the scan rate increases, the role of capacitive contribution further enlarges (Fig. 4b) with a maximum value of $\sim$95% at 5 mV s^{-1}. By similar analysis, the pseudocapacitive contribution is found more than 80% for the NF electrode, but only around 60% for the NW one at 0.8 mV s^{-1} (Supplementary Figs 12 and 13). This is unsurprising since the pseudocapacitive contribution should play a critical role for smaller particle size with high surface area ($\sim$154 m^2 g^{-1}) and/or high porosity (mesoporous from 7 to 37 nm, see Supplementary Fig. 14)[15,18]. Finally, the thin-film electric conductivity (Supplementary Fig. 15) and electrochemical impedance spectra (Supplementary Fig. 16) suggest that the NH electrode have a favourable charge transfer kinetics compared with NW electrode.

Comparison with Li-ion storage capability. It is widely believed that Na$^+$ transport and storage are more sluggish with more severe lattice expansion than the Li$^+$ one because of the larger radius of Na ions[1,29–32]. So far, the performance in SIB is generally worse than that in LIBs when the same electrode material is used, including capacity, high rate capability and polarization. Herein, we carefully compare the performance of SnS NH anode for both Na$^+$ and Li$^+$ tests.

To avoid the influence on Li uptakes in GF backbone, in our comparison experiment we used the SnS NH grown on Ni foam as the same electrode. Figure 5 shows the results. Strikingly, one can see the rate capacity for sodiation/desodiation is superior to Li$^+$ uptakes, particularly in the high-rate regions. At current of 30 A g^{-1}, the electrode delivers a Na$^+$ discharge capacity of $\sim$410 mAh g^{-1} compared with $\sim$105 mAh g^{-1} for the Li$^+$ electrode. In the galvanostatic charge/discharge processes (Fig. 5b), the Li$^+$ electrode shows two distinct plateaus at 1.2 − 1.4 and 0.01 − 0.4 V versus Li/Li$^+$ corresponding to their CV curves (Supplementary Fig. 17a), suggesting a lower fraction of capacitive contribution. The first discharge plateau during lithiation is attributed to the conversion reaction from SnS → Sn and the second one is the alloying reaction-forming Li$_{15}$Sn$_4$ phase[33,34]. The polarization from 30 mA g^{-1} to 7 A g^{-1} during lithiation ($\sim$340 mV) is twice to that in sodiation ($\sim$170 mV). In addition, the capacitive fraction for Li$^+$ storage is 75% (inset in Fig. 5b), whereas 85% for the Na$^+$ storage of the same electrode (Supplementary Fig. 18). Similar higher Na$^+$ capacitive contribution than that of Li$^+$ had also been observed in Li$_4$Ti$_5$O$_{12}$ spinel thin film electrode but with much lower discharge capacity[20]. Finally, the sodiation discharge curves have more moderate and continuous operation voltage than the lithiation ones (Fig. 5b), which is favourable to achieving high energy density of full cells and avoiding dendrite growth[19].

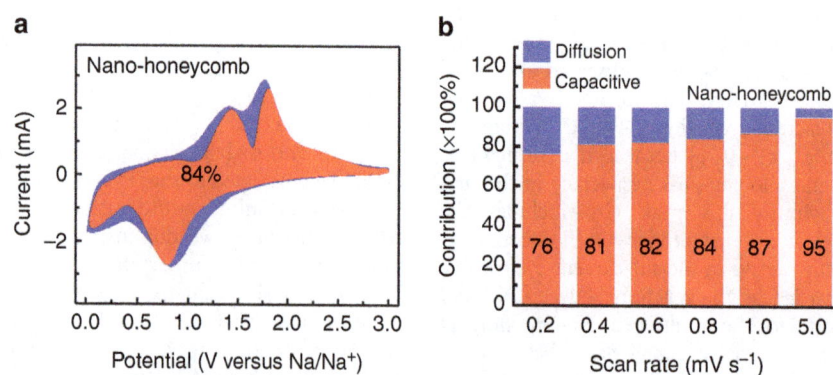

Figure 4 | Kinetics and quantitative analysis of the Na$^+$ storage mechanism. (**a**) Capacitive (red) and diffusion-controlled (blue) contribution to charge storage of nanohoneycomb at 0.8 mV s^{-1}. (**b**) Normalized contribution ratio of capacitive (red) and diffusion-controlled (blue) capacities at different scan rate.

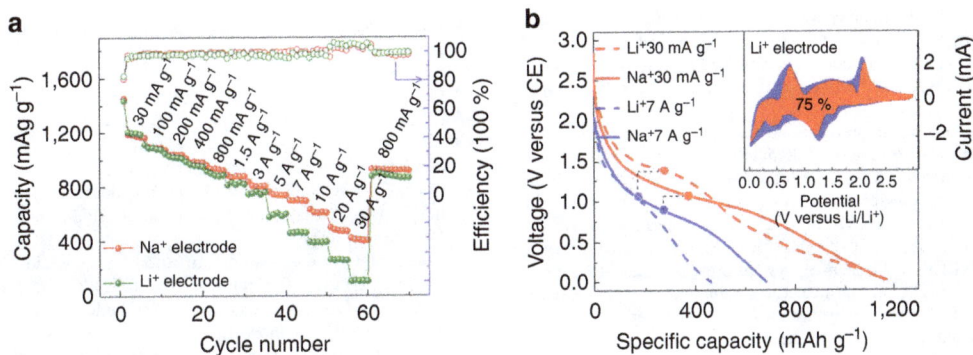

Figure 5 | Comparison between Na-ion and Li-ion storage. (**a**) Rate performance comparison of the nanohoneycomb electrode at various current densities from 30 to 30,000 mA g^{-1}. (**b**) Galvanostatic profiles of Na, Li electrodes at rates of 30 mA g^{-1} and 7 A g^{-1} after activation. Inset: capacitive (red) and diffusion-controlled (blue) contribution to charge storage at 0.8 mV s^{-1} during Li uptake. All of the batteries were tested in the same voltage range of 0.01–3 V versus Na/Na$^+$ and Li/Li$^+$ for Na-ion and Li-ion batteries, respectively.

Discussion

The results presented above demonstrate sodium-ion storage with both high capacity and high rate capability rendered by tunable extrinsic pseudocapacitance in our GF-supported SnS nanosheets. The electrode architecture provides, to the best of our knowledge, the highest reported reversible capacity of 1,100 mAh g^{-1} at 30 mA g^{-1}. Even at a high current density of 30 A g^{-1} (1,000-fold increase), the capacity is retained at 400 mAh g^{-1}, which is higher than that of Li$^+$ electrode ($\sim$105 mAh g^{-1} at 30 A g^{-1}). As the diffusion time of ions (t) is proportional to the square of the diffusion length (L), $t \approx L^2/D$, a short lithium diffusion time of $\sim$0.01 s is obtained on the basis of the ultrathin SnS architectures. As a consequence, similar to supercapacitors, the limiting factor for high rate charge/discharge is the transfer of ions and electrons to the surface of nanosheets rather than the conventional solid-state diffusion. As schematically shown in Fig. 6, the strongly solvating groups (carbonyl groups) in the organic electrolyte will arrange in an appropriate manner as a preferred solvation shell of the cations (Li$^+$ or Na$^+$) (refs 35–37), and it has been proven that sodium ion presents a weaker solvation shell with smaller de-solvation energy and lower activation barrier for sodiation transport compared with the lithium ion[35,38]. In additionally, higher mobility and conductivity of Na$^+$ solutions also contribute to the ion transfer in the electrolyte[39,40]. As a result, faster sodiation/desodiation kinetics are possible for SIBs as compared with LIBs[35,39–41].

The surface-dominated extrinsic pseudocapacitance is identified as a major energy-storage mechanism in favour of high capacity and fast Na$^+$ uptakes (Fig. 6). First, the chemically bonded GF–SnS hybrid demonstrates excellent structure stability and electronic/ionic conductivity through the network, which have also been shown to be a prerequisite for the extrinsic pseudocapacitance in nanosized MoO$_2$ (ref. 16). Moreover, the nanoscale dimension, especially thickness of the electrode materials, has been emphasized to be an important factor on the rate properties and corresponding redox capacitive contribution[20,27,42,43]. Compared with the SnS NW, the few layered architecture and mesoporous iso-oriented nanocrystals nature of the NH enables an interior Na$^+$ or electrolyte access into the van der Waals gaps of nanosheets, and thus results in both the exterior and interior parts participating in the electrochemical reaction[15]. This feature also facilitates ion access and shortens the ion diffusion.

Our kinetics analysis verifies the surface-dominated redox reaction mechanism in the Na-ion storage process, whereas the battery-type diffusion contribution is suppressed. This is closely correlated to the engineered thin-sheet structure of SnS NH, and

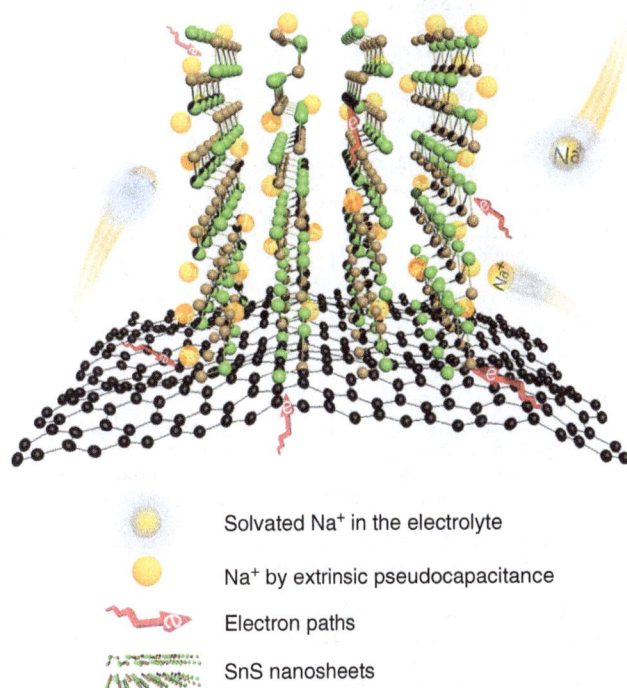

Figure 6 | Schematic illustration of high-rate charge storage of SnS architecture. During transfer of ions and electrons, the solvated Na$^+$/electron can easily enter into the open spaces between neighbouring ultrathin nanosheets on both the outside and inner surface of the graphene foam. After de-solvation of Na$^+$ on the surface of the layered SnS architecture, rapid sodiation takes place by the surface-dominated extrinsic pseudocapacitance.

may account for the superior Na$^+$ storage performance (high rate capacity) of the SnS NH to its Li$^+$ one. This strategy renders increased power density with maintained high energy density. This encouraging result may accelerate further development of SIBs by smart nanoengineering of the electrode materials.

Methods

Synthesis and characterization. SnS nanostructures were fabricated by a facile hot bath method. First, precursors with three sets of different concentrations, namely, high, medium and low concentrations of tin(II) chloride dehydrate: thioacetamide, respectively for nanowall (100:300 mM), nanoflake (50:150 mM) and nanohoneycomb (25:75 mM), were dissolved in 50 ml ethanol at 80 °C. Then 3D GFs(2 × 5 cm^2, ~0.8 mg cm^{-2}, prepared by chemical vapour deposition method according to our previous result[44]) or other type of substrates such as Ni foam, carbon cloth and ITO glass were immersed into the above reaction solutions and kept for 45 min. Finally, the samples were collected and rinsed with distilled water and ethanol in turn three times, and dried at 150 °C in vacuum to obtain 3D GF-supported SnS free-standing electrodes. The SnS loading was ~1.0 mg cm^{-2} for nanoflake and nanohoneycomb electrode, and 1.2 mg cm^{-2} for nanowall electrode.

The crystal structures of the samples were identified using X-ray diffraction (RigakuD/Max-2550 with Cu-Kα radiation). Raman spectra were obtained with a WITec-CRM200 Raman system (WITec, Germany) with a laser wavelength of 532 nm (2.33 eV). The morphologies of the samples were characterized by field emission scanning electron microscopy. The structures of the samples were investigated by HRTEM(JEOL JEM-2010F at 200 kV). The XPS measurements were performed by a VG ESCALAB 220i-XL system using a monochromatic Al Ka1 source (1,486.6 eV). The thin-film electric conductivity was measured by four-point probe sheet resistance. The surface area of the SnS electrode was determined by N$_2$ adsorption/desorption isotherms.

Electrochemical measurements. Standard CR2032-type coin cells were assembled in an argon-filled glove box (Mbraun, Germany) with the as-fabricated GF-supported SnS nanoarrays as the working electrode (without any binder or additives). For SIB fabrication, the metallic sodium foil as the counter-electrode, 1 M NaPF$_6$ in ethylene carbonate (EC)–diethyl carbonate (DEC)–fluoroethylene carbonate (FEC) (1:1:0.03 in volume) as the electrolyte, and glass fibre as the separator. For the LIB case, except for the metallic lithium foil as the counter-electrode and 1 M LiPF$_6$ as the solute, the other parameters are the same with the SIB fabrication. For full-cell testing, the cathode-active material was Na$_3$(VO)$_2$(PO$_4$)$_2$F nanoparticle. The one-time discharge/charge cycled SnS NH served as the anode. The weight ratio between anode and cathode active material was ~0.15:1. The specific capacity was calculated based on the mass of the cathode-active material. The CV measurements were carried out using CHI660 electrochemical workstation. Electrochemical impedance spectroscopy was recorded on Solartron 1470E, the amplitude of the sine perturbation signal was 5 mV, and the frequency was scanned from the highest (10 kHz) to the lowest (5 mHz). Galvanostatic charge discharge cycles were tested by Neware battery tester at different current densities at room temperature.

References

1. Larcher, D. & Tarascon, J. M. Towards greener and more sustainable batteries for electrical energy storage. *Nat. Chem.* **7**, 19–29 (2015).
2. Li, C. *et al.* An FeF3·0.5H$_2$O polytype: a microporous framework compound with intersecting tunnels for Li and Na batteries. *J. Am. Chem. Soc.* **135**, 11425–11428 (2013).
3. Yabuuchi, N. *et al.* P2-type Na(x)[Fe(1/2)Mn(1/2)]O2 made from earth-abundant elements for rechargeable Na batteries. *Nat. Mater.* **11**, 512–517 (2012).
4. Fang, Y. *et al.* Mesoporous amorphous FePO$_4$ nanospheres as high-performance cathode material for sodium-ion batteries. *Nano Lett.* **14**, 3539–3543 (2014).
5. Chao, D. *et al.* Graphene quantum dots coated VO$_2$ arrays for highly durable electrodes for Li and Na ion batteries. *Nano Lett.* **15**, 565–573 (2015).
6. Lim, S. Y. *et al.* Role of intermediate phase for stable cycling of Na$_7$V$_4$(P$_2$O$_7$)$_4$PO$_4$ in sodium ion battery. *Proc. Natl Acad. Sci. USA* **111**, 599–604 (2013).
7. Liu, H., Xu, J., Ma, C. & Meng, Y. S. A new O$_3$-type layered oxide cathode with high energy/power density for rechargeable Na batteries. *Chem. Commun.* **51**, 4693–4696 (2015).
8. Wu, L. *et al.* Electrochemical properties and morphological evolution of pitaya-like Sb@C microspheres as high-performance anode for sodium ion batteries. *J. Mater. Chem. A* **3**, 5708–5713 (2015).
9. Kohandehghan, A. *et al.* Activation with Li enables facile sodium storage in germanium. *Nano Lett.* **14**, 5873–5882 (2014).
10. Xie, X. *et al.* Sn@CNT nanopillars grown perpendicularly on carbon paper: a novel free-standing anode for sodium ion batteries. *Nano Energy* **13**, 208–217 (2015).
11. Zhou, T. *et al.* Enhanced sodium-ion battery performance by structural phase transition from two-dimensional hexagonal-SnS2 to orthorhombic-SnS. *ACS Nano* **8**, 8323–8333 (2014).
12. Augustyn, V., Simon, P. & Dunn, B. Pseudocapacitive oxide materials for high-rate electrochemical energy storage. *Energy Environ. Sci.* **7**, 1597–1614 (2014).
13. Augustyn, V. *et al.* High-rate electrochemical energy storage through Li+ intercalation pseudocapacitance. *Nat. Mater.* **12**, 518–522 (2013).
14. Simon, P., Gogotsi, Y. & Dunn, B. Where do batteries end and supercapacitors begin? *Science* **343**, 1210–1211 (2014).
15. Brezesinski, T., Wang, J., Tolbert, S. H. & Dunn, B. Ordered mesoporous alpha-MoO3 with iso-oriented nanocrystalline walls for thin-film pseudocapacitors. *Nat. Mater.* **9**, 146–151 (2010).
16. Kim, H. S., Cook, J. B., Tolbert, S. H. & Dunn, B. The development of pseudocapacitive properties in nanosized-MoO$_2$. *J. Electrochem. Soc.* **162**, A5083–A5090 (2015).
17. Li, S. *et al.* Surface capacitive contributions: towards high rate anode materials for sodium ion batteries. *Nano Energy* **12**, 224–230 (2015).
18. Chen, Z. *et al.* High-performance sodium-ion pseudocapacitors based on hierarchically porous nanowire composites. *ACS Nano* **6**, 4319–4327 (2012).
19. Chen, C. *et al.* Na($^+$) intercalation pseudocapacitance in graphene-coupled titanium oxide enabling ultra-fast sodium storage and long-term cycling. *Nat. Commun.* **6**, 6929 (2015).
20. Yu, P., Li, C. & Guo, X. Sodium storage and pseudocapacitive charge in textured Li$_4$Ti$_5$O$_{12}$ thin films. *J. Phys. Chem. C* **118**, 10616–10624 (2014).
21. Gou, X.-L., Chen, J. & Shen, P.-W. Synthesis, characterization and application of SnSx (x = 1, 2) nanoparticles. *Mater. Chem. Phys.* **93**, 557–566 (2005).
22. Sun, Y. *et al.* Direct atomic-scale confirmation of three-phase storage mechanism in Li$_4$Ti$_5$O$_{12}$ anodes for room-temperature sodium-ion batteries. *Nat. Commun.* **4**, 1870 (2013).
23. He, K. *et al.* Sodiation kinetics of metal oxide conversion electrodes: a comparative study with lithiation. *Nano Lett.* **15**, 5755–5763 (2015).
24. Ma, C. Z. *et al.* Investigating the energy storage mechanism of SnS$_2$-rGO composite anode for advanced Na-ion batteries. *Chem. Mater.* **27**, 5633–5640 (2015).
25. Zhu, C. B. *et al.* A general strategy to fabricate carbon-coated 3D porous interconnected metal sulfides: case study of SnS/C nanocomposite for high-performance lithium and sodium ion batteries. *Adv. Sci.* **2**, 1500200 (2015).
26. Luo, B. *et al.* Two dimensional graphene–SnS$_2$ hybrids with superior rate capability for lithium ion storage. *Energy Environ. Sci.* **5**, 5226–5230 (2012).
27. Muller, G. A., Cook, J. B., Kim, H. S., Tolbert, S. H. & Dunn, B. High performance pseudocapacitor based on 2D layered metal chalcogenide nanocrystals. *Nano Lett.* **15**, 1911–1917 (2015).
28. Bard, A. J. & Faulkner, L. R. *Electrochemical Method: Fundamentals and Applications* (John Wiley & Sons, 1980).
29. Kim, H. *et al.* Aqueous rechargeable Li and Na ion batteries. *Chem. Rev.* **114**, 11788–11827 (2014).
30. Kundu, D., Talaie, E., Duffort, V. & Nazar, L. F. The emerging chemistry of sodium ion batteries for electrochemical energy storage. *Angew. Chem. Int. Ed. Engl.* **54**, 3431–3448 (2015).
31. Xu, X. *et al.* In situ investigation of Li and Na ion transport with single nanowire electrochemical devices. *Nano Lett.* **15**, 3879–3884 (2015).
32. Wang, Y. *et al.* A zero-strain layered metal oxide as the negative electrode for long-life sodium-ion batteries. *Nat. Commun.* **4**, 2365 (2013).
33. Lu, J., Nan, C., Li, L., Peng, Q. & Li, Y. Flexible SnS nanobelts: facile synthesis, formation mechanism and application in Li-ion batteries. *Nano Res.* **6**, 55–64 (2012).
34. Tripathi, A. M. & Mitra, S. The influence of electrode structure on the performance of an SnS anode in Li-ion batteries: effect of the electrode particle, conductive support shape and additive. *RSC Adv.* **5**, 23671–23682 (2015).
35. Ponrouch, A. *et al.* Non-aqueous electrolytes for sodium-ion batteries. *J. Mater. Chem. A* **3**, 22–42 (2015).
36. Ponrouch, A., Marchante, E., Courty, M., Tarascon, J. M. & Palacin, M. R. In search of an optimized electrolyte for Na-ion batteries. *Energy Environ. Sci.* **5**, 8572–8583 (2012).
37. Abe, T., Fukuda, H., Iriyama, Y. & Ogumi, Z. Solvated Li-ion transfer at interface between graphite and electrolyte. *J. Electrochem. Soc.* **151**, A1120–A1123 (2004).
38. Jonsson, E. & Johansson, P. Modern battery electrolytes: ion-ion interactions in Li+/Na+ conductors from DFT calculations. *Phys. Chem. Chem. Phys.* **14**, 10774–10779 (2012).
39. Yabuuchi, N., Kubota, K., Dahbi, M. & Komaba, S. Research development on sodium-ion batteries. *Chem. Rev.* **114**, 11636–11682 (2014).
40. Zhang, Y., Narayanan, A., Mugele, F., Cohen Stuart, M. A. & Duits, M. H. G. Charge inversion and colloidal stability of carbon black in battery electrolyte solutions. *Colloids Surf. A* **489**, 461–468 (2013).
41. Ong, S. P. *et al.* Voltage, stability and diffusion barrier differences between sodium-ion and lithium-ion intercalation materials. *Energy Environ. Sci.* **4**, 3680–3688 (2011).
42. Come, J. *et al.* Electrochemical kinetics of nanostructured Nb$_2$O$_5$ electrodes. *J. Electrochem. Soc.* **161**, A718–A725 (2014).
43. McDowell, M. T. *et al.* In situ observation of divergent phase transformations in individual sulfide nanocrystals. *Nano Lett.* **15**, 1264–1271 (2015).
44. Chao, D. *et al.* A V$_2$O$_5$/conductive-polymer core/shell nanobelt array on three-dimensional graphite foam: a high-rate, ultrastable, and freestanding cathode for lithium-ion batteries. *Adv. Mater.* **26**, 5794–5800 (2014).

Acknowledgements

Z.X.S. acknowledge the financial supported by Ministry of Education, Tier 1 (Grant number: M4011424.110), Tier 2 (Grant number: M4020284.110); H.J.F. acknowledges the financial supported by MOE AcRF Tier 1 (RG104/14, RG98/15). We also acknowledge support from the Energy Research Institute @NTU (ERI@N). We thank Professor Bruce Dunn, University of California, Los Angeles for useful discussions.

Author contributions

D.L.C. and C.R.Z. conceived the experiment. D.L.C., C.R.Z., P.H.Y., X.H.X., J.L.L. and J.W. conducted the material synthesis, characterization and electrochemical measurement. X.F.F., S.V.S., J.Y.L., H.J.F. and Z.X.S. were involved in discussion on the kinetics and quantitative analysis. D.L.C., C.R.Z., H.J.F. and Z.X.S. wrote the manuscript.

Additional information

Competing financial interests: The authors declare no competing financial interests.

26

Discovery of abnormal lithium-storage sites in molybdenum dioxide electrodes

Jeong Kuk Shon[1,2], Hyo Sug Lee[1], Gwi Ok Park[2,3], Jeongbae Yoon[3], Eunjun Park[4], Gyeong Su Park[1], Soo Sung Kong[2], Mingshi Jin[5], Jae-Man Choi[1], Hyuk Chang[1], Seokgwang Doo[1], Ji Man Kim[2,3], Won-Sub Yoon[3], Chanho Pak[6,*], Hansu Kim[4] & Galen D. Stucky[7,*]

Developing electrode materials with high-energy densities is important for the development of lithium-ion batteries. Here, we demonstrate a mesoporous molybdenum dioxide material with abnormal lithium-storage sites, which exhibits a discharge capacity of 1,814 mAh g^{-1} for the first cycle, more than twice its theoretical value, and maintains its initial capacity after 50 cycles. Contrary to previous reports, we find that a mechanism for the high and reversible lithium-storage capacity of the mesoporous molybdenum dioxide electrode is not based on a conversion reaction. Insight into the electrochemical results, obtained by *in situ* X-ray absorption, scanning transmission electron microscopy analysis combined with electron energy loss spectroscopy and computational modelling indicates that the nanoscale pore engineering of this transition metal oxide enables an unexpected electrochemical mass storage reaction mechanism, and may provide a strategy for the design of cation storage materials for battery systems.

[1] Samsung Advanced Institute of Technology, Samsung Electronics Co., Ltd., Suwon 443-803, Republic of Korea. [2] Department of Chemistry, Sungkyunkwan University, Suwon 440-746, Republic of Korea. [3] Department of Energy Science, Sungkyunkwan University, Suwon 440-746, Republic of Korea. [4] Department of Energy Engineering, Hanyang University, Seoul 133-791, Republic of Korea. [5] Key Laboratory of Natural Resource of the Changbai Mountain and Functional Molecular (Yanbian University), Ministry of Education, Department of Chemistry, Park Road 977, Yanji City, Jilin Province 133002, China. [6] Fuel Cell Group, Corporate R&D Center, Samsung SDI Co., Ltd., Yongin 446-577, Republic of Korea. [7] Department of Chemistry and Biochemistry, University of California, Santa Barbara, California 93106, USA. * These authors jointly supervised this work. Correspondence and requests for materials should be addressed to J.M.K. (email: jimankim@skku.edu) or to W.-S.Y. (email: wsyoon@skku.edu) or to H.K. (email: khansu@hanyang.ac.kr).

Lithium (Li)-ion batteries (LIBs) are a key-enabling technology for addressing the power and energy demands of electric vehicles and stationary electrical storage for renewable energy as well as mobile electronics[1]. However, the energy density of currently commercialized LIBs is already close to its technological limit[2]. In order to achieve the battery performance that all applications expect, much effort has been made to develop new electrode materials to improve both the energy density and cycle performance of LIBs[3]. The main goal of these research efforts are to enable energy densities that are higher than the theoretical limit predicted for current Li-ion intercalation batteries. These new batteries would have the desired physicochemical properties, especially high reversible capacity, structural flexibility and stability, high rate capability, low cost and environmental benignity[4,5]. In order to improve the performance of anode parts replacing graphite (theoretical capacity of $372\,mAh\,g^{-1}$), there has been extensive research on developing anode materials such as transition metal oxides, silicon- or tin-based metal alloys, and related composite configurations[3-11]. Since they follow Li-storage mechanisms such as conversion and alloying reactions with Li, which are different from the Li intercalation reaction in the graphite anode, these newly developed anode materials show higher capacity than the current anode systems. In particular, there had been several reports that some transition metal oxides showed the reversible capacity exceeding their theoretical capacity based on the conversion reaction. This abnormal capacity of transition metal oxides has been explained by various mechanisms based on the interfacial reaction: (i) reversible formation/dissolution of organic film at the interface between the electrolyte and the transition metal oxide[9,12]; (ii) interfacial charge storage between metal nanocrystals and Li salts[13,14]; and (iii) generation of LiOH on the surface of metal oxide and subsequent reversible reaction between LiOH and Li[15]. Hence, even though the origins of extra capacity of transition metal oxides are different from each other, one of the effective approaches is nano-engineering of transition metal oxides for LIBs[16-18]. This approach has the potential to significantly increase the reversible capacity and the rate capability of anode materials by enlarging the interfacial area between the electrode and electrolyte. It is also expected that nanostructured transition metal oxides should show much improved cycle performance by providing the mechanical/structural integrity against huge changes in volume and crystal structure.

Mesoporous materials are excellent nanoscale-engineered candidates for numerous applications because of their high surface areas, tunable pore sizes, adjustable framework thickness and compositions, and diverse surface properties[19-23]. Recently, various novel mesoporous metal oxides have been widely investigated as electrode materials for LIBs[24-28]. These studies have opened up a possibility for the development of anode materials with significantly improved Li-storage performance. The mesostructure can be readily designed to exhibit hierarchical mesoporosity to promote facile and fast Li-ion diffusion, and a uniform framework thickness of 10 nm or less so that there is a reduced diffusion length for solid-state Li transport. Interestingly, we have found that the mesoporous MoO$_2$ anode presented here gives a high Li-storage capacity ($1,814\,mAh\,g^{-1}$ at first cycle and $1,607\,mAh\,g^{-1}$ after 50 cycles), which is much higher than its theoretical capacity based on the conversion reaction of MoO$_2$ with Li ($838\,mAh\,g^{-1}$). As aforementioned, several research groups have newly proposed Li-storage mechanisms[12-15,29]; however, this extra Li-storage mechanism of transition metal oxide is still unclear, suggesting that the high capacity of the mesoporous MoO$_2$ in the present work probably results from a different Li-storage mechanism than those previously reported[12-15].

Here we report a mechanism for the high and reversible Li-storage capacity of the mesoporous MoO$_2$ anode based on various physicochemical analyses and computational modelling. To the best of our knowledge, this is the first demonstration of a Li-storage oxide anode material that utilizes two different Li-storage mechanisms composed of both a Li-ion intercalation and a metallic Li storage. This enables a structural and electronic configuration that is, in a sense, a hybrid between the Li intercalation compound anode and the metallic Li anode, with the inherent safety features of the former and energy density approaching that of the latter. We believe that understanding the origin and mechanism of this excellent performance will enable important advances in the design and creation of high-energy storage devices.

Results

Li-storage characteristics of mesoporous MoO$_2$. The mesoporous MoO$_2$ material, which exhibits regular mesopores (18.2 nm in diameter), highly crystalline frameworks (~ 7 nm thickness) and high surface areas of $115\,m^2\,g^{-1}$, was successfully obtained from a mesoporous silica template (Fig. 1a–f and Supplementary Fig. 1). Electrochemical performances of the mesoporous MoO$_2$ electrode and a bulk MoO$_2$ (Aldrich, $S_{BET} = 0.23\,m^2\,g^{-1}$) electrode are shown in Fig. 1g,h. There is an obvious difference in the Li-storage behaviours of the mesoporous MoO$_2$ and bulk MoO$_2$ electrodes. For the bulk MoO$_2$ electrode, the discharge (Li insertion) and charge (Li removal) capacities are 385 and $183\,mAh\,g^{-1}$ (for the first cycle), respectively, indicating that 1.84 mol of Li per mol of MoO$_2$ can be stored into bulk MoO$_2$, and 0.87 mol of Li are reversibly released from the lithiated bulk MoO$_2$. On the other hand, the mesoporous MoO$_2$ delivered a reversible charge capacity of $1,308\,mAh\,g^{-1}$ (for the first cycle) with an initial coulombic efficiency of 72.1% (discharge capacity of $1,814\,mAh\,g^{-1}$), which is much higher than those of the bulk MoO$_2$ electrode and corresponds to the reversible removal of 6.24 mol of Li per mol of MoO$_2$ from the fully lithiated mesoporous MoO$_2$. More importantly, the reversible Li-storage capacity of mesoporous MoO$_2$ (6.24 mol of Li per mol of MoO$_2$) at the first cycle exceeds the theoretical limit of Li-storage capacity through conversion reaction of MoO$_2$ with Li (4 mol of Li per mol of MoO$_2$), suggesting that a new Li-storage mechanism should be introduced to explain this unexpected Li-storage performance.

As shown in Fig. 1i,j, Supplementary Figs 2 and 3 and Supplementary Table 1, the Li-storage capacities of the MoO$_2$ electrodes are linearly correlated with their surface areas, even though the shapes in the small-angle X-ray diffraction (XRD) patterns and N$_2$ sorption isotherms of mesoporous MoO$_2$ materials areas are very similar (Supplementary Figs 4 and 5). In the case of mesoporous MoO$_2$ with surface area of $39\,m^2\,g^{-1}$, the first and tenth reversible charge capacities are only $422\,mAh\,g^{-1}$ (discharge capacity of $1,022\,mAh\,g^{-1}$) and $814\,mAh\,g^{-1}$, respectively, which are much lower than those of the mesoporous MoO$_2$ with surface area of $115\,m^2\,g^{-1}$ ($1,308$ and $1,594\,mAh\,g^{-1}$, respectively). This result clearly indicates that the quality of the nanostructure of the mesoporous MoO$_2$ material is very crucial to achieve a high Li-storage capacity that exceeds the theoretical value. Differences in the synthesis of mesoporous MoO$_2$ materials with low and high surface areas are described in the Methods section. For more insight, scanning electron microscopic (SEM) images were obtained for the mesoporous MoO$_2$ materials with different surface areas (Supplementary Fig. 6). Interestingly, there are co-existing large particles in the cases of the low surface area materials (39 and $76\,m^2\,g^{-1}$; Supplementary Fig. 6c,e), whereas entire

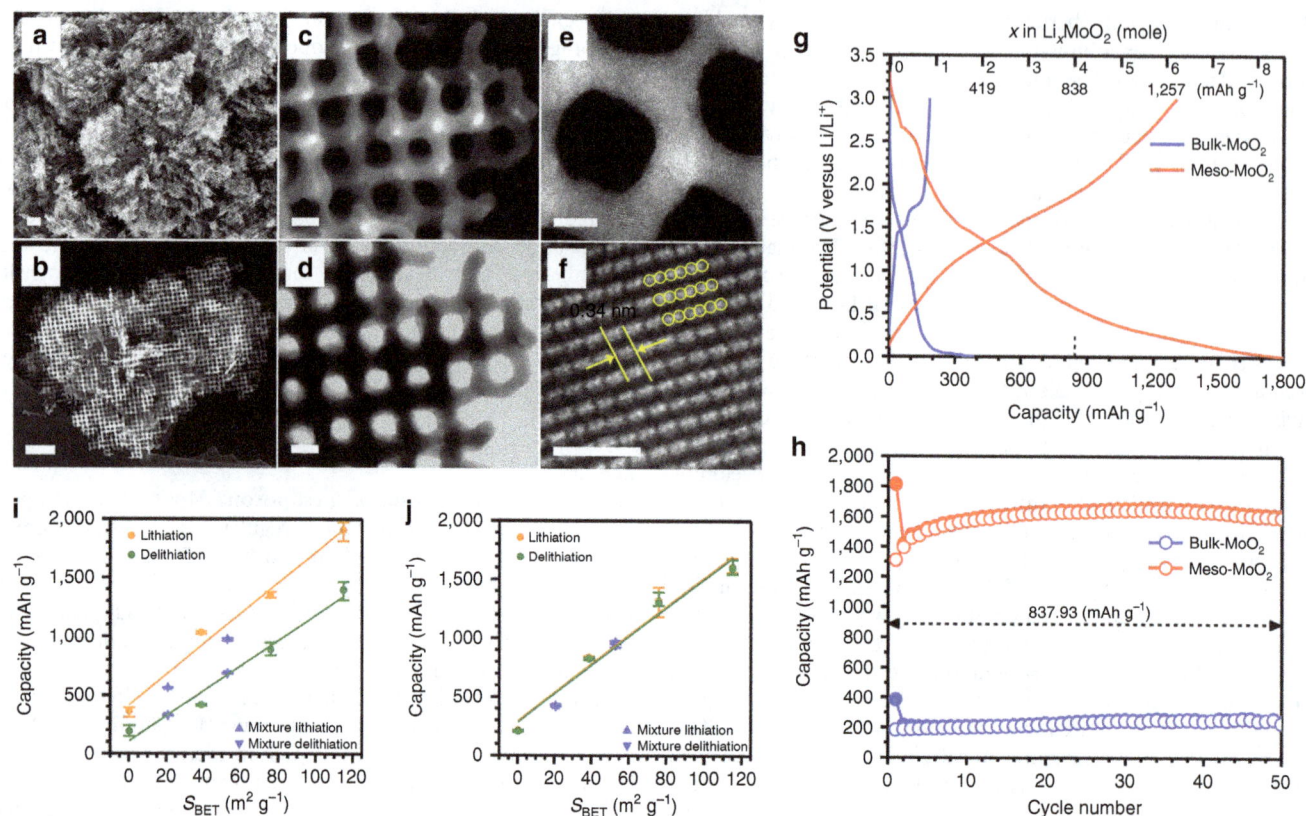

Figure 1 | Ordered mesoporous MoO₂ and Li-storage performance. (**a,b**) SEM, (**c,e**) STEM, (**d**) TEM and (**f**) HRTEM images of ordered mesoporous MoO₂ materials. Electrochemical performance of ordered mesoporous MoO₂ ($S_{BET} = 115\,m^2\,g^{-1}$) and bulk MoO₂ (Aldrich, $S_{BET} = 0.23\,m^2\,g^{-1}$): (**g**) voltage profiles and (**h**) cycle performances at current rate of 0.1 C in 1.3 M LiPF₆ (ethylene carbonate/diethyl carbonate (EC/DEC) = 3/7, by volume ratio). The value (837.93 mAh g⁻¹) represents the theoretical capacity for MoO₂, based on conversion reaction (4 mol of Li per 1 mol of MoO₂). Relationship of the Li-storage capacities of the MoO₂ electrodes with respective to their surface areas: (**i**) the first and (**j**) the tenth lithiation/delithiation. The galvanostatic lithiation/delithiation tests were replicated for three times. Error bars (**i,j**) represent the standard deviation of the mean about a series of the measured values. Straight line is a linear fit. Scale bars = 100 nm (**a,b**), 10 nm (**c,d**), 5 nm (**e**) and 1.5 nm (**f**).

particles of the mesoporous MoO₂ with the surface area of $115\,m^2\,g^{-1}$ exhibit highly ordered mesostructures (Fig. 1a and Supplementary Fig. 6f). For comparison, the Li-storage behaviours of two different mixtures of bulk MoO₂ and high-quality mesoporous MoO₂ materials (resulting in $S_{BET} = 21\,m^2\,g^{-1}$ and $53\,m^2\,g^{-1}$, respectively; Supplementary Figs 4, 5 and 6b,d and Supplementary Table 1) were also investigated, indicating that the mixtures give similar trends of Li-storage capacities depending on their surface areas.

In order to gain insight on the physics and material science of Li storage in the present ordered mesoporous MoO₂, we performed *ex situ* XRD analyses during the lithiation and delithiation of the ordered mesoporous MoO₂. Figure 2a shows that Bragg peaks corresponding to MoO₂ phase shifted to lower angles with an increase in the amount of Li stored, and returned to their initial scattering angles during the delithiation step, indicating highly reversible Li intercalation of the MoO₂ host without collapse of its crystalline structure during cycling. High-resolution transmission electron microscopic (HRTEM) images (Fig. 2b) also show reversible changes of crystalline structure during the first cycle. We also observed reversible shifts of the Bragg peaks of MoO₂ without the breakdown of the crystal structure even during the second cycle (Supplementary Fig. 7). It should be noted that we could not detect the evolution of any other phase related to the conversion reaction of MoO₂, such as metallic Mo and Li₂O, even after lithiation was electrochemically completed. These *ex situ* XRD analyses suggest that the present ordered mesoporous MoO₂ does not follow the conventional conversion reaction mechanism, in which there is

accommodation and release of Li ions by a Li intercalation mechanism. The XRD patterns obtained after the full lithiation (D − 0.0 V in Fig. 2a, >7 mol of Li per 1 mol of MoO₂) are very similar to the monoclinic Li$_x$MoO₂ phase, which means that the phase transition of crystal structures during the cycle is not the key for the unexpected high Li-storage capacity of the present mesoporous MoO₂ material. In the HRTEM images (Supplementary Fig. 8), we found that the highly crystalline framework of the pristine mesoporous MoO₂ material became a mixture of crystalline phase and amorphous phase upon the lithiation. This suggests that the amorphous phases formed between the crystalline domains may be critical for such high capacity of mesoporous MoO₂. *In situ* Mo *K*-edge X-ray absorption near edge structure (XANES; Fig. 2c and Supplementary Fig. 9) spectra for the ordered mesoporous MoO₂ electrode during lithiation show a clear shift of the absorption edge towards lower energy position, showing a change in the electronic structure of molybdenum from Mo⁴⁺ to a formal oxidation state close to Mo¹⁺. This result indicates formation of another domain (probably the amorphous phase shown in the HRTEM image; Supplementary Fig. 8), in addition to the Li-intercalated monoclinic Li$_x$MoO₂ phase during the lithiation. Furthermore, *in situ* Mo *K*-edge extended X-ray absorption fine structure (EXAFS; Fig. 2d) spectra for the ordered mesoporous MoO₂ show that the Mo–O and Mo–Mo peaks in the EXAFS spectra correspond to the coordination shells in MoO₂ structure, excluding any possibility of metallic Mo phase. Besides the extraordinarily high capacity, these contradictory XANES and EXAFS results against the conventional conversion

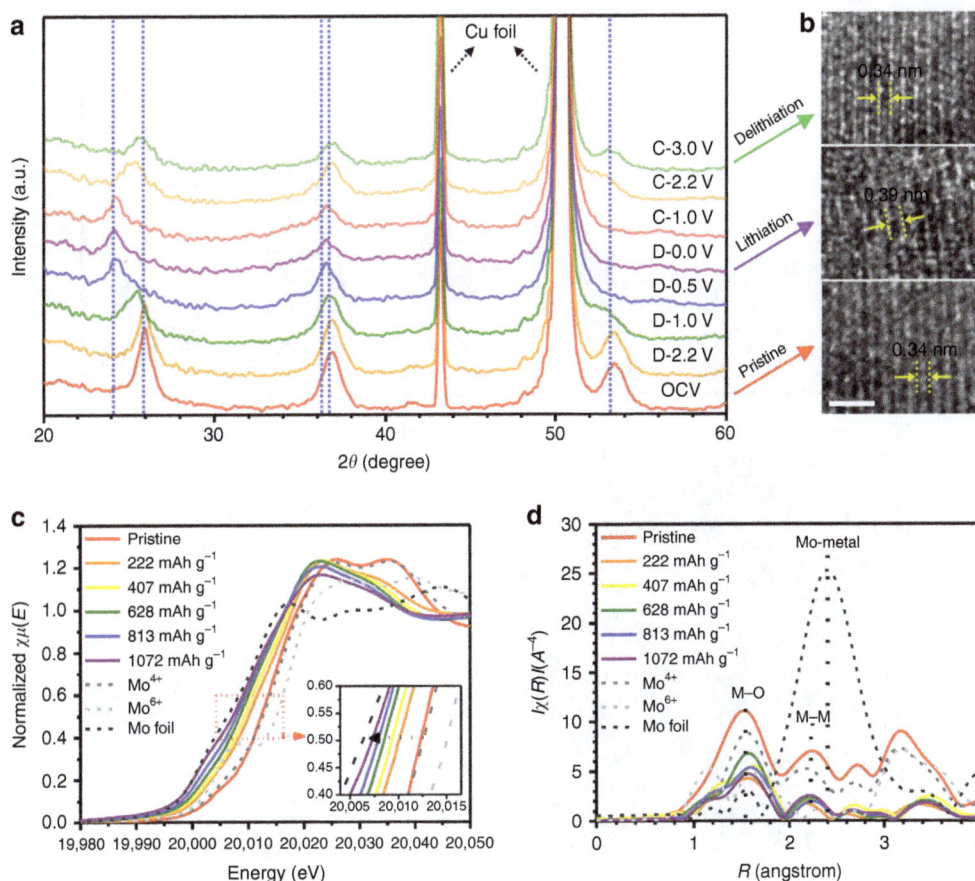

Figure 2 | Structure evolution of Li storage in ordered mesoporous MoO$_2$. (**a**) *Ex situ* XRD patterns of ordered mesoporous MoO$_2$ during the first cycle as a function of depth of discharge and charge. (**b**) HRTEM images of mesoporous MoO$_2$ before lithiation, after full lithiation and after full delithiation. Scale bar = 1.5 nm. (**c**) *In situ* XANES patterns and (**d**) *in situ* EXAFS patterns of ordered mesoporous MoO$_2$ electrode for the first lithiation.

reaction of transition metal oxide anode materials strongly suggest that another reaction model for Li storage into the ordered mesoporous MoO$_2$ should be considered in order to understand the origin of the extraordinary high Li-storage properties and the electrochemical reaction mechanism of the ordered mesoporous MoO$_2$ with Li.

Mechanism study on the Li-storage sites. In order to obtain a theoretical perspective into the atomic movement and redox behaviour that occur upon lithiation of MoO$_2$, we performed density functional theory (DFT) calculations on the Li-storage mechanism of the ordered mesoporous MoO$_2$. Figure 3a shows that the lithiation process of MoO$_2$ is almost the same for Li intercalation up to Li$_{1.5}$MoO$_2$ as evident from the small increase in unit cell volume ($\sim$20%) and the retention of initial crystal structure, that is, the monoclinic MoO$_2$ phase. However, further Li storage leads to phase separation of the lithiated MoO$_2$ into a Li-rich phase (the amorphous domain in Supplementary Fig. 8) and Li-intercalated MoO$_2$ crystalline phase (the crystalline domain in Supplementary Fig. 8). During the initial stage of this phase separation of lithiated MoO$_2$, we found that additional Li atoms are deposited as nearest neighbours to a position occupied by a former Li atom. This near-neighbour Li atom siting is more energetically favourable as a new position for a Li atom than is a random position (Supplementary Fig. 10). As shown in Fig. 3a, the further lithiation results in volume increase during cycling without expansion in the lithiated crystalline Li$_{1.5}$MoO$_2$ domain. This total volume expansion by formation of Li-rich phase is

probably an important characteristic that enables efficient Li storage in the ordered mesoporous MoO$_2$. With an increase in the depth of the lithiation layer, only the Li-rich phase (formed between crystalline Li$_x$MoO$_2$ domains) reacts with Li ion, and the interatomic distance between Li atoms tends to decrease, suggesting that the electronic structure of the Li atoms turns into a metallic state. As seen in Supplementary Fig. 11, the calculated partial density of state of Li s-band clearly shows the change of the electronic states of Li into metallic states when $x > 1.5$, which suggests that the metallic Li insertion between the crystalline Li$_x$MoO$_2$ domains causes the change in electronic structure.

It should be noted that this change in the electronic state of Li stored into a metal oxide differs from the interfacial Li-storage mechanism that has been used to explain the extra capacity of the transition metal oxide anode based on the conversion reaction mechanism[14]. According to the interfacial storage mechanism, extra Li ions are stored on the side of Li$_2$O at the interface between Li$_2$O and transition metal, which are formed through the fully conversion reaction of the transition metal oxide[14]. However, the Li-storage mechanism proposed in the present work for ordered mesoporous MoO$_2$ does not follow the conversion reaction of MoO$_2$ into Mo and Li$_2$O, but rather retains the crystalline phases (MoO$_2$ and Li$_x$MoO$_2$ during the cycle), with metallic Li accommodated between the Li-ion-intercalated MoO$_2$ crystalline domains. This unexpected Li-storage mechanism for the extraordinarily high capacity of ordered mesoporous MoO$_2$ can be attributed to the nanoscale pore-engineered structure, which offers a much higher surface area with a high metal oxide density. The thin MoO$_2$ frameworks

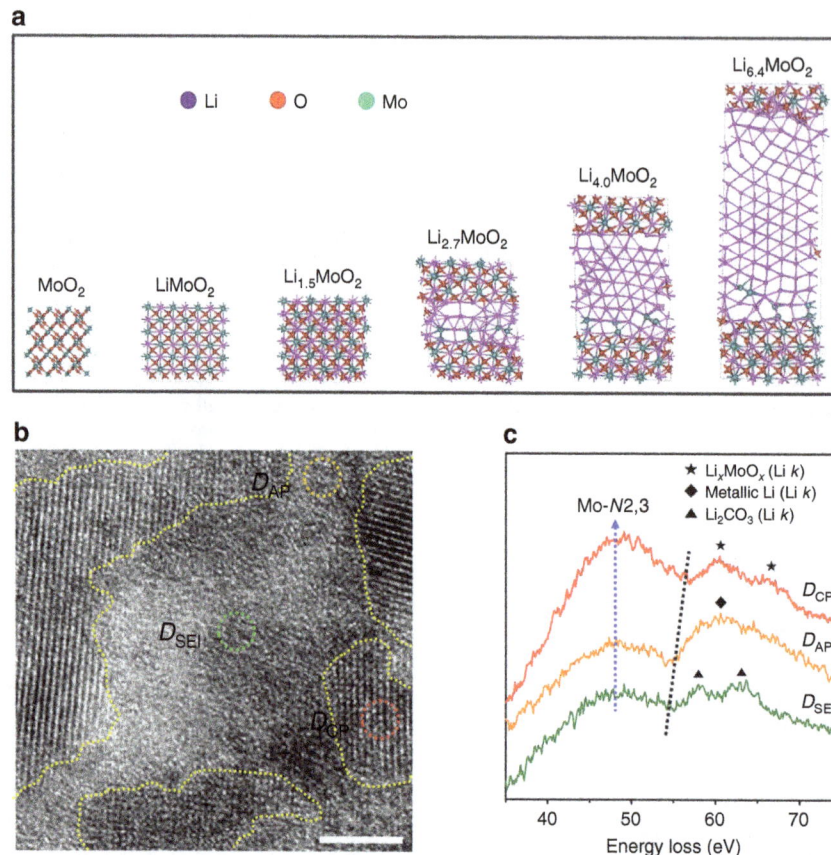

Figure 3 | DFT calculations and STEM-EELS spectra. (**a**) Snapshots of the ordered mesoporous MoO_2 electrode with the increase of Li inserted, calculated by DFT, and (**b**) top-view HRTEM image taken from a fully lithiated ordered mesoporous MoO_2, demonstrating the formation of three different phases: crystalline phase (D_{CP}), amorphous phase (D_{AP}) and SEI (D_{SEI}). Scale bar = 5 nm. (**c**) STEM-EELS spectra taken from areas D_{CP}, D_{AP} and D_{SEI} of the fully lithiated ordered mesoporous MoO_2. Dotted lines represent the peak position of Mo N edge (blue) and the onset of Li K-edge (black).

($\sim$7 nm) greatly shorten the diffusion length for the Li ions, thus enabling the electrochemical reaction that cannot occur in the bulk analogue.

In order to obtain additional direct experimental evidences for the two-phase reaction model proposed here (formation of both Li-ion-intercalated MoO_2 crystalline phase and metallic Li-rich phase) along with the change of the electronic state of Li, we performed HRTEM and electron energy loss spectroscopy (EELS) studies on the fully lithiated and delithiated MoO_2 electrode. The HRTEM image in Fig. 3b clearly shows that there are three different phases formed in the fully lithiated MoO_2, that is, crystalline phase (D_{CP}), amorphous phase (D_{AP}) and solid electrolyte interface (SEI) phase (D_{SEI}). The EELS data in scanning TEM (STEM) mode (STEM-EELS) taken at the crystalline area (D_{CP}) show that the Li K-edge corresponds to Li_xMoO_2 (ref. 30), and this is closely relevant to the phase observed by *ex situ* XRD analyses, that is, Li-ion-intercalated MoO_2 in Fig. 2a. However, STEM-EELS spectrum (Fig. 3c) taken at the amorphous area (D_{AP}) shows that energy loss near edge structures of Li species are different from those taken at other areas (D_{CP} and D_{SEI}) and corresponds to metallic Li[31], implying that some of Li ions are accumulated as a form of metallic Li at the interface between nanosized crystalline Li_xMoO_2 domains. This Li storage as a metallic cluster resembles Li-storage mechanism of disordered carbon in the microspaces located at the edges of carbon clusters[32–35].

In addition, as shown in Supplementary Fig. 12, the Z-contrast image and EELS spectra of O K-edge taken at the crystalline phase (D_{CP}) and amorphous phase (D_{AP}) strongly support this

finding because the area of amorphous phase (D_{AP}) shows darker contrast and lower oxygen intensity than the domain of the crystalline phase (D_{CP}). Another STEM-EELS result on the area of D_{SEI} clearly shows a peak for Li_2CO_3 (Supplementary Fig. 13), which is one of the main components of SEI formed on the anode in the LIBs. As expected, the delithiated mesoporous MoO_2 electrode does not contain any Li peaks, which suggests that it is reversibly returned to its initial state after the cycle (Supplementary Fig. 14). These EELS results strongly support the new reaction model of the ordered mesoporous MoO_2 described in the present work.

Cycle performance and rate capability. Recently, we have developed an *in operando* small-angle X-ray scattering (SAXS) technique to investigate the nanostructural changes of ordered mesoporous electrode materials[36]. As shown in Fig. 4a,b and Supplementary Fig. 15, there is no significant changes in the *in operando* SAXS data of mesoporous MoO_2 electrode during the lithiation and delithiation process ($\sim$20% reduction of peak intensity and $\sim$25% increase of net volume change). These small changes in nanostructural properties indicate that the present mesoporous MoO_2 electrode seems to follow the intercalation mechanism with additional Li-storage sites rather than the conventional conversion mechanism[36]. Figure 4c–e compares TEM images of the ordered mesoporous MoO_2 structure during the first cycle. After complete lithiation, the diameter of the MoO_2 frameworks increases from $\sim$7.8 to $\sim$11.8 nm. In addition to the diameter increase, there also should be a longitudinal volume

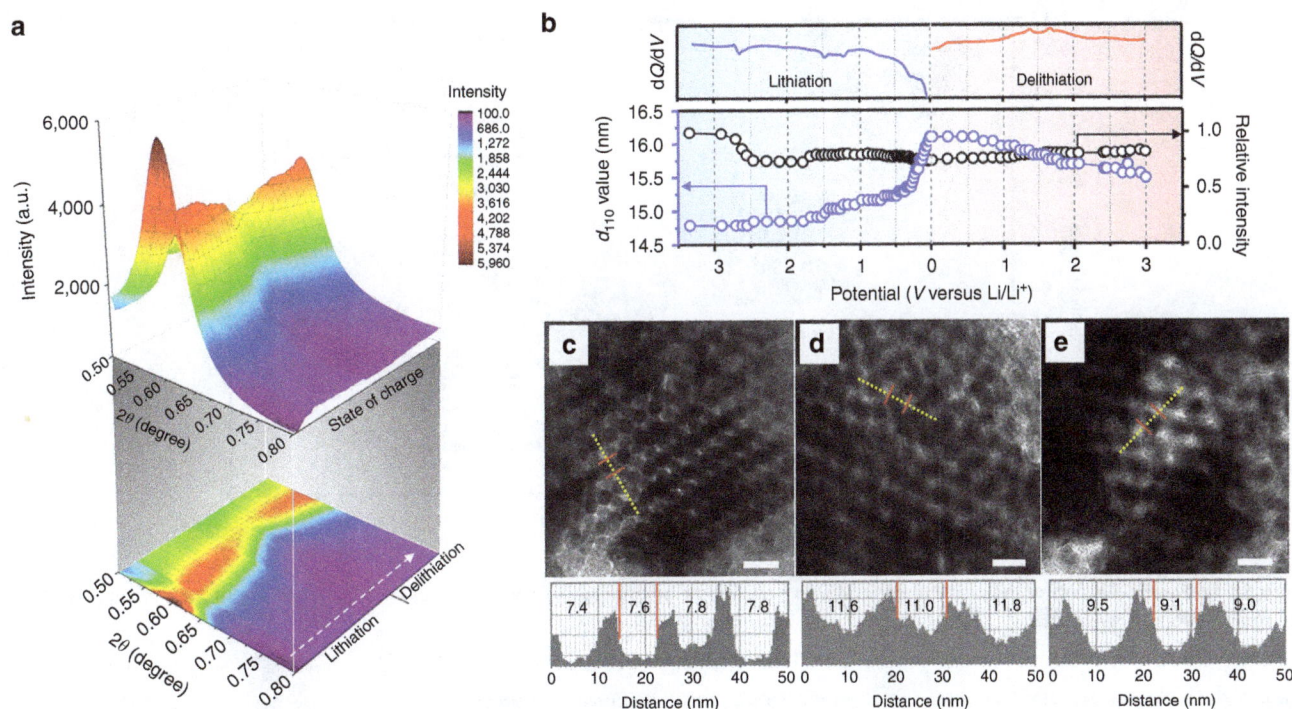

Figure 4 | Pore structure of mesoporous MoO₂ by *in operando* SAXS and *ex situ* TEM. (**a**) Color-coded 3D contour and projection map showing SAXS data collected from ordered mesoporous MoO₂ during *in operando* experiment. (**b**) The changes in lattice parameter and resolved peak relative intensity calculated from the (110) reflection with the corresponding dQ/dV plot. Representative TEM images and framework thickness of ordered mesoporous MoO₂ electrodes: (**c**) pristine, (**d**) lithiated and (**e**) delithiated. Scale bars = 20 nm.

expansion of MoO₂ frameworks upon the lithiation (not experimentally observable). As shown by the *ex situ* XRD patterns and the DFT calculation on the lithiated MoO₂ mesostructure (Figs 2a and 3a), the volume expansion can be attributed to the formation of both the Li-ion-intercalated crystalline phase and the amorphous metallic Li-rich phase during the lithiation. However, it is reasonable that the expansion associated with Li-ion intercalation is relatively smaller than that caused by the formation of the metallic Li-rich phase. This means that the volume expansion can be mainly attributed to the formation of the Li-rich phase, because the d_{110} spacing of crystalline MoO₂ in the XRD patterns and the HRTEM images of Fig. 2a,b expanded by 14.7% from the pristine state during the transformation from MoO₂ to Li$_x$MoO₂. It should be noted that, despite these volume changes during the first cycle, the ordered mesoporous MoO₂ retained its mesostructure even after complete lithiation and returned to the initial state after the complete delithiation (Fig. 4c–e). It is also important to note that this high capacity of ordered mesoporous MoO₂ electrode shows a significant improvement in capacity retention compared with the bulk material. As shown in Fig. 1h, the present mesoporous MoO₂ electrode retains 121.9% of the initial capacity after 50 cycles (from 1,308 to 1,594 mAh g^{-1}).

Cyclic voltammetry (CV) of the ordered mesoporous MoO₂ electrode reveals the same tendency of highly stable electrochemical storage and removal of Li up to 30 cycles (Supplementary Fig. 16). Li-storage capacity of the ordered mesoporous MoO₂ electrode shows the continuous increase with an increase of cycle numbers up to 20 cycles, and then reached the steady-state value (Supplementary Fig. 16b). Considering that this newly evolved couple of CV peaks could not be reported by other reports ever, the reduction (~0.2 V) and oxidation peak (~2.5 V), which can be also found at the differential capacity plots of the ordered mesoporous MoO₂ electrode (Supplementary Fig. 17), would be associated with Li-storage reaction mechanism

to account for high capacity of the ordered mesoporous MoO₂ electrode presented here. It should be also noted that the rectangular CV profiles took shape with an increase of CV cycle numbers, indicating that Li-storage reaction in the ordered mesoporous MoO₂ might resemble the electrochemical reaction observed in the supercapacitors rather than in the typical rechargeable batteries. Considering that under-potential deposition of metal ions is one of the typical super-capacitive electrochemical reactions, the rectangular shape of CV profiles observed in the mesoporous MoO₂ electrode might be a circumstantial evidence to prove the storage of Li as a metallic phase in the ordered mesoporous MoO₂ electrode.

Mo K-edge EXAFS spectra for MoO₂ during electrochemical cycles are shown in Supplementary Fig. 18. We found that Mo–O peak from oxide phase is still major contribution in EXAFS spectra during cycles, even though the peak intensities are gradually reduced due to the cycle-induced structural disorder. It is notable that Mo–Mo peak from Mo metallic phase does not appear even after 10 cycles, clearly confirming no possibility to have the conversion reaction in MoO₂ electrode.

Rate capability is also very important for electrode materials in addition to Li-storage capacity. We modified the mesostructure of the ordered mesoporous MoO₂ by reducing the wall thickness from 7.9 to 7.0 nm and 5.5 to 4.0 nm (but for similar mesopore size of ~18 nm, see Supplementary Figs 19–22 and Supplementary Tables 2 and 3), in order to examine the effects of framework thicknesses on the rate capability of mesoporous MoO₂. As shown in Supplementary Fig. 23, the rate capability was slightly enhanced by reducing the framework thickness of the mesoporous MoO₂ materials from 7.0 to 5.8 nm, which is probably by shortening the diffusion length for Li ion in the MoO₂ framework. However, neither further reduction (5.1 nm) nor increase (7.8 nm) gives a positive effect on the rate capability. Research is underway to find the optimized mesopore size and

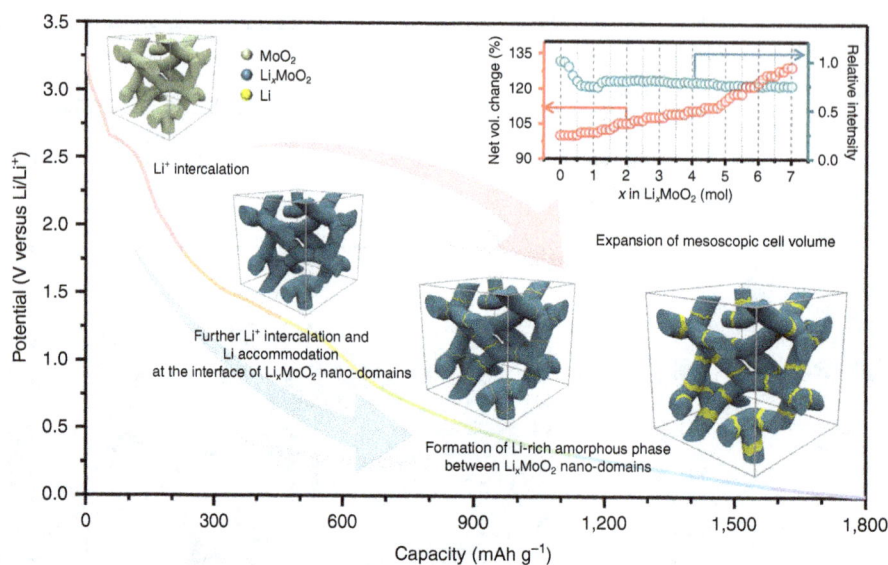

Figure 5 | Schematic diagram of the reaction pathways and mesoscale morphology. Schematic diagram of the reaction pathways and the resulting products of ordered mesoporous MoO_2 with respect to the amount of Li ion inserted.

framework thickness necessary to obtain the best electrochemical performance.

Discussion

We report a high-capacity Li-storage material, ordered mesoporous MoO_2, with a capacity that exceeds the storage capacity predicted by the conventional Li-storage mechanism such as Li intercalation and/or conversion reaction of transition metal oxide. We suggest a Li-storage mechanism consisting of a Li-ion intercalation reaction and the formation of a metallic Li-rich phase between the Li-ion-intercalated MoO_2 phase, based on first-principle calculations on the electronic structure of lithiated ordered mesoporous MoO_2, along with DFT calculations, *in situ* X-ray absorption spectroscopies and HRTEM analyses combined with EELS studies of the reacted ordered mesoporous MoO_2 (Fig. 5). In addition to the high capacity and excellent cycle performance of ordered mesoporous MoO_2, the proposed Li-storage mechanism clearly shows how the nanoscale engineering impacts the associated physicochemical properties of the material for electrochemical mass storage applications. The present results enable a structural and electronic configuration that is, in a sense, a hybrid between the Li-intercalated compound anode and the metallic Li anode, with the inherent safety features of the former and a high power density approaching to that of the latter. The results further provide a more complete understanding of possible Li-storage mechanisms for transition metal oxides, and thus make possible the further advancement of ultrahigh capacity anode materials for Li rechargeable batteries.

Methods
Mesoporous silica template. In the present work, mesoporous silica, KIT-6, with cubic $Ia3d$ mesostructure was used as the silica template for preparing mesoporous MoO_2 materials. Pluronic triblock copolymer P123 ($EO_{20}PO_{70}EO_{20}$, MW = 5,800) was utilized as the structure-directing agent for synthesis of KIT-6. Typically, 90.0 g of P123 was dissolved in a mixture of 3,255 g of distilled water, 90 g of 1-BuOH (99.7 wt%, Aldrich) and 177 g of c-HCl (35 wt%, Aldrich). After stirring at 35 °C for 10 min, 193.5 g of tetraethylorthosilicate (98 wt%, Aldrich) was added to this solution under vigorous stirring. The resulting mixture was stirred for 24 h at 35 °C and subsequently kept in static condition at 100 °C for 24 h in an oven. The solid product was filtered, washed with double distilled water and dried at 100 °C overnight. The white powder, thus obtained, was washed with EtOH, dried at 80 °C for 12 h, and finally calcined under static air conditions at 550 °C for 3 h in order to remove the structure-directing agent. Mesopore size of the KIT-6 template calculated by Barrett–Joyner–Halenda method was 7.1 nm. KIT-6 materials with

different pore sizes (5.1, 5.8 and 7.9 nm; see Supplementary Table 2 and Supplementary Fig. 19) were also synthesized using the above method, except for the hydrothermal treatment at 60, 80 and 140 °C, respectively.

Highly ordered mesoporous MoO_2 materials. The mesoporous MoO_2 materials were synthesized by the nano-replication method using the KIT-6 as the mesoporous silica templates. An amount of 6.49 g of ammonium molybdate tetrahydrate ($(NH_4)_6Mo_7O_{24} \bullet 4H_2O$, Aldrich) was dissolved in 8.4 g of distilled water. This precursor solution was infiltrated into 10.0 g of the calcined KIT-6 template by an incipient wetness method. After drying the composite at 80 °C for 24 h, the resulting material was heated to 500 °C under nitrogen atmosphere for 5 h for crystallization. After the heat treatment, the silica template was removed from the composite by a wet-etching process using a 20 wt% hydrofluoric acid (HF) solution. The resulting solid product was washed with distilled water and acetone several times, and then dried at 80 °C overnight in an oven. The mesoporous MoO_2 materials, synthesized from the KIT-6 with different pore sizes, are denoted as meso-$MoO_2 - x$ (where x means the synthesis temperature of KIT-6 templates). The mesoporous MoO_2 materials with surface area of 39 and 76 $m^2 g^{-1}$ were prepared by following the same synthesis method with the above procedure, except for the precursor infiltration conditions. The precursor solutions were prepared by dissolving 20.7 g of $(NH_4)_6Mo_7O_{24} \bullet 4H_2O$ in 15 g of distilled water and 8.11 g of $(NH_4)_6Mo_7O_{24} \bullet 4H_2O$ in 10.5 g of distilled water for the mesoporous MoO_2 with surface areas of 39 and 76 $m^2 g^{-1}$, respectively. Two mixed MoO_2 electrodes was simply prepared by physically mixing the bulk MoO_2 (Aldrich, $S_{BET} = 0.23 m^2 g^{-1}$) and the highly ordered mesoporous MoO_2 (this work, $S_{BET} = 115 m^2 g^{-1}$). The weight ratios between the bulk and mesoporous MoO_2 materials were 1:3 and 3:2 for the mixtures with surface areas of 21 $m^2 g^{-1}$ and 53 $m^2 g^{-1}$, respectively. The mixed MoO_2 materials were thoroughly hand-mixed with mortar and pestle.

Characterization. XRD patterns for the mesoporous silica and mesoporous MoO_2 powder were obtained using a Rigaku D/MAX-2200 Ultima equipped with Cu Kα radiation at 30 kV and 40 mA. *Ex situ* XRD patterns of electrodes during the cycles were recorded in reflection mode using an X'Pert PRO equipped with Cu Kα radiation at 40 kV and 40 mA in the 20–60° (2θ) angular range. *In situ* XANES and EXAFS spectra were collected on beamline 10 C at Pohang Accelerator Laboratory. Energy calibration was carried out using the first inflection point of the spectrum of Mo metal foil as a reference (that is, Mo K-edge = 20,000 eV). Reference spectra were simultaneously collected for each *in situ* spectrum using Mo metal foil. *In operando* SAXS experiments were carried out using BL 9 A U-SAXS beamline (Pohang Accelerator Laboratory) with a two-dimensional (2D) CCD detector (Rayonix SX165, USA) that was positioned 2 m away from the sample, which measured scattering in the 2θ range of 0.3–2.5 ($\lambda = 1.54$ Å). The size of the focused beam was 300 μm in diameter and the energy of the beam was 11 keV. The 2D patterns of the mesoporous samples during electrochemical cycling were recorded with a 1-s exposure time and 8 s detector readout time. The 2D patterns were scanned with the FIT2D software package to obtain the 1D patterns in the form of the intensity versus 2θ. The storage ring was operated at 3.0 GeV with a ring current of 300 mA. SEM images were taken using a LEO Supra 55 field-emission SEM operating at an accelerating voltage of 15 kV. Ultra-high-resolution SEM images were obtained using a Hitachi UHR S 5500 FE-SEM operating at 30 kV.

The main instruments we used in this study for microscopy were an aberration-corrected FEI TITAN field-emission TEM and a JEOL JEM 3010 TEM operated at an accelerating voltage of 300 kV. EELS analysis was performed in STEM mode using the field-emission TEM (TITAN) equipped with a high-resolution Gatan Tridieum 865ER imaging spectrometer. Because the MoO_2 electrodes during lithiation and delithiation were sensitive to air and moisture, all the sampling process was carried out in an argon-filled glove box. For TEM observations, all the specimens were moved to the instruments using argon-filled plastic tubes and carefully transferred to the instruments within 10 s in order to minimize the air-contacting time. N_2 adsorption–desorption isotherms were collected on a Micromeritics Tristar system at liquid N_2 temperature. All of the samples were completely dried under vacuum at 100 °C for 24 h before the measurement. The specific BET (Brunauer–Emmett–Teller) surface areas were calculated from the adsorption branches in the range of relative pressure $(p/p_0) = 0.05$–0.20. The pore size distribution curves were obtained by the Barrett–Joyner–Halenda method on the basis of the adsorption branches. Total pore volumes were measured at $p/p_0 = 0.99$.

Electrochemistry. Electrodes were prepared by coating the slurries that contain the mesoporous MoO_2 material as the active material powder, carbon black (Super-P, MMM) as the conducting agent and polyamideimide (Torlon 4,000 T, Solvay) as the binder in the weight ratio of 70:15:15 in N-methyl-2-pyrrolidone (Aldrich) solvent to produce an electrode slurry. The slurry was coated onto a Cu foil current collector using a Doctor-Blade technique. After the coating procedure, the electrodes were pressed and dried for 2 h at 200 °C under vacuum conditions. The electrodes thus obtained were cut into disks (12 mm in diameter). Coin-type cells (CR2016) were assembled in a dry room using Celgard 3,501 as the separator and Li foil as the counter and reference electrodes. On the basis of the current density of 0.1 C $(= 80\,mAh\,g^{-1})$, all cells were tested at within a fixed voltage window (0.001–3.0 V) using a battery cycle tester (TOSCAT 4,000 series, Tokyo, Japan).

Computational method. The first-principle calculations based on the DFT was performed using the VASP code. Initial structure of MoO_2 is the $3 \times 3 \times 1$ supercells including 36 Mo and 72 O atoms. The plane-wave cutoff energy was chosen to be 400 eV and appropriate k-points were chosen to ensure that the total energies are converged within a few meV. The exchange–correlation interactions between electrons were described by the generalized gradient approximation, and projector-augmented wave potentials were used for the description of ion–electron interactions. We used the conjugate gradient method for geometry optimizations, and the optimization procedure was truncated when the residual forces for the relaxed atoms were < 0.03 eV per Å. In the calculation of Li_xMoO_2, Li atoms were inserted to the position of the energetically most stable site. During the lithiation, the atomic position and cell structures are fully relaxed simultaneously.

References

1. Nishi, Y. The development of lithium ion secondary batteries. *Chem. Rec.* **1**, 406–413 (2001).
2. Tarascon, J.-M. & Armand, M. Issues and challenges facing rechargeable lithium battery. *Nature* **414**, 359–367 (2001).
3. Ellis, B. L., Lee, K. T. & Nazar, L. F. Positive electrode materials for Li-ion and Li-batteries. *Chem. Mater.* **22**, 691–714 (2010).
4. Jeong, G., Kim, Y.-U., Kim, H., Kim, Y.-J. & Sohn, H.-J. Prospective materials and applications for Li secondary batteries. *Energ. Environ. Sci.* **4**, 1986–2002 (2011).
5. Lee, K. T. & Cho, J. Roles of nanosize in lithium reactive nanomaterials for lithium ion batteries. *Nano Today* **6**, 28–41 (2011).
6. Huang, J. Y. *et al. In situ* observation of the electrochemical lithiation of a single SnO_2 nanowire electrode. *Science* **330**, 1515–1520 (2010).
7. Chan, C. K. *et al.* High-performance lithium battery anodes using silicon nanowires. *Nat. Nanotechnol.* **3**, 31–35 (2008).
8. Park, C.-M., Kim, J.-H., Kim, H. & Sohn, H.-J. Li-alloy based anode materials for Li secondary batteries. *Chem. Soc. Rev.* **39**, 3115–3141 (2010).
9. Poizot, P., Laruelle, S., Grugeon, S., Dupont, L. & Tarascon, J.-M. Nano-sized transition-metal oxides as negative-electrode materials for lithium-ion batteries. *Nature* **407**, 496–499 (2000).
10. Lee, Y. J. *et al.* Fabricating genetically engineered high-power lithium-ion batteries using multiple virus genes. *Science* **324**, 1051–1055 (2009).
11. Armand, M. & Tarascon, J.-M. Building better batteries. *Nature* **451**, 652–657 (2008).
12. Grugeon, S. *et al.* Particle size effects on the electrochemical performance of copoer oxides toward lithium. *J. Electochem. Soc.* **148**, A285–A292 (2001).
13. Li, H., Richter, G. & Maier, J. Reversible formation and decomposition of LiF clusters using transition metal fluorides an precursors and their application in rechargeable Li batteries. *Adv. Mater.* **15**, 736–739 (2003).
14. Jamnik, J. & Maier, J. Nanocrystallinity effects in lithium battery materials. *Phys. Chem. Chem. Phys.* **5**, 5215–5220 (2003).
15. Hu, Y.-Y. *et al.* Origin of additional capacities in metal oxide lithium-ion battery electrodes. *Nat. Mater.* **12**, 1130 (2013).
16. Cheng, F., Liang, J., Tao, Z. & Chen, J. Functional materials for rechargeable batteries. *Adv. Mater.* **23**, 1695–1715 (2011).
17. Bruce, P. G., Scrosati, B. & Tarascon, J.-M. Nanomaterials for rechargeable lithium batteries. *Angew. Chem. Int. Ed.* **47**, 2930–2946 (2008).
18. Liu, C., Li, F., Ma, L.-P. & Cheng, H.-M. Advanced materials for energy storage. *Adv. Mater.* **22**, E28–E62 (2010).
19. Kresge, C. T., Leonowicz, M. E., Roth, W. J., Vartuli, J. C. & Beck, J. S. Ordered mesoporous molecular sieves synthesized by a liquid-crystal template mechanism. *Nature* **359**, 710–712 (1992).
20. Zhao, D. *et al.* Triblock copolymer syntheses of mesoporous silica with periodic 50 to 300 angstrom pores. *Science* **279**, 548–552 (1998).
21. Joo, S. H. *et al.* Ordered nanoporous arrays of carbon supporting high dispersions of platinum nanoparticles. *Nature* **412**, 169–172 (2001).
22. Lee, H. I. *et al.* Spontaneous phase separation mediated synthesis of 3D mesoporous carbon with controllable cage and window size. *Adv. Mater.* **23**, 2357–2361 (2011).
23. Chang, H., Joo, S. H. & Pak, C. Synthesis and characterization of mesoporous carbon for fuel cell applications. *J. Mater. Chem.* **17**, 3078–3088 (2007).
24. Yue, W. *et al.* Syntheses, Li insertion and photoactivity of mesoporous crystalline TiO_2. *Adv. Func. Mater.* **19**, 2826–2833 (2009).
25. Ren, Y., Jardwick, L. J. & Bruce, P. G. Lithium intercalation into mesoporous anatase with an ordered 3D pore structure. *Adv. Mater.* **49**, 2570–2574 (2010).
26. Ren, Y., Armstrong, A. R., Jiao, F. & Bruce, P. G. Influence of size on the rate of mesoporous electrodes for lithium batteries. *J. Am. Chem. Soc.* **132**, 996–1004 (2010).
27. Shi, Y. *et al.* Ordered mesoporous metallic MoO_2 materials with highly reversible lithium storage capacity. *Nano Lett.* **9**, 4215–4220 (2009).
28. Sohn, J. K. *et al.* Nano-propping effect of residual silicas on reversible lithium storage over highly ordered mesoporous SnO_2 materials. *J. Mater. Chem.* **19**, 6727–6732 (2009).
29. Chen, C., Ding, N., Wang, L., Yu, Y. & Lieberwirth, I. Some new facts on electrochemical reaction mechanism for transition metal oxide electrodes. *J. Power Sources* **189**, 552–556 (2009).
30. Cosandey, F. in *Microscopy: Science, Technology, Applications and Education* (eds Méndez-Vilas, A. & Díaz, J.) Vol. 3, 1662–1666 (Formatex, 2010).
31. Hightower, A., Ahn, C. C., Fultz, B. & Rez, P. Electron energy-loss spectrometry on lithiated graphite. *Appl. Phys. Lett.* **77**, 238–240 (2000).
32. Mochida, I., Ku, C.-H. & Korai, Y. Anodic performance and insertion mechanism of hard carbons prepared from synthetic isotropic pitches. *Carbon N. Y.* **39**, 399–410 (2001).
33. Dahn, J. R., Zheng, T., Liu, Y. & Xue, J. S. Mechanisms for lithium insertion in carbonaceous materials. *Science* **270**, 590–593 (1995).
34. Mabuchi, A., Tokumitsu, K., Fujimoto, H. & Kasuh, T. Charge-discharge characteristics of the mesocarbon miocrobeads heat-treated at different temperatures. *J. Electrochem. Soc.* **142**, 1041–1046 (1995).
35. Mochida, I., Ku, C.-H., Yoon, S. H. & Korai, Y. Anodic performance and mechanism of mesophase-pitch-derived carbons in lithium ion batteries. *J. Power Sources* **75**, 214–222 (1998).
36. Park, G. O. *et al. In operando* monitoring of the pore dynamics in ordered mesoporous electrode materials by small angle X-ray scattering. *ACS Nano* **9**, 5470–5477 (2015).

Acknowledgements

This work was supported by Samsung Research Funding Center for Future Technology (SRFC-MA1401-03). We also thank partial supports obtained from the Energy Efficiency and Resources Core Technology Program of the Korea Institute of Energy Technology Evaluation and Planning (KETEP), which granted financial resource from the Ministry of Trade, Industry and Energy, Republic of Korea (no. 20132020000260), the research fund of Hanyang University (HY-2012-T), the US National Science Foundation (DMR 08–05148) and the National Research Foundation of Korea (NRF-2010-C1AAA001-2010-0029065 and the Mid-Career Researcher Program No. 2012R1A2A2A01010011).

Author contributions

J.M.K., W.-S.Y., C.P., H.K. and G.D.S. conceived, designed and coordinated the study. H.S.L. carried out theoretical calculations. J.K.S., G.O.P., J.Y., E.P., G.S.P., S.S.K., M.J., J.-M.C., H.C. and S.D. performed the experiment and acquired the data. J.M.K., H.S.L., J.K.S., W.-S.Y, C.P., H.K. and G.D.S. wrote the paper. All the authors participated in analysis of the experimental data and discussions of the results, as well as preparing the paper.

Additional information

27

Optical micromanipulation of nanoparticles and cells inside living zebrafish

Patrick Lie Johansen[1,*], Federico Fenaroli[1,*], Lasse Evensen[1], Gareth Griffiths[1] & Gerbrand Koster[1,†]

Regulation of biological processes is often based on physical interactions between cells and their microenvironment. To unravel how and where interactions occur, micromanipulation methods can be used that offer high-precision control over the duration, position and magnitude of interactions. However, lacking an *in vivo* system, micromanipulation has generally been done with cells *in vitro*, which may not reflect the complex *in vivo* situation inside multicellular organisms. Here using optical tweezers we demonstrate micromanipulation throughout the transparent zebrafish embryo. We show that different cells, as well as injected nanoparticles and bacteria can be trapped and that adhesion properties and membrane deformation of endothelium and macrophages can be analysed. This non-invasive micromanipulation inside a whole-organism gives direct insights into cell interactions that are not accessible using existing approaches. Potential applications include screening of nanoparticle-cell interactions for cancer therapy or tissue invasion studies in cancer and infection biology.

[1] Department of Biosciences, University of Oslo, Blindernveien 31, 0371 Oslo, Norway. * These authors contributed equally to this work. † Present address: Inven2 AS, Forskningsparken, Gaustadalléen 21, 0349 Oslo, Norway. Correspondence and requests for materials should be addressed to G.K. (email: gerbrand.koster@inven2.com).

Many cellular mechanisms depend on regulation of the physical contact between biological structures. A prominent example of structures rich in cellular interactions are the blood vessels where the lumen is surrounded by a layer of endothelial cells; these form a crucial barrier between the blood vessel lumen and the surrounding tissue. This barrier is important for many functions, such as preventing blood clotting, in inflammation, in formation of new blood vessels and in the control of blood pressure[1]. Endothelial cells can interact with different immune cells in the blood, such as neutrophils and macrophages; they can also interact with cancer cells that can pass through the endothelium in the process of extravasation, as well as with blood-borne pathogens[1].

As *in vivo* interactions are especially hard to characterize in detail since they are difficult to image and often occur at unpredictable positions and in short time windows, novel approaches are needed to increase the efficiency of 'catching' these interactions and to study them in a controlled manner. In addition, to properly understand why and how cellular interactions are established, we need analytical methods operating at the sub-micrometre scale of cells and on the time scale of bond formation.

Optical tweezers (OT)[2,3] are a highly sensitive and flexible micromanipulation tool that uses the force of light for non-invasive manipulation of nano- to micrometre sized particles. Moreover, by controlling the strength of the OT, interaction forces can be studied. Because of their versatility, OT have become a central tool for soft matter and life sciences in the last two decades[3]. Using this approach allows one to decipher when and where an interaction occurs, while parallel sensitive imaging and detection methods can be applied for detailed structural analysis. In order to achieve maximum control of the surrounding environment, and to simplify interpretation, a major part of OT experiments have been done *in vitro*, either in minimal reconstituted systems or with cells in culture. More recently, there have been developments towards optical trapping in the, technically more challenging, complex active environments inside living cells[4-7]. Ultimately, however, one would want to study interactions in a multicellular organism, where the interactions of interest occur deeper inside the organism, beneath layers of tissue. Towards this goal there has been one recent report describing optical trapping of erythrocytes in mouse[8]; however, these experiments were necessarily limited to red blood cells in thinner, superficial blood vessels in the ear.

Here we introduce the use of the zebrafish (ZF) larva for applying OT inside a living vertebrate. In the last years, the use of the ZF larva model has exploded in popularity for studying processes such as development and disease models[9]. An obvious advantage of using the ZF larva for OT is that it is optically transparent; moreover, many transgenic zebrafish lines are available with fluorescent cell types, such as macrophages and endothelial cells[10]. In a recent analysis of the biodistribution of different nanoparticles in the zebrafish we found that many of the nanoparticles bound tightly to the endothelium[11]. Here we take advantage of OT to follow and quantify these interactions. In another earlier project we have also followed the process of infection of *Mycobacterium marinum*, the organism of fish tuberculosis in the zebrafish and developed nanoparticle-based therapies against the disease[12].

Using a flexible optical tweezers-imaging system in combination with zebrafish lines, we demonstrate here that many different types of structures, such as microinjected nanoparticles or bacteria, or different cell types such as macrophages (fluorescent) or erythrocytes can be identified and manipulated; this allows their interactions to be determined inside the living zebrafish. We analyse the details of particle–endothelium

interactions and quantify 'stealth' properties of particles; such stealth particles are designed to avoid interactions with phagocytic cells, important for using nanoparticles against cancer[13]. Using multiplexing, we show that a number of traps can be activated to 'fish' out multiple nanoparticles from the blood stream simultaneously. We also demonstrate a procedure using OT where a region was first 'cleaned' of erythrocytes after which the interaction details of nanoparticles with the endothelium could be studied in detail in a cell-free environment, thereby simplifying the interpretation. This experiment reveals the formation and detection of tethering nanotubes that are formed when adhered particles are pulled away from endothelial cells, providing convincing evidence that the particles were indeed tightly bound to the cell plasma membrane. Collectively, these data establish the zebrafish larva as a powerful model for optical trap micromanipulation and for analysis of *in vivo* interactions under controlled experimental conditions.

Results and Discussion

Optical trapping of nanoparticles inside living zebrafish. The optical transparency of the thin zebrafish larvae (Supplementary Movie 1; Supplementary Fig. 1) is unique among *in vivo* verte-brate models and one of the main reasons for its immense popularity as a model system. With this in mind, we tested whether the zebrafish was also 'optically see-through' for the infrared laser beam of the optical tweezers and whether this approach could be used to manipulate different structures inside the living larva.

One recent emerging application of zebrafish as a model system is in the nanomedicine field[12]. For this, nanoparticles loaded with drugs and decorated on their surface with (potentially) targeting factors can be injected, enabling their biodistribution and blood circulation to be monitored using live imaging. To determine whether such injected particles can be trapped inside the complex environment of the fish, we injected latex particles into 2-day-old fish larvae (according to a recently developed protocol[12], see also Methods, Supplementary Movie 2 and Supplementary Fig. 2). The particles readily distributed throughout the circulation of the blood stream and with time an increasing portion of the nanoparticles either adhered to the endothelium lining of the blood vessels or were taken up by macrophages.

After mounting the fish in the optical tweezers microscope (Supplementary Fig 3), we used transmission microscopy to select particles of interest that had adhered to the endothelium lining the caudal vein (Fig. 1a). Next, the OT were turned on and we were able to carefully move away some particles off the endothelium (Fig. 1a at 3.5 s, see also Supplementary Movie 3). Subsequently, the particles could be displaced and also moved against the direction of fast blood flow ($200 \, \mu m \, s^{-1}$ in the vein and $\sim 700 \, \mu m \, s^{-1}$ in the artery[14]) indicating strong trapping. In the example shown (Fig. 1a), at time point 5.4 s an erythrocyte hits the optical trap, thereby dislodging the particle from the trap. Intriguingly, the dislodged particle was subsequently pulled back, spring-like, to the original adhesion point. This indicated that the particle was connected through a nanotube (as described previously[15]), which exerted a pulling force on the particle, that re-incorporated the tether into the endothelium after retraction. These manipulations could be done at a trapping laser power settings of 500 mW, about 10% of the maximum force available (corresponding to about max 75 mW in the sample). This demonstrates that statically adhering particles can be trapped in the zebrafish, making possible the investigating of details of interaction that are not possible with other methods.

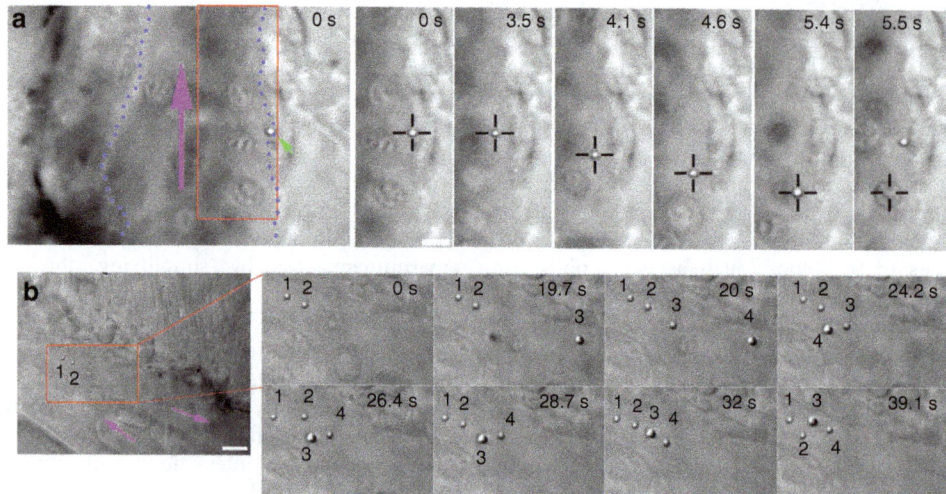

Figure 1 | *In vivo* optical micromanipulation of microinjected particles. (a) A particle (green arrowhead) adhered to the endothelium of the caudal vein (indicated with blue dotted lines) is pulled away from the endothelium into the fast blood flow (purple arrow) using optical tweezers (black crosshairs). At time 5.5 s an erythrocyte is drawn into the trap. This replaces the particle in the trap which is subsequently pulled back towards the original adhesion point of the endothelium, presumably due to a connecting nanotube. Experiment repeated at least 80 times. **(b)** Four separate particles (numbered) are fished out of the blood flow and moved towards a sheltered region at the tip of the tail. Purple arrows indicate flow direction. Experiment was repeated at least 10 times. Scale bar, 5 μm.

After having established that an already adhered particle could be trapped, we next investigated whether it was possible to catch particles that were injected into the blood stream and moving with the flow at high speed[14]. To do this, we made use of the available time-sharing multiplexing option of the optical tweezers system; by scanning the trapping laser at high speed over several positions, multiple traps can be created that can be used to trap in parallel a high number of objects[16]. We thus distributed several 'fishing' traps throughout the blood stream. In Fig. 1b (Supplementary Movie 4) using two traps, two particles (marked with '1' and '2') had already been moved towards the tail to a region with lower flow velocity outside the main blood flow (purple arrows). Next, another particle ('3') was caught (19.7 s) and moved together with the two other particles while another particle ('4') was trapped in the flow at $t = 20$ s. To demonstrate the positioning control that the multiplexed tweezers allow *in vivo*, the particles were positioned first in a straight line (32 s) and next repositioned into a square shape (39.1 s). The trapping of multiple particles within the blood flow and the subsequent reorganization of their relative positions demonstrates the robustness of the trapping, and the versatility of experiments that are possible; for example, by establishing simultaneous contact between several particles and specific cells (Supplementary Movie 5; Supplementary Fig. 4).

Trapping was possible throughout the fish and up to depths of 100 μm away from the bottom cover glass (including zebrafish and medium, Supplementary Fig. 3) and we for example trapped cells and particles inside the beating heart. However, we found that most straightforward trapping could be done in the caudal vein and artery in the (thinner) tail region of the fish, at a depth of about 50 μm, where the arterial blood flow turns from retrograde transport towards the tail to anterograde transport towards the head of the fish. The experiments demonstrate that the adhesive properties of particles can be tested inside the living fish using micromanipulation, even in the fast flowing blood in the caudal artery. We successfully trapped and moved polystyrene particles of 200 nm to 1 μm diameter, but conducted most experiments with particles of 840 nm diameter (see Methods section, fluorescence emission peak at ∼700 nm) or 1 μm (non-fluorescent) because of their clear visibility and higher

trapping stiffness[17]. Slowing down the blood flow using the anaesthetic tricaine[18] facilitates optical trapping, and allows also smaller particles down to 200 nm to be moved, although this was more difficult due to the lower trapping stiffness and the fact that smaller particles were difficult to discriminate using bright field transmission microscopy. However, using confocal fluorescence microscopy the particles could be identified. The larger particles could be trapped without obvious heating damage for extended periods of time inside the vasculature with laser powers 100 mW to 3 W (about 15 mW to ∼450 mW in the sample). However, a few scattered darker areas (possibly pigments that had developed despite use of phenylthiourea (PTU), see Methods) interacted strongly with the OT, and when the OT was moved transiently over these areas they became visibly damaged, presumably due to heating. However, by avoiding these areas no discernable damage was observed over the course of the experiments.

Trapping of cells in living zebrafish. Traditionally for experiments with optical tweezers spherical particles with high refractive index are used because they can be trapped with a high stiffness and they are especially suitable for trap calibration for quantitative experiments. However, for other applications, spatial micromanipulation is the crucial feature, as objects such as cells, particles or other nano–micron scale objects can be brought into contact with each other at well-defined time points and positions, opening up the possibility for analysing dynamics not accessible through passive observation.

As we found that particles can be easily and robustly trapped throughout the whole zebrafish, we next asked whether cells could also be micromanipulated inside the fish. In the absence of particles and using higher laser powers (about 2 W system setting, ∼250 mW in the sample), we activated the optical trap in the middle of the caudal vein, which almost instantly resulted in the immobilization of an erythrocyte in the trap. These cells could be stably held in position against the full flow of the blood stream (Fig. 2a; Supplementary Movie 6) and could even be moved in the direction against the blood flow. We found that the nucleus of the zebrafish erythrocytes[19] was trapped most strongly. In Fig. 2a, a membrane in a 'bag'-like shape can be observed behind the

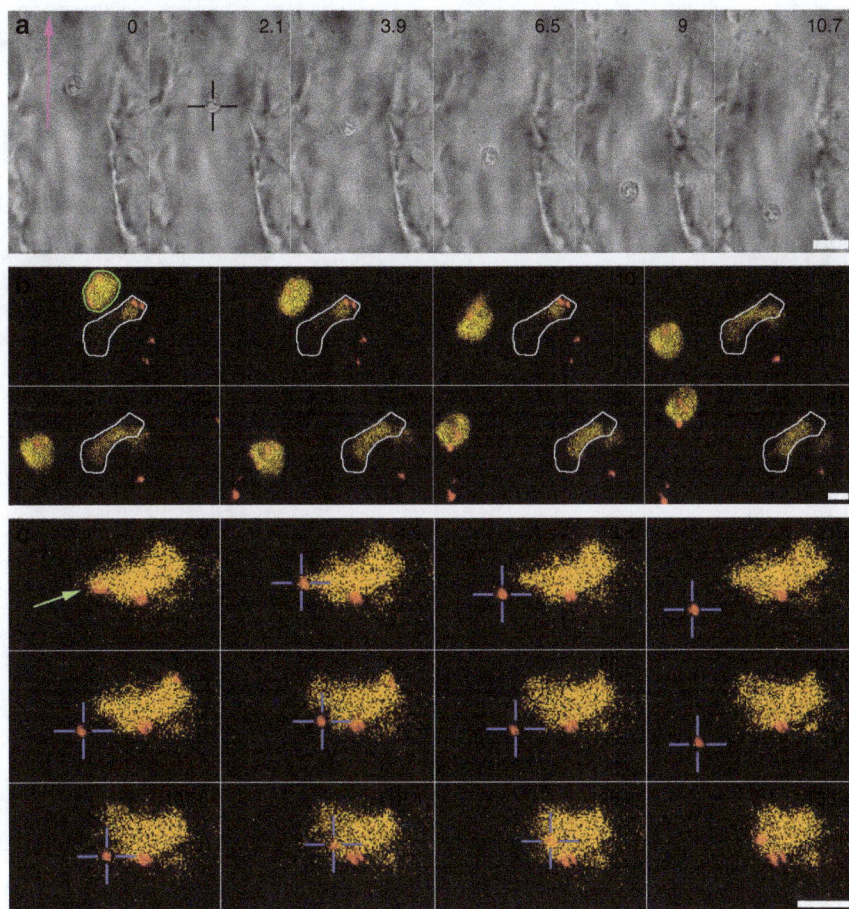

Figure 2 | Trapping of erythrocytes and macrophages. (**a**) An erythrocyte is trapped and moved in the blood flow. Scale bar, 10 µm. Experiment repeated at least 10 times. (**b**) A blood-resident fluorescent macrophage (yellow, green outline in $t = 0$ s) was micromanipulated and moved in 3D in a blood vessel. The white outline indicates another, non-mobile macrophage. The red dots are injected particles. Scale bar, 5 µm. Experiment repeated at least 5 times. (**c**) An injected particle (red colour) that was associated with a macrophage was tested for adhesion. First the particle was moved away ($t = 0$–21.7 s) after which the OT was briefly shut off. This did not result in the particle flowing away with the blood, suggesting that a nanotube (not visible) was tethering the particle; next the particle was carefully brought into contact and moved away again (52.3–101.6 s), indicating that no strong binding was established. Finally, the particle was moved further into the macrophage with a higher pushing force after which the particle could not be detached anymore ($t = 107.6$–136.1 s). Scale bar, 10 µm. Experiment was repeated at least 5 times.

stably trapped nucleus, this deformation is caused by the drag force that the blood exerts and resembles the blebs that form due to local heating of cells[20]. This experiment shows that in the zebrafish optical trapping of erythrocytes is strong and seems more robust than that achieved in mouse ear[8], where trapping had to be done gradually while making crucial use of the wall of the blood vessel. A possible explanation for this may be that erythrocytes in mice lack a nucleus.

Next, we tested whether other cell types could also be trapped in the zebrafish larva. However, it is not always trivial to identify different cell types based on solely using transmission imaging in the crowded and dynamic environment of a living organism. We therefore made use of a feature we implemented on the optical tweezers imaging system used here (adapted Nanotracker JPK Instruments AS, Berlin, see Methods), namely the possibility to do parallel trapping and confocal and transmission imaging (see also ref. 21). Using a transgenic zebrafish line with fluorescent macrophages, (Tg(mpeg1:mcherry))[22] we thus set out to test the trapping potential of these cells. In this fish line, a relatively high number of macrophages are resident in the tissue and these cells cannot be manipulated using OT. However, several macrophages were also present freely in the lumen of blood vessels. In Fig. 2b (see also Supplementary Movie 7) an area with two macrophages in the caudal vein was selected; these

macrophages could be identified using fluorescence imaging, one that was immobile (white contour) and another that could be moved using the OT. First, the macrophage was moved relative to the fish (using acousto-optic deflection-based repositioning; $t = 0$–18 s and 39–81 s), while at $t = 20$ s the whole stage, including the fish and the non-mobile macrophage, was moved (horizontally and axially), while the trap was held in a stable position. The combined trapping and confocal imaging in the zebrafish demonstrates that specific cell types can be identified and trapped using this versatile system. A plethora of transgenic zebrafish lines are available[10] and different cell types and physiological questions can thereby be addressed. As another example, in Supplementary Movie 8 (and Supplementary Fig. 5) a particle (red colour, emission 640 nm) was moved in zebrafish with (green) fluorescent endothelial cells (Tg(fli1:EGFP)[23].

We then asked whether particles that were associated with a macrophage could be moved using the tweezers; specifically, we wondered whether they were adhering to the macrophage surface or had been internalized by the cell. For this, we again used tricaine to slow the blood flow. In Fig. 2c (Supplementary Movie 9), a fluorescent macrophage was identified in the caudal vein of the zebrafish larva using confocal imaging. Next, a particle (red colour, indicated with a green arrow) was trapped with the tweezers and moved away, which resulted in its detachment from the macrophage ($t = 0$–21.7 s), thus indicating that it was not inside the macrophage. Shutting the OT off briefly did not result in the particle flowing away with the blood as other particles (the red dots) can be observed to do while they move past the stationary macrophage and stationary red nanoparticles. This suggests that the particle remained tethered to the macrophage through a nanotube (not visible either because of it being in a different plane than the confocal imaging plane or because it was very thin and not very bright). Next, the particle was moved towards the macrophage again; however, no interaction could be observed (52.3–101.6 s). Finally, the particle was moved with a higher pushing force against the macrophage after which it adhered robustly, and it was not possible to detach the particle anymore (even at 5 W, the highest power available on our system). Macrophages play a central role in clearing of larger objects[24], and it is crucial to understand this interaction for nanoparticle-based nanomedicine and delivery of drugs. The zebrafish-optical tweezers system makes it possible to study the dynamics of the interactions, make contact with a specific area of the cell, and may also be used to investigate the role of the contact forces with which particle interact when they are pushed against a macrophage in a controlled manner.

Multiple traps for *in vivo* nanotube formation. To study in more detail the particle adhesion to endothelial cells *in vivo*, a more complex system was tested involving multiple cells. In a zebrafish where the blood flow was slowed down (using tricaine), we first located a particle that adhered to the endothelium. Next, we applied several optical traps to clear the operation area of all cells: four erythrocytes that were present close to the adhered particle were removed (Fig. 3a; Supplementary Movie 10). First, cell '1' was moved away (behaving like a billiard ball pushing the other cells forward) and after moving it away it was kept trapped at a distance, allowing the assembly of a 'fence' together with several other traps placed in adjacent positions. Next, cell '2' was moved and placed behind cell '1'; the 'fence' formed by several optical traps prevented the cells from flowing and diffusing back into the operation region. In addition, cells that were out of focus experienced a scattering force that 'blew' them away from the area of the fence. At $t = 36.4$ s the earlier identified particle was next moved away from the endothelium. However, this particle did not fully detach, as a thin membrane protrusion was pulled out from the endothelial cell. Such membrane tethers or nanotubes have been studied in great detail *in vitro* since they can reveal information about, for example, membrane-cytoskeleton interactions, continuity of the membrane and its biophysical properties[25-27]. Our experiments using OT inside the live zebrafish larva show for the first time that nanotubes can be formed in a multicellular organism *in vivo* using active micromanipulation and that these nanotubes could function as a flexible adhesion, where adhered objects can be moved away while remaining tethered.

Quantification of adhesion and 'stealth' properties. Understanding how nanoparticles adhere to different cell types and how the dynamics of adhesion is regulated is crucial for nanoparticle-based drug delivery. We therefore wanted to test systematically the adhesion of particles to the endothelium cells lining the vessels. To do this we experimented with two types of particles: 'naked'—and polyethylene glycol (PEG)-coated polystyrene particles (1 μm diameter). The PEG should provide a coat that lowers the non-specific affinity of the particles for cells such as macrophages. PEG has been widely used to provide 'stealth' properties to nanoparticles designed for, for example, cancer therapy, facilitating a longer circulation time in the blood

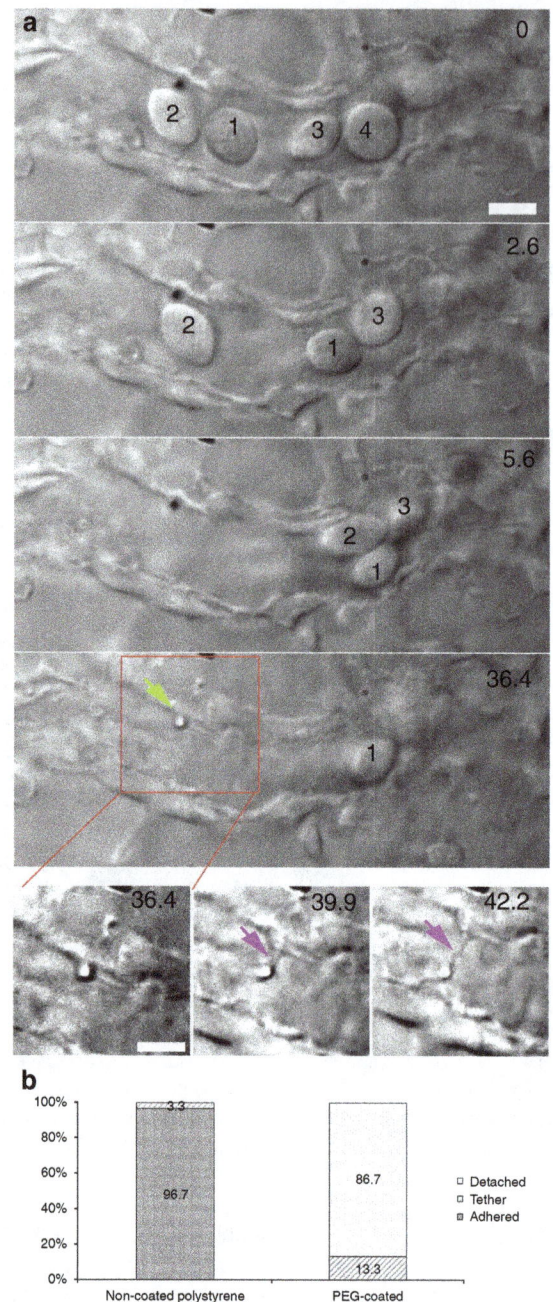

Figure 3 | Multiple traps for *in vivo* nanotube formation in cell-free blood vessels (a) In a smaller blood vessel several erythrocytes are cleared (0–36.4 s) and fenced off, after which an adhered particle was moved away from the endothelium, and a tethering nanotube was formed (36.4–42.2 s). Scale bar, 5 μm. (**b**) Quantification of the adhesion probability of naked and PEG-coated particles, classified as: detachable (white), strongly adhered (solid) and tethered (lines). Experiment was repeated at least 60 times.

stream. In another project we used PEG-coated particles in ZF and effectively monitored their biodistribution and characterized their targeting and 'stealth' properties to cancer cells[11]. Injected 1 μm particles adhering to the endothelium were detected and attempts were made to pull them away at a fixed laser power of 500 mW. For the naked polystyrene particles we found that the majority of the particles adhered strongly to the endothelium (Fig. 3b) and it was not possible to move them away at the laser power used (often also not at higher powers as they were strongly adhered). Out of 30 particles tested in the larvae, 29 could not be moved and only 1 could be pulled away but nevertheless remained connected through a tether and was pulled back after the trap was shut off (Fig. 3b, left bar). Next, we tested injected particles with a PEG-coating. These particles were also found to be adhered to the endothelium, and we determined the binding of these PEGylated particles (Fig. 3b, right). Out of the 30 PEG-coated particles that were found immobilized to the endothelium 26 could be detached, whereas four formed a tether. None of these particles were attached so strongly that it was impossible to move them away from the endothelium. This demonstrates that PEG lowers the binding affinity of the polystyrene particles for endothelial cells *in vivo*, even though they still adhered to some extent to the endothelium.

The fact that the PEG-coated particles more often form tethers indicates that the particles do adhere, but with a significantly lower strength than non-coated particles. The initial force required for tether formation depends strongly on the size of the adhesion zone and more frequent tether formation implies that the area of adhesion to endothelial cells is smaller in the PEGylated-particle-case. Naked-particles likely adhere through larger areas resulting in a higher threshold force for nanotube formation[15,28] than the tweezers can provide.

We then investigated whether heparin[29] would affect the binding of nanoparticles to the endothelial lining, given that proteins and drugs with heparin binding sites that bind the endothelium can be detached in the presence of heparin. We evaluated this first by comparing the circulation of fluorescent particles in ZF injected with mixtures (3 nl) of particles (non-coated polystyrene) together with or without heparin (concentrations 40 and 100 mg ml^{-1}). Confocal fluorescence imaging revealed no difference in adhesion properties in the presence or absence of heparin even at the highest drug concentration. We then used optical tweezers to evaluate whether the adhesion properties of particles had changed on a 'more subtle level' in the presence of heparin. As with the PEG-coated particles (Fig. 3b) we used the OT to pull on

nanoparticles adhered to the endothelium (500 mW laser power). These experiments show that there was no significant difference in the adhesion properties in the presence or absence of heparin (Supplementary Fig. 6).

Optical trapping and manipulation of injected bacteria. After having demonstrated trapping of injected particles and naturally occurring zebrafish cells inside the living zebrafish larva, we tested whether this approach can also be applied to study injected bacteria. Bacteria were amongst the first biological particles to be optically trapped *in vitro* in aqueous solutions[2], but as far as we know there are no reports of trapping bacteria inside living vertebrates. Rod-shaped bacteria are more difficult to trap than spherical (polystyrene) particles and they are smaller than macrophages and erythrocytes. As a proof of principle of optical manipulation of bacteria in zebrafish, we injected the fish bacterium *M. marinum*, which causes fish tuberculosis, and which has been widely and effectively used as a model for human tuberculosis in the zebrafish larva[12,30].

For this experiment we selected a region where the blood flow was very slow (again using tricaine), and using either fluorescence or transmission imaging, we could detect the presence of (red fluorescent) bacteria that had been injected (Fig. 4). Next, a bacterium was trapped and moved against the endothelium using the OT (Supplementary Movie 11). Before the trap was activated, the bacterium was seen moving in a Brownian manner through the blood vessel, rotating and diffusing. It could be imaged in a snapshot while oriented in the imaging plane (Fig. 4a, see also Supplementary Movie 11). Activating the optical tweezers (Fig. 4b) resulted in the bacterium becoming orientated in a direction perpendicular to the imaging plane (in the direction of the tweezer beam). This reorientation is expected for elongated objects. Next, the bacterium was pushed against the endothelium (Fig. 4c). Intriguingly, in Fig. 4d a cell (presumably a

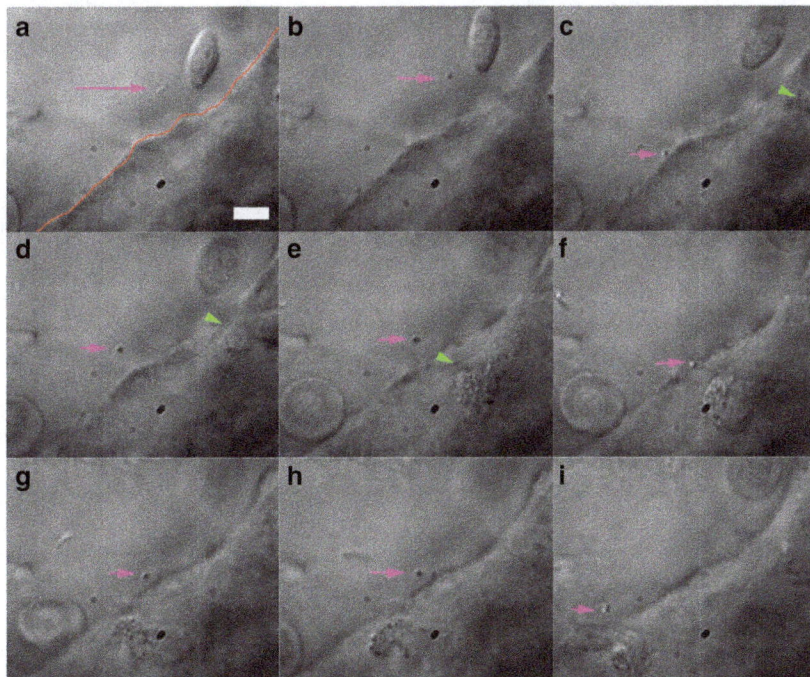

Figure 4 | Trapping of injected bacterium. (a) A diffusing bacterium (purple arrow) is **(b–d)** trapped, pushed against and moved away from the endothelium (red line in **a**). This seems to activate an immune cell (green arrowheads), which moves towards the contact point **(c–f)**. Repeated contact with the endothelium seems to again attract the attention of the crawling cell, which **(g–i)** finally moves into the vein. The total duration of experiment is ~2 min. Scale bar, 10 μm. Experiment was repeated at least 4 times.

macrophage[31]) could be observed crawling towards the contact point. After the bacterium was repeatedly moved against the endothelium, the immune cell was seen to arrest its movement and finally seemed to 'decide' to move across the endothelial barrier (Fig. 4g-i). Macrophages are known to collect bacteria[31], and controlling the position and timing of bacteria and interactions with different cells makes it possible to study this phenomenon in a controlled manner. The OT can for example be used to determine how long a bacterium needs to stay in contact to adhere to a cell or to evoke a response, or whether multiple bacteria will increase the recruitment of macrophages or influence phagocytosis dynamics.

We also tested the effects of anti-inflammatory drugs on the migration of immune cells to bacteria-invaded areas. Our preliminary experiments using fish lines with either fluorescent neutrophils or fluorescent macrophages confirmed that the anti-inflammatory drugs diclofenac and indomethacin[32] inhibited the recruitment of neutrophils (and macrophages to a lesser degree, unpublished results) to sites where bacteria had been injected[12]. However, these experiments need a systematic follow-up for their significance to be verified.

Collectively, the experiments we have described demonstrate for the first time active micromanipulation of a full scale of nano- to micron-sized structures inside a living vertebrate using the transparent zebrafish larva. The manipulated structures ranged from injected nanoparticles and bacteria to naturally occurring zebrafish cells as erythrocytes and macrophages.

We foresee many uses of this approach such as (but not limited to), the characterization of interaction properties of nanoparticles with specific cells for nanomedicine applications. In particular the properties of nanoparticles could be studied, for example, by functionalizing them with ligands for targeting to specific cells or with coats such as PEG to prevent interactions with other cells. Alternatively, optically manipulated nanoparticles releasing a specific compound could be brought in the proximity of the organismal structure of interest for testing of local cellular responses to chemicals (as has been already demonstrated elegantly with cells *in vitro*[33]) such as for studies of vascular function and endothelial integrity.

Controlled investigations of recruitment and activation of immune cells, by micromanipulating bacteria, other microorganisms or antigen-coated particles to specific regions in the organism will make possible to investigate adhesion to and activation/recruitment of immune cells, for example in the presence of anti-inflammatory drugs, especially in combination with imaging. Finally, quantitative optical tweezers have been instrumental to understanding cellular biomechanical properties and their regulatory role in function. However, this has been mostly done *in vitro* with cells in culture, and the work presented here opens up many possibilities to perform such experiments throughout a living vertebrate.

Methods

Zebrafish care and treatment.
Two lines of transgenic ZF larvae were used, Tg(fli1:EGFP) and Tg(mpeg1:mcherry) with green fluorescent endothelial cells and red fluorescent macrophages, respectively. The ZF larvae were kept in Petri dishes containing salt-containing water[18] with 0.003% w/v phenylthiourea (PTU, Sigma-Aldrich, St. Louis, USA) to keep the fish transparent by preventing pigmentation in retinal epithelium and melanophores[34]. All experiments were done at 28.5 °C. Experiments were conducted in agreement with the ethical provisions enforced by the Norwegian national animal research authority (NARA).

Microinjections of zebrafish larvae.
Injections were done using a glass micropipette (Harvard apparatus, Holliston, USA), with an outer diameter of 1.0 mm and inner diameter of 0.78 mm. The glass micropipettes were made using a micropipette puller Model P-97 (Sutter Instruments Co., Novato, USA). Manipulation of the micropipette was done using a Narishige MN-153 micromanipulator (Narishige, London, UK), and injection time and pressure was controlled using a FemtoJet

Express micro injector (Eppendorf, Hamburg Germany). Visualization of the ZF larvae during injections was done with a stereomicroscope (Leica DFC365FX with a × 1.0 Planapo lens).

For all injections, the ZF larvae were anaesthetized using (0.5–2 mg ml^{-1}) tricaine (Finquel, Argent Laboratories, Redmond, USA) in appropriate solution[18]. The ZF embryo was then placed on a gel made of hardened 2% agarose (Sigma-Aldrich) in water, and excess fluid was removed from around the ZF embryo using a pipette. This was done to immobilize the fish before injections.

Nanoparticles of five different sizes were used; four Fluoresbrite Microparticles (Polysciences Inc., Warrington, USA) with diameters of 100, 200, 500 and 1000 nm, and SPHERO particles (Spherotech Inc., Lake Forest, USA) with 840 nm diameter. The 200 nm Fluoresbrite particles contained fluorescein dye (yellow–green) with excitation maximum at 441 nm, while the 500 nm and 1 μm nanoparticles contained coumarin dye (bright blue) with excitation maxima 360 nm. The 840 nm SPHERO nanoparticles (Spherotech Inc.) contained sky blue, with excitation at 640 nm. For the experiments on the 'stealth' effect of PEG coating, the Fluoresbrite nanoparticles were modified with MPEG5000-NH2 on the surface[11]. The nanoparticles were diluted in PBS to a concentration of 2×10^8 nanoparticles per ml and loaded into a glass micropipette. Subsequently, 3–6 nl was injected into the posterior cardinal vein of the ZF larvae. *M. marinum* carrying the fluorescent reporter construct DsRed was injected (250 c.f.u.) in the posterior cardinal vein at 48 h post fertilization, as in ref. 12.

Sample preparation for optical tweezers experiments.
Following injection, the ZF larvae were moved to a Petri dish containing a tricaine solution. The concentration of tricaine was between 0.1 and 0.4 mg ml^{-1} in salt water[18] (depending on how slow blood flow was desired for the experiment).

Two parallel lines of silicone grease were applied to a 22 × 60 mm cover glass using a hypodermic needle and a syringe. The length of the grease lines was ~30 mm in length and the distance between the two lines was ~15 mm. Between these silicone lines the ZF embryo was placed in ~100 μl tricaine solution using a pipette. Using a small paint brush or hair loop, the embryo can be manipulated into a suitable position so that its body and tail are close to the cover glass (Supplementary Fig. 3, right), which facilitates optical trapping. A 22 × 22 mm coverslip (Karl Roth nr 1, thickness 0.13–0.16 μm) was carefully placed on top, resting on the two lines of silicone. The coverslip was pushed down onto the silicone, carefully, in order not to damage the embryo. It is important that the 22 × 22 mm coverslip is pushed far enough down to make sure the ZF embryo does not float around or move during the experiment. Excess fluid was removed from the edges using filter paper, or fill up the remaining space between the coverslips was filled with an embryo water-tricaine solution if necessary. The remaining openings between the cover glasses were sealed using clear nail polish. The sample was next mounted (with the 22 × 22 mm cover glass downwards) on the sample stage in the optical tweezers microscope.

Optical tweezers and imaging microscope.
An adapted version of the Nano-Tracker2 system (JPK Instruments AG, Berlin) was used. This custom-built system was developed in collaboration with JPK Instruments for parallel confocal, transmission and optical trapping (Supplementary Fig. 3, left). A 1,064-nm trapping laser (5 W) was split into two polarizations for independent trapping. One of these beams is controlled through a piezo-mirror and the other passes through Acousto-Optic Deflectors for position control and multiplexing. The trap position relative to the fish could also be controlled through movement of the whole sample with a piezo stage.

The optical trapping system was merged on a NIKON C2 confocal microscope with a × 60 (numerical aperture (NA) 1.2, WD 0.27 mm) water immersion objective for imaging and trapping, and we used Zeiss' 'Immersol' immersion fluid for water objectives ($n = 1.334$). Transmission light was focused in the sample through a × 60 water dipping condenser (NA 1, WD 2.5 mm). To be able to image in parallel in confocal mode, the ~700 to 900 nm band was used for transmission, which does not interfere with the confocal imaging.

Quantification of adhesion with and without PEG coating.
For the quantification of the binding affinities of the 1 μm polystyrene nanoparticles with and without PEG, we used three individual fishes for both PEG- and non-coated nanoparticles. In each of these 6 individuals 10 particles were trapped and manipulated using 500-mW laser power. Three possible outcomes were considered in the experiment: (1) the particle remained adhered to the endothelium; (2) the nanoparticle detached from the endothelial cell; or (3) a tether was pulled from the cell, allowing the particle to be moved away but maintaining the connection to the endothelium.

References

1. Pober, J. S. & Sessa, W. C. Evolving functions of endothelial cells in inflammation. *Nat. Rev. Immunol.* **7**, 803–815 (2007).
2. Ashkin, A. & Dziedzic, J. M. Optical trapping and manipulation of viruses and bacteria. *Science* **235**, 1517–1520 (1987).
3. Bendix, P. M., Jauffred, L., Norregaard, K. & Oddershede, L. B. Optical trapping of nanoparticles and quantum dots. *IEEE J. Sel. Top. Quantum Electron.* **20**, 15–26 (2014).

4. Blehm, B. H., Schroer, T. A., Trybus, K. M., Chemla, Y. R. & Selvin, P. R. *In vivo* optical trapping indicates kinesin's stall force is reduced by dynein during intracellular transport. *Proc. Natl Acad. Sci. USA* **110**, 3381–3386 (2013).

5. Hendricks, A. G., Holzbaur, E. L. & Goldman, Y. E. Force measurements on cargoes in living cells reveal collective dynamics of microtubule motors. *Proc. Natl Acad. Sci. USA* **109**, 18447–18452 (2012).

6. Oddershede, L. B. Force probing of individual molecules inside the living cell is now a reality. *Nat. Chem. Biol.* **8**, 879–886 (2012).

7. Hansen, P. M. & Oddershede, L. B. Optical trapping inside living organisms. *Proc. SPIE* **5930**, 1–9 (2005).

8. Zhong, M. C., Wei, X. B., Zhou, J. H., Wang, Z. Q. & Li, Y. M. Trapping red blood cells in living animals using optical tweezers. *Nat. Commun.* **4**, 1768 (2013).

9. Liu, S. & Leach, S. D. Zebrafish models for cancer. *Annu. Rev. Pathol. Mech. Dis.* **6**, 71–93 (2011).

10. Lieschke, G. J. & Currie, P. D. Animal models of human disease: zebrafish swim into view. *Nat. Rev. Genet.* **8**, 353–367 (2007).

11. Evensen, L. *et al.* Zebrafish as a model system for characterization of nanoparticles against cancer. *Nanoscale* **8**, 862–877 (2015).

12. Fenaroli, F. *et al.* Nanoparticles as drug delivery system against tuberculosis in zebrafish embryos: direct visualization and treatment. *ACS Nano* **8**, 7014–7026 (2014).

13. Kamaly, N., Xiao, Z., Valencia, P. M., Radovic-Moreno, A. F. & Farokhzad, O. C. Targeted polymeric therapeutic nanoparticles: design, development and clinical translation. *Chem. Soc. Rev.* **41**, 2971–3010 (2012).

14. Fieramonti, L. *et al.* Quantitative measurement of blood velocity in zebrafish with optical vector field tomography. *J Biophoton.* **8**, 52–59 (2015).

15. Koster, G., Cacciuto, A., Derenyi, I., Frenkel, D. & Dogterom, M. Force barriers for membrane tube formation. *Phys. Rev. Lett.* **94**, 068101 (2005).

16. Visscher, K., Gross, S. P. & Block, S. M. Construction of multiple-beam optical traps with nanometer-resolution position sensing. *IEEE J. Sel. Top. Quantum Electron.* **2**, 1066–1076 (1996).

17. Rohrbach, A. Stiffness of optical traps: quantitative agreement between experiment and electromagnetic theory. *Phys. Rev. Lett.* **95**, 168102 (2005).

18. Cosma, C. L., Swaim, L. E., Volkman, H., Ramakrishnan, L. & Davis, J. M. Zebrafish and frog models of Mycobacterium marinum infection. *Curr. Protoc. Microbiol.* Chapter 10, Unit 10B.2 (2006).

19. Zon, L. I. Developmental biology of hematopoiesis. *Blood* **86**, 2876–2891 (1995).

20. Oyama, K. *et al.* Directional bleb formation in spherical cells under temperature gradient. *Biophys. J.* **109**, 355–364 (2015).

21. Sorre, B. *et al.* Curvature-driven lipid sorting needs proximity to a demixing point and is aided by proteins. *Proc. Natl Acad. Sci. USA* **106**, 5622–5626 (2009).

22. Ellett, F., Pase, L., Hayman, J. W., Andrianopoulos, A. & Lieschke, G. J. mpeg1 promoter transgenes direct macrophage-lineage expression in zebrafish. *Blood* **117**, e49–e56 (2011).

23. Lawson, N. D. & Weinstein, B. M. In vivo imaging of embryonic vascular development using transgenic zebrafish. *Dev. Biol.* **248**, 307–318 (2002).

24. Desjardins, M. & Griffiths, G. Phagocytosis: latex leads the way. *Curr. Opin. Cell Biol.* **15**, 498–503 (2003).

25. Chaudhuri, O., Parekh, S. H., Lam, W. A. & Fletcher, D. A. Combined atomic force microscopy and side-view optical imaging for mechanical studies of cells. *Nat. Methods* **6**, 383–387 (2009).

26. Davis, D. M. & Sowinski, S. Membrane nanotubes: dynamic long-distance connections between animal cells. *Nat. Rev. Mol. Cell Biol.* **9**, 431–436 (2008).

27. Derényi, I. *et al.* in Controlled nanoscale motion Vol. 711 *Lecture Notes in Physics.* (eds Linke, H. & Månsson, A.) Ch. 7 141–159 (Springer Berlin Heidelberg, 2007).

28. Pontes, B. *et al.* Membrane elastic properties and cell function. *PLoS ONE* **8**, e67708 (2013).

29. Xu, D. & Esko, J. D. Demystifying heparan sulfate-protein interactions. *Annu. Rev. Biochem.* **83**, 129–157 (2014).

30. Ramakrishnan, L. Looking within the zebrafish to understand the tuberculous granuloma. *Adv. Exp. Med. Biol.* **783**, 251–266 (2013).

31. Herbomel, P., Thisse, B. & Thisse, C. Ontogeny and behaviour of early macrophages in the zebrafish embryo. *Development* **126**, 3735–3745 (1999).

32. Cordero-Maldonado, M. L. *et al.* Optimization and pharmacological validation of a leukocyte migration assay in zebrafish larvae for the rapid in vivo bioactivity analysis of anti-inflammatory secondary metabolites. *PLoS ONE* **8**, e75404 (2013).

33. Kress, H. *et al.* Cell stimulation with optically manipulated microsources. *Nat. Methods* **6**, 905–909 (2009).

34. Karlsson, J., von Hofsten, J. & Olsson, P.-E. Generating transparent zebrafish: a refined method to improve detection of gene expression during embryonic development. *Mar. Biotechnol.* **3**, 522–527 (2001).

Acknowledgements

JPK Instruments is acknowledged for excellent support and development of the optical tweezers imaging station, which is part of the NorMIC imaging platform at IBV (University of Oslo). We thank Dirk Linke for critical reading of the manuscript. G.K. acknowledges the Research Council of Norway for funding. We are grateful to the Norwegian Cancer Society for their generous funding (to F.F. and L.E.).

Author contributions

G.G. and G.K. conceived the project: P.L.J., F.F., G.G. and G.K. designed research; P.L.J., F.F., L.E., and G.K. conducted experiments: G.G. and G.K. supervised the project, P.L.J., G.G., and G.K. wrote the manuscript.

Additional information

Permissions

List of Contributors

Fabrizio Torricelli
Department of Information Engineering, University of Brescia, via Branze 38, Brescia 25123, Italy
Department of Electrical Engineering, Eindhoven University of Technology, Groene Loper 19, PO Box 513, Eindhoven 5600MB, The Netherlands

Luigi Colalongo and Zsolt Miklós Kovács-Vajna
Department of Information Engineering, University of Brescia, via Branze 38, Brescia 25123, Italy

Eugenio Cantatore and Daniele Raiteri
Department of Electrical Engineering, Eindhoven University of Technology, Groene Loper 19, PO Box 513, Eindhoven 5600MB, The Netherlands

Benjamin Pelz and Gabriel Žoldák
Physik Department E22, Technische Universita"t Mnchen, James Franck Strasse 1, 85748 Garching, Germany

Fabian Zeller and Martin Zacharias
Physik Department T38, Technische Universität München, 85748 Garching, Germany

Matthias Rief
Physik Department E22, Technische Universität München, James Franck Strasse 1, 85748 Garching, Germany
Munich Center for Integrated Protein Science, 81377 München, Germany

Li-Xian Wang and Cai-Zhen Li
State Key Laboratory for Mesoscopic Physics, Department of Physics, Peking University, Beijing 100871, China

Da-Peng Yu
State Key Laboratory for Mesoscopic Physics, Department of Physics, Peking University, Beijing 100871, China
Collaborative Innovation Center of Quantum Matter, Beijing, China
Institute of Physics and Electronic Information, Yunnan Normal University, Kunming 650500, China

Zhi-Min Liao
State Key Laboratory for Mesoscopic Physics, Department of Physics, Peking University, Beijing 100871, China
Collaborative Innovation Center of Quantum Matter, Beijing, China

Amirhossein Khalajhedayati
Department of Chemical Engineering and Materials Science, University of California, Irvine, California 92697, USA

Zhiliang Pan
Department of Mechanical and Aerospace Engineering, University of California, 4200 Engineering Gateway, Irvine, California 92697, USA

Timothy J. Rupert
Department of Chemical Engineering and Materials Science, University of California, Irvine, California 92697, USA
Department of Mechanical and Aerospace Engineering, University of California, 4200 Engineering Gateway, Irvine, California 92697, USA

Wenbin Li
Research Laboratory of Electronics, Massachusetts Institute of Technology, Cambridge, Massachusetts 02139, USA

Ju Li
Department of Nuclear Science and Engineering and Department of Materials Science and Engineering, Massachusetts Institute of Technology, Cambridge, Massachusetts 02139, USA

Mei-Ling Wu
Key Laboratory of Molecular Nanostructure and Nanotechnology, Institute of Chemistry, Chinese Academy of Sciences (CAS), Beijing 100190, China
Beijing National Laboratory for Molecular Sciences, Beijing 100190, China
University of CAS, Beijing 100049, China

Dong Wang and Li-Jun Wan
Key Laboratory of Molecular Nanostructure and Nanotechnology, Institute of Chemistry, Chinese Academy of Sciences (CAS), Beijing 100190, China
Beijing National Laboratory for Molecular Sciences, Beijing 100190, China

Haiping He, Qianqian Yu, Hui Li, Jing Li, Junjie Si and Zhizhen Ye
State Key Laboratory of Silicon Materials, School of Materials Science and Engineering, Zhejiang University, Hangzhou 310027, China

Yizheng Jin
Center for Chemistry of High-Performance and Novel Materials and State Key Laboratory of Silicon Materials, Department of Chemistry, Zhejiang University, Hangzhou 310027, China

Nana Wang and Jianpu Wang
Key Laboratory of Flexible Electronics (KLOFE) & Institute of Advanced Materials (IAM), Jiangsu National Synergetic Innovation Center for Advanced Materials (SICAM), Nanjing Tech University (NanjingTech), 30 South Puzhu Road, Nanjing 211816, China

Jingwen He, Xinke Wang and Yan Zhang
Department of Physics, Capital Normal University, Beijing Key Lab for Metamaterials and Devices, and Key Laboratory of Terahertz Optoelectronics, Ministry of Education, Beijing 100048, China

Eun Seon Cho, Anne M. Ruminski, Shaul Aloni and Jeffrey J. Urban
The Molecular Foundry, Materials Sciences Division, Lawrence Berkeley National Laboratory, 1 Cyclotron Road, Berkeley, California 94720, USA

Yi-Sheng Liu and Jinghua Guo
The Advanced Light Source, Lawrence Berkeley National Laboratory, Berkeley, California 94720, USA

Yuan Liu, Libo Gao, Jie Chen, Fuchi Liu, Yuanyuan Sun, Nujiang Tang and Youwei Du
Nanjing National Laboratory of Microstructures, Nanjing University, Nanjing 210093, China
Collaborative Innovation Center of Advanced Microstructures, Nanjing University, Nanjing 210093, China

Yuting Shen and Litao Sun
FEI Nano-Pico Center, Key Laboratory of MEMS of Ministry of Education, Collaborative Innovation Center for Micro/Nano Fabrication, Device and System, Southeast University, Nanjing 210096, China

Hui-Ming Cheng, Jincheng Li, Chang Liu, Wencai Ren and Feng Li
Advanced Carbon Division, Shenyang National Laboratory for Materials Science, Institute of Metal Research, Chinese Academy of Sciences, 72Wenhua Road, Shenyang, Liaoning Province 110016, China

Carson T. Riche
Mork Family Department of Chemical Engineering and Materials Science, University of Southern California, 925 Bloom Walk, HED 216, Los Angeles, California 90089, USA

Emily J. Roberts and Richard L. Brutchey
Department of Chemistry, University of Southern California, Los Angeles, California 90089, USA

Malancha Gupta and Noah Malmstadt
Mork Family Department of Chemical Engineering and Materials Science, University of Southern California, 925 Bloom Walk, HED 216, Los Angeles, California 90089, USA
Department of Chemistry, University of Southern California, Los Angeles, California 90089, USA

Guoxing Li, Jinhua Sun, Wenpeng Hou, Shidong Jiang, Yong Huang and Jianxin Geng
Technical Institute of Physics and Chemistry, Chinese Academy of Sciences, 29 Zhongguancun East Road, Haidian District, Beijing 100190, China

Calum Jack, Ryan Tullius , Marion Rodier, Brian Fitzpatrick, Laurence D. Barron, Adrian J. Lapthorn, Graeme Cooke and Malcolm Kadodwala
School of Chemistry, University of Glasgow, Joseph Black Building, Glasgow G12 8QQ, UK

Affar S. Karimullah
School of Chemistry, University of Glasgow, Joseph Black Building, Glasgow G12 8QQ, UK
School of Engineering, University of Glasgow, Rankine Building, Glasgow G12 8LT, UK

Larousse Khosravi Khorashad and Alexander O. Govorov
Department of Physics and Astronomy, Ohio University, Athens, Ohio 45701, USA

Nikolaj Gadegaard
School of Engineering, University of Glasgow, Rankine Building, Glasgow G12 8LT, UK

Vincent M. Rotello
Department of Chemistry, University of Massachusetts, 710 North Pleasant Street, Amherst, Massachusetts 01003, USA

Achim Woessner, Mark B. Lundeberg, Gabriele Navickaite and Davide Janner
ICFO—Institut de Ciencies Fotoniques, The Barcelona Institute of Science and Technology, 08860 Barcelona, Spain

Pablo Alonso-González
CIC nanoGUNE, 20018 Donostia-San Sebastian, Spain
Institute of Physics, Chinese Academy of Science, Beijing 100190, China

Yuanda Gao and James Hone
Department of Mechanical Engineering, Columbia University, New York, New York 10027, USA

Jose E. Barrios-Vargas and Aron W. Cummings
Catalan Institute of Nanoscience and Nanotechnology (ICN2), CSIC and The Barcelona Institute of Science and Technology, Campus UAB, 08193 Barcelona, Spain

Qiong Ma and Pablo Jarillo-Herrero
Department of Physics, Massachusetts Institute of Technology, Cambridge, Massachusetts 02139, USA

Kenji Watanabe and Takashi Taniguchi
National Institute for Materials Science, 1-1 Namiki, Tsukuba 305-0044, Japan

Valerio Pruneri and Frank H.L. Koppens
ICFO—Institut de Ciencies Fotoniques, The Barcelona Institute of Science and Technology, 08860 Barcelona, Spain
ICREA-Institució Catalana de Recerca i Estudis Avançats, 08010 Barcelona, Spain

Stephan Roche
Catalan Institute of Nanoscience and Nanotechnology (ICN2), CSIC and The Barcelona Institute of Science and Technology, Campus UAB, 08193 Barcelona, Spain
ICREA-Institució Catalana de Recerca i Estudis Avanats, 08010 Barcelona, Spain

Rainer Hillenbrand
CIC nanoGUNE and UPV/EHU, 20018 Donostia-San Sebastian, Spain
IKERBASQUE, Basque Foundation for Science, 48011 Bilbao, Spain

Maria Ibáñez
Department of Chemistry and Applied Biosciences, Institute of Inorganic Chemistry, ETH Zrich, Vladimir Prelog Weg 1, CH-8093 Zurich, Switzerland
Laboratory for Thin Films and Photovoltaics, Empa-Swiss Federal Laboratories for Materials Science and Technology, Dbendorf, berlandstrasse 129, CH- 8600 Dbendorf, Switzerland
Advanced Materials Department, Catalonia Energy Research Institute - IREC, Sant Adria de Besos, Jardins de les Dones de Negre n.1, Pl. 2, 08930 Barcelona, Spain

Zhishan Luo, Silvia Ortega, Doris Cadavid, Oleksandr Dobrozhan and Yu Liu
Advanced Materials Department, Catalonia Energy Research Institute - IREC, Sant Adria de Besos, Jardins de les Dones de Negre n.1, Pl. 2, 08930 Barcelona, Spain

Aziz Gen
Department of Advanced Electron Nanoscopy, Catalan Institute of Nanoscience and Nanotechnology (ICN2), CSIC and The Barcelona Institute of Science and Technology, Campus UAB, Bellaterra, 08193 Barcelona, Spain

Laura Piveteau and Maksym V. Kovalenko
Department of Chemistry and Applied Biosciences, Institute of Inorganic Chemistry, ETH Zrich, Vladimir Prelog Weg 1, CH-8093 Zurich, Switzerland
Laboratory for Thin Films and Photovoltaics, Empa-Swiss Federal Laboratories for Materials Science and Technology, Dbendorf,¡berlandstrasse 129, CH- 8600 Dübendorf, Switzerland

Maarten Nachtegaal
Paul Scherrer Institute, 5232 Villigen PSI, Switzerland

Mona Zebarjadi
Department of Mechanical and Aerospace Engineering, Rutgers University, 98 Brett Rd, Piscataway, New Jersey 08854-8058, USA

Jordi Arbiol
Department of Advanced Electron Nanoscopy, Catalan Institute of Nanoscience and Nanotechnology (ICN2), CSIC and The Barcelona Institute of Science and Technology, Campus UAB, Bellaterra, 08193 Barcelona, Spain
Institució Catalana de Recerca i Estudis Avanats, ICREA, Passeig de Lluís Companys, 23 08010 Barcelona, Spain

Andreu Cabot
Advanced Materials Department, Catalonia Energy Research Institute - IREC, Sant Adria de Besos, Jardins de les Dones de Negre n.1, Pl. 2, 08930 Barcelona, Spain
Institució Catalana de Recerca i Estudis Avanats, ICREA, Passeig de Lluís Companys, 23 08010 Barcelona, Spain

Pranab Kumar Das
Istituto Officina dei Materiali (IOM)-CNR, Laboratorio TASC, in Area Science Park, S.S.14, Km 163.5, I-34149 Trieste, Italy
International Centre for Theoretical Physics (ICTP), Strada Costiera 11, I-34100 Trieste, Italy

D. Di Sante
Consiglio Nazionale delle Ricerche—CNR-SPIN, I-67100 L'Aquila, Italy
Department of Physical and Chemical Sciences, University of L'Aquila, Via Vetoio, I-67100 L'Aquila, Italy

I. Vobornik, J. Fujii, G. Panaccione and R. Ciancio
Istituto Officina dei Materiali (IOM)-CNR, Laboratorio TASC, in Area Science Park, S.S.14, Km 163.5, I-34149 Trieste, Italy

T. Okuda
Hiroshima Synchrotron Radiation Center (HSRC), Hiroshima University, 2-313 Kagamiyama, Higashi-Hiroshima 739-0046, Japan

E. Bruyer and S. Picozzi
Consiglio Nazionale delle Ricerche — CNR-SPIN, I-67100 L'Aquila, Italy

A. Gyenis, B.E. Feldman and A. Yadzani
Joseph Henry Laboratories and Department of Physics, Princeton University, Princeton, New Jersey 08544, USA

J. Tao
Department of Condensed Matter Physics and Materials Science, Brookhaven National Laboratory, Upton, New York 11973, USA

G. Rossi
Istituto Officina dei Materiali (IOM)-CNR, Laboratorio TASC, in Area Science Park, S.S.14, Km 163.5, I-34149 Trieste, Italy
Dipartimento di Fisica, Universita` di Milano, Via Celoria 16, I-20133 Milano, Italy

R.J. Cava and M.N. Ali
Department of Chemistry, Princeton University, Princeton, New Jersey 08544, USA

Soonyoung Cha, Sangwan Sim, Jun Park and Hyunyong Choi
School of Electrical and Electronic Engineering, Yonsei University, Seoul 120-749, Korea

Ji Ho Sung, Hoseok Heo and Moon-Ho Jo
Center for Artificial Low Dimensional Electronic Systems, Institute for Basic Science (IBS), Pohang University of Science and Technology (POSTECH), Pohang 790-784, Korea
Division of Advanced Materials Science, Pohang University of Science and Technology (POSTECH), Pohang 790-784, Korea

Liguo Ma, Cun Ye, Yijun Yu, Xiaohai Niu, Donglai Feng and Yuanbo Zhang
State Key Laboratory of Surface Physics and Department of Physics, Fudan University, Shanghai 200433, China
Collaborative Innovation Center of Advanced Microstructures, Nanjing 210093, China

Xiu Fang Lu and Xian Hui Chen
Collaborative Innovation Center of Advanced Microstructures, Nanjing 210093, China
Hefei National Laboratory for Physical Science at Microscale and Department of Physics, University of Science and Technology of China, Hefei, Anhui 230026, China
Key Laboratory of Strongly Coupled Quantum Matter Physics, Chinese Academy of Sciences, School of Physical Sciences, University of Science and Technology of China, Hefei 230026, China

Sejoong Kim and Young-Woo Son
Korea Institute for Advanced Study, Hoegiro 85, Seoul 02455, Korea

David Tománek
Physics and Astronomy Department, Michigan State University, East Lansing, Michigan 48824, USA

Ricardo M.P. da Silva
Simpson Querrey Institute for BioNanotechnology (SQI), Northwestern University, Chicago, Illinois 60611, USA
Laboratory of Macromolecular and Organic Chemistry and Institute for Complex Molecular Systems, Eindhoven University of Technology, Eindhoven MB 5600, The Netherlands
Craniofacial Development & Stem Cell Biology, King's College London, London, SE1 9RT, UK

Daan van der Zwaag and E.W. Meijer
Laboratory of Macromolecular and Organic Chemistry and Institute for Complex Molecular Systems, Eindhoven University of Technology, Eindhoven MB 5600, The Netherlands

Lorenzo Albertazzi
Laboratory of Macromolecular and Organic Chemistry and Institute for Complex Molecular Systems, Eindhoven University of Technology, Eindhoven MB 5600, The Netherlands
Nanoscopy for Nanomedicine Group, Institute for Bioengineering of Catalonia (IBEC), Barcelona 08028, Spain

Sungsoo S. Lee
Department of Materials Science and Engineering, Northwestern University, Evanston, Illinois 60208, USA

Samuel I. Stupp
Simpson Querrey Institute for BioNanotechnology (SQI), Northwestern University, Chicago, Illinois 60611, USA

Department of Materials Science and Engineering, Northwestern University, Evanston, Illinois 60208, USA
Department of Chemistry, Northwestern University, Evanston, Illinois 60208, USA
Department of Medicine, Northwestern University, Chicago, Illinois 60611, USA
Department of Biomedical Engineering, Northwestern University, Evanston, Illinois 60208, USA

Woo Hyeon Jeong, Jae-Hoon Eom, Hee-Seung Lee and Jie-Oh Lee
Department of Chemistry, KAIST, Daejeon 34141, Korea

Haerim Lee and Sun Chang Kim
Department of Biological Sciences, KAIST, Daejeon 34141, Korea

Dong Hyun Song
Agency for Defense Development, Daejeon 34186, Korea

Hayyoung Lee
Institute of Biotechnology, Chungnam National University, Daejeon 34134, Korea

M.A.A. Hafiz, L. Kosuru and M.I. Younis
Physical Sciences and Engineering Division, King Abdullah University of Science and Technology, Thuwal 23955-6900, Saudi Arabia

Venkateshkumar Prabhakaran, B. Layla Mehdi, Nigel D. Browning, Grant E. Johnson, Julia Laskin and K. Don D. Gunaratne
Physical Sciences Division, Pacific Northwest National Laboratory, PO Box 999, MSIN K8-88, Richland, Washington 99352, USA

Jeffrey J. Ditto and David C. Johnson
Department of Chemistry, University of Oregon, Eugene, Oregon 97403, USA

Mark H. Engelhard and Bingbing Wang
Environmental Molecular Sciences Laboratory, Pacific Northwest National Laboratory, Richland, Washington 99352, USA

Yayuan Liu, Dingchang Lin, Zheng Liang, Jie Zhao and Kai Yan
Department of Materials Science and Engineering, Stanford University, Stanford, California 94305, USA

Yi Cui
Department of Materials Science and Engineering, Stanford University, Stanford, California 94305, USA
Stanford Institute for Materials and Energy Sciences, SLAC National Accelerator Laboratory, 2575 Sand Hill Road, Menlo Park, California 94025, USA

Yan Sun, Fang Guo, Tongfei Zuo, Jingjing Hua and Guowang Diao
College of Chemistry and Chemical Engineering, Yangzhou University, Yangzhou, Jiangsu 225002, China

Feng Zhang, Peng-Xiang Hou, Chang Liu, Bing-Wei Wang, Mao-Lin Chen, Dong-Ming Sun, Jin-Cheng Li and Hong-Tao Cong
Shenyang National Laboratory for Materials Science, Advanced Carbon Division, Institute of Metal Research, Chinese Academy of Sciences, 72 Wenhua Road, Shenyang 110016, China

Hua Jiang and Esko I. Kauppinen
Nano Materials Group, Department of Applied Physics and Center for New Materials, School of Science, Aalto University, PO Box 15100, FI-00076 Aalto, Finland

Hui-Ming Cheng
Shenyang National Laboratory for Materials Science, Advanced Carbon Division, Institute of Metal Research, Chinese Academy of Sciences, 72 Wenhua Road, Shenyang 110016, China
Faculty of Science, Chemistry Department, King Abdulaziz University, Jeddah 21589, Saudi Arabia

Dongliang Chao, Changrong Zhu, Peihua Yang, Jilei Liu and Hong Jin Fan
School of Physical and Mathematical Sciences, Nanyang Technological University, Singapore 637371, Singapore

Xinhui Xia
State Key Laboratory of Silicon Materials, Department of Materials Science and Engineering, Zhejiang University, Hangzhou 310027, China

Jin Wang and Jianyi Lin
Energy Research Institute @ NTU, Nanyang Technological University, Singapore 639798, Singapore

Xiaofeng Fan
College of Materials Science and Engineering, Jilin University, Changchun 130012, China

Serguei V. Savilov
Department of Chemistry, Moscow State University, Moscow 119992, Russia

Ze Xiang Shen
School of Physical and Mathematical Sciences, Nanyang Technological University, Singapore 637371, Singapore Energy Research Institute @ NTU, Nanyang Technological University, Singapore 639798, Singapore

Jeong Kuk Shon
Samsung Advanced Institute of Technology, Samsung Electronics Co., Ltd., Suwon 443-803, Republic of Korea Department of Chemistry, Sungkyunkwan University, Suwon 440-746, Republic of Korea

Hyo Sug Lee, Gyeong Su Park, Jae-Man Choi, Hyuk Chang and Seokgwang Doo
Samsung Advanced Institute of Technology, Samsung Electronics Co., Ltd., Suwon 443-803, Republic of Korea

Soo Sung Kong
Department of Chemistry, Sungkyunkwan University, Suwon 440-746, Republic of Korea

Gwi Ok Park and Ji Man Kim
Department of Chemistry, Sungkyunkwan University, Suwon 440-746, Republic of Korea
Department of Energy Science, Sungkyunkwan University, Suwon 440-746, Republic of Korea

Jeongbae Yoon and Won-Sub Yoon
Department of Energy Science, Sungkyunkwan University, Suwon 440-746, Republic of Korea

Eunjun Park and Hansu Kim
Department of Energy Engineering, Hanyang University, Seoul 133-791, Republic of Korea

Mingshi Jin
Key Laboratory of Natural Resource of the Changbai Mountain and Functional Molecular (Yanbian University), Ministry of Education, Department of Chemistry, Park Road 977, Yanji City, Jilin Province 133002, China

Chanho Pak
Fuel Cell Group, Corporate R&D Center, Samsung SDI Co., Ltd., Yongin 446-577, Republic of Korea

Galen D. Stucky
Department of Chemistry and Biochemistry, University of California, Santa Barbara, California 93106, USA

Patrick Lie Johansen, Federico Fenaroli, Lasse Evensen, Gareth Griffiths and Gerbrand Koster
Department of Biosciences, University of Oslo, Blindernveien 31, 0371 Oslo, Norway

Index